FRACTURE MECHANICS: FIFTEENTH SYMPOSIUM

Fifteenth National Symposium
on Fracture Mechanics
sponsored by ASTM
Committee E-24 on Fracture Testing
College Park, Md., 7–9 July 1982

ASTM SPECIAL TECHNICAL PUBLICATION 833
R. J. Sanford, University of Maryland, editor

ASTM Publication Code Number (PCN)
04-833000-30

 1916 Race Street, Philadelphia, Pa. 19103

Library of Congress Cataloging in Publication Data

National Symposium on Fracture Mechanics (15th : 1982 :
   College Park, Md.)
   Fracture mechanics.

   (ASTM special technical publication ; 833)
   "ASTM publication code number (PCN 04-833000-30)."
   Includes bibliographies and index.
   1. Fracture mechanics—Congresses. I. Sanford, R. J.
   II. ASTM Committee E-24 on Fracture Testing.
   III. Title. IV. Series.
   TA409.N38 1982   620.1′126   83-72816
   ISBN 0-8031-0208-9

NOTE

The Society is not responsible, as a body,
for the statements and opinions
advanced in this publication.

D

620.1126

NAT

Printed in Baltimore, Md. (b)
September 1984

Dr. George R. Irwin

Dedication

The dedication of this publication in honor of Dr. George R. Irwin on his 75th birthday recognizes his development of the basic theory of linear elastic fracture mechanics and its application in solving critical problems of national importance. In particular, we honor Dr. Irwin's continued counsel and guidance to ASTM Committee E-24 on Fracture Testing.

We wish Dr. Irwin many years of good health and happiness.

Foreword

The 15th National Symposium on Fracture Mechanics was held at the University of Maryland, College Park, on 7–9 July 1982. ASTM Committee E-24 on Fracture Testing was sponsor. R. J. Sanford, University of Maryland, served as symposium chairman and has edited this publication.

Related ASTM Publications

Environment-Sensitive Fracture: Evaluation and Comparison of Test Methods, STP 821 (1984), 04-821000-30

Fracture Mechanics: Fourteenth Symposium—Volume I: Theory and Analysis, STP 791 (1983), 04-791001-30

Fracture Mechanics: Fourteenth Symposium—Volume II: Testing and Applications, STP 791 (1983), 04-791002-30

Fractography of Ceramic and Metal Failures, STP 827 (1984), 04-827000-30

Elastic-Plastic Fracture: Second Symposium, Volume I—Inelastic Crack Analysis, STP 803 (1983), 04-803001-30

Elastic-Plastic Fracture: Second Symposium, Volume II: Fracture Resistance Curves and Engineering Applications, STP 803 (1983), 04-803002-30

Probabilistic Fracture Mechanics and Fatigue Methods: Applications for Structural Design and Maintenance, STP 798 (1983), 04-798000-30

Fracture Mechanics (Thirteenth Conference), STP 743 (1981), 04-743000-30

Fractography and Materials Science, STP 733 (1981), 04-733000-30

Elastic-Plastic Fracture, STP 688 (1979), 04-688000-30

A Note of Appreciation
to Reviewers

The quality of the papers that appear in this publication reflects not only the obvious efforts of the authors but also the unheralded, though essential, work of the reviewers. On behalf of ASTM we acknowledge with appreciation their dedication to high professional standards and their sacrifice of time and effort.

ASTM Committee on Publications

ASTM Editorial Staff

Contents

ELASTO-PLASTIC FRACTURE MECHANICS

Introduction

The George R. Irwin Anniversary Volume

The year 1982 marked a number of milestones in the history of fracture mechanics. In this year the National Symposium on Fracture Mechanics held its 15th annual forum to discuss a wide range of topics related to the fracture of materials. It also marked the 25th anniversary of the rocket motor fractures which led to the formation (December 1958) of the Special Committee on Fracture Testing of High-Strength Metallic Sheet Materials (in later years this committee was formally organized as ASTM Committee E-24 on Fracture Testing). Finally, in 1982, George R. Irwin, the major driving force in the early development of the theory of linear elastic fracture mechanics (LEFM), celebrated his 75th birthday. In commemoration of this latter event, the symposium subcommittee of E-24 assigned the University of Maryland the task of hosting this anniversary symposium and has dedicated this publication in Dr. Irwin's honor.

George Rankin Irwin was born in El Paso, Texas, in February 1907. His school years were spent in Springfield, Illinois, where he attended Springfield High School (1921–1925) and Knox College (1926–1931). Initially an English and journalism major, he earned his bachelor's degree in English but developed a keen interest in physics and took additional courses in this area. Continuing his studies, he attended the University of Illinois and obtained a master's degree in physics and then a doctorate in physics in 1937. During the latter stages of his doctorate study (1935–1936) he was an associate professor at Knox College.

In July 1937, George Irwin, with degree and wife, Georgia, moved to Washington, D.C., and joined the staff of the Ballistics Branch at the Naval Research Laboratory (NRL). He was assigned the task of investigating the cause of brittle failures of armor materials. Early in these studies he observed correlations between the energy absorbed in penetration and the appearance of the fractured area. These results would later be generalized to the strain energy release rate concept that was to become the cornerstone of the theory of LEFM prior to 1957.

During these early years, numerous conceptual advances in the theory of fracture were made by Irwin and his co-workers at NRL. Crack growth by advance nucleation was observed both in thin foils and brittle solids. Compliance calibration for characterizing fracture test specimens was developed. The role of fracture markings in postmortem analysis of fracture failures was demonstrated and catalogs of features and their origins compiled. Bifurcation as a mechanism for energy consumption in dynamic fracture was proposed.

In the mid 1950's Dr. Irwin turned his attention to the analytical aspects of fracture mechanics with particular emphasis on the characteristics of the stress field in the neighborhood of the crack tip. In 1957 he published a landmark paper in which the near-field stress equations were presented and the concept of the strength of the stress singularity (now referred to as K) was proposed. The use of the Westergaard method to determine the stress intensity factor for various geometric configurations followed. Later, the plastic zone correction concept was proposed as well as other conceptual ideas such as virtual crack extension. He was one of the founding members of the aforementioned ASTM Special Committee on Fracture Testing, and he continues to participate in ASTM Committee E-24.

After 30 years of federal service, George Irwin retired from the Naval Research Laboratory and assumed the position of University Professor at Lehigh University. During this period in his career, he placed his emphasis on the development of undergraduate and graduate courses in fracture mechanics.

Dr. Irwin retired from Lehigh in 1972 and accepted his current position as Visiting Professor at the University of Maryland, where his primary interests lie in guiding research in dynamic fracture. He continues to serve as adviser to government, university, and private industry, drawing on his vast experience to propose solutions to problems in fracture mechanics.

In recognition of his pioneering work he has received numerous awards and honors including:

> Navy Distinguished Civilian Service Award-1946
> ASTM Dudley Medal-1960
> ASME Thurston Lecturer-1966
> U.S. Navy Conrad Award-1969
> SESA Murray Lecturer-1973
> ASTM Honorary Member-1974
> ASM Sauveur Award-1974
> Société Française de Métallurgie Grande Médaille-1976
> National Academy of Engineering Membership-1977
> ASME Nadai Award-1977
> Honorary Doctor of Engineering, Lehigh University-1977
> SESA Lazan Award-1977
> ASTM (E-24) Irwin Medal-1978
> Franklin Institute Clauier Medal-1979

At the symposium banquet Dr. John S. Toll, President of the University of Maryland, presented to Dr. Irwin on behalf of the Governor of Maryland, the Honorable Harry Hughes, the Governor's Citation for distinguished service to the State of Maryland. In turn, George Irwin presented the 1982 medal named in his honor jointly to Drs. J. R. Rice and J. Hutchinson of Harvard University.

The Symposium Organizing Committee consisting of Prof. D. B. Barker, Prof. W. L. Fourney, Mr. John Gudas, Dr. John Merkle, Prof. R. J. Sanford, and Dr. H. H. Vanderveldt are pleased to have been involved in this effort to honor this truly remarkable scientist and educator. We would like to express our thanks to the staff and students of the Mechanical Engineering Department at the University of Maryland for their many efforts before and during the symposium. Finally, the committee is especially grateful to Mr. R. Chona, symposium secretary, for his invaluable assistance during the planning of the symposium and the preparation of this publication.

R. J. Sanford

Department of Mechanical Engineering, University of Maryland, College Park, Maryland; symposium chairman and editor

Linear Elastic Fracture Mechanics

A. F. Grandt, Jr.,[1] J. A. Harter,[2] and B. J. Heath[3]

Transition of Part-Through Cracks at Holes into Through-the-Thickness Flaws

REFERENCE: Grandt, A. F., Jr., Harter, J. A., and Heath, B. J., "**Transition of Part-Through Cracks at Holes into Through-the-Thickness Flaws,**" *Fracture Mechanics: Fifteenth Symposium, ASTM STP 833,* R. J. Sanford, Ed., American Society for Testing and Materials, Philadelphia, 1984, pp. 7-23.

ABSTRACT: This paper describes results of a numerical and experimental study of the behavior of part-through cracks located at holes as they transition into uniform through-the-thickness flaws. Fatigue crack growth tests are conducted with transparent polymer specimens which allow the crack plane to be photographed during the fatigue test. Stress intensity factors are computed by the three-dimensional finite-element-alternating method for the measured crack shapes. Both analysis and experiment indicate that the crack growth rate varies along the flaw perimeter in a manner which encourages the trailing edge of crack advance to "catch up" with the leading edge. Once a uniform through-thickness configuration is achieved, the trailing point then slows down to the growth rate occurring at the point of maximum crack advance.

KEY WORDS: fatigue cracks, surface cracks, stress intensity factors, fracture mechanics

Nomenclature

a Crack dimension measured along bore of hole as defined in Fig. 1

c Crack dimension measured perpendicular to bore of hole as defined in Fig. 1

c_L, c_R, c_t Free face crack dimensions as defined in Fig. 1

D Hole diameter

da/dN Fatigue crack growth rate

[1]Professor, School of Aeronautics and Astronautics, Purdue University, W. Lafayette, Ind. 47907.

[2]Former Graduate Assistant, Purdue University, W. Lafayette, Ind. 47907; currently with Northrop Corporation, Hawthorne, Calif. 92644.

[3]Former Graduate Assistant, Purdue University, W. Lafayette, Ind. 47907; currently with Garrett Turbine Engine Company, Phoenix, Ariz. 85010.

7

K Stress intensity factor
ΔK Cyclic stress intensity factor
N Number of elapsed cycles
P Applied force
R Hole radius
σ Remotely applied stress
T Plate thickness
W Specimen width
x,y Coordinate axes defined in Figs. 4 and 7

This paper describes results of an experimental and numerical study of the behavior of part-through cracks as they grow into through-the-thickness flaws. As indicated in Fig. 1, the paper is concerned with both corner and embedded surface cracks located along the bore of fastener holes. The transition period when the nonuniform through-the-thickness flaw grows to a stable through-thickness shape is of main concern here.

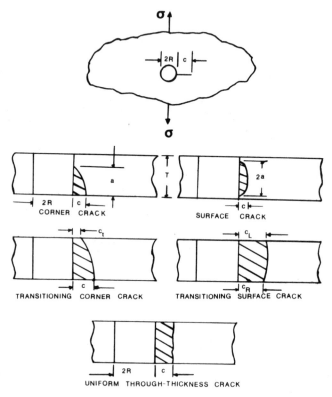

FIG. 1—*Schematic of flawed hole configurations.*

Several procedures have been described in the literature to analyze the transition from surface to uniform through-the-thickness cracks. Some authors, for example, have conservatively assumed that when the thickness dimension penetrates the back surface, the crack can be instantly treated as a uniform through-the-thickness crack [1,2]. Others have developed more elaborate criteria based on back surface yielding [3], imaginary equivalent surface cracks [4], or engineering weighting factors [5].

The goal of this paper is to examine the transition behavior in more detail. The cyclic growth and transition of initial part-through cracks into through-the-thickness flaws is documented through experiments with specimens made from polymethyl methacrylate (PMMA), a transparent polymer. Since the test specimens are transparent, it is possible to record crack growth, including changes in internal dimensions, by time-lapse photography. Stress intensity factors are computed for the observed crack shapes by the finite-element-alternating method. Fatigue crack growth rate changes at various points along the flaw border are correlated with the stress intensity factor solutions. Results are described for initial quarter-elliptical corner and embedded semi-elliptical surface cracks located at open holes in large plates loaded in remote tension, and for embedded surface cracks located along the bore of a simulated pin-loaded attachment lug.

Approach

This section reviews the experimental and numerical procedures used to characterize the transition of surface and corner cracked holes into through-the-thickness flaws. Since these methods have been employed before, only brief descriptions are given here.

Numerical Technique

The finite-element-alternating method (FEAM) developed by F. W. Smith and associates [6–9] was used to determine stress intensity factor variations along the perimeter of embedded surface and corner cracked holes. The FEAM approach involves iterative superposition of the three-dimensional finite-element solution for an uncracked body subjected to a specified surface loading [10] with the solution for a flat elliptical crack in an infinite body loaded with a nonuniform surface pressure [11]. Iteration between the cracked and uncracked solutions approximates the part-through crack boundary conditions and provides Mode I stress intensity factors and crack opening displacements for the three-dimensional flaw geometry.

The specific computer codes used here were developed by Smith and Kullgren for the analyses described in Ref 9. Other applications of this numerical method to cracks at holes are given in Refs 6 to 8 and 12 to 14. Prior comparisons of the FEAM results with other numerical and experimental techniques indicate that the finite-element alternating-method provides excellent stress intensity factor solutions for the part-through cracked fastener hole problem.

Experimental Procedure

The fatigue crack growth tests were conducted on specimens prepared from a single sheet of polymethyl methacrylate (PMMA), a transparent polymer. Since the specimens were transparent, it was possible to view internal crack dimensions directly with the aid of a mirror placed at an angle over the polished end of the specimen. The cracks were photographed with a 35-mm camera during the fatigue test. Subsequent measurement of the photographs gave crack dimensions as a function of applied load cycles. Figure 2 gives a schematic view of the test apparatus, showing the grips which were bolted and bonded to the tension specimen, the placement of the viewing mirror, and the camera.

Baseline fatigue crack growth tests [15] conducted with through-thickness edge cracks loaded in four-point bending at a frequency of 2 Hz gave the crack growth equation

$$\frac{da}{dN} = 1.0702 \times 10^{-31} \Delta K^{9.214} \tag{1}$$

Here the units of the fatigue crack growth rate da/dN are inch/cycle and the cyclic stress intensity factor ΔK is expressed in psi-in.$^{1/2}$. Equation 1 is restricted to crack growth rates which fall between 10^{-7} in./cycle and 2×10^{-3} in./cycle (2.54×10^{-7} and 5.08×10^{-3} cm/cycle). Equation 1 agrees well with other data collected from the same sheet of PMMA [16,17]. Additional details of the loading apparatus, specimen preparation procedures, and specimen material properties are given in Refs 15 to 17.

Discussion

This section discusses the behavior of the part-through cracks as they transition into through-the-thickness flaws. Experimental and numerical results are given for corner and embedded surface cracks located at open holes in plates loaded in remote tension. Experimental data are also described for embedded surface cracks located along the bore of a pin-loaded hole.

Corner Crack at Hole

This subsection describes the transition of corner cracks located at an open hole loaded in remote tension. The localized variation in stress intensity fac-

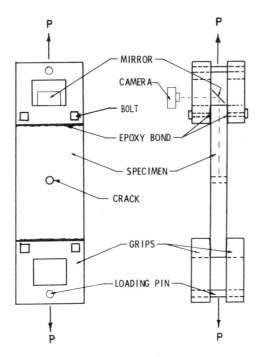

FIG. 2—*Schematic of experimental apparatus for flawed holes.*

tor during the transition period is demonstrated first with numerical results for a hypothetical problem, and then by measured fatigue crack growth rates and corresponding stress intensity factors for an actual transitioning fatigue crack.

Figure 3 presents results of a parameteric study of the transition of a hypothetical corner crack into a uniform through-the-thickness flaw. Here the front face dimension c of the corner crack was fixed at a value $c = D$ = hole diameter = plate thickness T. A series of quarter elliptical corner crack shapes was then analyzed by the FEAM method. The crack dimension a was allowed to increase along the hole bore from an initial value $a = 0.5\,c$, penetrate the back face, and then transition to a uniform through-the-thickness configuration ($a/c = 10$). After back face penetration, the dimension a represents the major axis of the portion of the quarter ellipse used to define the transitioning crack front, and is not an actual crack length dimension. The specific crack shapes analyzed are shown in Fig. 3.

Dimensionless stress intensity factors $K/\sigma\sqrt{D}$ are indicated directly on Fig. 3 at the fixed front face point and at the changing corner/back face location. Since the remote stress σ and hole diameter D are fixed for the present study, these dimensionless results provide the relative variation in the actual stress

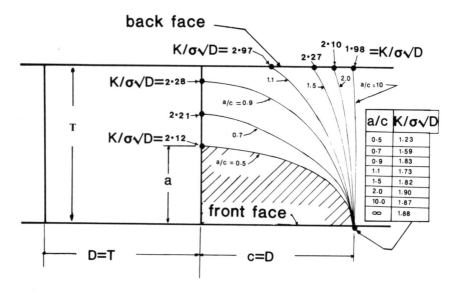

FIG. 3—*Hypothetical corner crack transition showing crack shape changes and dimensionless stress intensity factors at fixed front face point and changing hole bore/back face location.*

intensity factor as the crack shape changes. Although stress intensity factors at other points along the crack perimeter were also obtained from the FEAM analysis, the corner points bounded the K variation and are the only results given in Fig. 3.

Now compare the stress intensity factors at the fixed front face point and at the changing hole bore/back face location. Note that at the front face, $K/\sigma\sqrt{D}$ increases relatively uniformly from a value of 1.23 for the initial crack shape $a/c = 0.5$ to a value $K/\sigma\sqrt{D} = 1.87$ for an aspect ratio $a/c = 10$, although a slight oscillation is apparent around $a/c = 1$. As discussed in Ref 9, the present FEAM computer codes have a numerical programming instability at $a/c = 1$ which precludes analysis of flaw shapes $0.9 < a/c < 1.1$. Note that the front face stress intensity factors are relatively unaffected by the position of the back face penetration point ($1.73 \leq K/\sigma\sqrt{D} \leq 1.87$ for $0.9 \leq a/c \leq 10$). For comparison, the Bowie [18] analysis (as reported in Ref 19) for a through-cracked hole gives $K/\sigma\sqrt{D} = 1.88$ when $c/D = 1$. Thus, assuming a uniform through-the-thickness flaw of maximum length c during the transition period, the Bowie through-crack analysis gives a reasonable estimate for the stress intensity factor at the leading point of crack advance on the front face.

The stress intensity factor at the trailing point, where the crack penetrates the back face, varies significantly during the transition period. From Fig. 3, note that $K/\sigma\sqrt{D}$ at the hole bore/back face point increases from 2.12 at the initial crack shape $a/c = 0.5$, reaches a value 2.97 at $a/c = 1.1$, and then *decreases* to $K/\sigma\sqrt{D} = 1.98$ for the through-thickness shape $a/c = 10$. The fact that the back face value only decreases to 1.98 at $a/c = 10$, rather than to $K/\sigma\sqrt{D} = 1.87$ as seen at front face, indicates that there must still be a crack curvature effect even for this nearly uniform through-the-thickness flaw. From this change in K with increasing crack length, one would expect the back face crack dimension to initially extend quite rapidly after back face penetration, and then decrease to a slower growth rate as a uniform through-the-thickness geometry is achieved. As shown by the following discussion of experimental results, this increasing/decreasing growth rate is also observed in the fatigue tests.

The first experiment described here was originally conducted by Snow [17] and has subsequently been examined in more detail by the present authors. Snow subjected a 7.95 in. (20.2 cm) wide by 0.698 in. (1.77 cm) thick PMMA plate to a remote cyclic stress which varied between 18 and 630 psi (0.124 to 4.34 MPa) at a frequency of 2 Hz. A corner crack was initiated at a small notch machined at the edge of the 0.739 in. (1.88 cm) diameter hole. The loading apparatus shown schematically in Fig. 2 allowed the crack plane to be photographed during the fatigue test (recall the specimen material is transparent). Although Snow [17] measured only the major and minor axes of the corner crack, his original filmstrips have been re-examined with a photo interpreter/digitizer, and a much more detailed record of crack growth has been obtained [20]. Figure 4 shows the growth of the original corner crack (as measured 2000 cycles after the test began) into a fairly uniform through-the-thickness flaw at a cyclic life of 24 900 cycles. In Fig. 4, the origin of the x-y coordinate system is located at the intersection of the hole bore with the front face of the plate, the individual data points represent points measured along the crack front at various cyclic lives, and the solid lines are portions of ellipses fit through the measured points. (The actual coordinates of the digitized crack shape measurements are given in Ref 20.) For purposes of the present analysis, the ellipses were required to go through the point of front face penetration (crack position where $x = c, y = 0$) and through the hole bore/back face point (either the $x = 0, y = a$ or $x = c_t, y = T$ location). The ellipse origin coincides with the x-y origin in all cases. Note that the part-elliptical model represents the actual crack shapes quite well prior to back face penetration. During the transition/through-crack period, however, the actual crack fronts lead the elliptical shapes somewhat in the specimen interior.

Figures 5 and 6 present the change in crack dimensions and corresponding stress intensity factors as a function of applied load cycles during the transition period. Examining the Fig. 5 data first (open symbols), note the growth of the hole bore (a) or back face (c_t) crack dimension. Initially, the crack grows at a uniformly increasing rate along the bore of the hole until penetrat-

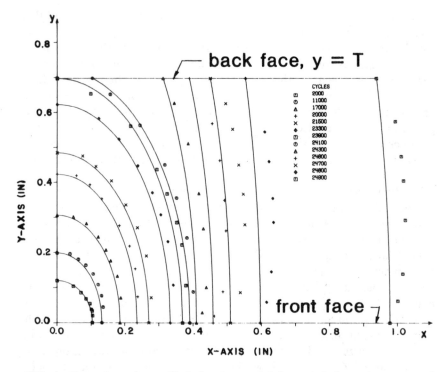

FIG. 4—*Comparison of part-elliptical and natural fatigue crack shapes for corner and through-cracked holes.*

ing the back face of the plate. The back face dimension c_t then increases very rapidly initially, but eventually slows down as a uniform through-thickness configuration is achieved. Finally, the back face length c_t approaches the front face dimension c, and both crack lengths grow at similar rates until final fracture (which occurred shortly after 24 900 cycles). Meanwhile, note that the front face crack length c grows at a uniformly increasing rate during the entire transition period, and is apparently little affected by the localized increase/decrease/increase in rate at the back face.

Now examine the dimensionless stress intensity factor curves given in Fig. 6 (solid symbols). Here $K/\sigma\sqrt{D}$ is plotted for the crack sizes and shapes corresponding to the current cyclic life N. (Note that the $K/\sigma\sqrt{D}$ curves in Fig. 6 are superimposed on the fatigue crack growth data given earlier in Fig. 5.) The finite-element-alternating method was used to compute K at the leading front face crack position and at the trailing crack bore/back face location. For the FEAM analysis, it was assumed that the hole diameter equaled the plate thickness, and that the fatigue cracks had the part-elliptical shapes shown in Fig. 4. For comparison purposes, the Bowie [18] analysis was also used to compute stress intensity factors for uniform through-thickness cracks

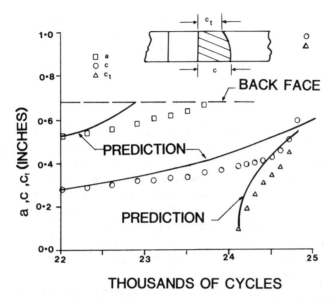

FIG. 5—*Fatigue crack growth during transition of corner crack into through-thickness flaw.*

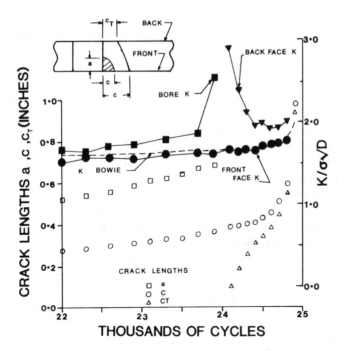

FIG. 6—*Comparison of stress intensity factors and crack length changes as function of elapsed cycles during transition of corner crack.*

of length c and is shown as the dashed line in Fig. 6. The following least-squares representation [21] of the Bowie solution was used here and is expected to be accurate within $\pm 3\%$ for $a/R \leq 10$.

$$K_I = \sigma\sqrt{\pi a}\left[\frac{0.8733}{0.3245 + a/R} + 0.6762\right] \qquad (2)$$

As seen in the hypothetical transition case considered in Fig. 3, K again increases at the hole bore position as the flaw grows, jumps to a large value along the back face immediately after transition, and then *decreases* along the back face as the crack transitions to the uniform through-thickness configuration. Meanwhile, the stress intensity factor at the front face increases smoothly as the front face crack length c slowly grows by fatigue. Again, the front face K is apparently little affected by the large changes occurring at the hole bore/back face location. Following back face penetration, the front face stress intensity factor can be closely approximated by the through-thickness solution for a crack of length c given by Eq 2.

The numerical and experimental results can also be compared by combining the predicted stress intensity factors with the fatigue crack growth law given by Eq 1 and integrating for cyclic life as a function of crack length. The predicted crack growth curves obtained in this manner are shown by the solid lines in Fig. 5. Here the stress intensity factor curves in Fig. 6 at the hole bore position (corresponding to crack dimension a), the back face location (corresponding to crack length c_t), and the front face position (crack length c) were treated independently to make the crack length predictions shown in Fig. 5. Thus the observed crack shape changes are incorporated in the stress intensity solutions and are implicit in the life calculations. The predictions begin with the actual flaw dimensions at 22 000 cycles of life. The c_t curve begins with the first measured flaw shape after back face penetration ($N = 24\ 100$), rather than at the predicted point where $a = T(22\ 900$ cycles), since the stress intensity factors were obtained for the observed crack shapes.

Although the crack shapes were constrained to the observed behavior and not a free parameter in the life prediction scheme, the predictions of Fig. 5 do verify the FEAM calculations. Note in Fig. 5 that the front and back face crack length predictions agree quite well with the experimental behavior. The predicted growth along the hole bore does, however, exceed the observed rate prior to back face penetration.

Embedded Surface Crack at Hole

Consider the results for the embedded surface crack hole configuration reported in Figs. 7 and 8. The PMMA specimen was 7.97 in. (20.3 cm) wide, 0.720 in. (1.83 cm) thick, and contained a hole with a 0.750 in. (1.91 cm)

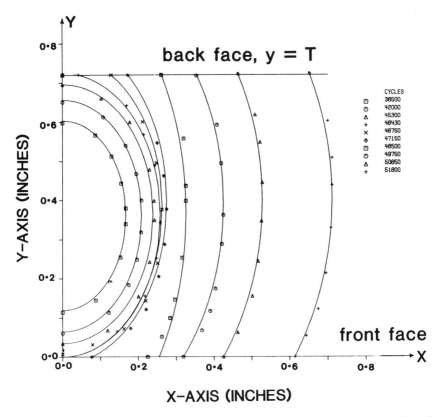

FIG. 7—*Comparison of digitized measurement of surface crack profiles and semielliptical model as function of elapsed cycles.*

diameter D. The specimen was machined from a piece of the original PMMA sheet tested by Snow [17] several years earlier. Care was taken to maintain the same crack orientation and to employ the same annealing cycle used by Snow to minimize potential residual stress effects. The specimen was subjected to a 558 psi (3.85 MPa) cyclic stress ($R = 0.01$) at a frequency of 2 Hz. As before, crack growth was recorded on film, and specific crack shapes were analyzed by the finite-element-alternating method.

Some of the digitized crack growth profiles are shown in Fig. 7. (Additional results for this specimen and for other similar tests are given in Ref 15.) In Fig. 7, the y-axis is oriented with the right edge of the hole bore and the x-axis corresponds to the front face of the plate. The back face of the plate is given by the line $y = 0.72$ in. The symbols in Fig. 7 represent digitized measurements of the crack profiles at various cyclic lives, while the solid lines are portions of ellipses used to model the crack shapes for the FEAM stress inten-

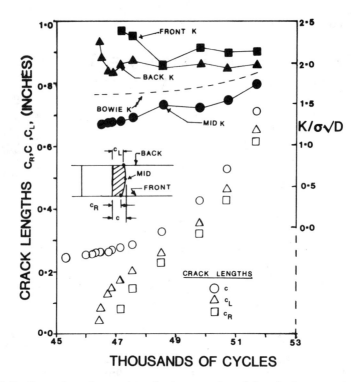

FIG. 8—*Comparison of stress intensity factors and crack length changes as function of elapsed cycles during transition of surface crack.*

sity factor analysis. For this case, the ellipse origin was allowed to shift along the *y*-axis in order to better approximate the actual crack shapes.

The fatigue crack growth curves and stress intensity factor results for the transition period are summarized in Fig. 8 in the same format employed earlier in Fig. 6 for the corner crack. As before, the crack lengths are represented by unconnected open symbols, while the dimensionless stress intensity factor curves are given by solid symbols connected with lines. The FEAM stress intensity factors were computed for the specific crack size and shape corresponding to the particular life in Fig. 8.

Examine the fatigue crack length measurements in Fig. 8 first. Here measurements of the back face dimension c_L, the midlength c, and front face crack length c_R are given as a function of elapsed cycles N. Note that, in this particular test, the initial embedded surface crack was not centered exactly along the hole bore, so that back face penetration (c_L versus N) occurred prior to front face break through (c_R versus N). Nevertheless, both c_L and c_R initially increase quite rapidly, but then grow at a slower rate as both dimensions approach the mid crack length c. This increase/decrease in crack growth rate, as a uniform through-thickness configuration is achieved, agrees with the corner crack results described earlier.

Now consider the dimensionless stress intensity factor $K/\sigma\sqrt{D}$ results given in Fig. 8. Note that again the stress intensity factors at the front and back face points are initially largest after penetration through the thickness, decrease as the crack dimension c_L or c_R increase, and eventually approach the stress intensity factor at the midpoint location along the crack perimeter as a uniform through-thickness shape is achieved. Note, however, that $K/\sigma\sqrt{D}$ at the front and back faces does not show the large changes seen earlier for the corner crack configuration. Perhaps this smaller K variation is due to the fact that transition from an embedded surface crack to a through-thickness flaw does not involve as dramatic a crack shape change as required for the corner crack configuration.

Two other points regarding the stress intensity factor results are of interest in Fig. 8. Firstly, note that the uniform through-thickness stress intensity factor solution labeled "Bowie K" (computed by Eq 2 for the maximum crack length c) gives a fairly good "average" value of the stress intensity factor across the perimeter of the transitioning flaw. Also note in Fig. 8 that at a particular cyclic life, the stress intensity factor is always smallest at the leading position of crack advance (midpoint location) and largest at the trailing point (crack dimension c_R). Thus the stress intensity factors vary along the crack perimeter in a manner which encourages the crack to grow to a uniform through-the-thickness shape.

Pin-Loaded Holes

Reference 15 describes several experiments conducted to simulate growth of surface cracks located along the bore of an attachment lug. As shown schematically in Fig. 9, the specimens considered here had a width $W = 6.75$ in. (17.1 cm), a hole diameter $D = 2.25$ in. (5.72 cm), and a thickness $T = 0.71$ in. (1.8 cm). The specimens were made from the same sheet of PMMA as before. One end of the specimen was loaded with the grip/mirror arrangement described earlier, while the other end was loaded through a close fit steel pin placed in the bore of the hole. The cyclic load P varied between 10 and 1100 lb (44 and 4900 N) for Specimen PT4 at a frequency of 2 Hz, while the load limits for specimen PT5 were 50 and 1200 lb (220 and 5300 N).

The transition portion of the surface crack growth for Specimens PT4 and PT5 are given in Figs.10 and 11. These crack lengths were measured from an image of the 35-mm negatives projected onto a screen.

Note in Fig. 10 that the surface crack transition is similar to the open hole case discussed in the last section. Initially, the dimensions c_L and c_R grow quite rapidly after free surface penetration, slow down as the midlength crack dimension c is reached, and then eventually all three crack dimensions accel-

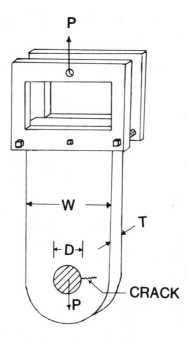

FIG. 9—*Schematic view of pin-loaded attachment lug specimen and loading apparatus.*

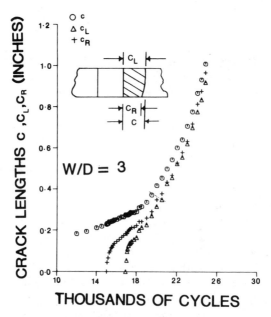

FIG. 10—*Fatigue crack growth curves showing changes in crack dimensions during transition of surface crack at pin-loaded attachment lug.*

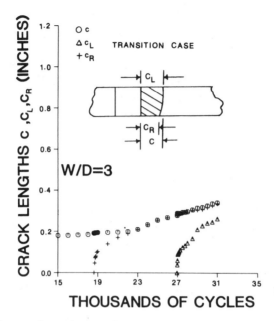

FIG. 11—*Fatigue crack growth curves showing changes in crack dimensions during transition of eccentric surface crack at pin-loaded attachment lug.*

erate at a similar rate until final fracture. Figure 11 presents much the same behavior, except in this case the initial embedded crack was located significantly away from the center of the hole bore, so that the crack penetrated through one free face much earlier than the other. Note that even for this nonsymmetric crack shape, the transitioning crack front grows locally at different rates in an apparent attempt to reach a stable through-the-thickness configuration. No stress intensity factor results are available for the pin-loaded configuration.

Summary and Conclusions

This paper has described experimental and numerical results from of the transition of initial corner and embedded surface cracks int the-thickness flaws. Fatigue crack growth tests conducted with a polymer provided a detailed record of the growth rates an changes during the transition period. A corresponding t' stress intensity factor analysis was performed by + alternating method.

Both experimental and numerical results indica' through cracks try to grow into a stable uniform + figuration. The stress intensity factor varies along

the maximum K occurs at the trailing, rather than the leading, point of crack advance. Moreover, at the trailing point, where the crack penetrates a free face, K varies significantly with crack advance.

The stress intensity factor reaches a large value initially after penetration, but then decreases locally at that point as the free face crack length grows to a uniform through-the-thickness shape. The stress intensity factor at the leading point of maximum crack advance usually has a smaller magnitude than at the trailing point, and is apparently unaffected by the large changes occurring at the trailing position during transition. The leading edge stress intensity factor may often be approximated by the two-dimensional analysis for a uniform through-the-thickness crack whose length equals the distance of maximum crack advance.

The measured fatigue crack growth rates agree with the computed stress intensity factors as the crack perimeter advances locally at rates corresponding to the K variation along the flaw perimeter. In particular, the free face crack dimension immediately grows quite rapidly after penetration, but then decreases to the rate seen by the point of maximum crack advance. Successful predictions for the corner crack transition crack growth using the computed stress intensity factors further verify the finite-element-alternating method stress intensity factor calculations.

Acknowledgments

Portions of this research were supported by AFWAL Flight Dynamics Laboratory Contract F33615-81-K-3206, with J. Rudd as technical monitor. Assistance was also provided by T. Nicholas of the AFWAL Materials Laboratory. The authors deeply appreciate the many discussions with T. E. Kullgren regarding the finite-element-alternating method, and gratefully acknowledge the computer support provided by G. Griffin and C. Malmsten, and the assistance of T. Myers in measuring the crack photographs.

References

[1] Rudd, J. L. in *Part-Through Crack Fatigue Life Prediction, ASTM STP 687*, J. B. Chang, Ed., American Society for Testing and Materials, 1979, pp. 96–112.
[2] Chang, J. B. in *Part-Through Crack Fatigue Life Prediction, ASTM STP 687*, J. B. Chang, Ed., American Society for Testing and Materials, 1979, pp. 156–167.
[3] Peterson, D. E. and Vroman, G. A. in *Part-Through Crack Fatigue Life Prediction, ASTM STP 687*, J. B. Chang, Ed., American Society for Testing and Materials, 1979, pp. 129–142.
[4] Johnson, W. S. in *Part-Through Crack Fatigue Life Prediction, ASTM STP 687*, J. B. Chang, Ed., American Society for Testing and Materials, 1979, pp. 143–155.
[5] Brussat, T. R., Chin, S. T., and Creager, M., "Flaw Growth in Complex Structure, Volume I—Technical Discussion," Technical Report AFFDL-TR-77-79, Vol. I, Air Force Flight Dynamics Laboratory, Wright-Patterson Air Force Base, Ohio, Dec. 1977.
Kullgren, T. E., Smith, F. W., and Ganong, G. P., *Journal of Engineering Materials and Technology*, Vol. 100, April 1978, pp. 144–149.

[7] Kullgren, T. E. and Smith, F. W., *International Journal of Fracture*, Vol. 14, 1968, pp. R319-R322.

[8] Kullgren, T. E. and Smith, F. W., *Journal of Engineering Materials and Technology*, Vol. 101, Jan. 1979, pp. 12-17.

[9] Smith, F. W. and Kullgren, T. E., "Theoretical and Experimental Analysis of Surface Cracks Emanating from Fastener Holes," Technical Report AFFDL-TR-76-104, Air Force Flight Dynamics Laboratory, Wright-Patterson Air Force Base, Ohio, 1977.

[10] Wilson, E. L., "Finite Element Analysis of Mine Structure," Technical Report Bureau of Mines OF 27-73, Denver Mining Research Center, Sept. 1972.

[11] Shaw, R. C. and Kobayashi, A. S., *Engineering Fracture Mechanics*, Vol. 3, No. 1, July 1971.

[12] Grandt, A. F., Jr., and Kullgren, T. E., *Journal of Engineering Materials and Technology*, Vol. 103, No. 2, April 1981, pp. 171-176.

[13] Grandt, A. F., Jr., *Engineering Fracture Mechanics*, Vol. 14, No. 4, 1981, pp. 843-852.

[14] Grandt, A. F., Jr., and Kullgren, T. E., "A Compilation of Stress Intensity Factor Solutions for Flawed Fastener Holes," Technical Report AFWAL-TR-81-4112, Air Force Wright Aeronautical Laboratory, Wright-Patterson Air Force Base, Ohio, Nov. 1981 (condensed version appears in *Engineering Fracture Mechanics*, Vol. 18, No. 2, 1983, pp. 435-451).

[15] Harter, J. A., "Fatigue Crack Growth of Embedded Flaws in Plate and Lug Type Fastener Holes," M.S. thesis, School of Aeronautics and Astronautics, Purdue University, W. Lafayette, Ind., May 1982.

[16] Grandt, A. F., Jr., and Hinnerichs, T. D., "Stress Intensity Factor Measurements for Flawed Fastener Holes," AMMRC MS 74-8, Army Materials and Mechanics Research Center, Watertown, Mass., Sept. 1974, pp. 161-176.

[17] Snow, J. R., "A Stress Intensity Factor Calibration for Corner Flaws at an Open Hole," Technical Report AFML-TR-74-282, Air Force Materials Laboratory, Wright-Patterson Air Force Base, Ohio, 1975.

[18] Bowie, O. L., *Journal of Mathematics and Physics*, Vol. 35, 1956, pp. 60-71.

[19] Paris, P. C. and Sih, G. C. in *Fracture Toughness Testing and its Applications, ASTM STP 381*, American Society for Testing and Materials, 1964, pp. 30-81.

[20] Grandt, A. F., Jr., and Macha, D. E., *Engineering Fracture Mechanics*, Vol. 17, No. 1, 1983, pp. 63-73.

[21] Grandt, A. F., Jr., *International Journal of Fracture*, Vol. 11, No. 2, April 1975, pp. 283-294.

C. R. Saff[1] *and K. B. Sanger*[1]

Part-Through Flaw Stress Intensity Factors Developed by a Slice Synthesis Technique

REFERENCE: Saff, C. R. and Sanger, K. B., **"Part-Through Flaw Stress Intensity Factors Developed by a Slice Synthesis Technique,"** *Fracture Mechanics: Fifteenth Symposium, ASTM STP 833,* R. J. Sanford, Ed., American Society for Testing and Materials, Philadelphia, 1984, pp. 24–43.

ABSTRACT: Part-through-the-thickness flaws are the most common type of flaw occurring in metal structure. Accurate prediction of their growth is vital to ensure adequate life. The majority of stress intensity factor solutions for these flaws are developed through finite-element analyses, iterative techniques, or less accurate superposition of simple solutions. Recently the authors have developed and extended a slice synthesis technique for computation of stress intensity factors for part-through flaws. This technique is far less expensive than finite-element methods, yet provides excellent accuracy. This paper presents the derivation of the method and recently obtained solutions for surface flaws, corner cracks at holes, and corner cracks at the edge of plates in tension and bending. Simple analytical expressions have been fit to these solutions which can be incorporated into computer routines for crack growth prediction.

KEY WORDS: stress intensity factors, surface flaws, corner flaws, mathematical models, crack propagation

Part-through-the-thickness flaws are the most common type of flaw occurring in metal structure. Accurate prediction of their growth is vital to ensure adequate life. The majority of stress intensity factor solutions for these flaws are developed through finite-element analyses, iterative techniques, or superposition of simple solutions [1]. Recently the authors have extended development of a slice synthesis technique for computation of stress intensity factors for part-through flaws. This technique is far less expensive than finite-element methods, yet provides excellent accuracy.

The slice synthesis technique was originally formulated by Fujimoto [2] and

---

[1]Technical Specialist and Engineer, respectively, Structural Research, McDonnell Aircraft Company, McDonnell Douglas Corporation, St. Louis, Mo. 63166.

is an extension of the line-spring model proposed by Rice and Levy [3]. Fujimoto's method was developed for analysis of flaws at fastener holes [4]. Later the method was extended by Dill and Saff to analysis of surface flaws in tension [4,5]. Recent improvements in the weight functions and solution technique have allowed the authors to use the same formulation to compute stress intensity factors for part-through flaws either centrally located or at holes in plates.

Derivation

As shown in Fig. 1, the part-through flaw is idealized as a system of horizontal slices in the $\bar{x}$-$\bar{y}$ plane, each containing a central through-crack (with or without a hole depending upon the solution desired) whose length is determined by the locations through the plate thickness at which the slice was taken. Each slice is considered to react independently to the applied stress (σ), but is coupled through the introduction of pressure distribution (p^*) acting on the faces of the cracks. The pressure p^* is determined by a second system of vertical slices in the $\bar{z}$-$\bar{y}$ plane. Each of the vertical slices contains an edge crack over which the pressure p^* acts in opposition to that applied to the center crack slices.

Thus there are two slice systems: center cracks and edge cracks. These sys-

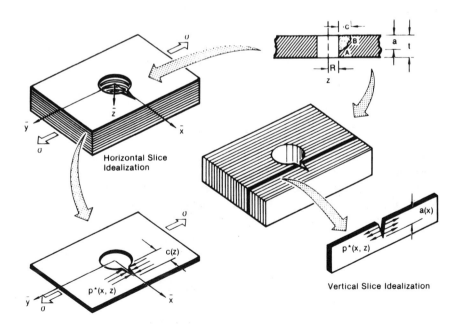

FIG. 1—*Slicing procedure for synthesizing three-dimensional crack solutions.*

tems are coupled by the pressure distribution p^* acting on the crack surfaces of each system and causing the displacements of the two systems to be equal.

Using the crack face pressure distribution, stress intensity factors at the maximum depth (A) and at the surface (B) can be determined from the weight functions

$$K_A = \int_0^{a_0} p^*(0, z)g(z, a_0)_{\text{v.s.}} dz \tag{1}$$

$$K_B = \int_0^{c_0} \{(\sigma - p^*(x, 0))\}g(x, c_0)_{\text{h.s.}} dx \tag{2}$$

where for the surface flaw we have assumed

$$p^*(x, z) = \sum_{i=0}^{2} \sum_{j=0}^{2} \alpha_{ij} x^i z^i \tag{3}$$

where $x = \bar{x} - R$ and $z = \bar{z}$.

The coefficients α_{ij} are determined by the requirements for continuity of displacements over the crack face:

$$v_{\text{h.s.}}(x, z) = v_{\text{v.s.}}(x, z) \tag{4}$$

These displacements can be determined from the pressure distributions and weight functions for each flawed slice system. For the horizontal slice:

$$v_{\text{h.s.}} = \frac{1}{E_{\text{h.s.}}} \int_x^{c(z)} K(\xi)_{\text{h.s.}} g(x, \xi)_{\text{h.s.}} d\xi \tag{5}$$

where ξ is radial crack length from 0 to $c(z)$, and $E_{\text{h.s.}}$ is the stiffness of the horizontal slice and is discussed later.

Substitution of the weight function representation for stress intensity factor (Eq 2) into the displacement expression produces more consistent computation of displacements for each crack face displacement. The resulting expression is

$$v_{\text{h.s.}} = \frac{1}{E_{\text{h.s.}}} \int_x^{c(z)} \int_0^{\xi} \{\sigma - p^*(u, z)\}g(u, \xi)_{\text{h.s.}} du\, g(x, \xi)_{\text{h.s.}} d\xi$$

Using a similar formulation for $v_{\text{v.s.}}$, Eq 4 is rewritten as

$$\frac{1}{E_{\text{h.s.}}} \int_x^{c(z)} \int_{-\xi}^{\xi} \{\sigma - p^*(u, z)\} g(u, \xi)_{\text{h.s.}} du \, g(x, \xi)_{\text{h.s.}} d\xi$$

$$= \frac{1}{E_{\text{v.s.}}} \int_z^{a(x)} \int_0^{\xi} p^*(x, u) g(u, \xi)_{\text{v.s.}} du \, g(z, \xi)_{\text{v.s.}} d\xi \quad (6)$$

The solution scheme used to find p^* is the same as that used by Fujimoto; p^* is expressed as a power series (from Eq 3):

$$p^*(x, z) = \sum_{i=0}^{2} \sum_{j=0}^{2} \alpha_{ij} x^i z^j \quad (7)$$

and Eq 6 can be rewritten as

$$Y(x, z) = \sum_{i=0}^{2} \sum_{j=0}^{2} \alpha_{ij} X(x, z)_{ij} \quad (8)$$

where

$$Y(x, z) = \sigma \int_x^{c(z)} \int_0^{\xi} g(u, \xi)_{\text{h.s.}} du \, g(x, \xi)_{\text{h.s.}} d\xi \quad (9)$$

and

$$X(x, z)_{ij} = \int_x^{c(z)} \int_0^{\xi} x^i z^j g(u, \xi)_{\text{h.s.}} du \, g(x, \xi)_{\text{h.s.}} d\xi$$

$$+ \frac{E_{\text{h.s.}}}{E_{\text{v.s.}}} \int_z^{a(x)} \int_0^{\xi} x^i z^j g(u, \xi)_{\text{v.s.}} du \, g(z, \xi)_{\text{v.s.}} d\xi \quad (10)$$

To assure that the coefficients α_{ij} represent the displacements over the entire crack face, the continuity expression is evaluated at the 13 points shown in Fig. 2. Then a multiple linear regression scheme is used to determine α_{ij}. Once α_{ij} are found, the stress intensity factors at A and B become (from Eq 1 and 2)

$$K_A = \sum_{j=0}^{2} \int_0^{a_0} \alpha_{0j} z^j g(z, a_0)_{\text{v.s.}} dz \quad (11)$$

$$K_B = \sum_{i=0}^{2} \int_0^{c_0} \{\sigma - \alpha_{i0} x^i\} g(x, c_0)_{\text{h.s.}} dx \quad (12)$$

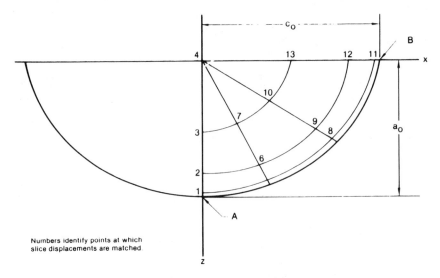

FIG. 2—*Surface flaw model.*

To determine the stiffness of the component slices, $E_{h.s.}$ and $E_{v.s.}$, we considered the displacements of an embedded flaw (Fig. 3) under uniform remote tension (p_0). Returning to a slice idealization we find that, because center cracks take on elliptical displacements, only a constant pressure (βp_0) will be required between slice systems to bring their displacements into agreement with the exact solution [6]. The crack surface displacement field of the horizontal slices is

$$v(x,\ +0,\ z) = \frac{2(1 - \beta)(1 - \mu^2)p_0 a_0}{E_s} \sqrt{1 - \left(\frac{z}{b_0}\right)^2 - \left(\frac{x}{a_0}\right)^2} \qquad (13)$$

and of the vertical slices is

$$v(x,\ +0,\ z) = 2\beta \frac{(1 - \mu^2)p_0 a_0}{E_s} \sqrt{1 - \left(\frac{z}{b_0}\right)^2 - \left(\frac{x}{a_0}\right)^2} \qquad (14)$$

where μ is Poisson's ratio.
Equating these displacements we find

$$\beta = a_0/(a_0 + b_0) \qquad (15)$$

Sneddon and Lowengrub [6] give the exact displacement field as

$$v(x,\ +0,\ z) = \frac{2(1 - \mu^2)p_0 a_0}{E\phi} \sqrt{1 - \left(\frac{z}{b_0}\right)^2 + \left(\frac{x}{a_0}\right)^2} \qquad (16)$$

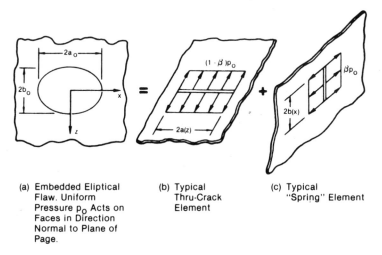

(a) Embedded Elliptical Flaw. Uniform Pressure p_0 Acts on Faces in Direction Normal to Plane of Page.

(b) Typical Thru-Crack Element

(c) Typical "Spring" Element

FIG. 3—*Synthesis idealization of fully embedded elliptical flaw.*

Thus

$$\frac{E_s}{E} = \frac{E_{v.s.}}{E} = \frac{E_{h.s.}}{E} = \phi\left(\frac{b_0}{a_0 + b_0}\right) \tag{17}$$

where

$$\phi = \int_0^{\pi/2} \sqrt{\sin^2\phi + \left(\frac{a_0}{b_0}\right)^2 \cos^2\phi}\, d\phi \tag{18}$$

Obviously, the accuracy of the slice model depends a great deal on the accuracy of the weight functions used. The weight functions used in this analysis are not exact but are accurate approximations. Their derivation is included in the Appendix. Three weight functions were developed: for single or double flaws emanating from a hole, for edge crack slices where out-of-plane deformation was allowed to occur, and for edge crack slices where out-of-plane deformation was restrained.

The weight function derived for the crack from hole slice was developed so that the hole radius could be set to zero for simulation of surface flaws or to infinity for simulation of plates having corner cracks at an edge. The two different solutions for edge cracks were required to account for the effect of plate continuity to restrain out-of-plane deformations occurring for deep flaws.

Analysis of Plate Restraint

Because shear stresses acting on the faces of the "free" edge crack slice are ignored in the slice synthesis model, these slices displace (Fig. 4). In reality,

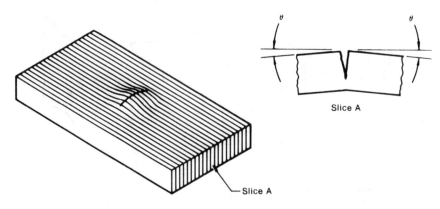

FIG. 4—*Deflection of edge crack slices without moment restraint.*

these types of displacements do not occur because of plate stiffness. Thus the model requires an estimate of the effect of plate stiffness in restraining out of plate deflection of "free" edge crack slices. The following paragraphs describe development of a restrained edge crack element based on analyses using NASTRAN three-dimensional finite elements.

Analyses were performed to determine the effect of rotational restraint offered by the plate to the edge crack slices. As shown in Fig. 5, the stress intensity, and hence displacements, for a "free" edge crack can be represented as

$$K_{\text{free}} = K_{\text{fixed}} + K_m \tag{19}$$

where K_{free} is the stress intensity for a "free" edge crack slice, K_{fixed} is the stress intensity for a "fixed" edge crack slice, and K_m is the stress intensity for an edge crack slice subjected to moments representing load eccentricity.

Values of stress intensity intermediate to free and fixed can be approximated by

$$K_{\text{restrained}} = K_{\text{fixed}} + K'_m \tag{20}$$

where K'_m is related to the amount of rotation (θ) due to the crack in the edge cracked slice, determined by a simple coupled beam analogy:

$$K'_m = \frac{K_{\text{free}} - K_{\text{fixed}}}{1 + \lambda\left(1 + 6\frac{t}{c}\alpha\right)} \tag{21}$$

where $\alpha = \theta E/4\sigma$. From Eq 21:

$$\alpha = \left(\frac{a/t}{1 - a/t}\right)^2 [5.93 - 19.69\,(a/t) + 37.14\,(a/t)^2 - 35.84\,(a/t)^3$$

$$+ 13.12\,(a/t)^4] \quad (22)$$

and

$$K_{\text{restrained}} = K_{\text{fixed}} + \frac{K_{\text{free}} - K_{\text{fixed}}}{1 + \lambda\left(1 + 6\frac{t}{c}a\right)} \quad (23)$$

If λ is zero, the stress intensity is that for a "free" edge crack slice; if λ is very large, the stress intensity approached that for a "fixed edge" crack slice. Analyzing the surface flaw geometry shown in Fig. 2 and using Eq 21 to determine the stress intensities and displacements for the edge crack slices, curves are obtained, labeled "FREE" and "FIXED", shown in Fig. 6.

To obtain an estimate of the restraint offered by the plate to the surface

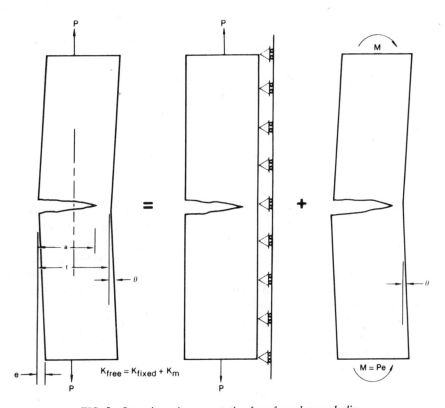

FIG. 5—*Stress intensity computation for a free edge crack slice.*

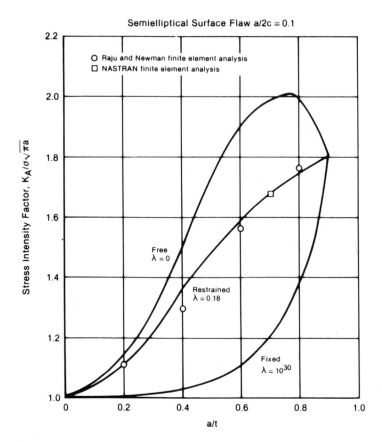

FIG. 6—*Selection of plate restraint coefficient to match finite-element results.*

flaw displacements, three finite-element analyses were performed. These three-dimensional analyses were performed for an embedded circular flaw in a large plate, a semicircular surface flaw in a large plate, and for a surface flaw having $a/2c = 0.1$ and $a/t = 0.7$. The first two analyses were used to show correlation with accepted stress intensity solutions for these problems; the third analysis was used to determine restraint to crack surface displacement provided by the plate and to allow λ to be determined.

Results of these analyses are summarized in Table 1. Estimates of stress intensity were developed through analyses of the crack surface displacements. A displacement function of the form

$$v = [Ax + Bx^2 + Cx^3]^{1/2} \tag{24}$$

TABLE 1—*Finite-element results normalized stress intensity.*

Flaw Geometry	Location	$K/\sigma\sqrt{\pi a}$	$K_{norm}/\sigma\sqrt{\pi a}$
Circular flaw in solid	. . .	0.608	0.637
Semicircular surface flaw ($a/t = 0.7$)	depth surface	0.618 0.741	0.647 0.776
Semielliptic surface flaw ($a/t = 0.7$, $a/2c = 0.1$)	depth surface	1.601 0.479	1.676 0.502

was assumed, and the terms A, B, and C were evaluated by least-squares linear regression analysis. The stress intensity was evaluated by use of the relationship

$$K = \frac{1}{2} \sqrt{\pi r_0} \frac{E}{(1 - \nu^2)} = \sqrt{\frac{\pi A}{8}} \frac{E}{(1 - \nu^2)} \tag{25}$$

where r_0 is the radius of curvature.

Estimates using this procedure are summarized in Table 1. The theoretical stress intensity for a circular flaw in a solid is $K/\sigma\sqrt{\pi a} = 2/\pi = 0.637$. Table 1 indicates the value obtained from the finite-element analysis is 0.608, an error of 4.77%. The analysis results for the other problems were normalized by $(\pi/2)/0.608$, bringing the stress intensity estimate for the circular flaw in agreement with the theoretical value. The estimate for the semicircular surface flaw, after normalizing by this factor, is in good agreement with other solutions for this problem. The estimate obtained from this analysis is $K/\sigma\sqrt{\pi a} = 0.647$, from Smith and Sorenson [7] is 0.653, from Hartranft and Sih [8] is 0.668, and from Tracey [9] is 0.662.

Analysis results for the surface flaw with $a/2c = 0.1$ and $a/t = 0.7$ were used as a basis for selecting λ in the slice synthesis model. When this was accomplished, the value $\lambda = 0.18$ was found by trial and error to yield results consistent with the finite-element analysis (Fig. 6). This value of λ was used in all other analyses, and good agreement with Raju and Newman's results [10] was obtained.

Surface flaw solutions were obtained by analyzing two symmetric corner flaws emanating from a hole in infinitesimal radius. These solutions are close to the results obtained by Raju and Newman [10], who used highly detailed finite-element models for tension and bending (Figs. 7 and 8).

Solutions for single and double flaws were obtained using the same analysis routine. The slice synthesis solutions for symmetric corner flaws were compared with those of Raju and Newman [11] for tension and bending. An example of this comparison is shown in Fig. 9. Agreement is good at intermedi-

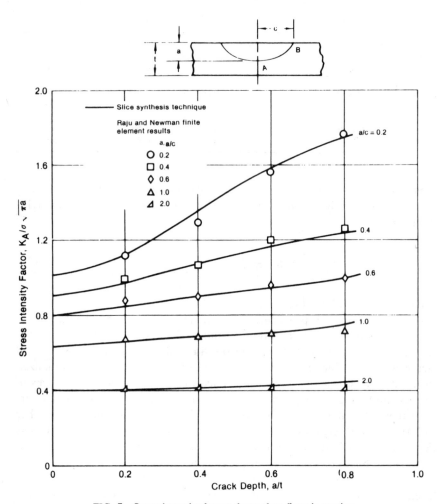

FIG. 7—*Stress intensity factors for surface flaws in tension.*

ate points ($\theta = \pi/6$ and $\theta = \pi/3$) but poor at the bore of the hole and the surface of the plate ($\theta = 0$ and $\theta = \pi/2$). The poorer agreement at the bore and plate surfaces is believed to be caused by a change in crack tip singularity at the junction of the crack and a free surface. This change is reflected in the finite-element results but not in the slice synthesis results. The change in singularity at the interface between a crack and a surface has been discussed by several authors [12-14], but currently the crack-surface interface is a zone of uncertainty. Correlation with test results using the current solutions has been good [15-17].

The corner flaw in an unnotched plate commonly occurs in caps of frames and spars of long-lived aircraft. Stress intensity factor solutions are not avail-

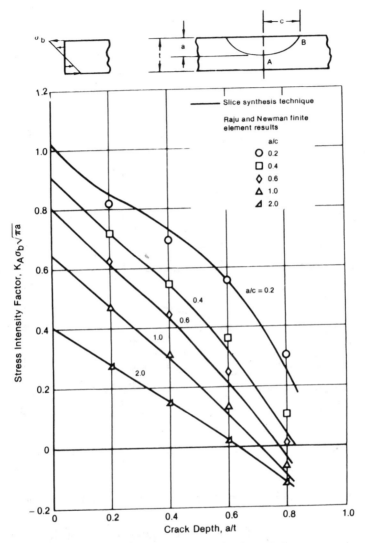

FIG. 8—*Stress intensity factors for surface flaws in bending.*

able for most of these cases. Slice synthesis solutions are obtained by letting the hole radius approach infinity. Results for tension are shown in Fig. 10. The only other analytical solution for a corner flaw in a plate was found from the finite-element results of Tracey [18] also shown in Fig. 10.

The authors have curve fit a spectrum of slice synthesis results for part-through flaws for both single and double flaws and tension or bending loads. These curve fits are shown in Figs. 7 to 10. Future results may be obtained through the use of these analytical expressions.

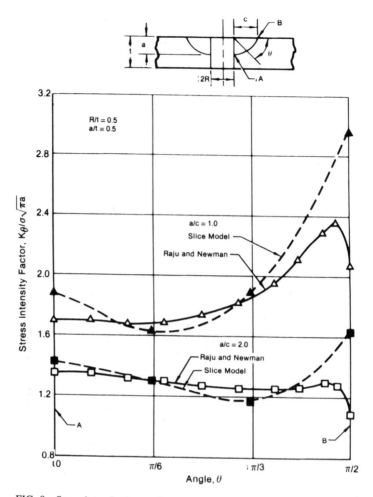

FIG. 9—*Stress intensity factors for symmetric corner flaws at holes in tension.*

Slice synthesis solutions can be used to analyze part-through flaws under combined tension and bending loads, through superposition. For example, consider a semicircular surface flaw under equal combined tension and bending stresses. Stress intensity factors at the surface and depth computed from superposition of tension and bending results are compared with results from direct analysis using the slice synthesis technique in Fig. 11. As expected, the correlation is excellent.

Summary

The primary advantages of the slice synthesis technique over finite-element solutions are in modeling time and cost and versatility. Modeling time is neg-

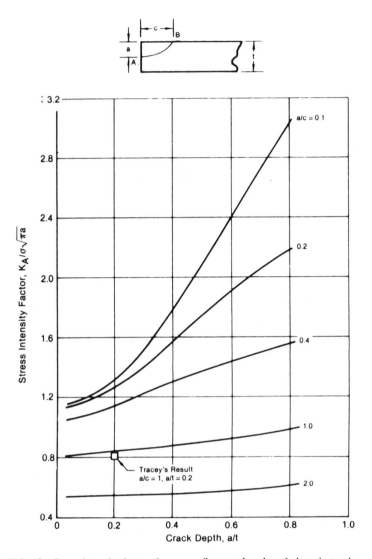

FIG. 10—*Stress intensity factors for corner flaws at the edge of plates in tension.*

ligible for the slice technique. Weight function derivation can be time consuming, particularly for checkout. However, many weight functions are available in the literature now and rapid techniques for development of weight functions are also being published [19].

Depending on the solution to be obtained, 10 to 50 slice synthesis solutions can be obtained for the cost of a single finite-element solution using 700 degrees of freedom. In addition, the slice technique solutions will always tend

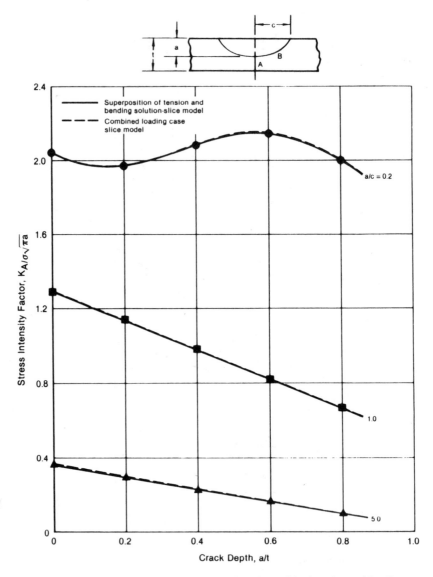

FIG. 11—*Stress intensity factors for surface flaws in combined tension and bending.*

toward accepted solutions for limiting cases since they are based upon solutions for such cases.

This technique is limited only by the weight functions available for modeling flaws. Using available weight functions, this technique can be extended to include embedded flaw solutions, flaws which have broken through the back face of the plate, and other geometries. Because of the reliance on weight function techniques, any applied or residual stress distribution can be analyzed, and, by iteration, contact problems such as the cracked pin-loaded hole can also be modeled.

Acknowledgments

The authors wish to acknowledge the work of Paul McAvoy, who, while a co-op student at MCAIR from Iowa State University, developed the weight function for a radial crack from a hole used herein.

APPENDIX

Weight Functions

The weight functions used in this analysis are not exact derivations but are felt to be accurate approximations. For the horizontal slices of Fig. 1, a weight function for single or double cracks from a hole was developed. This development was based on the methods developed by Fujimoto, Forman, and Hsu [2,20,21]. The new weight function is based primarily on that derived by Fujimoto for a short single flaw emanating from a hole. The work of Shivakumar and Forman was used to extend this solution to longer flaws and to the central crack solution by letting the hole radius go to zero. Hsu's work with the Green's theorem approach led to a simple correction which allowed the weight function to account for double cracks and finite width effects.

The resulting weight function used for analysis of central through cracks and cracks from holes is summarized herein.

For $c/R > 5$:

$$g_c(x, c) = \phi_{FW} \sqrt{\left(\frac{x}{c} + \frac{2R}{c}\right) \bigg/ \left[\left(\frac{1}{2} + \frac{R}{c}\right)\left(1 - \frac{x}{c}\right)\pi c\right]} \qquad \text{for single flaw}$$

$$g_c(x, c) = 2\phi_{FW} \sqrt{\{(c + R)/[(c + R)^2 - (x + R)^2]\}} \qquad \text{for double flaws}$$

For $c/R < 3$:

$$g_r(x, c) = \phi_{FW} [m_1(x, c) + m_2(x, c) + m_3(x, c)] \qquad \text{for single flaw}$$

$$g_r(x, c) = \phi_{DF}\phi_{FW}[m_1(x, c) + m_2(x, c) + m_3(x, c)] \qquad \text{for double flaws}$$

For $3 \leq c/R \leq 5$:

$$g(x, c) = \left(\frac{5}{2} - \frac{c}{2R}\right)g_r(x, c) + \left(\frac{c}{2R} - \frac{3}{2}\right)g_c(x, c) \qquad \text{for single or double flaws}$$

where $\phi_{FW} = \phi_1 \phi_2$ is a finite width correction.

$$\phi_1 = \frac{\sqrt{\pi(c + R)}}{w} \tan \frac{\pi(a + R)}{w}$$

$$\times \left[\frac{\sqrt{1 - \left(\frac{x + R}{c + R}\right)^2} + 0.297 \left[1 - \left(\frac{x + R}{c + R}\right)^2\right]\left[1 - \cos \frac{\pi(c + R)}{w}\right]}{\sqrt{1 - \left(\cos \frac{\pi(c + R)}{w} \middle/ \cos \frac{\pi(x + R)}{w}\right)^2}} \right]$$

$$\phi_2 = \left[1 + \frac{\frac{c}{2}\left(\frac{2R}{w}\right)}{\left(\frac{w}{2} - R\right)}\right]^{1.8}$$

$\phi_{DF} = \phi_3/(1 + \phi_4)$ is a single-to-double flaw correction

$$\phi_3 = 2 \sqrt{\frac{(c + R)(c/2 + R)(c - x)}{(c + R)^2 - (x + R)^2 (2R - x)}}$$

$$\phi_4 = \left[0.1612 \left(\frac{c/R}{1 + c/R}\right)^{1.24} - 0.008252 \left(\frac{c/R}{1 + c/R}\right)^{1.07}\right] \times$$

$$\sin \left[\pi \left\{2.31 \left(\frac{c/R}{1 + c/R}\right)^{0.57} - 1\right\}\right]$$

In the radial crack weight function, following the derivation of Fujimoto [2],

$$m_1(x, c) = \frac{\partial \beta_1(x, c)}{\partial c} \frac{\sqrt{c}}{\beta(c)} \left(A + 2C \frac{x}{c}\right) \sqrt{1 - \frac{x}{c}}$$

$$m_2(x, c) = \frac{\partial \beta_1(x, c)}{\partial c} \frac{\sqrt{c}}{\beta(c)} \left(\frac{x}{c}\right) \left(\frac{A}{2} + B\right) \ln \left[\frac{2c}{x} \left(\sqrt{1 - \frac{x}{c}} + 1\right) - 1\right]$$

$$m_3(x, c) = \frac{\beta_1(x, c)}{\beta_2(c)} \left[A + B \left(\frac{x}{c}\right) + C \left(\frac{x}{c}\right)^2\right] \middle/ \sqrt{c - x}$$

where

$A = 1.487975,$
$B = -0.889721,$ and
$C = 0.199631.$

$$\beta_1 = f + g\left(\frac{x}{c}\right) + h\left(\frac{x}{c}\right)^2$$

$$f = 2.7215\left(\frac{1}{1+c/R}\right) - 6.2066\left(\frac{1}{1+c/R}\right)^2 + 7.588\left(\frac{1}{1+c/R}\right)^3$$

$$- 3.1057\left(\frac{1}{1+c/R}\right)^4$$

$$g = 0.37547\left(\frac{c/R}{1+c/R}\right) + 0.109753\left(\frac{c/R}{1+c/R}\right)^2 + 0.450902\left(\frac{c/R}{1+c/R}\right)^3$$

$$h = -\left[0.083419\left(\frac{c/R}{1+c/R}\right) + 0.038756\left(\frac{c/R}{1+c/R}\right)^2 + 0.19357\left(\frac{c/R}{1+c/R}\right)^3\right]$$

$$\beta_2 = 0.4829 + 1.86\left(\frac{1}{1+c/R}\right) - 3.298\left(\frac{1}{1+c/R}\right)^2 - 2.246\left(\frac{1}{1+c/R}\right)^3$$

$$+ 10.106\left(\frac{1}{1+c/R}\right)^4 - 5.908\left(\frac{1}{1+c/R}\right)^5$$

For the vertical slices, edge crack weight functions were derived using Rice's expression [22]

$$g(z, a) = \frac{E}{K(a)}\frac{\partial v(z, a)}{\partial a}$$

The displacement of the edge cracks were assumed to be conic sections described by

$$v(z, a) = v_0(a)\sqrt{\frac{2}{2+m(a)}\left(1 - \frac{Z}{a}\right) + \frac{m(a)}{2+m(a)}\left(1 - \frac{Z}{a}\right)^2}$$

where $m(a)$ can be expressed in terms of the stress intensity factor (K) and the displacement at $z = 0$ as follows:

$$m(a) = \frac{\pi}{a}\left[\frac{v_0(a)E}{2K(a)}\right]^2 - 2$$

Displacements at $z = 0$ and stress intensity factors were determined from the handbook of Tada et al [23].
For the "free" edge crack:

$$K(a) = \sigma\sqrt{2t \tan\frac{\pi a}{2t}}\left[0.752 + 2.02\,(a/t) + 0.37\left(1 - \sin\frac{\pi a}{2t}\right)^3\right]\Bigg/\cos\frac{\pi a}{2t}$$

$$v_0(a) = \frac{4\sigma a}{E}\left[1.46 + 3.42\left(1 - \cos\frac{\pi a}{2t}\right)\right]\Bigg/\left(\cos\frac{\pi a}{2t}\right)^2$$

For the "fixed" edge crack we used the double-edge crack solution:

$$K(a) = \sigma \sqrt{2t \tan \frac{\pi a}{2t}} \left(1 + 0.122 \cos^4 \frac{\pi a}{2t} \right)$$

$$v_0(a) = \frac{4\sigma a}{E} \left(\frac{2t}{\pi a} \right) \left[0.459 \sin \frac{\pi a}{2t} - 0.065 \sin^3 \frac{\pi a}{2t} \right.$$

$$\left. -0.007 \sin^5 \frac{\pi a}{2t} + \cosh^{-1} \left(\sec \frac{\pi a}{2t} \right) \right]$$

References

[1] Newman, J. C., Jr., "Stress Intensity Factor Solution Techniques," presented at ASTM Symposium on Damage Tolerance Analysis, UCLA, Los Angeles, 29 June 1981.

[2] Fujimoto, W. T., "Determination of Crack Growth and Fracture Toughness Parameters for Surface Flaws Emanating from Fastener Holes," in *Proceedings, AIAA/ASME/SAE 17th Structures, Structural Dynamics, and Material Conference*, King of Prussia, Pa., 5-7 May 1976.

[3] Rice, J. R. and Levy, N., *Journal of Applied Mechanics*, Vol. 38, No. 1, March 1972, pp. 185-194.

[4] Dill, H. D. and Saff, C. R., "Environmental-Load Interaction Effects on Crack Growth," AFFDL-TR-78-137, Air Force Flight Dynamics Laboratory, Wright-Patterson Air Force Base, Ohio, Dec. 1978.

[5] Saff, C. R., "Environment-Load Interaction Effects on Crack Growth in Landing Gear Steels," NADC-79095-60, Naval Air Development Center, Warminster, Pa., Dec. 1980.

[6] Sneddon, I. N. and Lowengrub, M., *Crack Problems in the Classical Theory of Elasticity*, Wiley, New York, 1969.

[7] Smith, F. W. and Sorenson, D. R., *International Journal of Fracture*, Vol. 12, No. 1, Feb. 1976, pp. 47-58.

[8] Hartranft, R. J. and Sih, G. C., "Solving Edge and Surface Crack Problems by an Alternating Method," in *Mechanics of Fracture-1-Methods of Analysis and Solutions of Crack Problems*, G. C. Sih, Ed., Noordhoff, Leyden, The Netherlands, 1973.

[9] Tracey, D. M., *International Journal of Fracture*, Vol. 9, No. 3, Sept. 1973, pp. 340-343.

[10] Raju, I. S. and Newman, J. C., Jr., *Engineering Fracture Mechanics*, Vol. 11, No. 4, 1979, pp. 817-829.

[11] Raju, I. S. and Newman, J. C., Jr., in *Fracture Mechanics, ASTM STP 677*, American Society for Testing and Materials, 1979, pp. 411-430.

[12] Raju, I. S. and Newman, J. C., Jr., "Three-Dimensional Finite-Element Analysis of Finite-Thickness Fracture Specimens," NASA TN D-8414, National Aeronautics and Space Administration, Hampton, Va., May 1977.

[13] Hartranft, R. J. and Sih, G. C., *International Journal of Engineering Science*, Vol. 8, 1979, pp. 711-729.

[14] Schijve, J., *Engineering Fracture Mechanics*, Vol. 14, No. 4, 1981, pp. 789-800.

[15] Fujimoto, W. T. and Saff, C. R., *Experimental Mechanics*, Vol. 22, No. 4, April 1982, pp. 139-146.

[16] Saff, C. R., "F-4 Service Life Tracking Program (Crack Growth Gages)," AFFDL-TR-79-3148, Air Force Flight Dynamics Laboratory, Wright-Patterson Air Force Base, Ohio, Dec. 1979.

[17] Saff, C. R. and Rosenfeld, M. S. in *Design of Fatigue and Fracture Resistant Structures, ASTM STP 761*, American Society for Testing and Materials, 1982, pp. 234-252.

[18] From D. P. Rooke and D. J. Cartwright, *Compendium of Stress Intensity Factors*, Her Majesty's Stationery Office, London, 1976.

[19] Petroski, H. J. and Achenback, J. D., *Engineering Fracture Mechanics*, Vol. 10, No. 2, 1978, pp. 257–266.

[20] Forman, R. G. and Shivakumar, V., *International Journal of Fracture*, Vol. 16, No. 4, 1980, pp. 305–316.

[21] Hsu, T. M., McGee, W. M., and Aberson, J. A., "Extended Study of Flaw Growth at Fastener Holes, " AFFDL-TR-77-83, Air Force Flight Dynamics Laboratory, Wright-Patterson Air Force Base, Ohio, Vol. 1, April 1978.

[22] Rice, J. R., *International Journal of Solids and Structures*, Vol. 8, No. 6, 1972, pp. 751–758.

[23] Tada, H., Paris, P. C., and Irwin, G. R., *The Stress Analysis of Cracks Handbook*, Del Research Corp., Hellertown, Pa., 1973.

M. M. Ratwani[1] and H. P. Kan[1]

Analysis and Growth of Cracks in Skins with Variable Thickness

REFERENCE: Ratwani, M. M. and Kan, H. P., **"Analysis and Growth of Cracks in Skins with Variable Thickness,"** *Fracture Mechanics: Fifteenth Symposium. ASTM STP 833*, R. J. Sanford, Ed., American Society for Testing and Materials, Philadelphia, 1984, pp. 44–56.

ABSTRACT: Analysis of cracks in skins with variable thickness (skins with lands) was developed. The problem was reduced to the solution of an integral equation by using the Fourier transform technique. The integral equation was solved numerically and stress intensity factors were obtained. The influence of variable skin thickness ratios on stress intensity factors was investigated.

An experimental program was carried out to obtain crack growth data in skins with variable thickness (skins with lands). The crack growth behavior in those skins was predicted using analytical stress intensity factors and a crack growth model. A good correlation between predicted and observed crack growth behavior under constant amplitude as well as spectrum loading was found.

KEY WORDS: Fourier transforms, integral equations, crack growth, spectrum loading, variable thickness skins

Nomenclature

a	Half crack length
E_1, μ_1	Young's and shear moduli, respectively, of land material
E_2, μ_2	Young's and shear moduli, respectively, of skin material
f_1, g_1, ϕ_1	Unknown functions for the strip representing the land
f_2, g_2	Unknown functions for the half plane representing the skin
h	Half land width
u_1, v_1	Displacements in x and y directions, respectively, in the land
u_2, v_2	Displacements in x and y directions, respectively, in the skin
t_1	Land thickness
t_2	Skin thickness
$\sigma_{xx}^1, \sigma_{yy}^1, \sigma_{xy}^1$	Stresses in land
$\sigma_{xx}^2, \sigma_{yy}^2, \sigma_{xy}^2$	Stresses in skin
ν_1, ν_2	Poisson ratio for land and skin materials, respectively

[1] Manager and Engineering Specialist, respectively, Structural Life Assurance Research, Northrop Corporation, Aircraft Division, Hawthorne, Calif. 90250.

Typical structural details in aircraft construction often incorporate lands (local buildup in materials) around cutouts or at fastener holes at spars, ribs, or frames for riveted or bolted attachments to the skin. At these regions, stress gradients and peak stresses much greater than the gross design stress may occur. These stress concentrations generally represent fatigue critical areas in aircraft structures. At these locations, fatigue cracks, if not already present, may initiate and propagate during flight loads.

In order to predict crack growth behavior in skins with lands, it is necessary to know the stress intensity factors for the cracks occurring in lands, so that these can be used in a crack growth model for life predictions.

The majority of the analytical and experimental crack growth behavior studies reported in the literature are for skins of constant thickness. In Ref 1, crack growth behavior in skins with discontinuous thickness has been experimentally investigated. It was shown that the crack growth rate was discontinuous when the crack front reached the junction where thickness varied. No correlation with analysis was attempted. Crack growth behavior in skins containing lands and repaired with composite patches has been discussed in Refs 2 and 3.

The analytical techniques of Refs 4 and 5 have been extended here to obtain stress intensity factors for cracks occurring in the lands. The fatigue crack growth in the lands is then predicted using these stress intensity factors, and good correlation is shown with experimental crack growth data.

Analytical Development

Consider the skin with lands with a crack in the land as shown in Fig. 1. For the purposes of simplicity in the present analytical development, the influence of the radius of curvature at the junction of the land and the skin is neglected. It is also assumed that the land is symmetrically placed about the center line of the skin thickness as shown in Fig. 2. Thus the geometry of Fig. 1 will be replaced by that of Fig. 2 for developing analytical techniques.

Formulation of the Problem

The analysis of this problem is based on the following assumptions:

1. The skin and the land are either under generalized plane stress or plane strain conditions.

2. Displacements in the x and y directions (Fig. 2) are continuous at the junction of the skin with the land.

3. The total force in the x direction and the total shear force at the junctions are continuous. However, the stresses in the skin and the land at the junctions are discontinuous.

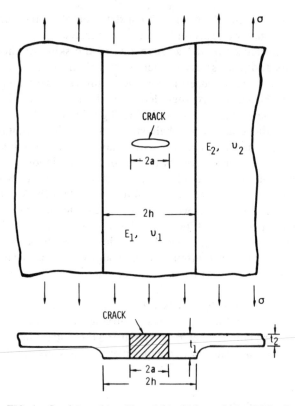

FIG. 1—*Crack in a skin with variable thickness (skin with land).*

In the generalized formulation of the problem, it will be assumed that the skin and the land have different elastic properties. The land and the skin are indicated by Indices 1 and 2, respectively.

The displacement field for a finite-width strip as derived by Sneddon and Lowengrub [6] is a superposition of well-known transform solutions for a body with $x = 0$ and $y = 0$ as planes of symmetry and an upper half plane symmetrical about the y-axis. The displacements for the land are given by

$$u_1(x, y) = -\frac{2}{\pi} \int_0^\infty \left\{ \frac{1}{\eta} \left[f_1(\eta) - \frac{\kappa_1 - 1}{2} g_1(\eta) \right] \sinh(\eta x) \right.$$

$$\left. + x g_1(\eta) \cosh(\eta x) \right\} \cos \eta y d\eta$$

$$- \frac{2}{\pi} \int_0^\infty \frac{\phi_1(\xi)}{\xi} \left(\frac{\kappa_1 - 1}{2} - \xi y \right) e^{-\xi y} \sin \xi x d\xi$$

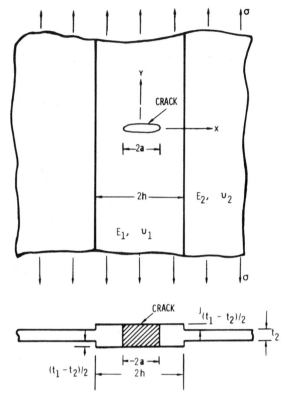

FIG. 2—*Idealized geometry of crack in a land.*

and

$$v_1(x, y) = \frac{2}{\pi} \int_0^\infty \left\{ \frac{1}{\eta} \left[f_1(\eta) + \frac{\kappa_1 + 1}{2} g_1(\eta) \right] \cosh(\eta x) \right.$$

$$\left. + xg_1(\eta) \sinh(\eta x) \right\} \sin \eta y \, d\eta$$

$$+ \frac{2}{\pi} \int_0^\infty \frac{\phi_1(\xi)}{\xi} \left(\frac{\kappa_1 + 1}{2} + \xi y \right) e^{-\xi y} \cos \xi x \, d\xi \quad (1)$$

where

$\kappa_1 = 3 - 4\nu_1$ for plane strain, and
$\kappa_1 = (3 - \nu_1)/(1 + \nu_1)$ for plane stress.

The corresponding stress field is given by the equations

$$\frac{\sigma_{xx}^1\,(x,\,y)}{2\mu_1} = -\frac{2}{\pi}\int_0^{\infty}\left[f_1(\eta)\cosh(\eta x) + \eta x g_1(\eta)\sinh(\eta x)\right]\cos\eta y d\eta$$

$$-\frac{2}{\pi}\int_0^{\infty}\phi_1(\xi)(1 - \xi y)e^{-\xi y}\cos\xi x d\xi$$

and

$$\frac{\sigma_{yy}^1\,(x,\,y)}{2\mu_1} = \frac{2}{\pi}\int_0^{\infty}\left\{\left[f_1(\eta) + 2g_1(\eta)\right]\cosh(\eta x)\right.$$

$$\left. + \eta x g_1(\eta)\sinh(\eta x)\right\}\cos\eta y d\eta$$

$$-\frac{2}{\pi}\int_0^{\infty}\phi_1(\xi)(1 + \xi y)e^{-\xi y}\cos\xi x d\xi$$

and

$$\frac{\sigma_{xy}^1\,(x,\,y)}{2\mu_1} = \frac{2}{\pi}\int_0^{\infty}\left\{\left[f_1(\eta) + g_1(\eta)\right]\sinh(\eta x)\right.$$

$$\left. + \eta x g_1(\eta)\cosh(\eta x)\right\}\sin\eta y d\eta$$

$$-\frac{2}{\pi}\int_0^{\infty}\xi y\phi_1(\xi)e^{-\xi y}\sin\xi x d\xi \qquad (2)$$

Similarly, displacement and stress fields for the skin (considered as a half plane) can be expressed as

$$u_2(x,\,y) = \frac{2}{\pi}\int_0^{\infty}\left\{\frac{1}{\eta}\left[f_2(\eta) + \frac{\kappa_2 - 1}{2}g_2(\eta)\right] + xg_2(\eta)\right\}e^{-\eta x}\cos\eta y d\eta$$

$$(3)$$

$$v_2(x,\,y) = \frac{2}{\pi}\int_0^{\infty}\left\{\frac{1}{\eta}\left[f_2(\eta) - \frac{\kappa_2 + 1}{2}g_2(\eta)\right] + xg_2(\eta)\right\}e^{-\eta x}\sin\eta y d\eta$$

and

$$\frac{\sigma_{xx}^2(x,\,y)}{2\mu_2} = -\frac{2}{\pi}\int_0^{\infty}\left[f_2(\eta) + \eta x g_2(\eta)\right]e^{-\eta x}\cos\eta y d\eta$$

$$\frac{\sigma_{yy}^2(x, y)}{2\mu_2} = \frac{2}{\pi} \int_0^\infty \left[f_2(\eta) + (\eta x - 2)g_2(\eta) \right] e^{-\eta x} \cos \eta y d\eta \qquad (4)$$

$$\frac{\sigma_{xy}^2(x, y)}{2\mu_2} = -\frac{2}{\pi} \int_0^\infty \left[f_2(\eta) + (\eta x - 1)g_2(\eta) \right] e^{-\eta x} \sin \eta y d\eta$$

where

$\kappa_2 = 3 - 4\nu_2$ for plane strain, and
$\kappa_2 = (3 - \nu_2)/(1 + \nu_2)$ for plane stress.

For the symmetrical problem shown in Fig. 2, the following boundary and continuity conditions have to be satisfied:

Continuity conditions at $x = h$:

$$u_1(h, y) = u_2(h, y); \qquad v_1(h, y) = v_2(h, y)$$

$$t_1 \sigma_{xx}^1(h, y) = t_2 \sigma_{xx}^2(h, y); \qquad t_1 \sigma_{xy}^1(h, y) = t_2 \sigma_{xy}^2(h, y) \qquad (5)$$

where

$t_1 = $ land thickness, and
$t_2 = $ skin thickness.

Homogeneous conditions at $y = 0$:

$$\sigma_{xy}^1(x, 0) = 0, |x| < h; \qquad \sigma_{xy}^2(x, 0) = 0, |x| > h$$

$$(6)$$

$$v_2(x, 0) = 0, |x| > h$$

Mixed boundary conditions at $y = 0$:

$$\sigma_{yy}^1(x, 0) = -p(x), |x| < a$$

$$(7)$$

$$v_1(x, 0) = 0, a < |x| < h$$

where $-p(x)$ is the traction on the crack surface and is obtained using the superposition principle.

Conditions $\sigma_{xy}^1(x, 0) = 0$ is identically satisfied by the representation of Eq 2. The homogeneous boundary conditions given by Eq 6 are satisfied by displacement and stress fields given by Eqs 3 and 4. The unknowns f_1, g_1, f_2, g_2 and ϕ_1 are solved by using the four continuity conditions given by Eq 5 and the mixed boundary conditions given by Eq 7.

Following the procedure of Ref 4, it can be shown that the problem of the crack in a land reduces to the integral equation given by

$$\int_{-a}^{a} \frac{G(t)}{t - x}\, dt + \int_{-a}^{a} G(t)K(t, x)dt = -\frac{\pi p(x)(1 + \kappa_1)}{4\mu_1}, \quad |x| < a \quad (8)$$

where $G(t)$ is an unknown function and

$$K(t, x) = \int_{0}^{\infty} k(t, x, \eta)e^{-\eta(h-t)}\, d\eta$$

$$k(t, x, \eta) = \frac{e^{-\eta h}}{[1 - 4\lambda_2\eta h e^{-2\eta h} - \lambda_1\lambda_2 e^{-4\eta h}]} \cdot \bigg[\lambda_1 \cosh(\eta x)$$

$$\cdot \{1 + \lambda_2(3 + 2\eta h)e^{-2\eta h}\} - 2\eta x \lambda_1\lambda_2 \sinh(\eta x)e^{-2\eta h}$$

$$+ \lambda_2\{1 - 2\eta(h - t)\}\}\{\cosh(\eta x)(3 - 2\eta h - \lambda_1 e^{-2\eta h})$$

$$+ 2\eta x \sinh(\eta x)\} \bigg] \quad (9)$$

$$\lambda_1 = \frac{\kappa_1\mu_2 t_2 - \kappa_2\mu_1 t_1}{\mu_2 t_2 + \kappa_2\mu_1 t_1}$$

$$\lambda_2 = \frac{\mu_2 t_2 - \mu_1 t_1}{\mu_1 t_1 + \kappa_1\mu_2 t_2}$$

The kernel $K(t, x)$ is a Fredholm's kernel for $a < h$, as it is bounded for all values of t and x in $(-a, a)$.

Solution of the Integral Equation

The unknown function $G(t)$ in Eq 8 has integrable singularities at the end points. Therefore the equation must be solved subject to the following single-valuedness condition

$$\int_{-a}^{a} G(t)dt = 0 \quad (10)$$

The singular-integral Eq 8 is solved by using the numerical procedure described in Ref 7. Using the technique described in this reference, the problem is reduced to a set of simultaneous equations for unknown function $G(t)$.

Evaluation of Stress Intensity Factors

The stress intensity factor K_I at the crack tips is defined as

$$K_I^1 = \lim_{x \to a} \sqrt{2(x - a)} \, \sigma_{yy}^1(x, 0) \tag{11}$$

The function $G(x)$ is written as

$$G(x) = \frac{\phi(x)}{\sqrt{a^2 - x^2}} \tag{12}$$

Equation 11 can be written as

$$K_I^1 = \frac{-2\mu_2}{(1 + \kappa_2)} \lim_{x \to a} \sqrt{2(a - x)} \, G(x) \tag{13}$$

$$= \frac{-2\mu_2}{(1 + \mu_2)} \frac{\phi(a)}{\sqrt{a}}, \, a \leqslant h \tag{14}$$

Thus the stress intensity factors are obtained by evaluating the function ϕ at the crack tips.

Numerical Results

Stress intensity factors have been obtained for several crack and structural configurations. In obtaining these stress intensity factors for the present work, it is assumed that the skin and land materials have the same Poisson ratio and elastic modulus. The ratio of land-to-skin thickness is varied in the calculations. The computer program developed is general, however, and can be used for cases where the skin and land have different Poisson ratios or elastic moduli. The results presented here correspond to a state of plane strain.

The variation of stress intensity factors with a/h (ratio of crack length to land width) is shown in Fig. 3 for t_1/t_2 (ratio of total land to skin thickness) = 1.5, 2.0, 3.0, and 4.0. It is seen that the stress intensity factors increase as the value of a/h increases or the crack approaches the edge of the land. This is due to the lesser stiffening provided by reduced skin thickness. As the value of a/h decreases—that is, as the width of land increases compared with the crack length—the stress intensity factors approach unity. Figure 3 also shows stress intensity factors for $t_1/t_2 = \infty$; that is, skin thickness is zero (finite width plate case). It is seen that stress intensity factors for $t_1/t_2 = \infty$ (finite width plate) are larger than those for $t_1/t_2 = 4$. This is due to the fact that the skin

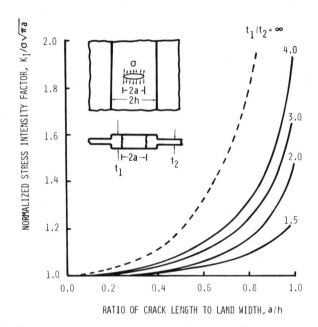

FIG. 3—*Variation of stress intensity factors with ratio of crack length to land width.*

adjacent to the land provides some stiffening to the crack in the land, and hence the stress intensity factors are lower than those for finite width (no stiffening at all).

Figure 4 shows the same data plotted as the variation of stress intensity factors with t_2/t_1 (ratio of skin to land thickness) for $a/h = 0.6, 0.7$, and 0.8. It is seen that the stress intensity factors decreases as the value of t_2/t_1 increases. This is due to the increased stiffening effect provided by the increase in skin thickness. At $t_2/t_1 = 1.0$ the normalized stress intensity factors approach unity (the stress intensity factor for an infinite plate). As t_2/t_1 approaches zero the stress intensity factors approach that for a finite-width plate.

Experimental Program

An experimental program was carried out to verify the analytical techniques discussed earlier. Crack growth data were obtained from a 7075-T6 center cracked specimen with 3.2 mm (0.125 in.) thickness at R (minimum to maximum stress ratio) = 0.1, and the crack growth law for this material thickness was established. The Forman equation was used to fit the data. The equation is given by

$$\frac{da}{dN} = \frac{C(\Delta K)^n}{(1 - R) K_c - \Delta K} \tag{15}$$

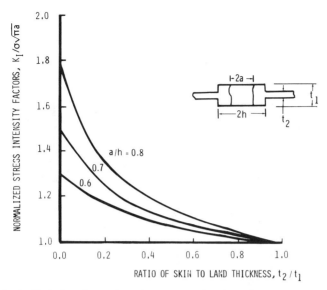

FIG. 4—*Variation of stress intensity factors with ratio of skin to land thickness.*

where

da/dN = crack growth rate,
ΔK = stress intensity factor range,
K_c = fracture toughness,
R = ratio of minimum stress to maximum stress, and
C and n = Forman constants which depend on the material, thickness, and environment.

The value of K_c for the material was found to be 1669 MPa$\sqrt{\text{mm}}$ (48 000 psi$\sqrt{\text{in}}$.). For 7075-T6 aluminum of 3.2 mm (0.125 in.) thickness, the Forman equation constants were found to be $C = 2.3953 \times 10^{-10}$ and $n = 2.3412$.

Next, crack growth data were obtained on a skin with variable thickness (land). The test specimen geometry is shown in Fig. 5. The specimen had a land thickness of 3.2 mm (0.125 in.) and a skin thickness of 1.6 mm (0.063 in.). The specimens were chem-milled from 3.2 mm (0.125 in.)-thick 7075-T6 aluminum plate. Two cracks were introduced in each specimen to obtain two sets of data. One specimen was tested under constant-amplitude loading at $R = 0.1$. The testing was conducted under room temperature environment in a MTS machine at 10-Hz frequency. Crack growth measurements were taken at frequent intervals.

The crack growth behavior in skins with lands (Fig. 5) was predicted using the analytical stress intensity factors (Fig. 3), and the crack growth rate was given by Eq 15. Predicted and observed crack growth behavior under con-

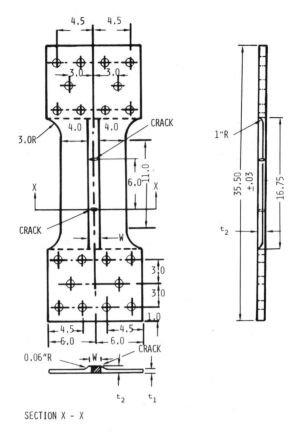

FIG. 5—*Test specimen geometry for cracks at lands (dimensions in inches).*

stant-amplitude loading are compared in Fig. 6. The figure shows excellent correlation between the predicted and observed behavior.

Another specimen, similar to that shown in Fig. 5 but with 6.4 mm (0.25 in.) holes drilled to simulate a realistic case, was tested under a typical fighter wing spectrum loading. In this specimen, the starter cracks emanated from both sides of a hole. The specimen had a land thickness of 3.0 mm (0.12 in.) instead of 3.2 mm (0.125 in.). Here the predictions were made using the analytical stress intensity factors (Fig. 3) with a Bowie correction [8] for the presence of a hole and the crack growth rate of Eq 15. The comparison of predicted and observed crack growth behavior is shown in Fig. 7. A Northrop-developed retardation model was used in predicted crack growth behavior. The correlation between observed and predicted crack growth behavior is reasonably good, particularly in view of the complexity of the spectrum loading.

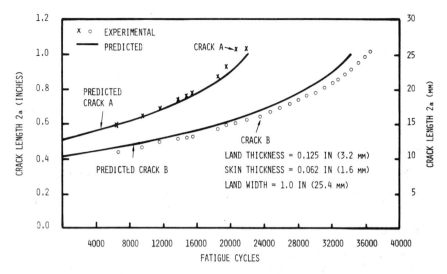

FIG. 6—*Comparison of observed and predicted crack growth behavior in skins with lands specimen; tested under constant-amplitude loading.*

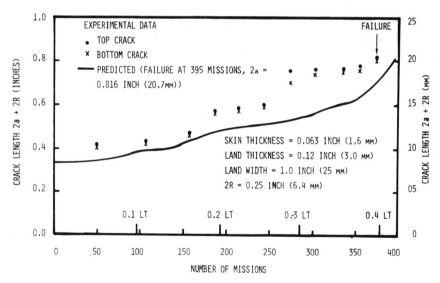

FIG. 7—*Comparison of observed and predicted crack growth in skins with lands specimen; tested under spectrum loading.*

Conclusions

Analytical techniques have been developed to obtain stress intensity factors for cracks in skins with variable thickness (lands). It was shown that the crack growth behavior in lands can be predicted reliably using the analytical stress intensity factors and the crack growth rate obtained from a plate having the same uniform thickness as the thickness of the land.

Acknowledgments

The work reported here was performed as part of a Northrop Corporation Independent Research Program.

References

[*1*] Chay, Y. S., Yang, W. H., and Kitgawa H., "Two and Three Dimensional Fatigue Crack Growth Behaviors in the Plate Having Discontinuous Thickness Interface," in *Proceedings*, International Conference on Fracture Mechanics and Technology, Hong Kong, March 1977, published by Sijthoff and Noordhoff, The Netherlands, pp. 81–88.

[2] Ratwani, M. M. and Kan, H. P., "Development of Composite Patches to Repair Complex Cracked Metallic Structures," Report NADC-80161-60, Vols. I and II, Naval Air Development Center, Warminster, Pa., Jan. 1982.

[3] Ratwani, M. M., Kan, H. P., and Fitzgerald, J. H., "Experimental Investigation of Fiber Composite Reinforcement of Cracks in Complex Metallic Structures," presented at Joint SESA–JSME Conference, Hawaii, May 1982.

[4] Gupta, G. D., *International Journal of Solids and Structures*, Vol. 10, 1973, pp. 1141–1146.

[5] Erdogan, F. and Bakieglu, M., *International Journal of Fracture*, Vol. 12, No. 1, Feb. 1976, pp. 71–84.

[6] Sneddon, I. N. and Lowengrub, M., *Crack Problems in the Classical Theory of Elasticity*, Wiley, New York, 1969, pp. 62–73.

[7] Erdogan, F. and Gupta, G. D., *Quarterly of Applied Mathematics*, Vol. 30, 1972, pp. 525–534.

[*8*] Bowie, O. L., *Journal of Mathematics and Physics*, Vol. 35, 1956, pp. 60–66.

Anthony P. Parker[1] and Christopher P. Andrasic[1]

Mode I Stress Intensity Factors for Point-Loaded Cylindrical Test Specimens with One or Two Radial Cracks

REFERENCE: Parker, A. P. and Andrasic, C. P., **"Mode I Stress Intensity Factors for Point-Loaded Cylindrical Test Specimens with One or Two Radial Cracks,"** *Fracture Mechanics: Fifteenth Symposium, ASTM STP 833*, R. J. Sanford, Ed., American Society for Testing and Materials, Philadelphia, 1984, pp. 57-71.

ABSTRACT: Weight (or Green's) functions are obtained, using an accurate modified mapping collocation numerical technique, for cylindrical specimens with either one or two internal or external radial cracks. Radii ratios (the ratio of external/internal radius) of 1.25, 1.5, 1.75, 2.0, 2.5, and 3.0 are considered. The weight functions are used to obtain stress intensity (K) factors for radial loading by point forces symmetrically located about the crack line and about an axis at right angles to the crack line, through the center of the specimen. The maximum errors associated with the solutions are estimated at 1%.

Notable features of the results obtained include a significant range of crack lengths with near-constant K-values, and configurations in which K changes sign as the crack length is increased. The point force configurations are suited to serve as cylindrical test specimens, either for fracture toughness or fatigue crack growth testing.

KEY WORDS: fracture (materials), stress intensity factor, cylinders, test specimens, residual stress, weight functions

Nomenclature

a Crack length

E Modulus of elasticity

H E (plane stress), $E/(1 - \nu^2)$ (plane strain)

K Mode I stress intensity factor

m Weight function

r General cylindrical coordinate

P Magnitude of radial point force per unit length

[1] Dean and Research Fellow, respectively, Faculty of Computing, Engineering and Science, North Staffordshire Polytechnic, Beaconside, Stafford, England.

R_1 Internal radius of cylinder
R_2 External radius of cylinder
r_0 Radius at which radial point force is applied
u Radial displacement
α Angle of application of radial point force
θ General cylindrical coordinate
ν Poisson's ratio

The cracked curved beam test specimen has gained some popularity, partly because the curved beam may be cut from real-life components such as pressure vessels and gun tubes. Numerous stress intensity solutions are available for the cracked curved beam, but a problem arises if the object is to test a complete ring. For instance, in the case of a cylinder with locked-in residual stresses such as an autofrettaged gun tube, the process of cutting radially through the tube actually releases most of the residual stress [1]. Furthermore, the current U.S. and U.K. practice of fatigue testing complete rings taken as specimens from pressure vessels and gun tubes, which involves cyclic hydraulic pressurization of the bore, is both time-consuming and expensive.

Thus there is a requirement for accurate stress intensity factor solutions for cylindrical specimens of the type illustrated in Fig. 1. This represents a plane cylindrical specimen with one radial crack or two diametrically opposed radial cracks of length a. Figure 1a represents the case of internal

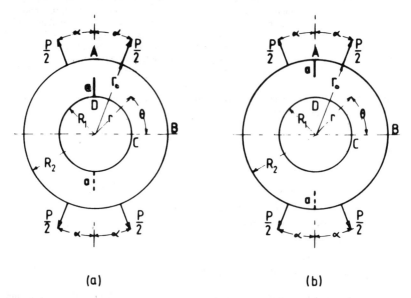

(a)   (b)

FIG. 1—*Point-loaded, radially cracked thick cylinder. (a) Internal crack(s). (b) External crack(s).*

cracks, whilst Fig. 1*b* represents external cracks. In order that the experimentalist may apply point loads at appropriate locations of his choice, either on the boundary of the specimen or at suitable internal points, we shall seek solutions for biaxially symmetric point forces applied anywhere on the line *ABCD* (Fig. 1*a*) or *ABCD* (Fig. 1*b*). High accuracy at short crack lengths is considered to be a particularly desirable feature of the required solutions.

The available solutions to these configurations are limited to those presented by Jones [2] for internal cracking, and the single set of results for an external crack which appears in a paper by Kapp [3]. Only a limited number of results are presented in Ref 2; furthermore, the results are reported to be of poor accuracy at short and long crack lengths, and some doubts over the results at intermediate crack lengths are indicated.

Modified Mapping Collocation

A particularly accurate method for determining stress intensity factors is the modified mapping collocation (MMC) technique [4]. The errors normally associated with this technique are only 0.5%, even at very short crack lengths. The MMC method is based upon a complex variable formulation, so that the state of stress and strain within a cracked elastic body may be completely defined by two analytic complex stress functions. By imposing stress-free conditions along a particular boundary the problem is further reduced to the determination of the coefficients in a single-series stress function.

By imposing resultant force or displacement conditions or both between nodal points on those parts of the boundary along which boundary conditions are not automatically satisfied, a set of linear simultaneous equations is obtained. The unknowns in these equations are the coefficients in the complex stress functions. Higher accuracy and improved efficiency in the numerical solution process is normally achieved by defining more equations than unknowns and obtaining a least-squares solution [4]. In addition, in order to solve some geometries it is necessary to partition the geometry into smaller subregions and then "stitch" them together. This is achieved by imposing appropriate conditions of equilibrium and compatibility along such "stitched" boundaries. The detailed MMC formulation for the cracked cylindrical geometries which are the subject of this paper is presented in Ref 5.

The Weight Function

Bueckner [6] and Rice [7] have demonstrated that a particular function, normally termed *the weight function*, is a property of a cracked geometry, and is independent of the loading. The weight function may be employed in the derivation of additional stress intensity solutions for the particular geometry, provided details of boundary loading are available. Thus the weight function may be regarded as being identical to the Green's function, which defines the contribution to stress intensity of a single, arbitrarily located point force. For the particular geometries under consideration here, we shall confine our attention to the weight function in the absence of body forces, defined by

$$m(r, \theta, a) = \frac{H}{2K^{(i)}(a)} \frac{\partial u}{\partial a}(r, \theta, a) \qquad (1)$$

where $H = E$ (plane stress), $H = E/(1 - \nu^2)$ (plane strain). $K^{(i)}$ is the stress intensity factor and u is the radial displacement field for loading system (i). Additional stress intensity solutions may be then generated from [7]

$$K^{(ii)} = \int_\Gamma \sigma_r \cdot m(r, \theta, a) d\Gamma \qquad (2)$$

where $K^{(ii)}$ is the stress intensity factor with boundary Γ loaded by a radial stress $\sigma_r(r, \theta)$. In the case of a point force P acting radially outwards at some point whose coordinates are (r_0, θ_0), the stress intensity is given by

$$K^{(ii)} = P \cdot m(r_0, \theta_0, a) \qquad (3)$$

In practice, all the information required to define $m(r, \theta, a)$ at a particular crack length (Eq 1) is obtained from two MMC computer runs [4]. This aspect is noteworthy, since the solution for a point force applied at any point on the body (including both boundary and internal locations) may be obtained from only two computer runs as opposed to the vast number which would be required if boundary loading were changed for each solution.

In the present work there are two significant combinations of geometrical and loading symmetries:

Biaxial Geometrical Symmetry and Biaxial Loading Symmetry—In this case there are two (external or internal) radial cracks. There is one axis of geometrical and loading symmetry about the crack line, and another at right angles to this axis through the center of the cylindrical geometry. Hence the weight function $m(r, \theta, a)$, which was obtained with an arbitrary loading applied to the crack line, gives directly the Green's function for the biaxially symmetric loading illustrated in Fig. 1.

Uniaxial Geometrical Symmetry and Biaxial Loading Symmetry—In this case there is one (external or internal) radial crack, and the only axis of symmetry is that about the crack line. In order to obtain the equivalent biaxial loading symmetry given above, the required weight function is given by

$$m(r, \theta, a) = \frac{H}{2K^{(i)}(a)} \left[\frac{\partial u}{\partial a}(r, \theta, a) + \frac{\partial u}{\partial a}(r, -\theta, a) \right] \qquad (4)$$

Results

The stress intensity factor results are presented in dimensionless form as K/K_0, where

$$K_0 = P/(\pi R_1)^{1/2} \tag{5}$$

In each case the cylinder is loaded by four symmetrical point forces acting radially outwards, each of magnitude $P/2$. R_1 and R_2 are the inner and outer cylinder radii respectively and a is the crack length (Fig. 1). Radii ratios (R_2/R_1) of 1.25, 1.5, 1.75, 2.0, and 3.0 are considered. Numerical (tabulated) results are contained in Ref. 8. Graphical results are presented for radii ratios of 1.25 and 2.0 in Figs. 2 to 9.

Figures 2 and 3 give the results for the case of one internal radial crack, whilst Figs. 4 and 5 are those for one external radial crack. Similarly Figs. 6 and 7 cover the results for two internal radial cracks, and Figs. 8 and 9 are results for two external radial cracks. Four sets of results are shown in each figure, two for loading applied to the inner boundary of the cylinder, and two for loading on the outer boundary. In each case the angle of inclination of the point force (α) is presented for the range 0 deg $\leqslant \alpha \leqslant$ 90 deg.

It is possible to compare the results presented in Fig. 3 with those in Ref 2 for the case $\alpha = 0$ deg, with two point loads of magnitude P on the outer boundary. Agreement is within 1%. Equally important is the accuracy at very short crack lengths. In order to check this we performed runs for crack lengths, $a/(R_2 - R_1)$, of 0.01 with $R_2/R_1 = 2.0$. These results may be compared directly with the accurate, unflawed stress analysis of Timoshenko and Goodier [9]. In each case the calculated stress intensity was compared with that predicted from

$$K = 1.12\sigma^*(\pi a)^{1/2} \tag{6}$$

where σ^* is the hoop (σ_θ) stress at the crack tip location given in Ref 9. The results are presented in Table 1. In each case, agreement is within 0.5%.

Discussion

A large number of results has been presented. It is possible, however, to make some general statements:

1. For thin-walled cylinders ($R_2/R_1 \leqslant 1.25$) the difference in stress intensity between external and internal loading of the cylinder, for a given value of α, is negligible.

2. In most cases for the configuration represented by $\alpha = 90$ deg there is a relatively small variation in K in moving the point force from the outer to the

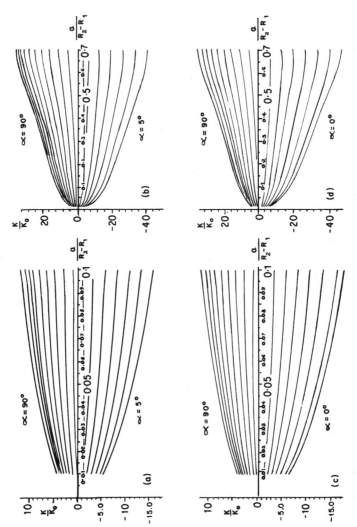

FIG. 2—*Stress intensity factors for one internal radial crack* ($R_2/R_1 = 1.25$). *(a) and (b). Point loads on inner boundary. (c) and (d) Point loads on outer boundary.*

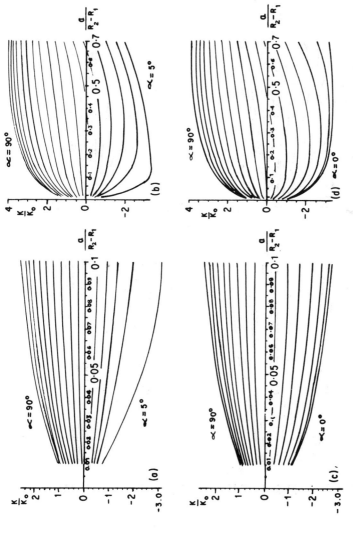

FIG. 3—*Stress intensity factors for one internal radial crack* ($R_2/R_1 = 2.0$). *(a) and (b) Point loads on inner boundary. (c) and (d) Point loads on outer boundary.*

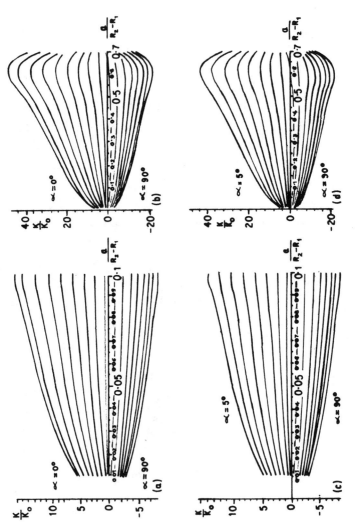

FIG. 4—Stress intensity factors for one external radial crack ($R_2/R_1 = 1.25$). (a) and (b) Point loads on inner boundary. (c) and (d) Point loads on outer boundary.

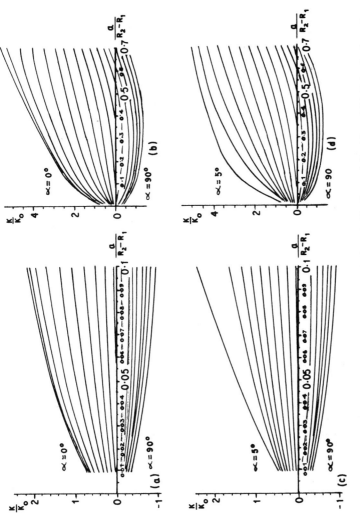

FIG. 5—Stress intensity factors for one external radial crack ($R_2/R_1 = 2.0$). (a) and (b) Point loads on inner boundary. (c) and (d) Point loads on outer boundary.

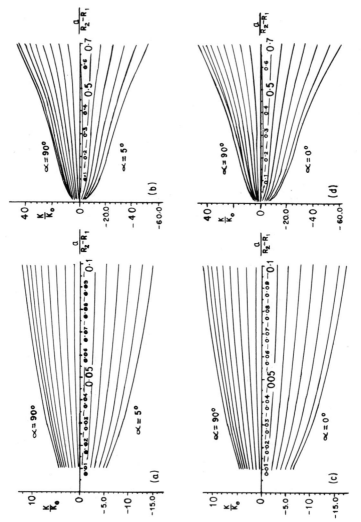

FIG. 6—*Stress intensity factors for two internal radial cracks* ($R_2/R_1 = 1.25$). *(a) and (b) Point loads on inner boundary. (c) and (d) Point loads on outer boundary.*

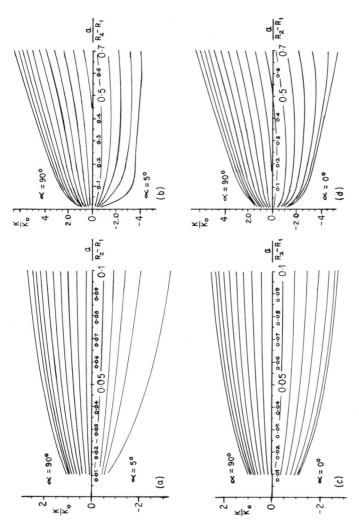

FIG. 7—*Stress intensity factors for two internal radial cracks ($R_2/R_1 = 2.0$). (a) and (b) Point loads on inner boundary. (c) and (d) Point loads on outer boundary.*

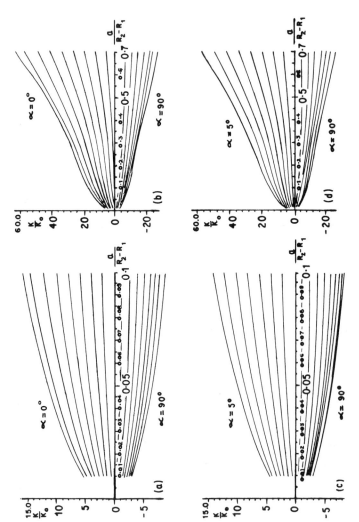

FIG. 8—*Stress intensity factor for two external radial cracks* ($R_2/R_1 = 1.25$). *(a) and (b) Point loads on inner boundary. (c) and (d) Point loads on outer boundary.*

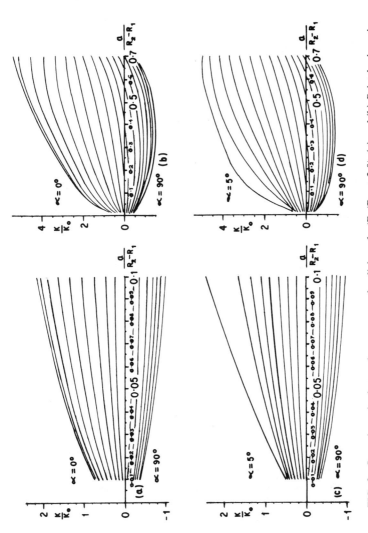

FIG. 9—Stress intensity factors for two external radial cracks ($R_2/R_1 = 2.0$). (a) and (b) Point loads on inner boundary. (c) and (d) Point loads on outer boundary.

TABLE 1—*Stress intensity factors,* $K(\pi a)^{1/2}/P$, *for a single short crack,* $a/(R_2 - R_1) = 0.01$, *with point load* P *applied to outer boundary.*

	$\alpha = 0$ deg		$\alpha = 90$ deg	
	Calculated Value (MMC)	Eq 6	Calculated Value (MMC)	Eq 6
Internal crack	−0.1098	−0.1093	0.0973	0.0973
External crack	0.8708	0.8689	−0.0325	−0.0324

inner boundary. (Results for loading on the line BC, as for all other loadings presented herein, are given numerically in Ref 8.)

3. In certain loading configurations there is a change in the sign of K as the crack extends from a short to a longer crack length (Fig. 5); that is, if the cylinder is loaded by compressive point forces on the outer boundary, the crack will propagate and then arrest at a predictable length. This feature may be of value in the accurate precracking of (for example) ceramic test specimens.

4. In several configurations there are significant ranges of crack length over which K is almost constant. (Note that because the nondimensionalizing term K_0 includes $(R_1)^{1/2}$ the ranges of constant K are easily identifiable.) This feature may be of value in the design of tests requiring constant K-values, such as stress corrosion cracking measurements.

5. Because these results apply to complete cylindrical specimens, it is not necessary to release any "locked-in" residual stresses during specimen preparation. Fracture toughness and fatigue crack growth testing may be performed in the presence of high residual stress fields if so desired.

Conclusions

Weight functions have been obtained, using an accurate MMC technique, for radially cracked ring specimens with a range of radii ratios, crack locations, and crack lengths. These weight functions give directly the solution for radial loading by point forces and may be used as Green's functions in reconstructing more complex boundary loadings. The point force configurations are ideally suited to serve as ring-type test specimens, either for fracture toughness testing of metals and nonmetals, or as fatigue crack growth specimens. In the latter form the solutions should be of value in assessing the effects of residual stress fields. For several rings there is a large range of crack lengths over which the stress intensity has a near-constant value. This feature should be of interest to workers seeking constant K specimens. In other cases there is a change of sign for K as the crack extends, which indicates that it may be possible to induce crack arrest.

In cases where comparison is possible, the solutions are within 0.5% of the limiting solution at very short crack lengths, and within 1% of a limited number of existing solutions.

Acknowledgments

This work was supported by a U.K. Ministry of Defence research contract.

References

[1] Parker, A. P., Underwood, J. H., Throop, J. F., and Andrasic, C. P. in *Fracture Mechanics: Fourteenth Symposium—Volume I: Theory and Analysis, ASTM STP 791*, J. C. Lewis and G. Sines, Ed., American Society for Testing and Materials, 1983, pp. I-216–I-237.

[2] Jones, A. T., *Engineering Fracture Mechanics*, Vol. 6, 1974, pp. 435–446.

[3] Kapp, J. A., "The Effect of Autofrettage on Fatigue Crack Propagation in Externally Flawed Thick-Walled Disks," Technical Report ARCLB-TR-77025, Benet Weapons Laboratory, Watervliet, N.Y., 1977.

[4] Andrasic, C. P. and Parker, A. P. "Weight Functions for Cracked Curved Beams," in *Proceedings*, 2nd International Conference on Numerical Methods in Fracture Mechanics, Swansea, 1980, pp. 67–82.

[5] Parker, A. P. and Andrasic, C. P., "Stress Intensity Factors for Externally and Internally Cracked, Pressurized Thick Cylinders with Residual and Thermal Stresses," presented at International Conference on Fracture Mechanics Technology, Melbourne, Australia, 1982.

[6] Bueckner, H. F., *Zeitschrift für Angewandte Mathematik und Mechanik*, Vol. 50, No. 9, 1970, pp. 529–546.

[7] Rice, J. R., *International Journal of Solids and Structures*, Vol. 8, No. 6, 1972, pp. 751–758.

[8] Parker, A. P. and Andrasic, C. P., "Mode I Stress Intensity Factors for Point Loaded Cylindrical Test Specimens with One or Two Radial Cracks," Technical Note MAT/34, Royal Military College of Science, Shrivenham, U.K., 1982.

[9] Timoshenko, S. P. and Goodier, J. N., *Elements of Elasticity*, 3rd ed., McGraw-Hill, New York, 1970.

K. Kathiresan,[1] T. M. Hsu,[2] and J. L. Rudd[3]

Stress and Fracture Analysis of Tapered Attachment Lugs

REFERENCE: Kathiresan, K., Hsu, T. M., and Rudd, J. L., "**Stress and Fracture Analysis of Tapered Attachment Lugs,**" *Fracture Mechanics: Fifteenth Symposium, ASTM STP 833*, R. J. Sanford, Ed., American Society for Testing and Materials, Philadelphia, 1984, pp. 72–92.

ABSTRACT: Attachment lugs are frequently used in aerospace applications to connect major structural components. It is important to assess the structural integrity of such lugs with cracks present. These assessments are necessary to prevent abrupt failure of the lugs before their intended service lives are reached. The estimation of the associated fatigue life of a lug, which consists of crack initiation and crack growth periods, demands stress and fracture analyses of the lug. In this paper, the results of two-dimensional stress and fracture analyses of tapered attachment lugs subjected to symmetric and off-axis loadings are presented. The effects of various parameters such as the outer-to-inner diameter ratio, crack length, and crack location on the tangential stress distribution, pin contact pressure distribution, stress concentration factor, fatigue critical location, and stress intensity factor are investigated. The finite element method utilizing a crack-tip singularity element is used in the analyses.

KEY WORDS: tapered lug, stress and fracture analyses, stress concentration factors, stress intensity factors, off-axis loadings, residual strength, crack propagation

Nomenclature

R_i Hole radius of lug

R_o Outer radius of lug head

θ Angle measured from axis of lug in clockwise direction

β Angle between two edge surfaces of tapered head

σ_θ Tangential stress along lug hole

σ_{max} Maximum σ_θ

p Pin-bearing pressure between pin and lug

t Thickness of lug

[1] Specialist Engineer, Lockheed-Georgia Company, Marietta, Ga. 30063.

[2] Staff Engineer, Lockheed-Georgia Company, Marietta, Ga. 30063; presently with Gulf E & P Company, Houston, Tex. 77036.

[3] Aerospace Engineer, Air Force Wright Aeronautical Laboratories, Wright-Patterson AFB, Ohio 45433.

P Total applied pin-load
σ_{br} Average bearing pressure $= P/(2R_i t)$
K_{tb} Stress concentration factor $= \sigma_{max}/\sigma_{br}$
K Mode I stress intensity factor
a Crack length (through-the-thickness)

Flaws are present in structural components for many reasons. Some such reasons are manufacturing deficiencies and tool marks. Fatigue cracks often nucleate from these flaws. Cracks can also develop due to service fatigue loading. Since the elastic gross section stress concentration for lugs is relatively high, crack initiation and crack growth lives for lugs can be unusually short. In spite of improved damage tolerance design practices and the use of sophisticated nondestructive inspection techniques, small cracks can still be present in lugs which cannot be detected during regular maintenance inspections. If undetected, such small cracks can grow under service fatigue loading and result in failure of the lug when the crack reaches a critical dimension. If the cracks are detected, the part can either be repaired or replaced. The engineer is posed with a practical design problem of estimating the crack growth lives and residual strengths of lugs. In order to assure aircraft safety, the U.S. Air Force has imposed the damage tolerance design requirements of MIL-A-83444 [1] for aircraft structural components. This document requires the prediction of fatigue crack growth and residual strength of the structure with small initial cracks present at critical locations. These cracks are assumed to exist due to various material and structural manufacturing and processing operations.

There have been a number of publications on stress, fatigue, and fracture analyses of attachment lugs using analytical or experimental procedures or both. Extensive studies have been made on fatigue of straight and tapered attachment lugs [2-9]. Several methods have been proposed for estimating stress intensity factors for cracked attachment lugs. A simple compounding solution method in which known solutions are superposed was used by Liu and Kan [10], Kirkby and Rooke [11], Brussat [12], and Cartwright and Rooke [13]. Hsu [14] used the finite element method with the inclusion of a crack-tip singularity element to analyze straight and tapered lugs subjected to symmetric axial loading. He also investigated the effects of bearing pressure distribution change due to crack extension and relative rigidity of the pin on stress intensity factors. A hybrid finite element procedure was used by Pian et al [15] to study cracks oriented at various angles to the axial direction of straight lugs. Energy release rate concepts and finite element methods were employed by Zatz et al [16] to study the crack growth behavior in attachment lugs. The popular James-Anderson [17] back-tracking method of reducing crack growth rate data to stress intensity factors was used by Schijve and Hoeymakers [18] and Wanhill [19] to derive empirical stress intensity factor formulas for lugs subjected to constant amplitude loading. The weight function method, commonly known as Green's function method, was derived by Bueckner [20] for edge crack problems in semi-infinite plates. These solutions were modified using a series of geometric correction factors for lug problems by Impellizzeri and Rich [21]. Since attachment lugs are some of the most fatigue and fracture critical components, it is also very important to study methods of improving fatigue and crack growth lives of lugs. Repairs of lugs with a bonded steel sleeve was investigated by Jones and Callinan [22]. The significance of residual stresses in lugs introduced by the split-sleeve cold working method, with reference to the depth of penetration of cold work and the relaxation of residual stresses upon subsequent cyclic loading, was analyzed by Schijve et al [23]. The idea of installing an interference-fit bushing to introduce residual stresses around the hole of the lug prior to pin fitting, to improve the fatigue and fracture performance of the lug, was investigated by Hsu and Kathiresan [24].

This paper will assess the structural integrity of attachment lugs from the point of view of predicting their residual strengths and crack growth lives. The related task of determining the stress distributions and stress intensity factors for various structural and crack geometries and loading complexities is a formidable one. Generally, in the open literature, the attachment lug is considered to be a male lug with a straight shank and subjected to a loading which is axially symmetric. If an off-axis or transverse loading is present, however, the lug is generally tapered in order to carry the load effectively. In this paper the results of two-dimensional stress and fracture analyses of tapered attachment lugs are presented. Using the developed stress analysis solutions, fatigue critical locations are identified and the results of fracture analysis can be used to predict the residual strengths and crack growth lives of lugs. Effects of various parameters such as lug outer-to-inner radii ratios, loading directions, crack lengths and crack locations on tangential stress distributions, pin contact pressure distributions, stress concentration factors, and stress intensity factors are also presented. The problems were analyzed using the finite element method with a crack-tip singularity element to calculate the stress intensity factors.

Stress Analysis

Tapered attachment lugs have been used frequently in aircraft structural fittings to provide strength against off-axis loading. In order to determine the critical location and direction where a fatigue crack may initiate and subsequently grow, it is necessary to determine the stress distribution for the unflawed lug. This is especially true for the circumferential stress along the edge of the hole for a tapered lug subjected to a pin loading applied in various in-plane loading directions. Because of the complexity of the lug geometry and off-axis loading, the finite element method was used to determine the tangential stress distributions along the edge of the hole in the unflawed lug. This analysis was conducted for a pin load applied at 0, 90, 135, 180, 270, and 315 deg measured in the clockwise direction from the axis of the lug. Figure 1 depicts the geometry and typical two-dimensional finite element model used in the stress

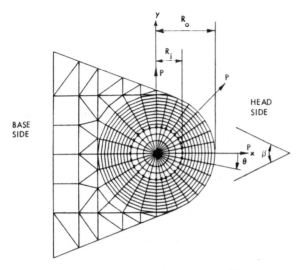

FIG. 1—*Finite-element model for a tapered attachment lug having* R_o/R_i *ratio of 2.25.*

analysis of the unflawed lug. The angle between the two edge surfaces of the tapered head (β) is 45 deg. Constant strain triangular and quadrilateral elements were used to represent both the lug and the pin. Spring elements were used to connect the pin and the lug at each pair of nodes having identical nodal coordinates along the contact surface. Contact conditions were determined through an iterative analysis by removing the spring elements which were in tension and imposing very high stiffnesses for the spring elements which were in compression. The finite element model shown in Fig. 1 consists of 429 nodes, 72 triangular elements, 348 quadrilateral elements, 28 spring elements, and a total of 850 degrees of freedom. A concentrated force was applied at the center of the pin to simulate pin loading and was reacted at the base of the lug. The analysis was carried out to determine the stresses in the lug and the pin-bearing pressure distributions at the pin-lug contact area for the six loading directions mentioned above. Three outer-to-inner radius ratios of the tapered head— that is, R_o/R_i = 1.50, 2.25, and 3.00—were evaluated. In all cases of the analysis, the rigidity of the pin was assumed to be three times the rigidity of the lug which simulates an aluminum lug loaded by a steel pin. Within the range of practical application, the effect of the relative rigidity between the pin and the lug on the stress and pin-bearing pressure distributions is negligible as reported in a previous study for straight attachment lugs [14].

For the conventional finite element analysis using constant strain elements, the values of stress and strain obtained for a given element were assigned to the centroid location of that element. In order to determine the stress at the edge of the hole, the stresses at the centroids of a series of elements located in the same radial direction were used to extrapolate to the edge location. This procedure was used to determine the tangential stresses along the edge of the hole of tapered lugs having R_o/R_i ratios of 1.50, 2.25, and 3.00. The results are shown in Figs. 2, 3, and 4 for pin loadings applied in three principal directions, namely 0, −45 (315), and −90 (270) deg, respectively. In each case, the tangential stresses were normalized with the average pin bearing pressure, σ_{br}, which is defined as $P/(2R_it)$ where t is the thickness of the lug.

As can be seen from Figs. 2 to 4, for each loading direction there are two local maximum tangential stresses located at each side of the loading direction. The locations of these maximum stresses depend upon the loading direction and R_o/R_i ratio. For a pin loading applied in the axial direction of the lug (Fig. 2), the maximum stress locations are found at about ±85 to ±90 deg away from the loading direction. The maximum stress locations do not change significantly with the change of the R_o/R_i ratio. When the pin loading is applied in the −45 deg direction (Fig. 3), for $R_o/R_i = 1.50$, the absolute maximum stress occurs at about 65 deg measured from the axis of the lug (or 110 deg away from the load direction), with the other local maximum stress located about 180 deg from the first. When the R_o/R_i ratio increases, the locations of

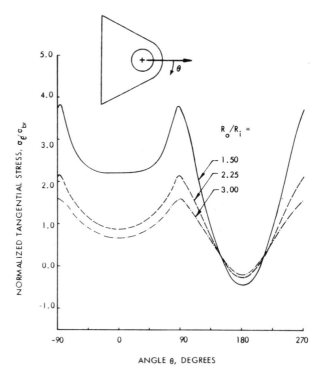

FIG. 2—*Normalized tangential stress distributions along the edge of the hole for tapered lugs subjected to a pin loading applied in 0 deg direction.*

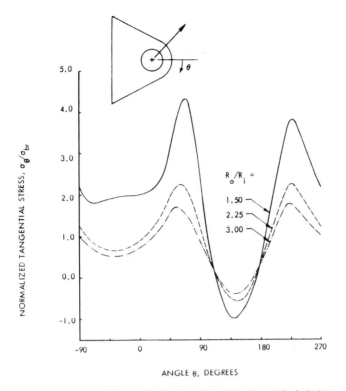

FIG. 3—*Normalized tangential stress distributions along the edge of the hole for tapered lugs subjected to a pin loading applied in −45 deg direction.*

the maximum stresses change only slightly. However, the absolute maximum stress location switches from the head side of the lug to the base side of the lug. When the pin loading is applied in the direction perpendicular to the axis of the lug (Fig. 4), for all R_o/R_i ratios, the maximum stress occurs at the location in the base of the lug at a θ-value of about 200 to 210 deg. Figure 5 summarizes the locations of the local maximum tangential stresses at the edge of the hole for each loading direction. These are the most critical locations, where one would anticipate that a fatigue crack would initiate. For each lug geometry and loading condition, Numbers 1 and 2 in Fig. 5 indicate the probable order of crack initiation. They are chosen based upon the relative magnitude of the two computed local maximum stresses.

Larsson [7,8] conducted fatigue testing of axially and transversely loaded aluminum lugs having a R_o/R_i ratio of 2.2. He observed and tabulated the locations of fretting and crack initiation for three different lug configurations. His results of crack initiation locations are summarized in Figs. 6 and 7 for a pin loading applied in the 0 and 90 deg directions, respectively. As can be seen

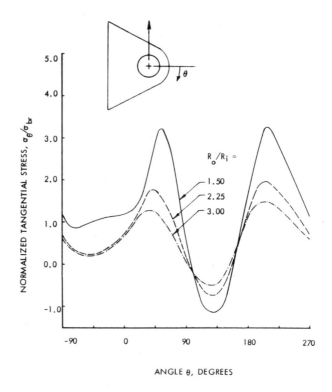

FIG. 4—*Normalized tangential stress distributions along the edge of the hole for tapered lugs subjected to a pin loading applied in −90 deg direction.*

from these figures, the current predicted fatigue critical locations as well as the possible sequence of crack initiation agree well with these experimental data. These predicted critical locations will be used in the modeling of the cracked lugs in the subsequent fracture analysis.

The fatigue critical area of an attachment lug for a given loading direction is not necessarily subjected to compression when the loading direction is reversed. Thus, for some load orientations, the reversed fatigue loading might have a significant effect on crack growth behavior. Figures 8 to 10 show the tangential stress distributions along the edge of the hole for a pin loading applied in the reversed direction of the three primary load orientations presented in Figs. 2 to 4, respectively. As can be seen from the figures, in most cases the reversed loading stretches the critical area in tension but the magnitude is reduced. It should be noted here that the result in Fig. 10 is essentially the same as that in Fig. 4 except for the definition of the angle θ. Plots of the stress concentration factors at the edge of the hole in logarithmic scales, shown in Figs. 11 and 12 for symmetrically loaded straight and tapered attachment lugs, respectively, reveal

SYMBOL	LOAD DIRECTION
--O--	$0°$
—△—	$-45°$
·—□—·	$-90°$

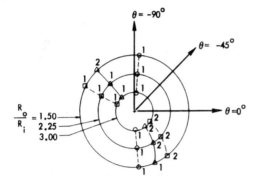

FIG. 5—*Fatigue critical locations of tapered lugs subjected to various load orientations.*

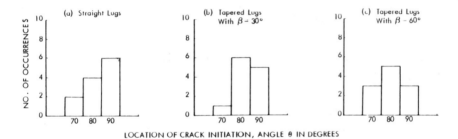

FIG. 6—*Locations of crack initiation for various attachment lugs subjected to a pin-loading applied in 0 deg direction.*

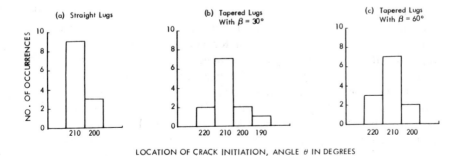

FIG. 7—*Locations of crack initiation for various attachment lugs subjected to a pin loading applied in −90 deg direction.*

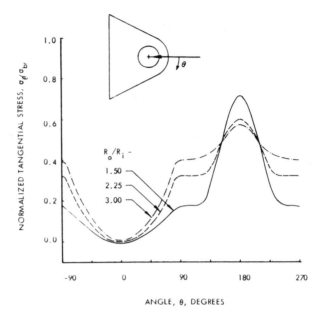

FIG. 8—*Normalized tangential stress distributions along the edge of the hole for tapered lugs subjected to a pin loading applied in 180 deg direction.*

that a simple empirical formula can be derived. The relationship is given by the equation

$$K_{tb} = \frac{\sigma_{max}}{\sigma_{br}} = \left(2.75 - \frac{\beta^{\circ}}{135}\right)\left(\frac{R_o}{R_i} - 1\right)^{-(0.675 - \beta^{\circ}/1000)}$$

where β° is the taper angle of the attachment lug in degrees.

The stress concentration factors were obtained by normalizing the peak tangential stresses with the average pin bearing pressure σ_{br}. The equation of the logarithmic straight lines shown in Figs. 11 and 12 compares with the finite element solution within 0.5% for straight lugs and 0.8% for tapered lugs. The values computed by the above equation and the finite element method are also listed in tabular form in Figs. 11 and 12 for comparison. This simple empirical equation may be used for interpolating for taper angles less than 45 deg or extrapolating for R_o/R_i values outside the range of 1.5 and 3.0. The values of the stress concentration factors presented in Figs. 11 and 12 correspond to the $\theta = \pm 90$ deg locations. The actual maximum tangential stress location is between ± 85 and ± 90 deg. From a knowledge of the stress distributions, it is anticipated that there will be no appreciable difference in the solutions. Also, in the fracture analysis of tapered lugs subjected to a pin loading in the axial direction, the crack surface will be assumed to be in the plane perpendicular to the loading direction. For off-axis loadings of ± 45 and ± 90 deg, however, the

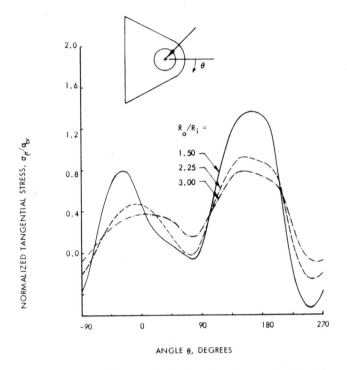

FIG. 9—*Normalized tangential stress distributions along the edge of the hole for tapered lugs subjected to a pin loading applied in 135 deg direction.*

crack surface will be assumed to be in the critical location as predicted by the stress analysis.

The computed pin bearing pressure distributions along the contact surfaces with no cracks present are presented in Figs. 13 to 15 for pin loadings applied in the three primary directions, that is, 0, −45, and −90 deg, respectively. Similar results obtained for the reversed loadings of 180 and 135 deg are presented in Figs. 16 and 17, respectively. For the reversed loading of 90 deg, the contact pressure can be obtained from Fig. 15 by redefining the angle θ as previously discussed.

Fracture Analysis

Based on the stress analysis of unflawed tapered attachment lugs, the most critical locations were selected for the fracture analysis to obtain the stress intensity factors for various crack lengths. A special high-order crack tip singularity element [25] was used in the present analysis to calculate the stress intensity factors. In the analysis, it is assumed that, for a given tapered lug subjected

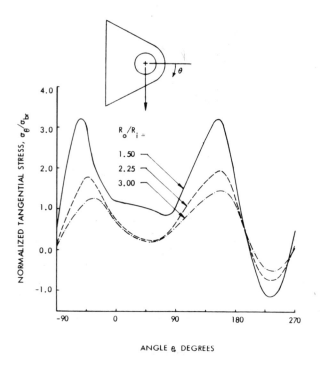

FIG. 10—*Normalized tangential stress distributions along the edge of the hole for tapered lugs subjected to a pin loading applied in 90 deg direction.*

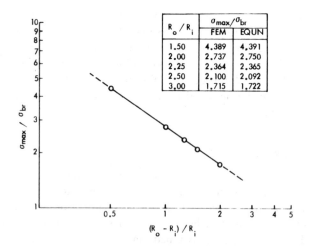

FIG. 11—*Elastic stress concentration factors for straight attachment lugs.*

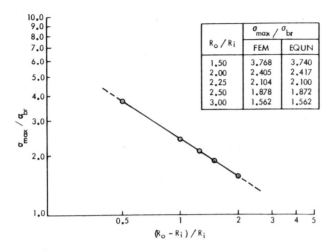

FIG. 12—*Elastic stress concentration factors for tapered attachment lugs.*

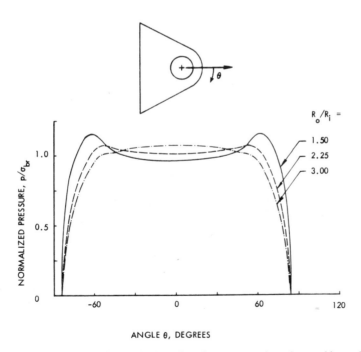

FIG. 13—*Pin-bearing pressure distributions along the contact surface of tapered lugs subjected to a pin loading applied in 0 deg direction.*

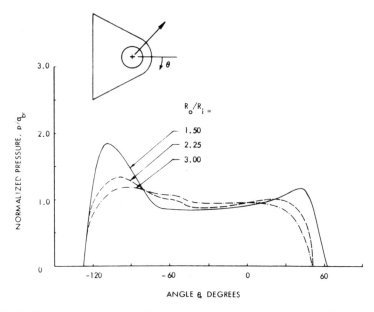

FIG. 14—*Pin-bearing pressure distributions along the contact surface of tapered lugs subjected to a pin loading applied in −45 deg direction.*

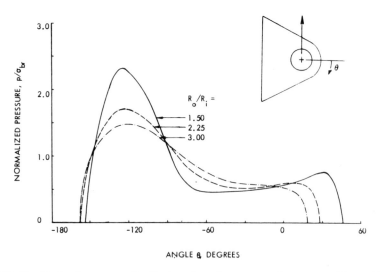

FIG. 15—*Pin-bearing pressure distributions along the contact surface of tapered lugs subjected to a pin loading applied in −90 deg direction.*

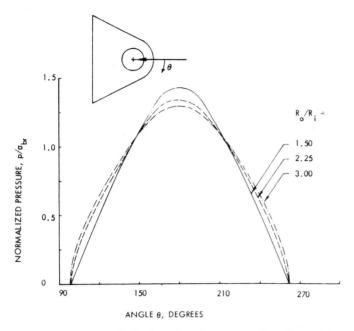

FIG. 16—*Pin-bearing pressure distributions along the contact surface of tapered lugs subjected to a pin loading applied in 180 deg direction.*

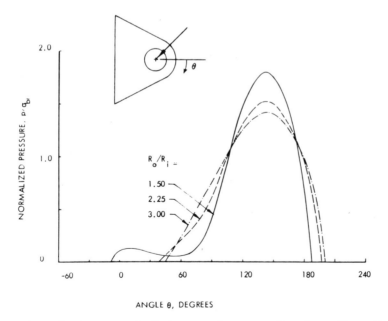

FIG. 17—*Pin-bearing pressure distributions along the contact surface of tapered lugs subjected to a pin loading applied in 135 deg direction.*

to a specific direction of pin loading, the crack will initiate from the maximum tangential stress location at the hole and propagate radially. Figure 18 shows a typical finite element model which was used for a single through-the-thickness crack emanating from a tapered attachment lug subjected to a pin-loading applied in the 0 and 180 deg directions. The crack surface is assumed to be in the plane perpendicular to the loading direction. The finite element breakdown consists of 366 nodes, 150 triangular elements, 228 quadrilateral elements, 32 spring elements, 1 crack-tip element, and a total of 724 degrees of freedom. The computed normalized opening mode stress intensity factors, as a function of normalized crack length, for single through-the-thickness cracks emanating from the holes of tapered attachment lugs having R_o/R_i ratios ranging from 1.5 to 3.0 are presented in Fig. 19. The normalized stress intensity factor values at the lug hole, that is, at $a/R_i = 0$, were obtained by multiplying the concentration factors determined from the unflawed stress analysis by 1.12, which was derived by Gross et al [26] for a straight edge crack in a large plate subjected to remote tension. The trends of the above tapered lug solutions are very similar to those obtained for straight lugs [14]. Similar solutions of normalized stress intensity factors as a function of normalized crack length for tapered attachment lugs subjected to pin loadings applied in the 180 deg direction are presented in Fig. 20.

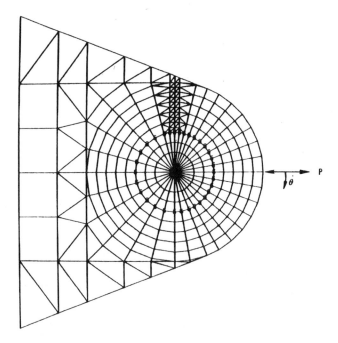

FIG. 18—*Finite element model for a cracked tapered lug subjected to a pin loading applied in 0 and 180 deg directions.*

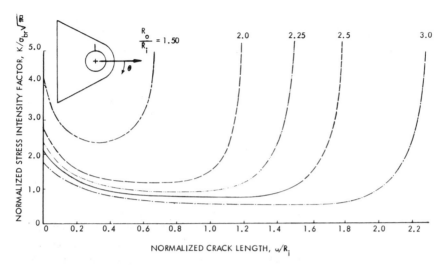

FIG. 19—*Normalized stress intensity factors for single through-the-thickness crack emanating from tapered attachment lugs subjected to a pin loading applied in 0 deg loading direction.*

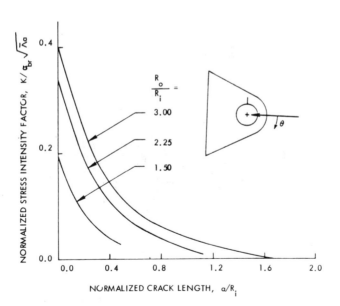

FIG. 20—*Normalized stress intensity factors for single through-the-thickness crack emanating from tapered attachment lugs subjected to a pin loading applied in 180 deg loading direction.*

The analysis was then extended to tapered attachment lugs subjected to off-axis loadings of −45, 135, −90, and 90 deg. A R_o/R_i ratio of 2.25 was considered for the off-axis loading fracture analysis. Figure 21 shows the finite element model used for analyzing a single through-the-thickness crack emanating from a tapered lug subjected to a pin loading applied in −45 deg and its reversed (135 deg) direction. The model contains 401 nodes, 224 triangular elements, 218 quadrilateral elements, 34 spring elements, 1 crack-tip element, and a total of 794 degrees of freedom. It should be noted that although there are two critical locations modeled, 58 and 227 deg measured from the axis of the lug, only one crack was analyzed at a time. The computed normalized stress intensity factors are shown in Fig. 22 as a function of the normalized crack length. As seen from this figure, when the crack length is small, say $a/R_i <$ 0.15, the stress intensity factor for a crack located closer to the base of the lug is higher than the one located at the head side of the lug. When the crack length increases ($a/R_i > 0.15$), the stress intensity factor for a crack located at the head side becomes larger than the one located at the opposite side of the hole. The difference between the two computed K-values increases as the crack length increases. For the case when the direction of the applied pin loading is reversed, the stress intensity factors are also computed and included in the fig-

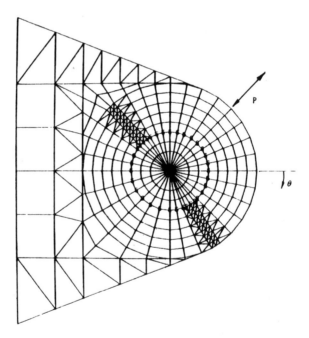

FIG. 21—*Finite element model for a cracked tapered lug subjected to a pin loading applied in −45 and 135 deg directions.*

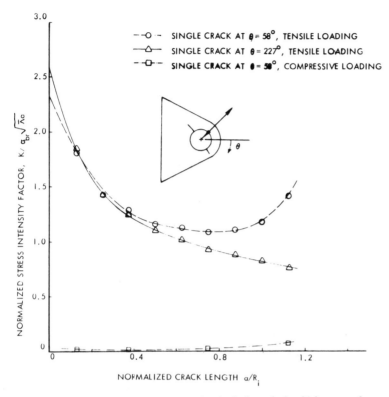

FIG. 22—*Normalized stress intensity factors for single through-the-thickness crack emanating from a tapered lug subjected to a pin loading applied in* −45 *deg and its reversed directions* ($R_o/R_i = 2.25$).

ure for a crack located at the head side of the lug. The surfaces of a crack located closer to the base of the lug are completely closed during the reversed loading. The computed K-values corresponding to the reversed loading are much smaller than the corresponding ones obtained under the primary tensile loading. In the above analysis, only one crack was assumed to exist at a time. In reality, however, both cracks may exist and grow at the same time. In such a case, the influence of one crack on the stress intensity factor of the other crack may be significant. This can be accounted for by developing a matrix of stress intensity factor solutions for various crack lengths of both cracks. This matrix of solutions may then be used to accurately estimate the stress intensity factor for either crack. However, such an effort is not made in the present analysis.

A similar finite element model to the one shown in Fig. 21 was used for a tapered lug subjected to a pin-loading applied in the ±90 deg directions. Two critical locations, 43 and 205 deg measured from the axis of the lug, were determined from the results of the stress analysis of an unflawed tapered lug subjected to a pin loading applied in the −90 deg direction. Similar to the −45 deg loading case, only one crack was analyzed at a time, though two critical locations were modeled. The computed normalized stress intensity factors are shown in Fig. 23 as a function of the normalized crack length. As can be seen from the figure, when the crack length is smaller than $0.85R_i$ the stress intensity factor for a crack located closer to the base of the lug is higher than the one located at the head side of the lug. When the crack length increases ($a/R_i >$ 0.85), the stress intensity factor for a crack located at the head side becomes larger than the one located at the opposite side of the hole. For the case when the direction of the applied pin loading is reversed, the stress intensity factors are also computed and included in the figure for a crack located at the head side of the lug. The surfaces of a crack located closer to the base of the lug are completely closed during the reversed loading. The K-values corresponding to the reversed loading are smaller than the corresponding ones obtained under primary tensile loading. However, these magnitudes are significantly larger than the corresponding ones obtained for a case where the pin loading was applied in the 135 deg direction.

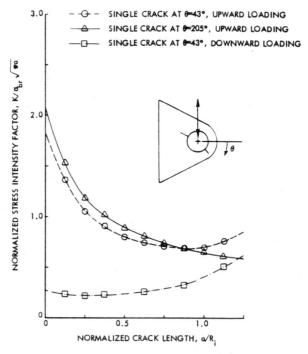

FIG. 23—*Normalized stress intensity factors for a single through-the-thickness crack emanating from a tapered lug subject to a pin loading applied in* −90 *deg and its reversed directions* ($R_o/R_i = 2.25$).

Conclusions

The finite element method, with the inclusion of a high-order crack tip singularity element, has been used to obtain the unflawed stress distributions and stress intensity factors for cracks in tapered attachment lugs subjected to symmetric and off-axis loadings. The change in pin-bearing pressure distribution was shown [14] to have a significant effect on stress intensity factors. The current analysis accounts for such a change in the pin bearing distribution as a function of the change in crack length which results in accurate determinations of stress intensity factors. A simple empirical formula was derived for stress concentration factors for straight and tapered attachment lugs subjected to symmetric loading. As one would anticipate, for a constant R_o/R_i and a constant pin load, the tapered lug provides lower stress concentration factors and lower stress intensity factors than the straight lug when subjected to symmetric loading. Among the principal pin loading directions of 0, -45, and -90 deg, the loading of -45 deg is the most severe loading case from the point of view of fatigue crack initiation and fatigue crack propagation. Based on the current unflawed stress and fracture analysis, one may conclude that, when a tapered lug having a R_o/R_i ratio of 2.25 is subjected to cyclic fatigue pin loading in -45 and -90 deg directions, a crack will probably initiate at the critical location closer to the base of the lug first. When this crack propagates, a second crack will initiate at the head side of the lug. Eventually, the growth of the base-side crack will slow down, while the growth rate of the head-side crack will increase and exceed that of the base-side crack. The lug will finally fail from the head side. Failure modes experimentally studied by Larsson [7] for tapered lugs loaded in the -90 deg direction are very similar to those discussed above. An experimental program will be conducted to evaluate and verify the present analysis for tapered lugs subjected to symmetric and off-axis loadings and the results will be reported in the near future.

Acknowledgments

This work was performed under Contract F33615-80-C-3211 for the Flight Dynamics Laboratory, Air Force Wright Aeronautical Laboratories. The authors would like to thank Dr. T. R. Brussat for his valuable discussions of the results presented in this paper.

References

[1] "Airplane Damage Tolerance Requirements," MIL-A-83444, Air Force Aeronautical Systems Division, July 1974.
[2] Larsson, S. E., "The Development of a Calculation Method for the Fatigue Strength of Lugs and a Study of the Test Results for Lugs of Aluminum Alloys," in *Fatigue Design Procedures*, ICAF Symposium, International Committee of Aeronautical Fatigue, Munich, Germany, 1969, pp. 309–339.

[3] Moon, J. E. and Edwards, P. R., "Fatigue Behavior of Pin Loaded Lugs in BS 2L65 Aluminum Alloy," Reports and Memoranda No. 3834, RAE Farnborough, Hants, Nov. 1977.

[4] Moon, J. E., "Fatigue Behavior of BS 2L65 Aluminum Alloy Pin Loaded Lugs with Interference Fit Bushes," Reports and Memoranda No. 3835, RAE Farnborough, Hants, Nov. 1977.

[5] Buch, A., "Comparison of Fatigue Behavior of 2024-T3 and 7075-T6-Al-Alloy Lugs—Part I," TAE No. 365, Technion Israel Institute of Technology, May 1979.

[6] Schijve, J., "Fatigue of Lugs," in *Contributions to the Theory of Aircraft Structures*, Prof. A. Van der Neut Anniversary Volume, Nijgh-Wolters Noordhoff United Press, 1972, pp. 423–440.

[7] Larsson, N., "Fatigue Testing of Transversely Loaded Aluminum Lugs," TN FFA HU-1673, The Aeronautical Research Institute of Sweden, Stockholm, Sweden, October 1977.

[8] Larsson, N., "Fatigue Testing of Transversely Loaded Aluminum Lugs—Sequence II," TN FFA HU-1848, The Aeronautical Research Institute of Sweden, Stockholm, Sweden, April 1978.

[9] Larsson, N., "Fatigue Testing of Slender Aluminum Lugs," TN FFA HU-1990, The Aeronautical Research Institute of Sweden, Stockholm, Sweden, Aug. 1978.

[10] Liu, A. F. and Kan, H. P., "Test and Analysis of Cracked Lugs," in Fracture 1977, Vol. 3, ICF4, Waterloo, Canada, 19–24 June 1977, pp. 657–664.

[11] Kirkby, W. T. and Rooke, D. P., "A Fracture Mechanics Study of Residual Strength of Pin-Loaded Specimens," *Fracture Mechanics in Engineering Practice*, Applied Scientific Pub., London, 1977, p. 339.

[12] Brussat, T. R., "Stress Intensity Formulas for Structural Applications," LR 29412, Lockheed-California Co., Burbank, Calif., April 1981.

[13] Cartwright, D. J. and Rooke, D. P., *Engineering Fracture Mechanics*, Vol. 6, 1974, pp. 563–571.

[14] Hsu, T. M., *Journal of Aircraft*, Vol. 18, No. 9, Sept. 1981, pp. 755–760.

[15] Pian, T. H. H., Mar. J. W., Orringer, O., and Stalk, G., "Numerical Computation of Stress Intensity Factors for Aircraft Structural Details by the Finite Element Method," AFFDL-TR-76-12, Air Force Flight Dynamics Laboratory, Wright Patterson AFB, Ohio, May 1976.

[16] Zatz, I. J., Eidinoff, H. L., and Armen, H., "An Application of the Energy Release Rate Concept to Crack Growth in Attachment Lugs," in *Proceedings, 22nd AIAA/ASME/ASCE Structures, Structural Dynamics and Materials Conference*, Atlanta, Ga., April 1981.

[17] James, L. A. and Anderson, W. E., *Engineering Fracture Mechanics*, Vol. 1, 1969, pp. 565–568.

[18] Schijve, J. and Hoeymakers, A. H. W., "Fatigue Crack Growth in Lugs and the Stress Intensity Factor," Report LR-273, Delft University of Technology, Delft, The Netherlands, July 1978.

[19] Wanhill, R. J. H., "Calculation of Stress Intensity Factors for Corner Cracking in a Lug," *Fracture Mechanics Design Methodology*, AGARD CP221, Feb. 1978, p. 8.

[20] Bueckner, H. F., *Zeitschrift für Angewandte Mathematik und Mechanik*, Vol. 51, 1971, pp. 97–109.

[21] Impellizzeri, L. F. and Rich, D. L. in *Fatigue Crack Growth under Spectrum Loads, ASTM STP 595*, American Society for Testing and Materials, 1976, pp. 320–336.

[22] Jones, R. and Callinan, R. J., *International Journal of Fracture*, Vol. 17, 1981, pp. R53–R55.

[23] Schijve, J., Jacobs, F. A., and Meulman, A. F., "Flight Simulation Fatigue Tests on Lugs with Holes Expanded According to the Split-Sleeve Cold Working Method," NLR TR 78131 U, National Aerospace Laboratory ULR, The Netherlands, Sept. 1978.

[24] Hsu, T. M. and Kathiresan, K., in *Fracture Mechanics: Fourteenth Symposium—Volume I: Theory and Analysis, ASTM STP 791*, American Society for Testing and Materials, 1983, pp. I-172–I-193.

[25] Chu, C. S. et al, "Finite Element Computer Program to Analyze Cracked Orthotropic Sheets," NASA-CR-2698, Washington, D.C., July 1976.

[26] Gross, B., Srawley, J. W., and Brown, W. F., "Stress Intensity Factors for a Single-Edge-Notch Tension Specimen by Boundary Collocation of a Stress Function," NASA-TN-D-2395, Washington, D.C., Aug. 1964.

J. C. Newman, Jr.[1]

An Elastic-Plastic Finite Element Analysis of Crack Initiation, Stable Crack Growth, and Instability

REFERENCE: Newman, J. C., Jr., **"An Elastic-Plastic Finite Element Analysis of Crack Initiation, Stable Crack Growth, and Instability,"** *Fracture Mechanics: Fifteenth Symposium, ASTM STP 833*, R. J. Sanford, Ed., American Society for Testing and Materials, 1984, pp. 93–117.

ABSTRACT: An elastic-plastic (incremental and small strain) finite element analysis was used with a crack growth criterion to study crack initiation, stable crack growth, and instability under monotonic loading to failure of metallic materials. The crack growth criterion was a critical crack-tip-opening displacement (CTOD) at a specified distance from the crack tip, or equivalently, a critical crack-tip-opening angle (CTOA). Whenever the CTOD (or CTOA) equaled or exceeded a critical value, the crack was assumed to grow. Single values of critical CTOD were found in the analysis to model crack initiation, stable crack growth, and instability for 7075-T651 and 2024-T351 aluminum alloy compact specimens. Calculated and experimentally measured (from the literature) CTOD values at initiation agreed well for both aluminum alloys. These critical CTOD values from compact specimens were also used to predict failure loads on center-crack tension specimens and a specially designed three-hole-crack tension specimen made of the two aluminum alloys and of 304 stainless steel. All specimens were 12.7 mm thick. Predicted failure loads for 7075-T651 aluminum alloy and 304 stainless steel specimens were generally within ±15% of experimental failure loads, while the predicted failure loads for 2024-T351 aluminum alloy specimens were generally within ±6% of the experimental loads. The technique presented here can be used as an engineering tool to predict crack initiation, stable crack growth, and instability for cracked structural components from laboratory specimens such as compact specimens.

KEY WORDS: fracture mechanics, fracture strength, finite element method, aluminums, steels, displacement, cracks, plastic deformation

Nomenclature

a Crack length defined in Fig. 1, m

a_0 Initial crack length, m

d Minimum element size along crack line, m

E Young's modulus, N/m^2

[1]Senior Scientist, NASA Langley Research Center, Hampton, Va. 23665.

K Stress intensity factor, $N/m^{3/2}$
M Number of finite elements
N Number of nodes
n Ramberg-Osgood strain-hardening power
P Applied load, N
P_f Failure load, N
t Specimen thickness, m
W Specimen width, m
Δa_p Physical crack extension, m
δ_c Critical crack-tip-opening displacement, m
δ_i Crack-tip-opening displacement at initiation, m
ϵ Uniaxial strain
κ Ramberg-Osgood strain-hardening coefficient, N/m^2
σ Uniaxial stress, N/m^2
σ_{ys} Uniaxial yield stress (0.2% offset), N/m^2
σ_u Uniaxial tensile strength, N/m^2

Experiments on metals have shown that, under monotonic loading to failure, a crack goes through three stages of behavior: (1) a period of no crack growth, (2) a period of stable crack growth, and (3) crack growth instability under load control (or stable crack growth with decreasing load under displacement control). In the past decade, the phenomenon of stable crack growth has been studied extensively using elastic-plastic finite element methods [1-8]. These studies were conducted to develop efficient techniques to simulate crack extension and to examine various local and global fracture criteria. Some of these criteria

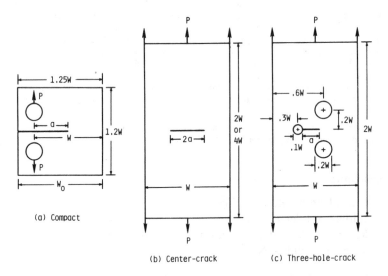

FIG. 1—*Specimen configurations tested and analyzed.*

were crack tip stress or strain, crack-tip-opening displacement or angle, crack tip force, energy release rates, J-integral, and tearing modulus. Of these, the crack-tip-opening angle (CTOA) or displacement (CTOD) at a specified distance from the crack tip was shown to be most suited for modeling stable crack growth and instability during the fracture process [3,7,8]. Some discrepancies among the various analyses, however, have been observed at initiation of stable crack growth. de Koning [3] showed that CTOA was nearly constant from initiation. Shih et al [7] and Kanninen et al [8] showed that CTOA at initiation was larger, and in some cases much larger, than the value needed for stable crack growth. On the other hand, Luxmoore et al [9] have experimentally shown that CTOA (or CTOD) was constant from the onset of stable crack growth in two aluminum alloys, but have found different values for different crack configurations (center-crack and double-edge crack tension specimens). These results show the necessity for studying different crack configurations when assessing the validity of any fracture criteria.

One of the objectives of the present paper was to critically evaluate the CTOD growth criterion using an elastic-plastic finite element analysis under monotonic loading to failure. In particular, the analysis was conducted to see whether or not the critical CTOD was constant during crack initiation, stable crack growth, and instability. The second objective was to determine whether or not fracture data from laboratory specimens, such as the compact specimen, could be used to predict failure loads on other crack configurations. To assess the CTOD criterion, experiments and analyses were conducted on several crack configurations made of various materials.

Fracture tests were conducted on compact, center-crack tension, and specially designed three-hole-crack tension specimens (Fig. 1) made of 7075-T651 aluminum alloy, 2024-T351 aluminum alloy, and 304 stainless steel. The compact specimens were tested by NASA Langley Research Center and Westinghouse Research and Development Laboratory [10] to provide basic fracture data (applied load against "physical" crack extension and failure loads). Center-crack and three-hole-crack specimens were tested only by NASA Langley. The three-hole-crack specimen has a complicated stress intensity factor solution, like that for a cracked stiffened panel.

In the finite element analysis, a critical value of CTOD at a specified distance from the crack tip was chosen as the fracture criterion. During incremental loading to failure, whenever the CTOD equaled or exceeded a preset critical value (δ_c), the crack tip node was released and the crack advanced to the next node. This process was repeated until crack growth became unstable at the failure load. Comparisons between experimental and calculated load against physical crack extension data were made on the compact specimens to determine the critical CTOD values. These critical CTOD values were then used to predict failure loads on other compact specimens, center-crack tension specimens, and three-hole-crack tension specimens. Comparisons were made between predicted and experimental failure loads for all specimen types. Comparisons were also made between calculated and experimentally measured [11] CTOD values at initiation for the two aluminum alloys.

Experimental Procedure

The experimental test program was conducted by NASA Langley Research Center and Westinghouse Research and Development Laboratory [10] as part of an ASTM Committee E-24 round robin on fracture. Tests were conducted on compact specimens (with initial crack-length-to-width ratios, a_0/W, of 0.5) to obtain load against physical crack extension data and failure loads. The NASA Langley Research Center also conducted fracture tests on other compact specimens (with a_0/W equal to 0.3 and 0.7), center-crack tension specimens, and a "structurally-configured" specimen (with three circular holes and a crack emanating from one of the holes) subjected to tensile loading. The specimen configurations tested are shown in Fig. 1. In addition, tension specimens were tested to obtain uniaxial stress-strain curves.

Materials

The three materials tested were 7075-T651 aluminum alloy, 2024-T351 aluminum alloy, and 304 stainless steel. These materials were selected because they exhibit a wide range in fracture toughness behavior. They were obtained in plate form (1.2 by 3.6 m) with a nominal thickness of 12.7 mm.

Specimen Configurations and Loadings

Four types of specimens were machined from one plate of each material. The specimens were: (1) tensile, (2) compact, (3) center-crack tension, and (4) three-hole-crack tension. A summary of specimen types, nominal widths, and nominal crack-length-to-width ratios is shown in Table 1.

TABLE 1—*Test specimen matrix and number of specimens for 7075-T651, 2024-T351, and 304 stainless steel.*

Specimen Type	Nominal Width, mm	Nominal Crack-Length-to-Width Ratio			
		0.3	0.4	0.5	0.7
Compact	51	2	...	5[a]	2
Compact	102	2	...	5[a]	2
Compact	203	2	...	5[a]	2
Center-crack	127	...	2	...	...
Center-crack	254	...	2	...	...
Three-hole crack	254	8($0.05 \leq a_0/W \leq 0.4$)			
Tensile[a]	12.7	...			

[a]Data used to determine δ_c.

Tensile Specimens—Eight tension specimens (ASTM E 8) with square cross sections (12.7 by 12.7 mm) were machined from various locations in each plate of material. The specimens were machined to obtain tensile properties perpendicular to the rolling direction. Full engineering stress-strain curves were obtained from each specimen. The initial load rate was 45 kN/min, but after yielding, the load rate was set at 4.5 kN/min. Average tensile properties (E, σ_{ys} and σ_u) are given in Table 2.

Compact Specimens—The compact specimen configuration is shown in Fig. 1a. The planar configuration is identical to the "standard" compact (ASTM E 399) specimen, but the nominal thickness was 12.7 mm. Twenty-seven specimens were machined from each plate of material, and the cracks were oriented in the same direction (parallel to the rolling direction). The nominal widths (W) were 51, 102, and 203 mm, and the nominal crack-length-to-width ratios were 0.3, 0.5, and 0.7. All specimens were fatigue precracked in accordance with ASTM E 399 requirements.

The specimens tested by Westinghouse ($a_o/W = 0.5$) were loaded under displacement-control conditions and periodically unloaded (about 15% at various load levels) to determine crack lengths from compliance [10,12]. The specimens tested by NASA Langley were loaded under load-control conditions to failure. The initial load rates on the NASA Langley tests were about the same as those tested by Westinghouse. Load against crack extension data were obtained from visual observations and from unloading compliance data (at both the crack mouth and the load line). Initial crack lengths (a_o) and failure loads (P_f) were also recorded. The initial crack lengths were measured from broken specimens and were three-point weighted averages through the thickness ($3a_o = a_1 + 2a_2 + a_3$) where a_1 and a_3 were surface values and a_2 was the value in the middle of the specimen. Specimen dimensions, initial crack lengths, and experimental failure loads are given in Tables 3 to 5 for the three materials. These experimental results will be presented and discussed later.

Center-Crack and Three-Hole-Crack Tension Specimens—The center-crack and three-hole-crack specimen configurations are shown in Figs. 1b and 1c, respectively. Again, all specimens were machined so that the cracks were oriented parallel to the rolling direction. Four center-crack specimens ($W = 127$ and 254 mm) were machined from each plate of material. The nominal crack-length-to-width ratio was 0.4. Eight three-hole-crack specimens ($W =$

TABLE 2—*Average tensile properties of the three materials.*[a]

Material	E, MN/m^2	σ_{ys}, MN/m^2	σ_u, MN/m^2	κ, MN/m^2	n
7075-T651	71 700	530	585	640	30
2024-T351	71 400	315	460	550	10
304 stainless steel	203 000	265	630	745	5

[a]Average values from eight tests.

TABLE 3—*7075-T651 aluminum alloy data.*

t, mm	W, mm	a_o, mm	Experimental P_f, kN	P_{pred}/P_{exp}
Compact Specimens				
12.4	51	16.1	16.1	1.00
12.5	51	15.4	16.0	1.01
12.7[a]	51	25.6	8.73	1.10
12.8[a]	51	25.6	8.85	1.08
12.6	51	25.4	8.54	1.12
12.6	51	25.9	8.85	1.08
12.6	51	35.4	3.75	1.03
12.5	51	36.3	3.34	1.15
12.7	102	31.8	27.4	0.95
12.7	102	30.6	27.2	0.95
12.8[a]	102	50.8	15.5	0.95
12.7[a]	102	50.7	15.5	0.95
12.8[a]	102	51.4	14.5	1.02
12.7	102	50.9	15.1	0.98
12.8	102	51.4	15.1	0.98
12.6	102	71.2	5.78	1.08
12.7	102	71.0	5.65	1.11
12.7	203	60.6	47.4	0.90
12.8	203	60.4	46.3	0.92
12.8[a]	203	102.0	24.1	1.00
12.7[a]	203	102.2	24.1	1.00
12.8[a]	203	100.8	25.4	0.95
12.8	203	101.2	25.7	0.94
12.8	203	101.2	26.2	0.92
12.8	203	142.0	10.2	0.97
12.7	203	142.2	10.5	0.95
Center-Crack Tension Specimens				
12.8	127	26.4	209	0.97
12.8	127	24.9	200	1.02
12.8	254	49.8	365	0.89
12.7	254	49.1	356	0.91
Three-Hole Crack Specimens				
12.8	254	11.9	696	0.83
12.8	254	25.5	685	0.84
12.7	254	39.7	698	0.82
12.7	254	50.5	651	0.88
12.8	254	64.8	620	0.92
12.7	254	75.6	578	0.98
12.8	254	90.1	462	0.96
12.8	254	100.8	362	0.95

[a]Tested at Westinghouse Research Laboratory [10].

TABLE 4—*2024-T351 aluminum alloy data.*

t, mm	W, mm	a_o, mm	Experimental P_f, kN	P_{pred}/P_{exp}
Compact Specimens				
12.4	51	16.1	29.8	0.98
12.3	51	16.0	29.5	0.99
12.6[a]	51	26.5	14.2	1.01
12.5[a]	51	26.3	14.7	0.97
12.5[a]	51	26.1	14.8	0.97
12.3	51	26.1	14.5	0.98
12.4	51	26.4	14.7	0.97
12.3	51	36.2	5.22	0.95
12.3	51	36.3	5.29	0.93
12.5	102	31.4	54.7	1.06
12.5	102	31.2	54.7	1.06
12.5[a]	102	51.9	28.8	0.98
12.6[a]	102	51.6	28.9	0.98
12.6[a]	102	51.4	29.8	0.95
12.5	102	51.6	28.2	1.00
12.5	102	51.9	28.7	0.99
12.6	102	71.2	10.1	0.99
12.5	102	71.4	10.1	0.98
12.5	203	61.8	98.5	0.93
12.6	203	61.7	100.3	0.92
12.6[a]	203	102.4	52.1	1.04
12.6[a]	203	102.5	51.9	1.04
12.5	203	102.2	52.3	1.04
12.6	203	102.2	52.0	1.04
12.5	203	142.9	18.6	1.03
12.5	203	143.0	18.9	1.02
Center-Crack Tension Specimens				
12.6	127	26.2	302	1.06
12.6	127	25.2	311	1.03
12.6	254	51.2	581	1.05
12.6	254	52.1	574	1.06
Three-Hole Crack Specimens				
12.6	254	13.9	754	1.05
12.5	254	25.7	738	1.06
12.5	254	38.6	735	1.04
12.5	254	51.8	718	1.04
12.6	254	64.3	696	1.02
12.6	254	75.8	660	1.01
12.5	254	90.0	580	1.04
12.5	254	101.5	505	1.04

[a] Tested at Westinghouse Research Laboratory [10].

TABLE 5—*304 stainless steel data.*

t, mm	W, mm	a_o, mm	Experimental P_f, kN	P_{pred}/P_{exp}
Compact Specimens				
13.1	51	16.5	52.7	0.95
13.3	51	16.3	53.6	0.94
12.8[a]	51	25.6	27.3	0.87
12.8[a]	51	26.1	25.9	0.91
12.8[a]	51	25.8	26.8	0.88
13.1	51	25.8	27.5	0.88
13.1	51	26.2	26.9	0.90
13.2	51	36.2	9.56	0.85
13.1	51	36.2	9.61	0.85
13.4	102	34.1	93.4	1.09
13.3	102	31.1	104	0.98
13.0[a]	102	49.4	55.1	0.93
13.0[a]	102	50.7	50.8	1.01
13.0[a]	102	51.4	47.8	1.07
13.0	102	50.5	51.8	0.99
13.3	102	51.8	50.6	1.03
13.4	102	72.1	17.7	0.90
13.3	102	72.3	17.3	0.92
13.5	203	62.0	195	[b]
13.5	203	62.0	192	[b]
12.8[a]	203	102.0	86.8	1.13
12.8[a]	203	102.3	85.4	1.15
12.8[a]	203	102.0	85.3	1.15
13.4	203	101.4	96.3	1.02
13.4	203	102.2	96.1	1.03
13.3	203	142.6	34.1	0.97
13.4	203	142.8	32.9	1.01
Center-Crack Tension Specimens				
13.6	127	26.1	458	1.12
13.6	127	26.2	469	1.09
13.5	254	50.1	882	1.11
13.6	254	50.8	878	1.11
Three-Hole Crack Specimens				
13.6	254	13.5	1260	[b]
13.6	254	26.3	1220	[b]
13.5	254	39.1	1180	[b]
13.4	254	51.6	1150	[b]
13.5	254	64.4	1120	[b]
13.5	254	77.7	999	1.11
13.6	254	89.8	895	1.09
13.6	254	102.7	790	1.08

[a]Tested at Westinghouse Research Laboratory [10].
[b]Prediction not made on this specimen.

254 mm) were also machined from each plate of material. The nominal crack lengths in the three-hole-crack specimen ranged from 13 to 102 mm. All center-crack and three-hole-crack specimens had 510 mm between griplines. The initial load rates were selected such that the initial stress intensity factor rates were roughly the same (30 MN/m$^{3/2}$/min) for all crack specimens. Again, initial crack lengths (three-point weighted average through the thickness) and failure loads were recorded. Specimen dimensions, initial crack lengths, and experimental failure loads are given in Tables 3 to 5 for the three materials. These results will be presented and discussed later.

Finite-Element Analysis of Fracture

The elastic-plastic analysis of the three specimen types (Fig. 1) employed the finite element method and the initial stress concept [13,14]. The elastic-plastic analysis was based on incremental flow theory with a small strain assumption. The finite element models for these specimens were composed of two-dimensional, constant-strain, triangular elements under assumed plane-stress conditions. Several mesh patterns were used to model different size specimens so that the minimum element size (d) along the line of crack extension would be the same in all specimens. Fictitious springs were used to change boundary conditions associated with crack extension. The use of springs was found to be computationally efficient. For free nodes along the crack line, the spring stiffnesses were set equal to zero; for fixed nodes, the stiffnesses were assigned extremely large values. See Ref 5 for details of the elastic-plastic finite element analysis with crack extension.

Although the finite element analysis (constant-strain elements) used here does not contain a singularity at the crack tip, such as the Hutchinson-Rice-Rosengren (HRR) singularity [15,16], the use of singularities may not be necessary. Singularities, which do not exist in real materials, are only mathematical consequences of continuum mechanics. The finite element analysis with very small elements gives a high, but finite, strain concentration at the crack tip, like a crack in a real material.

The average material stress-strain curves used in the finite element analysis were approximated by the Ramberg-Osgood equation [17] as

$$\epsilon = \frac{\sigma}{E} + \left(\frac{\sigma}{\kappa}\right)^{n} \tag{1}$$

where κ and n are the strain-hardening coefficient and power, respectively. Values of the constants are given in Table 2 for the three materials.

Crack Growth Criterion

The crack growth criterion used here was based on a critical crack-tip-opening displacement (δ_c) at a specified distance (d) from the crack tip. The distance d

was the element size along the crack line; in other words, d is the distance between the first free node and the crack tip. (The critical CTOD criterion is also equivalent to a critical CTOA criterion, since CTOA $= 2 \tan^{-1} (\delta_c/2d)$.) During incremental loading to failure, whenever the CTOD equaled or exceeded a preset critical value (δ_c), the crack tip node was released (see Ref 5 for details) and the crack advanced to the next node. This process was repeated until crack growth continued without any increase in load. The use of the CTOD (or CTOA) criterion does require that the absolute size (d) and arrangement of elements in the crack tip region and along the line of crack extension be the same in all crack configurations considered.

The procedure used to establish the critical δ_c value and mesh size (d) is as follows. The mesh size in the crack tip region of the large compact specimen ($W = 203$ mm) was systematically reduced until the calculated loads at initiation and at failure were reasonably close to the experimental values using a given value of δ_c. In other words, the mesh size was used as a variable to fit the experimental load against crack length data. After the mesh size d was determined, the final δ_c value was selected so that the mean of the calculated-to-experimental failure load ratio on the various size compact specimens ($W = 51$, 102, and 203 mm) with $a_o/W = 0.5$ was about unity. The critical δ_c value was then used to predict failure loads on other compact, center-crack tension, and three-hole-crack tension specimens.

The typical crack growth behavior that was obtained from the finite element analysis using the critical CTOD criterion is shown in Fig. 2. Applied load is plotted against crack length. During initial loading, a plastic zone developed

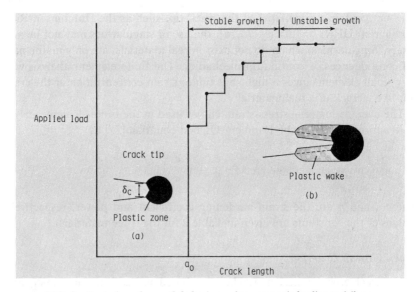

FIG. 2—*Typical crack growth behavior under monotonic loading to failure.*

at and to the right of the crack tip as illustrated in Insert (a). At a certain load, the CTOD became critical (δ_c) and the crack moved forward one element size (d) while the applied load was held constant. The CTOD for the new crack tip was found to be less than the critical value. The dashed curve in Insert (b) shows what the crack surface displacements would have been if residual plastic deformations had not been retained in the material to the left of the crack tip. The plastic deformations remaining in the wake of the advancing crack (lightly shaded region to the left of the crack tip) caused the new displacements to be substantially less than that for the crack without the "plastic wake." The corresponding crack tip strains were also less for the crack with the plastic wake than for the crack without the plastic wake [5]. Thus an increase in applied load was required to make the CTOD critical and to advance the crack further. This process was repeated until crack growth became unstable (continuous crack growth with no further increase in load). At the instability load, the CTOD value at the new crack tips were always equal to or greater than the δ_c value and the crack continued to grow from node to node. If the analysis had been conducted under displacement control, instead of load control, then a reduction in applied load would have been required to maintain a constant δ_c.

Compact Specimens

A typical finite element model (Mesh A) for one half of the compact specimen is shown in Fig. 3. The crack line was a line of symmetry. The minimum element size along the crack line was 0.00625 W_o for this mesh pattern, where W_o is the overall width of the compact specimen. (The large number of elements around the pin-loaded hole was not necessary for this study. This particular mesh was used in a previous analysis to study the deformations around the hole and was used here only for convenience.)

Critical CTOD—A comparison between experimental and calculated load

Mesh A: M = 3654   N = 1985

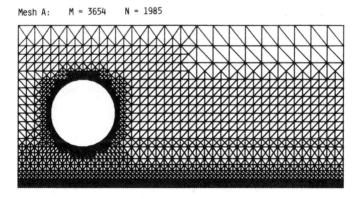

FIG. 3—*Finite element idealization of one half of the compact specimen.*

against physical crack extension data on a 7075-T651 aluminum alloy compact specimen is shown in Fig. 4. The symbols show the average experimental crack extension measurements made using visual and unloading compliance (load-line and crack mouth) methods [10]. The solid lines show calculations made using three different mesh sizes (Meshes A, B, and C) in the crack tip region. The critical CTOD values (δ_c) for each mesh size was selected (by trial and error) to give about the same failure load as measured on the compact specimen. The δ_c values, shown on the figure, were lower for smaller mesh sizes. Smaller mesh sizes also gave lower loads at initiation of crack growth. The calculated crack growth behavior from Mesh C ($d = 0.4$ mm) agreed well with the experimental data up to maximum load.

The final δ_c value for the 7075-T651 aluminum alloy was selected so that the mean of the calculated-to-experimental failure load ratio on the various size compact specimens with $a_o/W = 0.5$ was about unity. Mesh C was used for the 203-mm-wide specimen; Meshes B and A with all coordinates scaled by a factor of 0.5 and 0.25, respectively, were used for the 102- and 51-mm-wide specimens, so that the element size along the crack line was 0.4 mm in all meshes. A comparison between experimental and calculated load against crack extension data on two compact specimen sizes ($W = 51$ and 203 mm) is shown in Fig. 5. The large solid symbols show the experimental data from a singe test, and the bars indicate the range and mean of failure (or maximum) load on four or five tests. The bars are placed at the average value of crack extension at maximum load. The solid lines show the calculated crack growth behavior

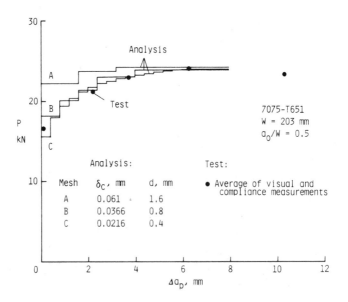

FIG. 4—*Effect of mesh size and critical CTOD on crack growth.*

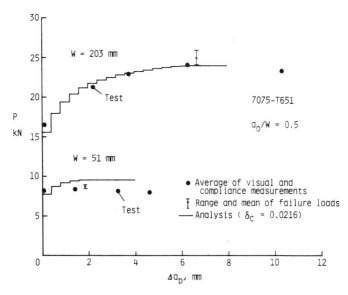

FIG. 5—*Comparison of calculated and experimental crack growth behavior for 7075-T651 aluminum alloy compact specimens.*

from the finite element analysis with $\delta_c = 0.0216$ mm and $d = 0.4$ mm for both specimens. The tests were conducted under displacement-control conditions, whereas the analysis was conducted under load-control conditions. Thus calculations were not made beyond maximum load. The calculated failure load on the large specimen was about 5% lower than the average experimental failure load. For the small specimen, however, the calculated failure load was about 10% higher than the average experimental failure load. The calculated failure loads for all compact specimens with a_0/W equal to 0.5 are given in Table 3. The mean of the ratios of calculated-to-experimental failure load was about 1.01.

For convenience, the same mesh size ($d = 0.4$ mm) used for the 7075-T651 aluminum alloy specimens was also used for the 2024-T351 aluminum alloy and 304 stainless steel specimens.

A comparison between experimental and calculated load against physical crack extension data 2024-T351 aluminum alloy compact specimens is shown in Fig. 6. Again, the symbols show average experimental crack extension measurements made on two compact specimens using visual and unloading compliance data [10]. The respective bars indicate the range and mean of failure loads on four or five tests. A critical CTOD value was selected so that the mean of the calculated-to-experimental failure load ratio on the various size compact specimens with $a_0/W = 0.5$ was about unity. Again, the solid lines show the calculated crack growth behavior from the finite element analysis with $\delta_c = 0.0457$ mm and $d = 0.4$ mm. The calculated behavior at initiation and at insta-

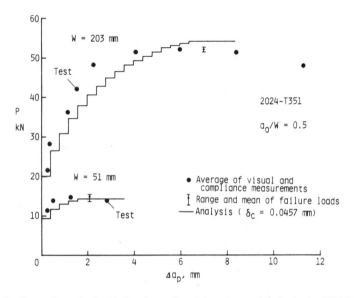

FIG. 6—*Comparison of calculated and experimental crack growth behavior for 2024-T351 aluminum alloy compact specimens.*

bility (maximum load) agreed well with the experimental data. Some discrepancies between calculated and experimental behavior were observed for the stable crack growth region; however, the calculated failure loads on the large and small compact specimens were within 5% of the average experimental failure loads.

As with the other materials, the critical CTOD value for 304 stainless steel specimens was selected so that the mean of the calculated-to-experimental failure load ratio on the various size compact specimens ($W = 51$, 102, and 203 mm) with $a_o/W = 0.5$ was, again, about unity. The critical CTOD value (δ_c) was 0.356 mm with $d = 0.4$ mm. The 304 stainless steel compact specimens exhibited large deformations and rotations. Because the finite element analysis was based on a small strain assumption, no comparisons were made between calculated and experimental crack growth.

Comparison of Calculated and Experimental CTOD Values—Paleebut [*11*] used a laser-interferometric displacement technique to measure crack-opening displacements near the tip of a crack in 7075-T651 and 2024-T351 aluminum alloy compact specimens ($W = 51$ mm). The specimen thicknesses ranged from 3 to 25 mm. Only the results for 13-mm-thick specimens were of interest here. The specimens were fatigue precracked in accordance with ASTM E 399 requirements to a specified crack length ($a_o/W = 0.5$), and then two indentations 0.1 mm apart were placed across the fatigue crack about 0.1 mm from the crack tip. The specimens were then statically pulled to failure and load against CTOD traces were obtained from the test. The first indication of a major "popin" was taken to be the CTOD at initiation (δ_i). These experimental values are given in Table 6.

The critical CTOD values (δ_c) determined from fitting the finite element analysis to load against crack extension data for the two materials are also shown for comparison. Despite the fact that the δ_c values in the analyses were measured at 0.4 mm from the crack tip, instead of 0.1 mm as in the experiments, the agreement is extremely good. This good agreement may be caused by crack tunneling. Cracks in 13-mm-thick compact specimens tend to tunnel in the interior. Consequently, the "effective" distance from the crack tip to the measurement point would have been larger than 0.1 mm. For the 7075-T651 material, the average tunneling was about 0.6 mm, whereas for the 2024-T351 material the average tunneling was about 1 mm. The 0.4 mm used in the analysis was less than the average tunneling for both materials.

Failure Load Predictions—Having determined the critical CTOD values for the three materials, the finite element analysis was used to predict failure loads on compact specimens with $a_o/W = 0.3$ and 0.7. Figures 7, 8, and 9 show experimental (symbols) and predicted failure loads plotted against a_o/W for 7075-T651 aluminum alloy, 2024-T351 aluminum alloy, and 304 stainless steel, respectively. The predicted failure loads were calculated for $a_o/W = 0.3$, 0.5, and 0.7, and a curve was drawn through the results. A failure load prediction was not made on the large 304 stainless steel specimen with $a_o/W = 0.3$ because of the excessive computer cost. All predictions were within $\pm 10\%$ of the experimental failure loads, except for some results on the 304 stainless steel specimens. The predicted results on the steel specimens showed some systematic errors in failure load predictions with a_o/W and W, but all predicted failure loads were within $\pm 15\%$ of experimental loads. The larger and systematic errors on the steel specimens were probably caused by using a small strain analysis for a large deformation problem.

Experimental and predicted failure loads for compact specimens are given in Tables 3 to 5 for the three materials.

Center-Crack Tension Specimens

Because the critical CTOD values were determined from compact specimens, the good failure load predictions on other compact specimens may have

TABLE 6—*Comparison of experimental and calculated values of CTOD.*

Material	t, mm	Experimental δ_i, mm	Calculated δ_c, mm
7075-T651	13	0.021	0.0216
2024-T351	13	0.048	0.0457

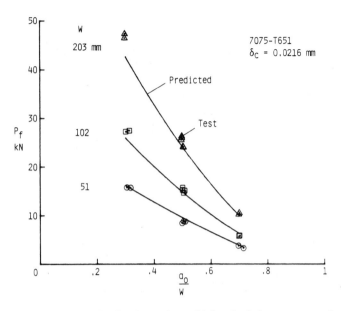

FIG. 7—*Comparison of predicted and experimental failure loads for compact specimens made of 7075-T651 aluminum alloy.*

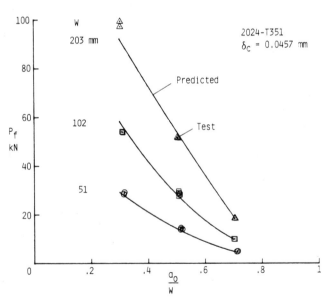

FIG. 8—*Comparison of predicted and experimental failure loads for compact specimens made of 2024-T351 aluminum alloy.*

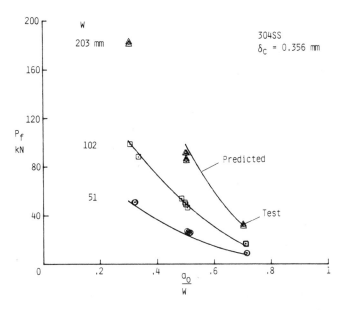

FIG. 9—*Comparison of predicted and experimental failure loads for compact specimens made of 304 stainless steel.*

been fortuitous. To test the CTOD criterion and to determine whether or not fracture data from compact specimens could be used to predict failure of other crack configurations, center-crack tension specimens were tested and analyzed.

Figure 10 shows the finite element meshes used to model the two center-crack tension specimens ($W = 127$ and 254 mm). Because of symmetry, only one quarter of the specimen was modelled. Insert (a), along the crack line, shows the typical element sizes and patterns in the crack tip region for the two specimen sizes. The minimum element size along the crack line for both specimens was 0.4 mm.

The finite element analysis with the CTOD criterion was used to predict failure loads on the center-crack tension specimens made of the three materials (Tables 3 to 5). Figure 11 shows experimental (symbols) and predicted failure loads plotted against specimen width for a nominal $a_o/W = 0.4$. The predicted failure loads were made at $W = 127$ and 254 mm, and a line was drawn through the results. All predictions were within about $\pm 10\%$ of experimental failure loads for all materials.

Three-Hole-Crack Tension Specimens

To verify the general applicability of the CTOD criterion to complex cracked components, a structurally configured specimen, the three-hole-crack tension specimen (Fig. 1c) was tested and analyzed. The three-hole-crack specimen

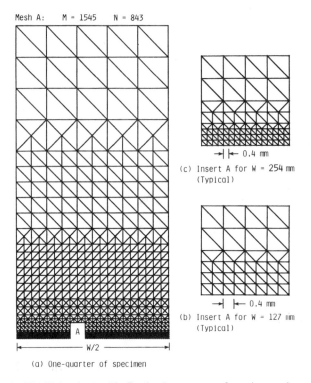

(c) Insert A for W = 254 mm
(Typical)

(b) Insert A for W = 127 mm
(Typical)

(a) One-quarter of specimen

FIG. 10—*Finite element idealization for center-crack tension specimens.*

was designed to give a complicated stress intensity factor solution, like that for a cracked stiffened panel. Although the stress intensity factor solution was not used in the fracture analysis, the stress intensity factor solution will be used in the discussion of results. In the following sections, the stress intensity factor solution and failure load predictions on the three materials are presented.

Stress Intensity Factor Solution—The stress intensity factor solution for the three-hole-crack tension specimen was obtained by using a two-dimensional elastic finite element analysis [14]. The finite-element mesh for one half of the specimen is shown in Fig. 12. The minimum element size was 0.00625 W. Stress intensity factors as a function of crack length were obtained by using a local-energy approach proposed by Irwin [18]. To verify the local-energy approach and mesh pattern in the crack tip region, compact and center-crack tension specimens were analyzed. The details of the approach are given in Appendix B of Ref 19.

The stress intensity factor solution for the three-hole-crack tension specimen (Fig. 1c) is shown in Fig. 13. The symbols show the finite element results. Stress intensity factors were normalized by gross applied stress (P/Wt) and are plotted against crack-length-to-width ratio (a/W) with $W = 254$ mm.

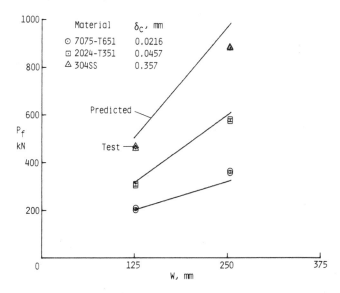

FIG. 11—*Comparison of predicted and experimental failure loads for center-crack tension specimens.*

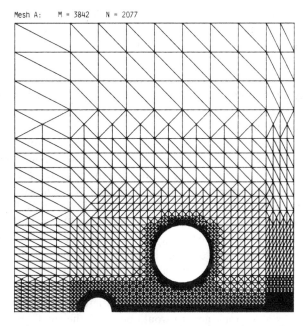

FIG. 12—*Finite element idealization of one half of the three-hole-crack specimen.*

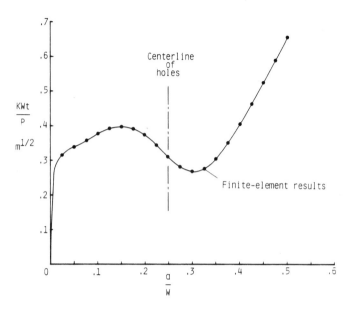

FIG. 13—*Stress intensity factors for the three-hole-crack tension specimen.*

Crack length is measured from the edge of the small hole. The dash-dot line shows the centerline of the two large holes. This solution simulates the solution for a cracked stiffened panel [20] when the centerline of the stiffener is located at $a/W = 0.25$ (dash-dot line). The stress intensity factors have a minimum value near the centerline of the two large holes, or a stiffener, in the case of a stiffened panel.

Failure Load Predictions—During the failure of the three-hole-crack specimens with a_o/W less than 0.3, large amounts of stable crack growth were expected (as much as 65 mm for $a_o/W = 0.05$) because of the stress intensity factor solution shown in Fig. 13. During monotonic loading, as the crack stably tears toward the centerline of the large holes, the stress intensity factor decreases. An increase in load is required to cause further stable crack growth. Crack growth instability should only occur when the crack has passed the minimum stress-intensity factor at an a/W value of about 0.3. The same general behavior might intuitively be expected for elastic-plastic materials. Fracture tests on the three materials and the finite element analysis will test this hypothesis.

To save computer time and cost, the small elements ($d = 0.4$ mm) were used only between a/W values of 0.275 and 0.375, for $a_o/W \leq 0.25$ because the crack should stably tear to an a/W value of about 0.3. At this point, the crack will be located in the small element region. For $a_o/W = 0.3$, 0.35, and 0.4, the small element pattern was placed at the appropriate location along the crackline so that all stable crack growth would occur in the small element region. But for a_o/W less than 0.275 ($a_o < 70$ mm) stable crack growth would occur in a region with larger element sizes than 0.4 mm. To use the larger element sizes, the CTOD criterion was modified as follows. The large compact specimen ($W = 203$ mm) with $a_0/W = 0.5$ was reanalyzed with three different element sizes along the line of crack extension. Critical CTOD values needed to predict only the experimental failure loads on the three materials were determined. These results are shown in Fig. 14. The critical CTOD (δ_c') for a given element size (d') normalized by the critical CTOD (δ_c) for an element size (d) of 0.4 mm is plotted against the element size ratio (d'/d). Symbols show the finite element results for the three materials. The solid line is an equation fit to the results. The equation

$$\delta_c' = \delta_c \left[1 + \frac{2}{3}\left(\frac{d'}{d} - 1\right) \right] \qquad (2)$$

was used to determine the δ_c' value for any value of element size d'. Figure 14 shows δ_c and d for the three materials. (The dashed line indicates the relationship between δ_c' and d' needed for a constant CTOA.)

Motion pictures (200 frames per second) were taken of some of the three-hole crack specimen tests. A voltmeter was used to indicate applied load in the movie. Load against crack length measurements taken from these motion pictures are shown in Fig. 15. The initial crack lengths (a_0) were about 25.4 mm. The symbols show experimental data on 7075-T651 and 2024-T351 aluminum alloy specimens, and the solid lines show the prediction from the finite element analysis. The solid symbols show the final crack lengths near maximum load condi-

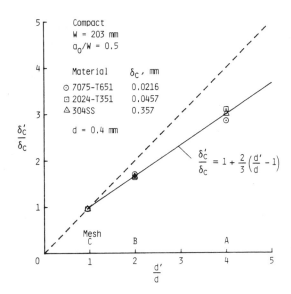

FIG. 14—*Effect of mesh size on critical CTOD needed to predict failure loads on the large compact specimens.*

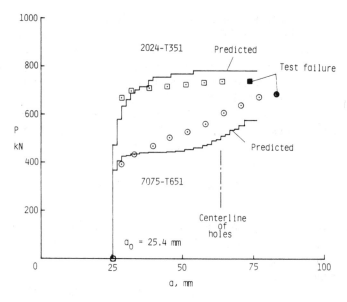

FIG. 15—*Comparison of predicted and experimental crack growth behavior on the three-hole-crack tension specimens.*

tions. As expected, the final crack lengths were pass the centerline of the large holes and were very near the minimum stress intensity factor location in Fig. 13. Predictions on the 7075-T651 aluminum alloy specimen were in fair agreement with the experimental results. The predicted failure load was about 16% lower than the experimental failure load. Predictions on the 2024-T351 aluminum alloy specimen, on the other hand, agreed well with the experimental results; here the predicted failure load was only 6% higher than the experimental failure load. The predicted crack lengths at instability (crack length at first attainment of maximum load), however, were less than the experimental values, especially on the 2024-T351 specimen. This may have been caused by the size of the load step used in the analysis, since the predicted curve is nearly horizontal near maximum load. Photographs from motion picture frames near maximum (failure) load conditions are shown in Fig. 16 for the three materials. Figure 16c illustrates why finite deformation analyses may be necessary to improve the predictions for the 304 stainless steel specimens.

A comparison between predicted and experimental failure load for various initial crack lengths for the three materials is shown in Fig. 17. Again, symbols show the experimental failure loads and solid curves show the predicted results from the finite element analysis. Predictions were not made on most of the 304 stainless steel specimens because of the high computer cost. The predicted failure loads on 7075-T651 aluminum alloy specimens were 2 to 18% lower than the experimental failure loads. The largest errors occurred at the smallest

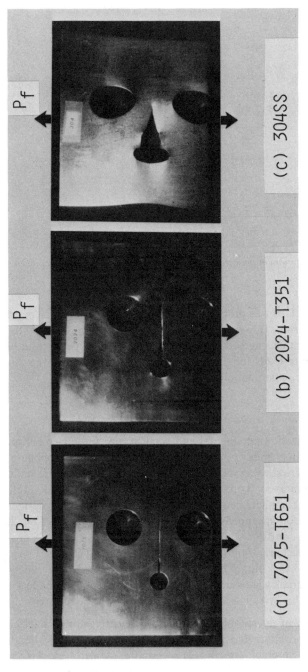

FIG. 16—*Three-hole-crack tension specimens near maximum load (failure) condition* ($a_o = 25.4$ mm).

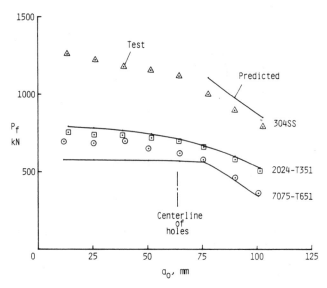

FIG. 17—*Comparison of predicted and experimental failure loads on three-hole-crack tension specimens.*

crack lengths. This was probably caused by using the larger element sizes for a_0 less than 70 mm. However, predicted failure loads on the 2024-T351 aluminum alloy specimens agreed well (1 to 6%) with the experimental loads. The predicted failure loads for the 304 stainless steel specimens were about 10% higher than the experimental results.

Conclusions

An elastic-plastic (incremental and small strain) finite element analysis, in conjunction with a crack growth criterion based on crack-tip-opening displacement, was used to study crack growth behavior under monotonic loading to failure for three crack configurations and for three materials. Fracture tests were conducted on compact, center-crack tension, and specially designed three-hole-crack tension specimens made of 7075-T651 aluminum alloy, 2024-T351 aluminum alloy, and 304 stainless steel. In the analysis, the crack growth criterion was a critical crack-tip-opening displacement (CTOD) at a specified distance from the crack tip. Whenever the CTOD equaled or exceeded a critical value (δ_c), the crack was assumed to grow. Comparisons were made with experimental data. This investigation supports the following conclusions:

1. The present elastic-plastic analysis predicted three stages of crack growth behavior, characteristic of metals, under monotonic loading to failure: (1) a period of no crack growth, (2) a period of stable crack growth, and (3) crack growth instability under load-control conditions.

2. Single values of critical CTOD (δ_c) were found to reasonably model crack

initiation, stable crack growth, and instability for 7075-T651 and 2024-T351 aluminum alloy compact specimens.

3. Calculated CTOD values agreed well with experimental measured (from the literature) values for crack growth initiation for the two aluminum alloys.

4. Critical CTOD (δ_c) values from compact specimens were used to predict failure loads on center-crack tension and three-hole-crack tension specimens generally within $\pm 15\%$ of experimental failure loads on 7075-T651 aluminum alloy and 304 stainless steel specimens, and within $\pm 6\%$ of experimental failure loads on 2024-T351 aluminum alloy specimens.

References

[1] Anderson, H., *Journal of the Mechanics and Physics of Solids*, Vol. 21, 1973, pp. 337–356.

[2] Kobayashi, A. S., Chiu, S. T., and Beeuwkes, R., *Engineering Fracture Mechanics*, Vol. 5, No. 2, 1973, pp. 293–305.

[3] de Koning, A. U., "A Contribution to the Analysis of Slow Stable Crack Growth," National Aerospace Laboratory Report NLR MP 75035U, The Netherlands, 1975.

[4] Light, M. F., Luxmoore, A., and Evans, W. T., *International Journal of Fracture*, Vol. 11, 1975, pp. 1045–1046.

[5] Newman, J. C., Jr. in *Cyclic Stress-Strain and Plastic Deformation Aspects of Fatigue Crack Growth*, ASTM STP 637, American Society for Testing and Materials, 1977, pp. 56–80.

[6] Rousselier, G., "A Numerical Approach for Stable-Crack-Growth and Fracture Criteria," in *Proceedings*, Fourth International Conference on Fracture, University of Waterloo, Ontario, Canada, Vol. 3, 1977.

[7] Shih, C. F., de Lorenzi, H. G., and Andrews, W. R. in *Elastic-Plastic Fracture*, ASTM STP 668, American Society for Testing and Materials, 1979, pp. 65–120.

[8] Kanninen, M. F., Rybicki, E. F., Stonesifer, R. B., Broek, D., Rosenfield, A. R., and Halin, G. T. in *Elastic-Plastic Fracture*, ASTM STP 668, American Society for Testing and Materials, 1979, pp. 121–150.

[9] Luxmoore, A., Light, M. F., and Evans, W. T., *International Journal of Fracture*, Vol. 13, 1977, pp. 257–259.

[10] McCabe, D. E., "Data Development for ASTM E24.06.02 Round Robin Program on Instability Prediction," NASA CR-159103, National Aeronautics and Space Administration, Aug. 1979.

[11] Paleebut, S., "CTOD and COD Measurements on Compact Tension Specimens of Different Thicknesses," M.S. thesis, Michigan State University, East Lansing, Mich., 1978.

[12] Clarke, G. A., Andrews, W. R., Paris, P. C., and Schmidt, D. W. in *Mechanics of Crack Growth*, ASTM STP 590, American Society for Testing and Materials, 1976, pp. 27–42.

[13] Zienkiewicz, O. C., Valliappan, S., and King, I. P., *International Journal for Numerical Methods in Engineering*, Vol. 1, 1969, pp. 75–100.

[14] Newman, J. C., Jr., "Finite-Element Analysis of Fatigue Crack Propagation—Including the Effects of Crack Closure," Ph.D. thesis, Virginia Polytechnic Institute and State University, Blacksburg, Va., May 1974.

[15] Hutchinson, J. W., *Journal of the Mechanics and Physics of Solids*, Vol. 16, 1968, pp. 13–31.

[16] Rice, J. R. and Rosengren, G. F., *Journal of the Mechanics and Physics of Solids*, Vol. 16, 1968, pp. 1–12.

[17] Ramberg, W. and Osgood, W. R., "Description of Stress-Strain Curves by Three Parameters," NACA TN-902, National Advisory Committee for Aeronautics, 1943.

[18] Irwin, G. R., "Analysis of Stresses and Strains Near the End of a Crack Traversing a Plate," *Journal of Applied Mechanics, Transactions of ASME*, Vol. 79, 1957.

[19] Newman, J. C., Jr., "Finite-Element Analysis of Initiation, Stable Crack Growth, and Instability Using a Crack-Tip-Opening Displacement Criterion," NASA TM-84564, National Aeronautics and Space Administration, Oct. 1982.

[20] Poe, C. C., Jr., "Stress Intensity Factor for a Cracked Sheet with Riveted and Uniformly Spaced Stringers," NASA TR-358, National Aeronautics and Space Administration, May 1971.

C. William Smith[1] and George C. Kirby[1]

Stress Intensity Distributions and Width Correction Factors for Natural Cracks Approaching "Benchmark" Crack Depths

REFERENCE: Smith, C. W. and Kirby, G. C., **"Stress Intensity Distributions and Width Correction Factors for Natural Cracks Approaching "Benchmark" Crack Depths,"** *Fracture Mechanics: Fifteenth Symposium, ASTM STP 833*, R. J. Sanford, Ed., American Society for Testing and Materials, Philadelphia, 1984, pp. 118–129.

ABSTRACT: A series of frozen stress photoelastic experiments were conducted on natural surface flaws in both wide and finite width flat plates under uniform uniaxial tension. The flaws grown under monotonic load were semielliptic and retained that shape (but with varying aspect ratio) up to 75% of the plate depth. The stress intensity distributions compared favorably with the predictions of the Newman-Raju analysis, but were slightly higher at the points of maximum flaw penetration for the deeper flaws. The crack growth path of the photoelastic models differed from that resulting from fatigue tests on aluminum up to $a/T \approx 0.3$. The difference is conjectured to be caused by crack closure in the fatigue tests.

KEY WORDS: fracture mechanics, stress intensity factors, photoelasticity, fatigue crack growth, frozen stress

G. R. Irwin, in a classic paper in 1962 [1], characterized surface cracks in large bodies as being semielliptic in shape and provided a stress intensity (SIF) distribution around the flaw border. This characterization has been generally found to be valid, but has been questioned for use when the crack width occupies a substantial portion of a finite width plate or the crack depth is a substantial portion of plate depth. In 1976, a meeting of workers in the field of three-dimensional fracture problems was convened at Battelle Columbus Laboratories and a set of "benchmark" problem geometries were prescribed [2] for use by analysts and experimentalists for use in checking and

---

[1] Alumni Professor and Graduate Research Assistant, respectively, Department of Engineering Science and Mechanics, Virginia Polytechnic Institute and State University, Blacksburg, Va. 24061.

comparing their results. One of these geometries was for a surface flaw in a finite depth, wide plate under remote uniaxial tension.

Several numerical solutions utilizing finite elements, alternating methods, boundary integrals, and hybrid methods were subsequently compared and showed reasonable agreement for cracks in wide plates [3]. Further studies [4] revealed that the specified geometries could not be obtained by growing cracks subcritically from mechanical defects where the aspect ratio of the starter crack is approximately 0.9. An experimental study was then undertaken in order to (1) find the aspect ratios of cracks grown in finite plates under uniaxial extension when they reach "benchmark flaw depths", (2) determine if flaw shapes deviate from semiellipses during growth, (3) assess finite width effects experimentally, and (4) identify a mathematical model which could provide reasonable estimates of all these effects. Results of the first phase of this study, reported in Ref 5, identified the Newman-Raju analysis [6] as yielding good correlation with limited experimental results.

The present work is an extension of Ref 5 and seeks to address the first three items mentioned above in some detail and to assess the validity of the Newman-Raju analysis for finite width plates.

Analytical Background (Mode I Loading)

Sih and Kassir [7] have shown that the Irwin near tip stress field equations for the plane case apply also for plane cracks with curved crack borders. For the problem geometry and notation of Fig. 1, these equations may be written as

$$\sigma_{ij} = \frac{K_{\mathrm{I}}}{(2\pi r)^{1/2}} f_{ij}(\theta) + \sigma_{ij}^0(r, \theta) \qquad (i, j = n, z) \tag{1}$$

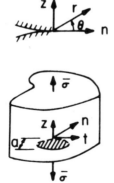

FIG. 1—*General problem geometry and notation.*

where the second set of terms are included to account for the contribution of the nonsingular part of the stress field to the stresses in the measurement zone, since said zone is a short distance from the crack tip. These terms (σ_{ij}^0) may be thought of as Taylor Series Expansions of the regular stresses in the measurement zone. It is important to make measurements outside the very near tip region, since a nonlinear zone exists here. If one takes data just outside the nonlinear zone, however, then σ_{ij}^0 can be adequately represented by only the leading terms in the Taylor Series for the stress components of interest. Thus σ_{ij}^0 become a set of constants for a given point along the flaw border but vary from point to point. Moreover, σ_{nn}^0 and σ_{nz}^0 need not vanish to satisfy crack free surface conditions, since Eq 1 is employed only in a narrow zone away from the crack surfaces. In fact, since Mode I fringes tend to spread in a direction approximately normal to the crack tip (Fig. 2), data are taken only along $\theta = \pi/2$, which is close to the direction of the minimum fringe gradient. Thus, if we compute the maximum shearing stress along $\theta = \pi/2$,

$$\tau_{nz}^{max} = \frac{1}{2}[(\sigma_{zz} - \sigma_{nn})^2 + 4\sigma_{nz}^2]^{1/2} \tag{2}$$

and truncate to the same order as Eq 1, we obtain

$$\tau_{nz}^{max} = \frac{K_I}{(8\pi r)^{1/2}} + f(\overline{\sigma}_{ij}^0) \tag{3}$$

which may be normalized and rewritten as

$$\frac{K_{AP}}{\overline{\sigma}(\pi a)^{1/2}} = \frac{K_I}{\overline{\sigma}(\pi a)^{1/2}} + \frac{(8\pi)^{1/2}}{\overline{\sigma}} f(\overline{\sigma}_{ij}^0)\left(\frac{r}{a}\right)^{1/2} \tag{4}$$

where

$K_{AP} = \tau_{nz}^{max}(8\pi r)^{1/2}$, the "apparent" SIF,
$\overline{\sigma}$ = load parameter such as remote stress,
K_I = Mode I SIF,
$\overline{\sigma}_{ij}^0$ = set of constant values at a point, and, from the stress-optic law,

$$\tau_{nz}^{max} = \frac{Nf}{2t'} \tag{5}$$

where

N = isochromatic fringe order,
f = material fringe value, and
t' = slice thickness in t direction.

FIG. 2—*Spreading of fringes normal to crack plane (Mode I).*

Equation 4 describes the stress state in a zone where the normalized apparent SIF $\{K_{AP}/\overline{\sigma}(\pi a)^{1/2}\}$ varies linearly with the square root of the normalized distance from the crack tip $(r/a)^{1/2}$ where r is measured along $\theta = \pi/2$. A typical set of raw photoelastic data for this zone is shown in Fig. 3 for one point on the border of a semielliptic surface flaw. Except when cracks are approaching other boundaries, this zone normally lies between 0.1 and 1.0 mm from the crack tip along $\theta = \pi/2$. However, the exact location of the zone is not obtained from a single point such as represented in Fig. 3, but rather by comparing K_{AP} curves from all points along a given crack front from which measurements were made. It should be common to all points along the same flaw border. Once the linear zone is so located, data in that zone are fitted with a straight line which is then extrapolated to the origin in order to obtain $K_I/\overline{\sigma}(\pi a)^{1/2}$. This procedure allows one to pass over the near tip nonlinear zone.

The foregoing approach was used in order to estimate SIF distributions in the experiments described in the sequel.

The Experiments

Beginning with fringe-free flat plates of a stress-freezing photoelastic material PSM-9 with 15 parts hardener cast by Photolastic Incorporated, natural starter cracks were inserted into a series of specimens by striking a sharp

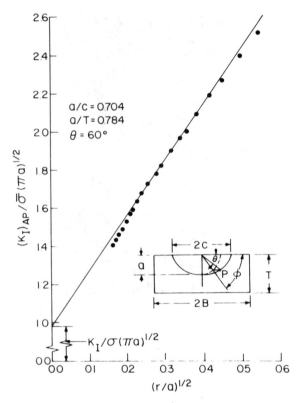

FIG. 3—*Typical data for estimating SIF.*

round-tipped blade held normal to the plates with a hammer. A single starter crack emanated dynamically from the blade tip and quickly arrested. The plane of each crack was normal to the plate surface. The cracked plates were then placed in an oven in a deadweight loading rig consisting of nylon lines connected through fishing swivels to pins through equally spaced holes near the top and bottom of dog-boned-shaped wide plate specimens. The cracked models were then heated to critical temperature, soaked thermally, and loaded with sufficient load to enlarge the starter cracks. When the desired crack size was achieved, as viewed through a glass port in the oven wall, the load was reduced to stop flaw growth and models were slowly cooled to room temperature, freezing in the fringe and deformation fields associated with the reduced load above critical temperature. Upon load removal (with negligible recovery), slices parallel to the nz plane about 2.0 mm thick were removed at intervals along the flaw border, coated with matching index fluid, and analyzed in a crossed circular polariscope using the Tardy fractional fringe order method. Data so obtained were plotted on graphs such as Fig. 3 for each slice, and SIF values were estimated therefrom. Dimensions and geometry of the eleven plates studied are given in Table 1.

Results

Bearing in mind that stress-freezing photoelastic materials are incompressible above critical temperature (that is, $v \approx 0.5$), the experimental SIF distributions are compared with the predictions of the Newman-Raju analysis for wide plates in Fig. 4. This analysis consists of a finite element model which utilizes isoparametric linear strain elements away from the crack tip and pentahedron-shaped singular elements at the crack tip with displacement terms which were proportional to the square root of the distance from the crack tip. SIF values were calculated using a nodal force method so as to avoid the assumptions of plane stress or plane strain. Convergence studies on the

TABLE 1—*Dimensions and geometry of the eleven plates.*

	Infinite Plates					
Moderately Shallow Flaws						
Test	$\bar{\sigma}$, MPa	a, mm	C, mm	a/C	a/T	C/B
S-1	0.10	4.12	4.69	0.88	0.30	0.13
S-2	0.10	4.33	5.56	0.78	0.32	0.15
S-3[a]	0.19	5.33	5.54	0.96	0.39	0.15
S-4[a]	0.24	5.26	5.33	0.99	0.39	0.15
Intermediate Flaws						
Test	$\bar{\sigma}$, MPa	a, mm	C, mm	a/C	a/T	C/B
I-5	0.08	7.04	9.42	0.75	0.52	0.21
I-6	0.08	6.86	8.92	0.77	0.53	0.20
Deep Flaws						
Test	$\bar{\sigma}$, MPa	a, mm	C, mm	a/C	a/T	C/B
D-7	0.08	11.00	15.62	0.70	0.78	0.21
D-8	0.07	9.54	14.00	0.68	0.72	0.20
D-9	0.09	10.41	17.28	0.60	0.78	0.25
	Finite Width Plates					
Test	$\bar{\sigma}$, MPa	a, mm	C, mm	a/C	a/T	C/B
D-10	0.16	7.67	9.78	0.78	0.62	0.51
D-11	0.07	6.28	10.24	0.61	0.50	0.55

[a]7 parts hardener (all others 15 parts hardener). The equation is

$$\Phi = \int_0^{\pi/2} \left[\left(\frac{a}{C} \right)^2 \cos^2\phi + \sin^2\phi \right]^{1/2} d\phi$$

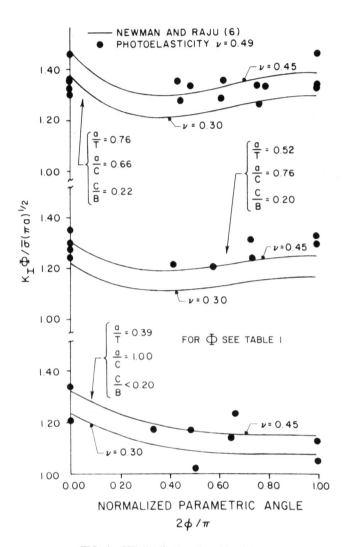

FIG. 4—*SIF distributions for wide plates.*

SIF showed agreement with closed form three-dimensional solutions to within 1 or 2%. The models had 4300 to 4800 degrees of freedom.

Since the Newman-Raju analysis predicted an elevation of the SIF of about 6% (or the order of the experimental scatter) for the deep flaws for $\nu = 0.45$ above results for $\nu = 0.30$, the correlation between theory and experiment in Fig. 4 appears quite reasonable. Moreover, the experimental flaw shapes remained semielliptic, even for $a/T \approx 0.75$. It is noted, however, that for the deeper flaws slightly higher experimental SIFs were obtained at maximum flaw depth than predicted by theory. Figure 5 provides similar comparisons

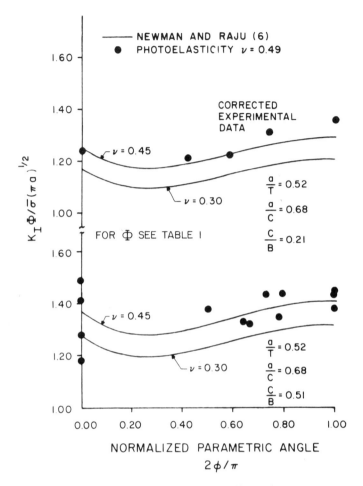

FIG. 5—*Influence of width effect on SIF distribution.*

between wide $(C/B \approx 0.2)$ and finite $(C/B \approx 0.5)$ width plate results. In making these comparisons, results from wide plate models were used to obtain a correction for thickness effect for the wide plates, and this correction was used on the wide plate models to correct them to the same a/T as the finite width plates. These results suggest that the Newman-Raju analysis provides very reasonable finite width effect assessments. Again, however, the measured SIF values at maximum flaw depth were slightly higher than predicted analytically.

By applying a modified Paris crack growth rate law at the plate surface and at maximum flaw depth, and assuming a semielliptic flaw shape, Newman and Raju were able to predict crack growth paths on plots of a/C versus a/T. Figure 6 shows a comparison between such a prediction and some crack

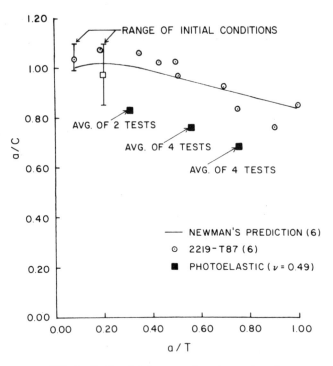

FIG. 6—*Predicted and measured crack growth paths.*

growth test data on 2219-T87 aluminum plates which exhibited crack closure during fatiguing under a tension-tension load spectrum. The modification used by Newman and Raju was to adjust the growth rates at the surface and maximum crack depth so that $a/C \approx 1.0$ up to $a/T \approx 0.3$. Averaged results from the current test program are also included in Fig. 6. In prior studies [8,9], the authors have found that natural cracks grown under monotonic load in thick-walled photoelastic models above critical temperature would virtually overlay those produced under tension-tension fatigue of geometrically similar metal structures. In Fig. 6, however, we see a significant departure of the photoelastic data from the metal fatigue test data for $a/T < 0.3$. This reduction in aspect ratio for the photoelastic models, or increased growth rate along the surface of the body relative to depth for relatively shallow flaws, is believed due to an absence of crack closure, which retards flaw growth near the surface for metals under fatigue load. Referring to the simple Paris law in the form

$$\frac{da}{dN} = C(\Delta K)^m \qquad (6)$$

for stress ratios ($R = \sigma_{min}/\sigma_{max}$) without closure, Newman [10] has computed load paths for various values of m. For metals, m usually lies between 2 and 4. For polymeric materials, however, values ranging from 5.7 for Plexiglas to 14 for polyvinyl chloride [11] have been obtained. Figure 7 shows crack growth paths predicted by different values of m. The photoelastic results (without closure) are also shown. Moreover, Jolles and Tortoriello [12] have obtained similar curves for aluminum alloy at R ratios near unity without closure. On the basis of this evidence, we conjecture that the divergence of the photoelastic results from the fatigue test results in Fig. 6 is due to the closure effect for $a/T < 0.3$. Beyond this depth the closure effects are apparently small and the photoelastic crack growth path appears parallel to the metal fatigue crack growth path. However, these observations should be strengthened by further study.

Summary

After briefly reviewing the analytical background for an experimental technique consisting of a combination of frozen stress photoelasticity and the equations of linear elastic fracture mechanics for estimating SIF distributions, a series of frozen stress experiments on flat plates containing natural

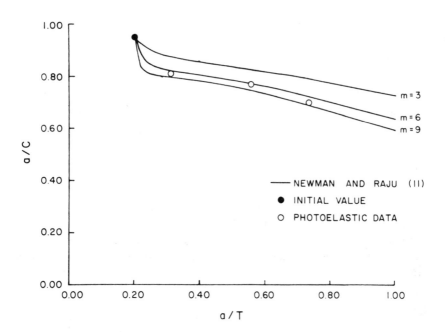

FIG. 7—*Influence of crack growth rate exponent on crack path with no closure present using Newman-Raju model.*

surface flaws under uniaxial tension were described. Measured SIF distributions were compared with results predicted by the Newman-Raju analysis, and crack growth paths were compared with both predicted values and results from fatigue crack growth tests on aluminum plates.

It was concluded that:

1. Surface flaws grown under uniaxial tension in flat plates under monotonic load retain a semielliptic shape of varying aspect ratio a/C up to at least 75% of the depth of the plate.

2. SIF distributions predicted by the Newman-Raju analysis agree with experiments to within experimental scatter. However, experimental values at maximum flaw depth were slightly higher than predicted.

3. The influence of the elevated Poisson's ratio in the photoelastic experiments elevates SIF values by the order of the experimental scatter (about 6%).

4. Crack growth paths in the monotonically loaded photoelastic models diverge from those in fatigue-loaded metal models for relatively shallow flaws. This is believed to be caused by crack closure during fatigue.

The experimental method used has its own limitations if it is to predict both flaw shape and SIF distribution due to fatigue. They include restriction to:

1. Gross elastic behavior of an incompressible elastic material with slightly elevated SIF values.

2. Cracked areas should be sufficiently thick in net section that fatigue enhancement (crack closure) is negligible.

3. Starter crack geometries of photoelastic and metal models should be identical.

4. Some limitation on the fatigue load spectrum (no fatigue overloads).

Despite these restrictions, the method has potential for opening the way to code verification of numerical solutions for three-dimensional cracked body problems where neither flaw shape nor SIF distributions are known *a priori*.

Acknowledgments

The authors wish to acknowledge the useful work of J. C. Newman and I. S. Raju, the staff and facilities of the Department of Engineering Science and Mechanics at VPI & SU, and the support of the Division of Mechanical Engineering and Applied Mechanics of the National Science Foundation under Grant MEA-8113565.

References

[1] Irwin, G. R., *Journal of Applied Mechanics, Transactions of ASME*, Series E, Vol. 84, Dec. 1962, pp. 651–654.
[2] Hulbert, L. E., Ed., *Proceedings of a Workshop on Three Dimensional Fracture Analysis*, Battelle-Columbus Laboratories, Columbus, Ohio, Sept. 1966.

[3] McGowan, J. J., Ed., *Journal of Experimental Mechanics*, Vol. 20, No. 8, Aug. 1980, pp. 253-264.

[4] Smith, C. W., Peters, W. H., Kirby, G. C., and Andonian, A. T. in *Fracture Mechanics: Thirteenth Conference, ASTM STP 743*, American Society for Testing and Materials, 1981, pp. 422-437.

[5] Smith, C. W. and Kirby, G. C., in *Fracture Mechanics: Fourteenth Symposium—Volume I: Theory and Analysis, ASTM STP 791*, J. C. Lewis and G. Sines, Eds., American Society for Testing and Materials, 1983, pp. I-269-I-280.

[6] Newman, J. C., Jr., and Raju, I. S., "Analyses of Surface Cracks in Finite Plates under Tension or Bending Loads," NASA Technical Paper 1578, Washington, D.C., Dec. 1979.

[7] Sih, G. C. and Kassir, M. K., *Journal of Applied Mechanics*, Vol. 33, 1966, pp. 601-611.

[8] Smith, C. W. and Peters, W. H., "Prediction of Flaw Shapes and Stress Intensity Distributions in Three Dimensional Problems by the Frozen Stress Method," in *Sixth International Conference on Experimental Stress Analysis Preprints*, Sept. 1978, pp. 861-865.

[9] Smith, C. W., *Journal of Experimental Mechanics*, Vol. 20, No. 4, April 1980, pp. 126-133.

[10] Newman, J. C., Jr., private communication, 1982.

[11] Radon, J. C., *International Journal of Fracture*, Vol. 16, No. 6, Dec. 1980, pp. 533-551.

[12] Jolles, M. and Tortoriello, V., this publication, pp. 300-311.

M. Ramulu,[1] *A. S. Kobayashi,*[1] *and B. S.-J. Kang*[1]

Dynamic Crack Branching— A Photoelastic Evaluation

REFERENCE: Ramulu, M., Kobayashi, A. S., and Kang, B. S. -J., **"Dynamic Crack Branching—A Photoelastic Evaluation,"** *Fracture Mechanics: Fifteenth Symposium, ASTM STP 833*, R. J. Sanford, Ed., American Society for Testing and Materials, 1984, pp. 130–148.

ABSTRACT: A necessary and sufficient condition for crack branching based on a crack branching stress intensity factor, K_{Ib}, accompanied by a minimum characteristic distance of r_c is proposed. This crack branching criterion is evaluated by dynamic photoelastic experiments involving crack branching of six single-edged notch specimens and six wedge-loaded rectangular double cantilever beam specimens. Consistent crack branching at $K_{Ib} = 2.04$ MPa$\sqrt{m}$ and $r_c = 1.3$ mm verified this crack branching criterion. The crack branching angle predicted by this crack branching criterion agreed well with those measured in the crack branching experiments.

KEY WORDS: dynamic fracture, crack curving, crack branching, branching angle, critical radial distance, dynamic stress intensity factor, nonsingular stress, finite element method, fracture (materials), crack propagation

Literature on crack branching criteria can be grouped into two categories: dynamic crack tip stress field [1–3] and initiation of secondary cracks [4–7]. While the former relates only to the singular stress field at the crack tip, the latter incorporates the nonsingular stress components. Studies on the crack tip stress field can also be divided into pre- and post-branching analyses. Pre-branching analysis normally leads to a branching criterion, while direction of the branched crack and its propagation are studied in post-branching analysis. An excellent review of such crack branching analysis can be found in Ref 7.

Crack branching has been frequently observed during the ten-plus years of dynamic fracture research at the University of Washington [8] and at the University of Maryland [9]. Earlier attempts to evaluate these crack branching results were hampered by the lack of adequate data reduction procedures

[1]Research Assistant Professor, Professor, and Graduate Student, respectively, Department of Mechanical Engineering, University of Washington, Seattle, Wash. 98195.

as well as by the paucity of theoretical understanding on elasto-dynamic crack propagation. Many of these obstacles have been removed today and thus it appears appropriate to re-evaluate these photoelastic data on crack branching in view of the available new data reduction procedure [10]. This data analysis will be preceded by a brief review of existing crack branching criteria, after which a new crack branching criterion will be presented.

Review of Crack Branching Criteria

The most popularly held cause of dynamic crack branching is the pre-branching distortion of the crack tip stress field at a critical crack velocity. Yoffe's theoretical analysis [1] of a constant velocity crack showed that at a crack velocity of about $c/c_1 = 0.33$, the maximum circumferential stress, $\sigma_{\theta\theta}$, shifted away from its original location of $\theta = 0$ at a lower crack velocity.[2] This crack branching criterion based on dynamic crack kinking was followed by that of Craggs [11], who derived a critical crack velocity of $c/c_1 = 0.40$ for a propagating semi-infinite crack. Unfortunately, experimentally measured crack velocities never attained the high velocity predicted by this critical crack velocity criterion. Although Döll measured a branching crack velocity of $c/c_1 = 0.28$ and 0.3 in glass [12], the crack branching velocities in steels reported by Irwin [6], Hahn et al [13], Congleton et al [5], and in photoelastic polymers reported by Kobayashi et al [8] and by Kobayashi and Dally [14] were less than $c/c_1 = 0.25$. Also, the precise ultrasonic ripple marking techniques used to mark instantaneous crack front by Kerkhoff [15] showed only a 10% decrease in crack speed in glass immediately after branching, while Schardin [16] observed no change in crack velocity in plate glass. Acloque [17] observed only a 6% decrease in crack velocities immediately after branching in prestressed glass. Thus, the experimentally observed lower branching velocities, which hardly decreased after crack branching, showed that the postulated critical crack velocity could not be a prerequisite to crack branching in these materials.

Since crack branching is also observed at extremely low crack velocity, such as that in stress corrosion cracking, other crack tip parameters, such as the stress intensity factor, which could trigger branching of a crack propagating at any crack velocity must be sought. For example, attempts have been made to determine experimentally a critical crack branching stress intensity factor, K_{Ib}. Kobayashi et al [8] showed that crack branching occurred in Homalite-100 single-edge notch (SEN) specimens when K_I reached a maximum value of 3.6 times its fracture toughness, K_{Ic}. Dally et al [9,14] obtained a $K_{Ib} = 3.8\ K_{Ic}$ from SEN, double cantilever beam (DCB), and compact specimens when the cracks are propagating at the terminal velocity in Homalite-100.

---

[2]c and c_1 are crack velocity and dilatational stress wave velocity, respectively. The maximum circumferential stress, $\sigma_{\theta\theta}$, is in terms of polar coordinate (r, θ) with origin at crack tip.

A crack kinking criterion based on the development of secondary cracks in a region off-axis to the primary crack is also an attractive alternative, since the crack kinking angle is governed by the dynamic crack tip state of stress. Clark and Irwin [18] concluded that branching occurs by advanced off-axis cracking under a critical stress intensity factor, K_{Ib}, at a limiting crack velocity smaller than those of Yoffe and Craggs. These advanced cracks created crack surfaces of increasing roughness which were associated with increasing stress and velocity and which usually terminated after crack branching.

Crack Branching Angle

A characteristic feature of a branched crack is the crack branching angle. Many attempts have been made to predict this angle. Sih [19] used the prebranching minimum strain energy density to predict a branching angle of 15 to 18 deg which varies with Poisson's ratio. Kitagawa [20] and Kalthoff [21] used the static post branching state of stress of symmetrically branched edged cracks and postulated that a small initial wedge angle between two branched cracks was governed by a vanishing Mode II stress intensity factor; that is, $K_{II} = 0$. Kitagawa et al predicted a branching angle of 30 to 40 deg, while Kalthoff predicted a branching angle of 28 deg which was in agreement with his measured angle in fracturing glass.

The branching angles measured by Christie [22] in an SEN specimen impacted by stress waves was about 25 deg. Congleton [5] observed branching angles of about 30 to 40 deg in center and edge-notched steel plates and 70 to 80 deg in bursting steel tubes. It will be shown later that these variations in measured crack branching angles can be attributed to the influence of nonsingular stress terms which govern the direction of crack branching in various fracture specimen geometry.

Crack Branching Criterion

As described previously, experimental evidence indicates that dynamic crack branching at a terminal crack velocity is accompanied by a critical dynamic stress intensity factor and that the crack branching angles associated with each specimen configuration are very similar. A plausible crack branching criterion would be to postulate that the crack branching stress intensity factor, K_{Ib}, is a necessary condition accompanied by a sufficient condition for crack kinking which governs the crack branching angle. The former necessary condition is supported by crack branching data which show that K_{Ib} is about four times its fracture toughness in Homalite-100.

As for the latter sufficient condition, either of the two dynamic crack curving criteria [23] advanced by the authors can be used to estimate the crack branching angle. These dynamic crack kinking criteria are derived from the near field, mixed-mode elasto-dynamic state of stress associated with a crack tip propagating at constant velocity. The dynamic state of crack tip stress field is given by Freund [24] in terms of local rectangular and polar coordinates of (x, y) and (r, θ), respectively, and the Mode I and II dynamic stress intensity factors, K_I and K_{II}, respectively.[3] The second-order term σ_{0x}, which is acting parallel to the direction of crack extension, is also included in the above crack tip state of stress so that crack kinking can be triggered at crack velocities lower than those of Yoffe [1] and Craggs [11]. The two crack kinking criteria based on this dynamic crack tip stress are the maximum circumferential stress and the minimum strain energy density criteria, both of which will predict nearly identical crack kinking angles in the crack velocity range of $c/c_1 < 0.15$. Thus, for brevity, only the crack kinking criterion based on maximum circumferential stress will be discussed in this paper.

The angle θ_c at which circumferential stress $\sigma_{\theta\theta}$ is maximum when evaluated in conjunction with a pure Mode I dynamic crack tip state of stress will yield a transcendental relation between the critical values of θ and r as

$$r = \frac{1}{4\pi}\left[\left(\frac{K_I}{\sigma_{0x}}\right)\frac{B_1(c)}{\sin 2\theta}\left\{((S_1^2 - S_2^2) - (1 + S_1^2)\cos 2\theta)\frac{\partial f_{11}}{\partial\theta}\right.\right.$$
$$+ 2(1 + S_1^2)\sin 2\theta\, f_{11} + \frac{4S_1 S_2}{1 + S_2^2}\cos 2\theta\,\frac{\partial f_{22}}{\partial\theta}$$
$$-2\frac{4S_1 S_2}{1 + S_2^2}\sin 2\theta\, f_{22} - (2S_1 \sin 2\theta)\left(\frac{\partial g_{11}}{\partial\theta} - \frac{\partial g_{22}}{\partial\theta}\right)$$
$$\left.\left.- (4S_1 \cos 2\theta)(g_{11} - g_{22})\right\}\right]^2 \tag{1a}$$

where

$$\left.\begin{array}{l} f_{11} = [f(c_1) + g(c_1)]^{1/2} \\ g_{11} = [f(c_1) - g(c_1)]^{1/2} \\ f_{22} = [f(c_2) + g(c_2)]^{1/2} \\ g_{22} = [f(c_2) - g(c_2)]^{1/2} \end{array}\right\} \quad \text{for } 0 < \theta \leq \pi \tag{1b}$$

$$f(c_1) = \frac{1}{\left(1 - \frac{c^2}{c_1^2}\sin^2\theta\right)^{1/2}} ; \quad g(c_1) = \frac{\cos\theta}{\left(1 - \frac{c^2}{c_1^2}\sin^2\theta\right)} \tag{1c}$$

$$f(c_2) = \frac{1}{\left(1 - \frac{c^2}{c_2^2}\sin^2\theta\right)^{1/2}} ; \quad g(c_2) = \frac{\cos\theta}{\left(1 - \frac{c^2}{c_2^2}\sin^2\theta\right)}$$

[3]The superscript "dyn" to identify dynamic stress intensity factor will not be used in this paper, since all quantities refer to dynamic values.

$$B_I(c) = \frac{1 + S_2^2}{4S_1S_2 - (1 + S_2^2)^2} \tag{1d}$$

$$S_1^2 = 1 - \frac{c^2}{c_1^2} \; ; \qquad S_2^2 = 1 - \frac{c^2}{c_2^2} \tag{1e}$$

where c_2 is the distortional wave velocity.

The condition for a self-similar propagation of straight crack was assumed in Ref 23 by setting $\theta = 0$, which yielded the characteristic distance r_0 of

$$r_0 = \frac{1}{128\pi} \left[\frac{K_I}{\sigma_{0x}} V_0(c, c_1, c_2) \right]^2 \tag{2a}$$

where

$$V_0(c, c_1, c_2) = B_1(c) \{ -(1 + S_2^2)(2 - 3S_1^2)$$

$$- \frac{4S_1S_2}{1 + S_2^2}(14 + 3S_2^2) - 16S_1(S_1 - S_2) + 16(1 + S_1^2) \} \tag{2b}$$

The directional instability of a rapidly running crack is assumed to occur when $r_0 \leq r_c$, where r_c is the critical radial distance ahead of the crack tip. This instability criterion implies that off-axis cracks within this critical distance r_c are actuated by a critical crack tip stress field and deflect the crack from its otherwise self-similar propagation path. The critical radial distance was postulated to be a unique material property which was found to be $r_0 = r_c = 1.3$ mm for Homalite-100 at the onset of directional instability of a propagating crack [23], regardless of shape and geometry of the specimen.

It can be easily shown that for zero crack velocity or $c = 0$, Eq 2 reduces to Streit and Finnie's [25] characteristic radial distance of

$$r_0 = \frac{9}{128\pi} \left[\frac{K_I}{\sigma_{0x}} \right]^2$$

for crack kinking of an initially stationary crack. This static r_0 is always larger than the dynamic characteristic distance r_0 for crack velocities of $0 < c/c_1 \leq 0.33$. The crack curving angle θ_c for a stationary crack can be obtained from Eq 1a as

$$\theta_c = \cos^{-1} \left[\frac{1 \pm \sqrt{1 + \frac{1024\pi}{9} r_0 \left(\frac{\sigma_{0x}}{K_I} \right)^2}}{\frac{512\pi}{9} r_0 \left(\frac{\sigma_{0x}}{K_I} \right)^2} \right] \tag{3}$$

Thus this crack kinking criterion can also be used to estimate the crack branching angle for quasi-static crack branching under stress corrosion cracking conditions, provided a static counterpart of the necessary crack branching stress intensity factor can be established.

The foregoing crack kinking criterion for $\sigma_{0_x} > 0$, when applied to crack branching, should estimate a crack kinking angle which is one half of the included crack branching angle. The high crack branching stress intensity factor will result in sufficient energy release rate to create two kinked cracks simultaneously.

To recapitulate, then, crack branching will occur when the dynamic stress intensity factor reaches K_{Ib} and $r_0 = r_c$. In the following sections, this crack branching criterion will be tested by re-evaluating previous dynamic experiments in which crack branching was observed. Results of eleven dynamic photoelastic results involving SEN and wedge-loaded rectangular DCB (WL-RDCB) fracture specimens are also reported in the following sections.

Crack Branching in Homalite-100 Fracture Specimens

Homalite-100 SEN Specimens

The SEN specimens considered are 3.2 and 9.5 mm thick Homalite-100 plates with a 254 by 254 mm test section loaded in fixed grip configuration. The prescribed boundary conditions included both uniform and linearly decreasing displacements along the fixed grip edges of the specimen. At fracture load, the crack propagated from the SEN starter crack which was saw cut and chiseled. Further details of the test setup and the test conditions can be found in Ref 26. Figure 1 shows three frames out of a 16-frame dynamic photoelastic record of a crack propagating and branching in a 3.2 mm thick, 254 by 254 mm Homalite-100 plate loaded under fixed grip, linearly varying tension load.

Figure 2 shows the dynamic K_I and K_{II} variations obtained from the dynamic photoelastic patterns preceding and after crack branching of Fig. 1. By extrapolating the dynamic K_I associated with two branch cracks, an after-branching dynamic stress intensity factor, $K_I = 1.2$ MPa$\sqrt{m}$ and $K_{II} = 0.45$ MPa$\sqrt{m}$, is obtained. The branching stress intensity factor immediately prior to branching is estimated to be $K_{Ib} = 2.03$ MPa$\sqrt{m}$. Also shown in Fig. 2 are the variations in the r_0 values as computed from Eq 2. Note that r_0 reached a minimum value of $r_c = 1.2$ mm at crack branching.

Figure 3 shows another set of K_I, K_{II} and r_0 for two branch cracks in a similar dynamic photoelastic experiment. By extrapolating the K_I associated with the two branch cracks, an after-branching $K_I = 1.2$ MPa$\sqrt{m}$ and $K_{II} = -0.1$ MPa$\sqrt{m}$ are obtained. Immediately prior to branching, the instantaneous dynamic stress intensity factor reached its maximum value of 2.0 MPa$\sqrt{m}$; this is consistent with previous results. The estimated minimum r_0 at crack branching was $r_c = 1.3$ mm. Evaluations of four other SEN tests

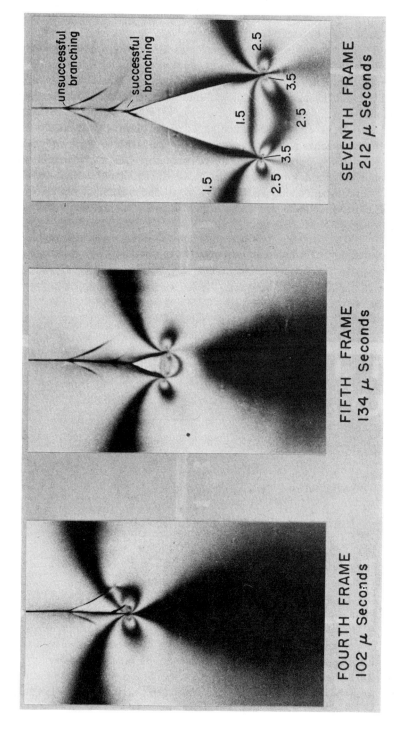

FIG. 1—*Typical crack branching dynamic photoelastic patterns of Homalite-100 single edge notched specimen (fixed grip loading): Specimen B8.*

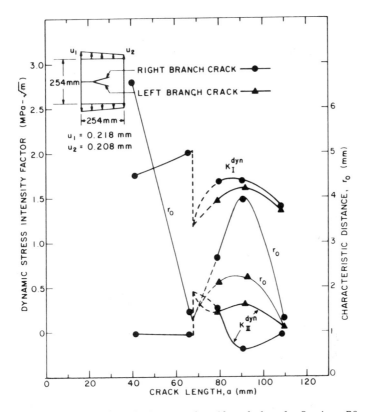

FIG. 2—*Dynamic stress intensity factors and r_0 of branched cracks; Specimen B8.*

yielded branching stress intensity factors of $K_I = 2.00$ and 2.09 MPa√m (Table 1). The r_c values ranged from 1.2 to 1.4 mm.

The K_I, K_{II} and σ_{0x} presented were evaluated by using isochromatics within a radial distance of $r_m = 4.5$ mm in Fig. 2. The extracted data, K_I, K_{II} and σ_{0x}, were used to compare the calculated and recorded fringe patterns and to check the interaction effects of the two branched cracks immediately after branching. The theoretical and recorded isochromatics in Fig. 1 agreed reasonably well for the region inside a radial distance r_m of 5 mm. The interaction effect of the two adjoining isochromatics in Fig. 2 on the K_I, K_{II} and σ_{0x} presented is thus considered negligible.

The crack velocities in the six tests were essentially constant at about 15 ± 5% of the dilitational wave velocity, $c_1 = 2400$ mps. Nevertheless, the crack velocities before and after crack branching were very close to the maximum velocity observed in all dynamic fracture tests involving Homalite-100. This so-called terminal velocity varied from test to test in a range of 0.15 to 0.20 c_1 where the crack always accelerated slightly just prior to crack branching.

The variations in the characteristic distance, r_0, which was computed from

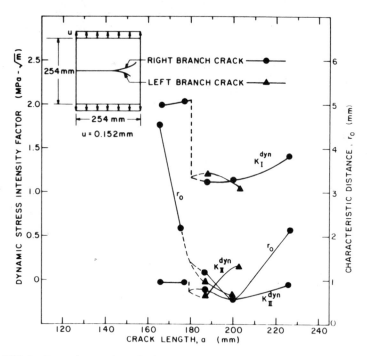

FIG. 3—*Dynamic stress intensity factors and* r_0 *of branched cracks; Specimen B9.*

Eq 2, for the branching cracks in the six tests all reached a minimum value prior to and at crack branching. This minimum value, which was obtained by interpolation at crack branching, averaged 1.3 mm and is consistent with the previously measured r_c values for crack curving [23], and is further evidence that r_c is a material property. Since minimum r_0 or r_c is derived through σ_{0x}, this r_c value indicates that σ_{0x} has a significant effect on crack branching.

Table 1 also shows the measured and estimated crack branching angles in the six tests. The crack branching angles, which were computed by Eq 1, for a known r_c, K_I and σ_{0x} are within 10% of the measured values, thus validating the use of this crack kinking criterion.

As an interesting sideline, Fig. 4 shows the enlarged view of Test B5 where an isochromatic pattern of a pure Mode II crack tip deformation—that is, nearly pure shear state of stress—is generated at branched cracks. The Mode II stress intensity factor K_{II} and remote stress σ_{0x} associated with these isochromatics are listed in Table 2. Figure 5 shows that within the 49-μs interval the propagating crack turned about 81 deg and arrested. The mixed-mode stress intensity factors prior to this severe crack kinking were $K_I = 0$, $K_{II} = 0.41$ MPa$\sqrt{m}$, and $\sigma_{0x} = 0.18$ MPa. The predicted theoretical kinking angle of 84 deg agreed well with the experimentally measured angle. After

TABLE 1—*Summary of experimental crack branching data at the onset of branching in a single edge notched specimen under fixed grip loading.*

Test	Plate Thickness (h), mm	Initial Crack Length (a_0), mm	Crack Length Branching (a_b), mm	At Branching					Measured Branch Angle ($2\theta_c$), deg	Estimated Branch Angle ($2\theta_c$), deg
				c/c_1	K_{Ib}, MPa$\sqrt{\text{m}}$	σ_{0x}, MPa	r_c, mm	K_{Ib}/K_{Ic}		
B8	3.18	5.6	66.0	0.18	2.08	−6.93	1.2	4.95	23	26
B9	3.18	4.3	177.0	0.18	2.03	−5.55	1.3	4.83	30	24
W082270[a]	3.58	5.8	139.7	0.18	2.03	−6.75	1.4	4.83	26	26
B7[b]	9.53	5.1	52.6	0.18	2.00	−6.80	1.4	4.76	30	28
B5	9.53	13.5	19.1	0.18	2.08	−7.08	1.2	4.95	30	28
B6	9.53	13.5	28.7	0.18	2.09	−7.60	1.3	4.98	28	32
Average				0.18	2.05	−6.79	1.3	4.88	27.8	27.3

[a]Second branching.
[b]Crack blunted.

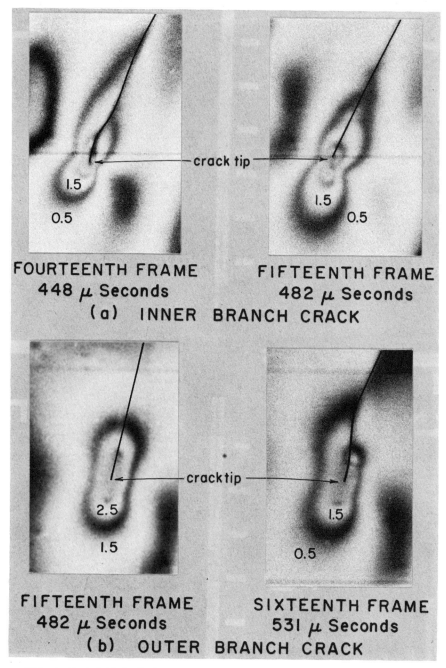

FIG. 4—*Typical Mode II dynamic isochromatic patterns of arresting branched cracks; Specimen B5.*

TABLE 2—K_{11} and σ_{0x} of arrested branch
cracks in Fig. 4.

| | Inner Branch Crack | |
	14th Frame	15th Frame
K_{11}	0.4 MPa$\sqrt{m}$	0.44 MPa$\sqrt{m}$
σ_{0x}	0.32 MPa	−0.04 MPa
	Outer Branch Crack	
	15th Frame	16th Frame
K_{11}	0.44 MPa$\sqrt{m}$	0.41 MPa$\sqrt{m}$
σ_{0x}	0.08 MPa	0.18 MPa

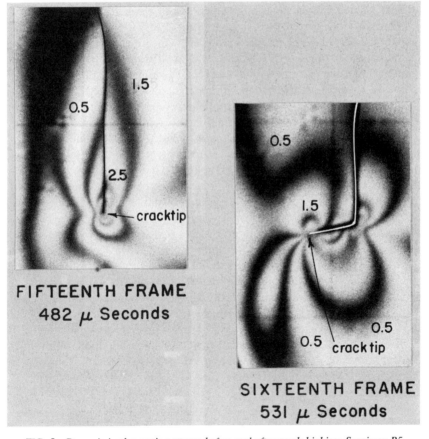

FIG. 5—*Dynamic isochromatic patterns before and after crack kinking; Specimen B5.*

crack kinking, the crack arrested with $K_I = 0.34$ MPa$\sqrt{m}$, $K_{II} = 0.08$ MPa$\sqrt{m}$, and $\sigma_{0x} = 1.4$ MPa. These results show that crack kinking can also occur under the high K_{II} state of stress.

Homalite-100 WL-RDCB Specimen

As mentioned previously, the proposed crack branching criterion should be applicable to quasi-static crack branching where inertia effects in the prebranched crack are negligible or nonexistent. Experimental data of the former were found in Homalite-100 WL-RDCB specimens where the crack immediately branched after initiating at a blunt starter crack tip. The necessary condition for branching is satisfied by the high K_{IQ} due to the blunt crack tip.[4] The crack branching angle (Eq 3) is a function of σ_{0x}/K_{IQ} and is thus a function of the specimen geometry.

The WL-RDCB fractured specimen considered is 76 by 152 by 9.5 mm thick and is of the geometry shown in Fig. 6. The crack immediately branched and propagated from a single-edge notch starter crack 24.3 to 29.3 mm long with a crack tip blunted by a drilled hole 2.2 to 5.0 mm in diameter. The branched crack paths of six fractured specimens are also shown in Fig. 6.

In all six tests of the WL-RDCB specimens, the crack branched at initiation, forming two or three branches. Table 3 summarizes the experimental data along with the measured branching angle in six WL-RDCB specimens. The angles of deviation of the post-branched cracks were measured along the crack path by averaging the measured crack curving angle on the front and back surfaces of the fractured specimen. Included angles for all major branches averaged 53.4 deg, and each is approximately twice the branching angle in a SEN specimen. This averaged branch angle agrees with the experimental results of Nakasa and Takei [27], where bending of the SEN specimens due to cantilever loading resulted in a positive σ_{0x} which in turn caused larger branching angles.

Although reliable data on the crack initiation condition were lacking for this series of experiments, the crack branching angle can be estimated from standard finite element analysis. Equation 3 shows that θ_c involves only the ratio of σ_{0x}/K_{IQ} and the predetermined r_c; thus the exact applied loading condition need not be known for estimating the branch angle of an initially stationary crack. In other words, the crack branching angle in this WL-RDCB specimen is governed only by the specimen geometry, provided sufficient driving force is given to branch the crack upon initiation. With a unit vertical wedge loading displacement applied to the specimen, K_I and σ_{0x} were calculated by least-square fitting the following plane stress crack tip

---

[4]K_{IQ} is the crack initiation stress intensity factor which is larger than the fracture toughness K_{Ic}.

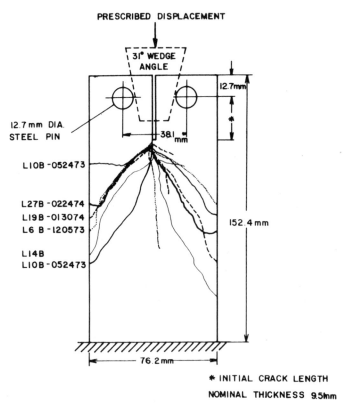

FIG. 6—*Branched crack paths in wedge-loaded rectangular double cantilever beam specimen (WL-RDCB).*

TABLE 3—*Summary of crack branching angle distribution in a wedge-loaded rectangular double cantilever beam specimen.*

Test	Specimen Thickness (h), mm	Diameter of Blunt Notch (ρ), mm	Measured Branch Angle (1st Branching) ($2\theta_c$), deg	Calculated 1st Branch Angle ($2\theta_c$), deg
L6B-120573	9.5	2.2	52	52
L10B-052473	9.5	2.2	52	52
L14B	9.5	5.0	55	52
L19B-013074	9.5	4.0	54	52
L27B-022474	9.5	2.4	54	52
		Average	53.4	52

displacement field of three to four sets of nodal displacements on the crack surface.

$$u_x = \frac{\sigma_{0x}r}{2(1 + \nu)G} \tag{4a}$$

and

$$u_y = \frac{2K_1\sqrt{r}}{(1 + \nu)G\sqrt{2\pi}} \tag{4b}$$

where G and ν in Eq 4 are shear modulus and Poisson's ratio, respectively. An average $K_{1Q}/\sigma_{0x} = 0.223\ \sqrt{m}$ was obtained from the finite element analysis using Eq 4 and a half branch angle of $\theta_c = 26$ deg was obtained using Eq 3. This value is in good agreement with the averaged branch angle of 54 deg shown in Table 3. Figure 7 shows two frames out of a 16-frame dynamic photoelastic record of branched cracks in a WL-RDCB specimen of 9.5-mm-thick Homalite-100 plate. Experimental details of this series of tests can be found in Ref 28. Figure 8 shows the dynamic fracture parameters K_1 and σ_{0x} obtained from the dynamic photoelastic pattern of the three branched cracks shown in Fig. 7. K_{II}, which oscillated between ± 0.3 MPa$\sqrt{m}$, was not plotted in order to avoid cluttering of Fig. 7. The decreasing stress intensity factor as well as the fluctuations in σ_{0x} (and K_{II}) along the post-branching curved cracks are noted. Crack 2 arrested at $K_1 = 0.4$ MPa$\sqrt{m}$. This arrest stress intensity factor is close to the arrest stress intensity factor for Homalite-100 determined by Dally [29].

Discussion

Table 1 shows that at the onset of branching, the instantaneous dynamic stress intensity factor and nonsingular stress reached an average maximum of 2.05 MPa$\sqrt{m}$ and -6.79 MPa, respectively, regardless of specimen thickness, loading condition, and initial crack geometry. This branching stress intensity factor, K_{Ib}, is approximately 4.88 times the fracture toughness and is in agreement with that of Dally [29]. Figs. 2 and 3 show that while K_1 hovers about K_{Ib}, crack branching will not occur prior to the precipitous drop in r_0. At the onset of branching, the characteristic r_0 value reaches its average minimum, $r_c = 1.3$ mm, for this material. These results show that K_{Ib} is a necessary condition for crack branching. The sufficiency condition involves the characteristic distance r_0, which is a function of the crack velocity, K_1 and σ_{0x}. The ratio of K_1 values before and after crack branching averages 2.2. Although this value is consistent with the postulate that crack branching occurs to dissipate fracture energy along two pro-

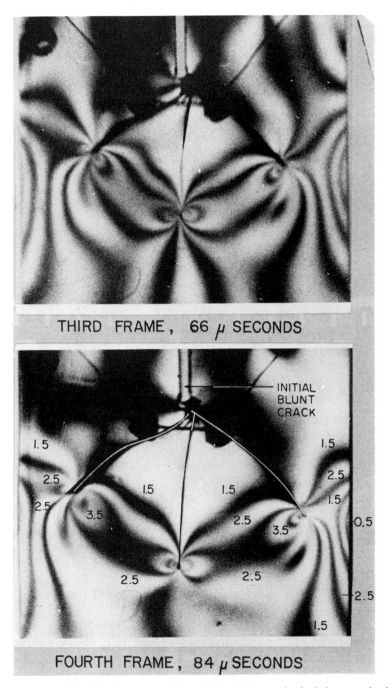

FIG. 7—*Typical photoelastic patterns of branched cracks in a wedge-loaded rectangular double cantilever beam (WL-RDCB) Homalite-100 specimen; Specimen L6B-120573.*

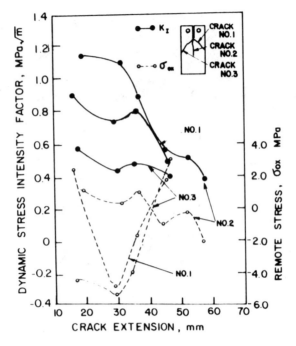

FIG. 8—*Modes I and II dynamic stress intensity factors of branched cracks shown in Fig. 7.*

pagating cracks, it is higher than the expected $\sqrt{2}$ value. Similar observations were also made by Romaniv et al [*30*] in structural steel.

It is also interesting to note that $K_{II} = 0$ prior to crack branching increases a small amount immediately after crack branching consistent with the postulated directional stability model [*23*]. Irrespective of the crack geometry and specimen thickness, the crack branched when it reached $K_I = K_{Ib}$ and $r_0 = r_c$, regardless of crack traveling length.

Of a total of 31 dynamic fracture tests involving WL-RDCB, 14 cracks curved and 6 branched at initiation. These results imply that crack branching in WL-RDCB specimens is observed only in a few cases. This is attributed to the fact that the crack propagates in a decreasing K_I field, a situation which does not promote crack branching beyond the initiation of crack extension.

The crack branching angles of Kobayashi [*8*], Kalthoff [*21*], and Christie [*22*] all converged to about 25 to 28 deg. This agreement is not surprising, since the loading conditions and the specimen geometries are quite similar in all three cases and resulted in a negative σ_{0x} value which reduces the fracture angle.

Conclusions

1. A crack branching criterion based on the directional stability of a running crack under pure Mode I loading is developed.

2. A necessary and sufficient condition for dynamic crack branching is a crack branching stress intensity factor, K_{Ib}, accompanied by minimum characteristic distance $r_0 \leq r_c$.

3. Dynamic photoelastic experimental results of crack branching in Homalite-100 are used to verify the proposed criterion. A crack branching stress intensity factor, $K_{Ib} = 2.04$ MPa$\sqrt{m}$, and the characteristic distance, $r_c = 1.3$ mm, were obtained.

4. The computed branching angles by the directional instability model agreed well with the experimentally measured branching angles in Homalite-100 SEN and WL-RDCB specimens. The crack branching angle in WL-RDCB specimens was found to be greater than the branching angle in SEN specimens.

Acknowledgments

The work reported here was obtained under ONR Contract 0014-76-C000 NR 064-478. The authors wish to acknowledge the support and encouragement of Dr. Y. Rajapakse of the Office of Naval Research during the course of this investigation

References

[1] Yoffe, E. H., *Philosophical Magazines*, Vol. 42, 1951, pp. 739-750.
[2] Achenbach, J. D., "Elasto Dynamic Stress Intensity Factors for a Bifurcated Crack," in *Prospects of Fracture Mechanics*, G. C. Sih et al, Eds., Noordhoff, Leyden, The Netherlands, 1974, pp. 319-336.
[3] Achenbach, J. D., *Journal of Elasticity*, Vol. 9, No. 2, 1979, pp. 113-129.
[4] Anthony, S. R. and Congleton, J., *Metal Science Journal*, No. 2, 1968, pp. 158-160.
[5] Congleton, J., "Practical Application of Crack Branching Measurements," in *Dynamic Crack Propagation*, G. C. Sih, Ed., Noordhoff, Leyden, The Netherlands, 1973, pp. 427-438.
[6] Irwin, G. R. in *Fast Fracture and Crack Arrest, ASTM STP 627*, American Society for Testing and Materials, 1977, pp. 7-18.
[7] Rossmanith, H. P., "Crack Branching in Brittle Materials, Part I," Photomechanics Laboratory Report, University of Maryland, College Park, 1977-1980.
[8] Kobayashi, A. S., Wade, B. G., Bradley, W. B., and Chiu, S. T., *Engineering Fracture Mechanics*, Vol. 6, No. 1, 1974, pp. 81-92.
[9] Irwin, G. R., Dally, J. W., Kobayashi, T., Fourney, W. L., Etheridge, M. J., and Rossmanith, H. P., *Experimental Mechanics*, Vol. 19, No. 4, 1979, pp. 121-128.
[10] Kobayashi, A. S. and Ramulu, M., *Experimental Mechanics*, Vol. 21, No. 1, 1981, pp. 41-48.
[11] Craggs, J. W., *Journal of the Mechanics and Physics of Solids*, Vol. 8, 1960, pp. 66-75.
[12] Döll, W., *International Journal of Fracture*, Vol. 11, 1975, pp. 184-186.
[13] Hahn, G. T., Hoagland, R. G., and Rosenfield, A. R., "Crack Branching in A 533B Steel," in *Fracture 1977*, Vol. 2, University of Waterloo Press, Ontario, 1977, pp. 1333-1338.

[14] Kobayashi, T. and Dally, J. W. in *Fast Fracture and Crack Arrest, ASTM STP 627*, American Society for Testing and Materials, 1977, pp. 257-273.
[15] Kerkhoff, F., *Dynamic Crack Propagation*, G. C. Sih, Ed., Noordhoff, Leyden, The Netherlands, 1973, pp. 3-35.
[16] Schardin, H., "Velocity Effects in Fracture," in *Fracture*, B. L. Averbach et al, Eds., Wiley, New York, 1959, pp. 297-330.
[17] Acloque, P., *Silicate Industrials*, Vol. 28, 1963, p. 323.
[18] Clark, A. B. J. and Irwin, G. R., *Experimental Mechanics*, Vol. 6, 1966, pp. 321-330.
[19] Sih, G. C., "Dynamic Crack Problems: Strain Energy Density Fracture Theory," *Elastodynamic Crack Problems*, Vol. 4, G. C. Sih, Ed., Noordhoff, Leyden, The Netherlands, 1977, pp. 17-37.
[20] Kitagawa, H., Yuuki, R., and Ohira, T., *Engineering Fracture Mechanics*, Vol. 7, 1975, pp. 515-529.
[21] Kalthoff, J. F., "On the Propagation of Bifurcated Cracks," in *Dynamic Crack Propagation*, G. C. Sih, Ed., Noordhoff, Leyden, The Netherlands, 1973, pp. 449-458.
[22] Christie, D. G., *Journal of the Society of Glass Technology*, Vol. 36, 1952, pp. 74-89.
[23] Ramulu, M. and Kobayashi, A. S., *Experimental Mechanics*, Vol. 23, No. 1, 1983, pp. 1-9.
[24] Freund, L. B., "Dynamic Crack Propagation," in *Mechanics of Fracture*, Vol. 19, F. Erdogan, Ed., ASME, 1976, pp. 105-134.
[25] Streit, R. and Finnie, I., *Experimental Mechanics*, Vol. 20, No. 1, 1980, pp. 17-23.
[26] Bradley, W. B., "A Photoelastic Investigation of Dynamic Brittle Fracture," Ph.D. dissertation, University of Washington, Seattle, 1969.
[27] Nakasa, K. and Takei, H., *Engineering Fracture Mechanics*, Vol. 11, 1979, pp. 739-751.
[28] Lee, M. H., "Dynamic Photoelastic Analysis of a Compression Double Cantilever Beam Specimen," M.Sc. Thesis, University of Washington, Seattle, 1975.
[29] Dally, J. W., *Experimental Mechanics*, Vol. 19, No. 10, 1979, pp. 349-367.
[30] Romaniv, O. N., Nikiforchin, G. N., Student, A. G., and Tsirul'nik, A. T., *Soviet Materials Science*, July 1982, pp. 30-40.

A. R. Rosenfield,[1] P. N. Mincer,[1] C. W. Marschall,[1] and A. J. Markworth[1]

Recent Advances in Crack-Arrest Technology

REFERENCE: Rosenfield, A. R., Mincer, P. N., Marschall, C. W., and Markworth, A. J., "**Recent Advances in Crack-Arrest Technology,**" *Fracture Mechanics: Fifteenth Symposium, ASTM STP 833,* R. J. Sanford, Ed., American Society for Testing and Materials, Philadelphia, 1984, pp. 149-164.

ABSTRACT: Statistical analysis of the variability of crack-arrest toughness (K_a) is reviewed. The standard deviation of K_a is 12 MPa $\cdot$ m$^{1/2}$ for an individual heat of low-alloy steel and 15 MPa $\cdot$ m$^{1/2}$ when data for all heats are referred to the Charpy 41-J (30 ft-lb) temperature. These results aid in the definition of acceptable size limitations for crack-arrest test specimens. Miniature specimens (50 by 50 by 12.7 mm) are shown to give acceptably conservative results when tested in the vicinity of the reference transition temperature.

KEY WORDS: crack arrest, reactor pressure vessel steels, fracture statistics, lower-bound toughness, specimen-size effects

Nomenclature

a Crack length
B Crack-arrest specimen thickness
B_n Crack-arrest specimen thickness at the roots of the side grooves
CV30   41-J (30 ft-lb) Charpy V-notch impact temperature
$2H$ Crack-arrest specimen height
K_{IA} Crack-arrest toughness
K_a Stress intensity shortly after crack arrest
K_D Propagating-crack toughness
K_{Id} Rapid-loading fracture toughness
K_{IR} Lower-bound fracture toughness
K_0 Stress intensity at crack initiation
N Crack-arrest specimen notch width

---

[1]Battelle-Columbus Laboratories, Columbus, Ohio 43201.

RT_{NDT} Reference transition temperature as defined in WRC 175 [2]
   T Temperature
   w Crack-arrest specimen width
   Δa Crack-jump length
   2δ Displacement measured in a crack-arrest test
   σ_{yd} Dynamic yield strength
   σ_{ys} Static yield strength

During the past decade, there has been a steady advance in crack-arrest technology. An ASTM task group is close to agreement on a proposed standard, and a number of laboratories have acquired the necessary expertise to conduct crack-arrest tests. At the same time, research has begun to improve the efficiency of testing and to assess the reliability of the results. This paper reports recent progress at Battelle-Columbus Laboratories (BCL) in these areas.

The object of crack-arrest measurements is to determine the stress intensity at arrest (K_A). (Note that the subscript I is not used here because plane-strain requirements have not been established.) Since it would be very difficult to make precise measurements at the instant of arrest, the experiments actually measure the stress intensity shortly after arrest, denoted K_a. It is generally assumed that K_a is a lower bound and thus a conservative estimate of K_A [1].

Past research at Battelle-Columbus focused on K_D, the propagating-crack toughness. Because K_A is the zero-velocity value of K_D, and because K_D generally increases with increasing velocity, measured K_D values are upper-bound estimates of K_A [1]. For this reason, this paper concentrates on measurements of K_a.

Lower-Bound Toughness Values

The K_{IR} curve in WRC 175 [2], which was conceived as a lower-bound toughness value, was determined partly by K_a. No data available at the time (1972) fell below that curve. Since then, both occasional rapid-initiation toughness (K_{Id}) [3] and K_a [4,5] values have been found to fall below K_{IR}. As a result, it has been concluded that K_{IR} has to be redefined [3,5,6] in statistical terms. Oldfield et al [3,6] have assumed that the transition portion of the K_{Id} curve has a hyperbolic-tangent form which is essentially indistinguishable from a straight line in the approximate temperature range between RT_{NDT} and $(RT_{NDT} + 100)°C$. The slope of this line is typically on the order of 3 MPa·m$^{1/2}$/°C.

Most of the K_a data reported in the literature have been obtained between RT_{NDT} and $(RT_{NDT} + 60)°C$. Crosley and Ripling [7] found essentially linear temperature dependence in this region, with an average slope of 0.8 MPa·m$^{1/2}$/°C, much smaller than the slope of K_{Id} versus temperature. One implication of this result is that K_a should dominate the K_{IR} curve to an increasingly greater extent as the test temperature is raised within the ductile/brittle transition region.

The initial Battelle-Columbus statistical analysis involved fitting the K_a/temperature (T) curve to an arbitrary polynomial and assessing the scatter about the mean curve [4,5,8]. With the exception of one point, the data lay between RT_{NDT} and (RT_{NDT} + 65)°C. Expressing the curve in a higher order than linear did not improve the goodness of fit. It was also found that the points had a normal distribution around the best straight line. The K_{IR} curve was shown to represent data between the 90th and 99th percentile with approximately 95% confidence [5].

Based on the foregoing results, a linear-regression curve was used to predict crack arrest in the thermal-shock experiments at Oak Ridge National Laboratory, in which the inner wall of a large axially precracked cylinder is rapidly quenched [9]. The K_a data, generated at Battelle-Columbus using compact specimens, were fitted to a linear relation. The values of K_a at the CV30 temperature were measured on the steel from each cylinder, and a regression line for the thermal-shock data was suggested:

$$K_a = K_a(CV30) + 4.89 + 0.47(T - CV30) \qquad (1)$$

where K_a (CV30) is the value of K_a at the temperature where the Charpy V-notch energy is 41-J (30 ft-lb), with the units being MPa · m$^{1/2}$ and °C. The use of CV30 instead of RT_{NDT} is based on its likely adoption in nuclear standards. The standard deviation of K_a determined from Eq 1 is 16 MPa · m$^{1/2}$. For any single heat, the standard deviation is about 12 MPa · m$^{1/2}$ [7,8], and this observation will be used later in this paper.

ASTM A508 steel with three different heat treatments was used in the four most recent ORNL thermal-shock experiments, and the Eq 1 parameters of the various materials are given in Table 1. Figure 1 is a test of the equation. The solid line is Eq 1, which is based solely on BCL compact-specimen data. The points are the Oak Ridge National Laboratory (ORNL) thermal-shock results, and it is seen that the results of both types of experiments are in good agreement. Also shown in Fig. 1 are allowable lower limits for these K_a data modeled on the same considerations as for aerospace structural reliability [10].

TABLE 1—*Crack-arrest toughness parameters for steels used in thermal-shock experiments.*

Experiment	CV30, °C	K_a(CV30), MPa · m$^{1/2}$
TSE4	60	97
TSE5	40	72
TSE5A	−7	66
TSE6	40	68

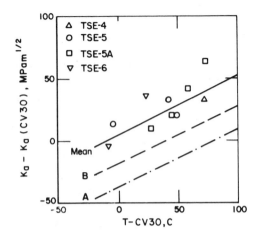

FIG. 1—*Comparison between predicted and measured* K_a *values for the ORNL-TSE data.*

It should be noted that the temperature dependence of K_a of Eq 1 is somewhat less than was found in the Crosley and Ripling data [7]. The question of most representative slope is unresolved and is related to the choice of reference temperature, because the use of different reference temperatures results in different slopes [4,5,8]. Currently a more advanced statistical technique, recently developed at Battelle-Columbus [11], is being applied to the Battelle-Columbus fracture-toughness data. It is expected that this technique will provide a better fixing of reference temperature as well as an improved description of lower-bound toughness.

Specimen Miniaturization

In addition to providing information about lower-bound toughness, the statistical analysis makes it possible to judge whether systematic changes in experimental procedures produce systematic changes in K_a. As was stated previously, the material variability within a single heat can be characterized by a standard deviation of 12 MPa $\cdot$ m$^{1/2}$ [8,12]. This observation has been used in judging the limiting specimen size for essentially linear elastic behavior as part of an effort in designing miniature specimens. Data outside of these limits are considered to have resulted from the use of too small a specimen.

The research at Battelle-Columbus on miniature specimens is motivated by the need to obtain irradiated-steel data efficiently. The large specimens (200 by 200 by 50 mm) used in the Cooperative Test Program [12] would be impractical for surveillance testing because of space limitations in reactors. Somewhat smaller specimens (150 by 150 by 25 mm) were used to duplicate the ORNL thermal-shock experiment crack-arrest data, with the result shown in Fig. 1. Marschall et al [13] further downsized the specimens to 100 by 100 by 12.7 mm and had a successful test of a specimen 6.4 mm thick. These data were obtained on steel from the Cooperative Test Program and agreed with the large-specimen values.

A new series of experiments was undertaken to verify the results reported by Marschall et al [13] and to determine whether even smaller specimens could be used. Specimens of several sizes were machined from a prolongation of the ASTM A508 steel cylinder used in ORNL thermal-shock experiment TSE5A [9]. They were tested at 0°C, a temperature at which BCL compact specimen data are consistent with the large-cylinder data [14]. The specimens, which had starting-notch root radii of 0.25 mm, were of the design used by Marschall et al [13] with two exceptions:

1. Two of the three 6.3-mm-thick specimens exhibited extreme crack tunnelling, with crack travel being considerably less at the roots of the side grooves than at the midplane. Marschall et al [13] did not have this problem in the single specimen of this thickness that they tested. To resolve the problem in the current tests, a 40% side-groove depth was used in a second set of 6.3-mm-thick specimens, instead of the usual 25%. Data for the 25% side-grooved specimens are listed in Table 2 but are not plotted in the following summary graphs.

2. As is shown in Fig. 2, specimens with the smallest face dimension required modification of the starting-notch tip and of the displacement-measurement point. Since the usual welding rod used for notch-tip embrittlement (Hardex N) was too wide to fit a notch of usual proportions, the TIG process was used instead to harden the notch-tip region without depositing a weld bead. In addition, the clip gage was placed 0.33 w behind the load line instead of at the usual 0.25 w position. To correct for this difference, the displacement behind the load line was considered to vary linearly with position in accord with the calculations of Newman [15] and of Saxena and Hudak [16]. As a result, the measured displacement was multiplied by 0.93 for calculation of stress intensity via the ASTM Task Group E24.01.06 stress-intensity relation of Ripling [1]. The cyclic-load test procedure [13], designed to minimize the effect of plastic deformation on measured K_a values, was used to produce unstable fracture. Values of K_a were calculated using the displacement at crack arrest.

The results of the size-effect study are given in Tables 2 and 3.[2] As will be seen subsequently, with the exception of the specimens with the smallest face dimensions, the data were reasonably consistent. These small specimens not only had significantly lower K_a values than did the balance of the specimens, but they tended to have lower K_0 (stress intensity at initiation) values. Although the lower K_0 values probably resulted from use of the TIG weld, this factor should not invalidate the K_a data. Nakano and Tanaka [17] also reported low

---

[2]The K_0 and K_a values previously reported [14] have been recalculated using the ASTM task group relation. This decreased K_a for Specimen 97 by 10%. The other specimens changed by less than 5%.

TABLE 2—Crack-arrest data (TSE5A steel tested at 0°C).

Specimen No.	Specimen Dimensions, mm[a]					Crack Length, mm		Cycles to Failure	Displacement (2δ), mm	
	2H	w	B	B_n	N	Initiation	Arrest		Initiation	Arrest
5A-91	152.4	127.0	50.9	38.1	10.67	53.1	103.2	16	1.17	1.31
92	152.4	127.0	50.9	38.1	10.67	51.3	107.7	14	1.13	1.18
93	101.6	84.68	25.3	18.92	7.62	30.0	67.1	10	0.74	0.83
94	101.6	84.58	25.3	19.30	7.62	33.3	72.1	19	1.02	1.09
95	101.6	84.58	25.3	18.80	7.62	33.0	70.6	16	0.94	1.08
96	101.6	84.68	12.6	9.27	7.62	33.6	73.0	17	1.08	1.19
97	101.6	84.84	12.7	9.53	7.62	29.7	72.9	15	0.98	1.06
98	101.6	84.71	12.6	9.27	7.62	35.4	71.3	10	1.00	1.07
99	101.6	84.84	6.31	4.70	7.62	31.7	46.4[b] 57.1[c]	16	1.03	1.06
100	101.6	84.58	6.31	4.70	7.62	30.0	46.0[b] 52.5[c]	13	0.86	0.89
101	101.6	84.71	6.31	4.70	7.62	29.7	69.1	16	0.92	0.94
102	101.6	84.71	6.31	3.81	7.62	29.3	67.8	8	0.71	0.82
103	101.6	84.71	6.31	3.81	7.62	29.3	70.2	10	0.74	0.85
104	101.6	84.58	6.37	3.56	7.62	29.7	62.0	10	0.70	0.77
45	152.4	127.0	26.7	20.57	10.67	44.2	91.4	21	0.88	0.96
55	152.4	127.0	26.7	20.32	10.67	45.1	99.4	17	0.93	1.02
1	50.8	42.29	12.7	9.65	3.81	16.9	35.7	9	0.47	0.55
2	50.8	42.29	12.7	9.65	3.81	16.9	35.5	11	0.60	0.61
3	50.8	42.29	12.7	9.65	3.81	16.9	36.6	15	0.57	0.67

[a] Terminology of Marschall et al [13].
[b] Average length.
[c] Greatest extent.

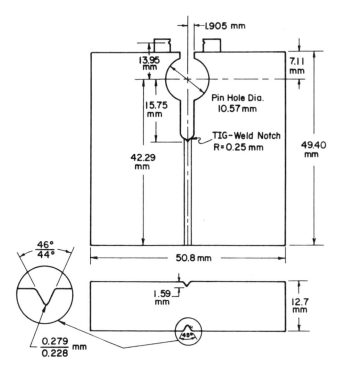

FIG. 2—*Miniature crack-arrest specimen.*

K_a values for steel specimens that were embrittled at the notch tip by means of an electron-beam weld and were tested at and below NDT.[3] However, those data were limited to two points obtained on double-cantilever-beam specimens and may not be relevant.

The effect of thickness on K_a is shown in Fig. 3. The breadth of the horizontal scatter band is two standard deviations for typical crack-arrest data. The 12.7-mm-thick specimens at the lower edge of the band are those with the smallest face dimension. The single positive K_a deviation from the scatter band is for one of the thinnest specimens used. The dashed line represents the likely ASTM-recommended size limit [1,14], which is seen to be conservative with respect to thickness requirements.

When the data are ordered according to the length of the ligament remaining after crack arrest (Fig. 4), the different behavior of the specimens with the smallest face size becomes more evident. Even so, those specimens do not raise the apparent toughness, but lower it. This is, of course, in contrast to typical small specimen behavior, where plasticity causes an increase in the apparent toughness when the stress intensity is calculated from displacement instead of

---

[3]The steel was designated KD32. Its composition and mechanical properties resemble those of ASTM A 537-1 steel.

TABLE 3—*Crack-arrest toughness (TSE5A steel tested at 0°C).*

Specimen No.	Stress Intensity and Toughness, MPa · m$^{1/2}$		
	K_0	K_D	K_a
5A-91	173.9	120.9	82.7
92	172.0	113.9	65.7
93	149.7	100.0	68.6
94	191.6	126.1	72.4
95	179.8	119.0	77.7
96	205.4	133.1	77.1
97	198.7	123.8	69.0
98	183.8	124.4	75.6
99[a]	202.3	171.6[b]	154.1[b]
		153.3[c]	119.0[c]
100[a]	175.0	146.7[b]	130.0[b]
		136.5[c]	111.7[c]
101	187.2	121.7	72.1
102	162.5	106.4	73.8
103	169.4	108.1	69.6
104	164.6	115.5	86.6
45	144.4	102.8	77.6
55	151.9	102.1	70.1
1	123.1	81.5	53.4
2	158.3	105.4	60.3
3	150.4	97.5	59.8

[a]Data invalid due to extreme tunnelling. These points are not plotted.
[b]Based on average crack length.
[c]Based on greatest extent of crack growth.

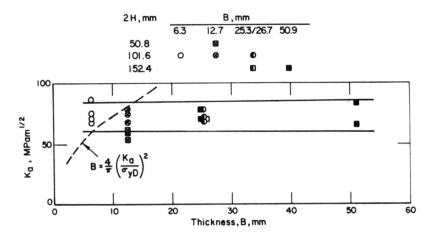

FIG. 3—*Effect of thickness on* K_a. *Data are for TSE5A steel tested at 0°C.*

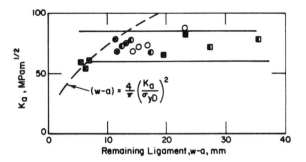

FIG. 4—*Effect of length of the remaining ligament on* K_a. *Data are for TSE5A steel tested at 0°C. See Fig. 3 for key to symbols.*

load. The cause of the low K_a values for the specimens with small remaining ligaments is not completely clear. As was noted in the Cooperative Test Program Report [12], there is a tendency for K_a to decrease with increasing crack-jump length (Δa). However, the miniature specimens experienced relatively short jumps (that is, low Δa values) as a result of their small size. There is a slight tendency in the K_a data of Table 3 for K_a to decrease with relative jump length ($\Delta a/w$) and this would be consistent with the Cooperative Test Program result. The significance of this observation is not clear.

Finally, no trend is apparent in the relation between K_a and normalized initiation stress intensity (Fig. 5). This figure also contains a suggested limitation on face dimension to prevent excessive plasticity during loading.

A possible explanation for the size independence of K_a can be found with the aid of fracture surface observations. Figure 6 shows specimen fracture surfaces. Note that, for the specimens of the largest size, the one with the larger K_a value (5A-91) shows an extreme example of unbroken ligaments of the kind reported by Hahn et al [4]. These ligaments are believed to be nucleated when the advancing crack encounters a tough region in the steel. As was suggested

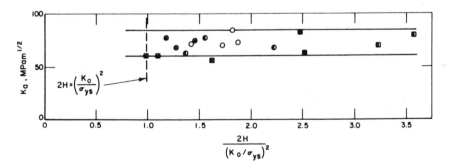

FIG. 5—*Effect of face dimensions on* K_a. *Data are for TSE5A steel tested at 0°C. See Fig. 3 for key to symbols.*

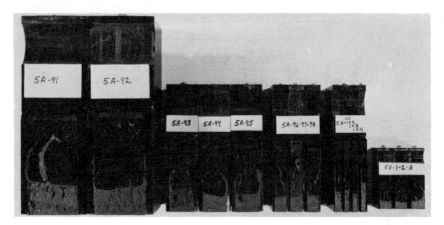

FIG. 6—*Fracture surfaces of crack-arrest specimens of TSE5A steel tested at 0°C.*

by Hahn et al [4], the crack front develops a step in order to bypass the tough region. As the step advances, it leaves behind a layer of unbroken material that serves as a line of pinching forces imposing a retarding contribution to the crack tip stress intensity and raising the apparent value of K_a. Actually, visual observation underestimates the ligamentation, since, on a microscopic scale, many fine ligaments are observed. Further visual observation of specimens of any given size shows that the higher K_a values are associated with more prominent ligaments. In addition, the tendency for visual ligamentation decreases with decreasing thickness (Fig. 6).

The material savings made possible by the use of miniature specimens is shown in Fig. 7. The volume of the smaller sample in the figure is $1/36$ of that of the larger one.

FIG. 7—*Comparison of largest and smallest TSE5A specimens used in the size-effect study.*

To determine whether the crack-arrest data for TSE5A steel are representative, a second series of miniature specimens was fabricated. These were made from broken halves of a specimen of Cooperative Test Program (CTP) steel. The specimen preparation and experimental procedures were identical to those used on the miniature specimens made from steel from TSE5A.

The results are given in Tables 4 and 5. Of the 12 specimens tested, 10 developed rapid cracks that arrested. The other two exhibited stable tearing. The highest temperature for a successful result was 5°C, which was 22°C above CV30, the Charpy 41-J (30 ft-lb) temperature. For comparison, the miniature-specimen experiments on TSE5A were carried out 7°C above CV30.

Examination of the data obtained at 0 to 4°C shows that the experimental results are comparable with those obtained on TSE5A steel. In particular, both K_0 and K_a for these miniature specimens were on the low end of the range of CTP data [12,13] (Fig. 8). Specifically, these miniature specimens are in approximately the tenth percentile of all CTP data at 0°C (including duplex-specimen points [4] that are not plotted in Fig. 8). The miniature-specimen K_a values also were much lower than the K_a value of the specimen from which they were cut (Specimen EG-7). The difference may be due, at least in part, to differences in the degree of ligamentation. Specimen EG-7 was particularly heavily ligamented, an effect that can raise K_a significantly, whereas ligamentation on the miniature specimens was much lighter.

The data also were examined in the light of size requirements likely to be adopted by the ASTM task group on crack arrest [1,14]. The CTP steel has a much lower yield strength than the TSE5A steel. Accordingly, the miniature CTP specimens would be expected to be more likely to display high K_a values

TABLE 4—*Crack-arrest data obtained with miniature specimens of CTP steel.*[a]

Specimen No.	Test Temperature, °C	Crack Length, mm		Cycles to Failure	Displacement (2δ), mm	
		Initiation	Arrest		Initiation	Arrest
EG-7-6	29	16.6	[b]	17	[b]	[b]
9	13	16.6	[b]	20	[b]	[b]
11	5	16.6	34.0	15	0.57	0.65
10	4	16.6	33.9	8	0.50	0.55
4	0	16.6	34.5	15	0.57	0.64
5	0	16.6	34.2	15	0.57	0.64
8	−18	16.6	34.1	7	0.57	0.62
7	−19	16.6	36.4	13	0.57	0.63
12	−39	16.6	35.7	14	0.61	0.69
13	−39	16.6	36.1	15	0.63	0.63
14	−78	16.6	37.6	12	0.59	0.71
15	−78	16.6	38.1	10	0.52	0.65

[a]Specimens were cut from broken halves of Specimen EG-7.
[b]Stable crack growth.

TABLE 5—*Crack-arrest toughness for miniature specimens of CTP steel.*

Specimen No.	Test Temperature, °C	Stress Intensity and Toughness, MPa · m$^{1/2}$		
		K_0	K_D	K_a
EG-7[a]	0	171	131	101[b]
EG-7-6	29	c	c	c
9	13	c	c	c
11	5	154.0	105.3	73.4
10	4	133.1	91.3	62.5
4	0	152.1	103.1	69.3
5	0	151.8	103.2	70.9
8	−18	155.0	105.9	70.2
7	−19	151.8	99.1	57.8
12	−39	163.4	107.5	67.4
13	−39	168.2	109.8	59.1
14	−78	157.3	99.1	56.9
15	−78	138.2	85.7	48.9

[a]Large specimen (197.6 by 203.2 by 50.8 mm) from whose broken halves the miniature specimens were cut.
[b]Large value due to excessive ligamentation.
[c]Stable crack growth.

as a result of excess plasticity. Comparison with the data of Marschall et al [13], however, indicates that this is not the case, giving credence to the idea that the cyclic-loading procedure minimizes plasticity effects. Figures 9 and 10 show that the miniature specimens fail both the remaining-ligament and face-dimension criteria likely to emerge from the ASTM task group. A thickness plot (not shown) is similar to the one for remaining ligament. However, the small departure in K_a values from large-specimen results gives credence to the conservatism of the criteria.

In addition to the 0°C data, crack-arrest values for miniature specimens were measured down to −78°C. These data are plotted in Fig. 11 along with the Cooperative Test Program results and the duplex-CT-specimen data generated by Hahn et al [4] in preparation for that program. The lower limit of the data is parallel to the K_{IR} curve (CV30 was used as RT_{NDT} in constructing the graph). Future plans include adding these data to the BCL crack-arrest data bank to help extend its temperature range.

Discussion

Crack arrest was originally viewed as the inverse of crack initiation [18], and to some extent the view is apt. For crack initiation, it has been argued that the large data scatter reflects the existence of widely distributed weak spots that trigger cleavage [19]; large specimens are more likely to have a weak spot near the tip of the preexisting fatigue crack and to exhibit lower toughness than

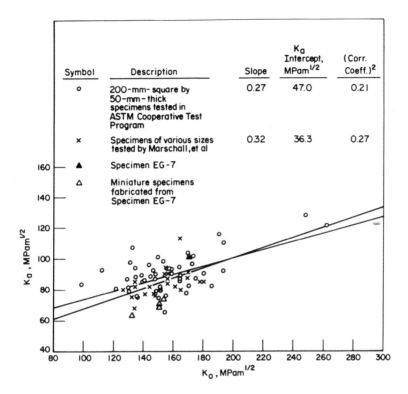

FIG. 8—*Relation between initiation and arrest stress intensity for cooperative test program steel (after Marschall et al [13]).*

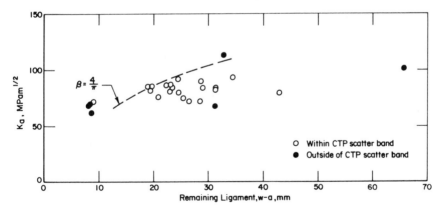

FIG. 9—*Effect of variation in the size of the remaining ligament on K_a at $0°C$ for cyclic loading of the CTP steel. Data of Marschall et al [13] and miniature-specimen results.*

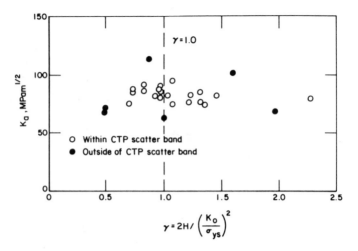

FIG. 10—*Effect of variation in face size on* K_a *at 0°C for cyclic loading of the CTP steel. Data of Marschall et al [13] and miniature-specimen results.*

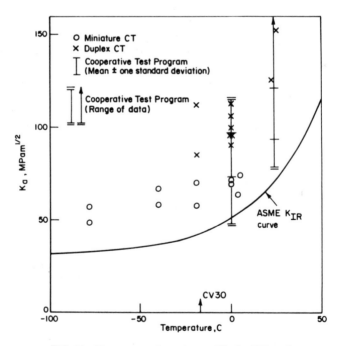

FIG. 11—*Temperature dependence of* K_a *for CTP steel.*

small specimens. In crack arrest, the existence of widely distributed tough spots that nucleate ligaments can be hypothesized. In turn, the ligaments exert a retarding effect on the propagating crack and lead to premature arrest [4]. If this explanation is correct, the miniature specimens, which appear to exhibit a lesser tendency to ligamentation, should be particularly useful for surveillance purposes, since they are more likely to approach lower-bound K_a values than the larger specimens previously used.

One aspect where initiation and arrest differ is in the extent of data scatter; the variability of K_{Ic}, which can be seen in Refs 20 to 23 for example, is greater than that of K_a. While the reasons for this are not completely clear, the practical effect is that it should be possible to define reliable lower-bound K_a values with many fewer specimens than would be required for K_{Ic}.

The results of the Cooperative Test Program suggest that much of the scatter in K_a may arise from differences in crack velocity during propagation in different specimens and the resulting variation in propagating-crack toughness K_D. It has been suggested [4] that the aforementioned ligaments are the major contribution to K_D, and the results in this paper are consistent with such an explanation. It is now generally agreed [12] that K_D is an upper bound estimate of the actual crack-arrest material property K_{IA}. In turn, K_a is a lower-bound estimate and is accordingly more useful for safety analysis, despite greater scatter than for K_D.

Acknowledgments

This research is part of a program being carried out under UCCND Subcontract 85X-13876C between Union Carbide Nuclear and Battelle-Columbus. We are grateful to the following persons for their many helpful discussions of the program: R. D. Cheverton, G. D. Whitman, W. H. Corwin, and J. G. Merkle, Oak Ridge National Laboratory; M. Vagins and J. Strosinder, Nuclear Regulatory Commission; G. R. Irwin, W. L. Fourney, T. Kobayashi, University of Maryland; and R. G. Hoagland, The Ohio State University.

References

[1] Fourney, W. L., "Proposed Test Method for Crack Arrest Fracture Toughness of Metallic Materials," submitted to ASTM Task Group E24.01.06 on Crack Arrest, 1983.
[2] "PVRC Recommendations on Toughness Requirements for Ferritic Materials," Bulletin 175, Welding Research Council, 1972.
[3] Oldfield, W., *Journal of Engineering Materials Technology*, Vol. 102, 1980, pp. 107–117.
[4] Hahn, G. T. et al in *Crack Arrest Methodology and Applications, ASTM STP 711*, American Society for Testing and Materials, 1980, pp. 289–320.
[5] Rosenfield, A. R. et al, "Critical Experiments, Measurments, and Analyses to Establish a Crack-Arrest Methodology for Nuclear-Pressure-Vessel Steels," NUREG/CR-1887, BMI-2071, Nuclear Regulatory Commission, Washington, D.C., 1981.
[6] Wullaert, R. A., Oldfield, W., and Server, W. L., *Journal of Engineering Materials Technology*, Vol. 102, 1980, pp. 101–106.

[7] Crosley, P. B. and Ripling, E. J. in *Crack Arrest Methodology and Applications, ASTM STP 711*, American Society for Testing and Materials, 1980, pp. 321-337.

[8] Rosenfield, A. R. et al, "BCL HSST Support Program," in NUREG/CR-2141, Nuclear Regulatory Commission, Vol. 3, 1981, pp. 10-43.

[9] Cheverton, R. D. et al, "Fracture Mechanics Data Deduced from Thermal-Shock and Related Experiments with LWR Pressure Vessel Material," ASME Publication PVP-Vol. 58, American Society of Mechanical Engineers, 1982, pp. 1-16.

[10] "Military Standardization Handbook," Metallic Materials and Elements for Aerospace Vehicle Structures," MIL-HDBK 5C, Wright-Patterson AFB, Ohio, 1978.

[11] Bishop, T. A., Markworth, A. J., and Rosenfield, A. R., *Metallurgical Transactions A*, Vol. 14A, 1983, pp. 687-693.

[12] Crosley, P. B. et al, "Cooperative Test Program on Crack Arrest Toughness Measurements," NUREG/CR-3261, Nuclear Regulatory Commission, 1983.

[13] Marschall, C. W., Mincer, P. N., and Rosenfield, A. R. in *Fracture Mechanics: Fourteenth Symposium—Volume II: Testing and Applications, ASTM STP 791*, American Society for Testing and Materials, 1983, pp. II-295-II-319.

[14] Rosenfield, A. R., *Journal of Engineering Materials Technology*, Vol. 106, 1984, pp. 207-208.

[15] Newman, J. C., Jr., in *Fracture Analysis, ASTM STP 560*, American Society for Testing and Materials, 1974, pp. 105-121.

[16] Saxena, A. and Hudak, S. J., Jr., *International Journal of Fracture*, Vol. 14, 1978, pp. 453-468.

[17] Nakano, Y. and Tanaka, M., *Transactions of the Iron and Steel Institute of Japan*, Vol. 22, 1982, pp. 147-153.

[18] Irwin, G. R. and Wells, A. A., *Metallurgical Reviews*, Vol. 10, 1965, pp. 223-270.

[19] Rosenfield, A. R. and Shetty, D. K., *Engineering Fracture Mechanics*, Vol. 17, 1983, pp. 461-470.

[20] Landes, J. D. and Shaffer, D. H. in *Fracture Mechanics: Twelfth Symposium, ASTM STP 700*, American Society for Testing and Materials, 1980, pp. 368-382.

[21] Andrews, W. R., Kumar, V., and Little, M. M. in *Fracture Mechanics: Thirteenth Symposium, ASTM STP 743*, American Society for Testing and Materials, 1981, pp. 576-598.

[22] Iwadate, T. et al in *Elastic-Plastic Fracture: Second Symposium, Volume II—Fracture Resistance Curves and Engineering Applications, ASTM STP 803*, American Society for Testing and Materials, 1983, pp. II-531-II-561.

[23] Rosenfield, A. R. and Shetty, D. K., "Cleavage Fracture of Steel in the Ductile-Brittle Transition Region," paper presented at ASTM Symposium on User's Experience with Elastic-Plastic Fracture Toughness Methods, Louisville, Ky., 20-22 April 1983; to be published in *Elastic-Plastic Fracture Test Methods: The User's Experience, ASTM STP 856*, American Society for Testing and Materials, 1985.

J. M. Bloom[1] and S. N. Malik[1]

A Failure Assessment Approach for Handling Combined Thermomechanical Loading

REFERENCE: Bloom, J. M. and Malik, S. N., **"A Failure Assessment Approach for Handling Combined Thermomechanical Loading,"** *Fracture Mechanics: Fifteenth Symposium, ASTM STP 833,* R. J. Sanford, Ed., American Society for Testing and Materials, Philadelphia, 1984, pp. 165–189.

ABSTRACT: This paper presents a failure assessment approach for the handling of combined thermomechanical loading of nuclear pressure vessels. The proposed approach is based on a failure assessment curve that accounts for the thermal and mechanical stresses in terms of the elastic stress intensity factor with a plastically corrected crack length. The square of this stress intensity factor divided by the effective elastic modulus is added to a *J*-integral deformation plasticity solution term. The result is valid for the full range of material/structural behavior, from linear elastic to fully plastic. The fracture behavior parameter, the total *J*-integral, can then be formulated in terms of a failure assessment curve to account for the combined loading states.

The proposed approach was benchmarked using a modified Babcock & Wilcox version of the ADINA computer program. The program generated incremental plasticity finite element crack solutions for a single edge cracked plate and a circumferentially flawed cylinder subjected to thermomechanical loadings. The ADINA finite element results agreed quite well with the proposed failure assessment approach.

The failure assessment approach is illustrated by a sample problem of a postulated accident condition, based on the occurrence of an overcooling transient (thermal loading) combined with an uncontrolled repressurization (mechanical loading) of a typical nuclear reactor vessel.

KEY WORDS: fracture, plastic collapse, failure assessment diagram, thermomechanical loading, nuclear pressure vessels, ADINA

There is a need for a simple, viable engineering method for assessing ductile fracture in nuclear pressure vessels. Present safety margins based on linear elastic fracture mechanics (LEFM) analyses for normal operating or accident conditions are likely to be overly conservative. These overly conservative

[1]Technical Advisor and Research Engineer, respectively, Applied Mechanics Section, Babcock & Wilcox, A McDermott Company, Research and Development Division, Alliance, Ohio 44601.

safety margins may lead to conclusions that could result in the unnecessary shutdown of some nuclear power plants. Therefore the availability of a simple, accurate, and conservative engineering assessment procedure that recognizes the ductile nature of nuclear pressure-boundary components would be most valuable in avoiding costly repair procedures and in establishing realistic safety margins.

This need has been addressed in part through Electric Power Research Institute (EPRI) funded work by General Electric [1] and Nuclear Regulatory Commission (NRC) funded work reported in NUREG-0744 [2]. Neither program, however, has addressed the accident condition of an overcooling transient (thermal loading) combined with an uncontrolled repressurization (pressure loading) of the reactor vessel. This situation is of prime concern to both the NRC and utilities.

The failure assessment diagram (FAD) approach originally developed by the Central Electricity Generating Board of the United Kingdom (CEGB), known as the R-6 failure assessment procedure [3], was extended under EPRI Contract RP 1237-2 [4] to include both strain hardening of the material and geometric effects of the pressure vessel.

Both the original R-6 approach and the extension work recognize brittle fracture and plastic collapse of the pressure vessel in the form of a FAD. The FAD is a safety/failure plane defined by the stress intensity factor/fracture toughness ratio (K_r) as the ordinate and the applied stress/plastic collapse ratio (S_r) as the abscissa (shown in terms of the CEGB R-6 failure assessment diagram in Fig. 1).

An extensive review of CEGB's R-6 FAD approach for handling thermal

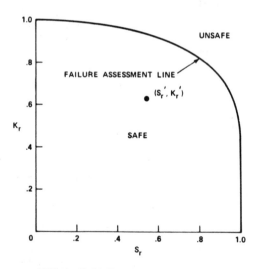

FIG. 1—*R-6 failure assessment diagram.*

stresses revealed that it is overly conservative for most applications. This is because the R-6 curve does not account for the actual strain hardening effects of the material or the actual geometry of the pressure vessel.

The proposed procedure discussed in this paper is based on a failure assessment curve that accounts for the thermal-plus-mechanical stresses in terms of the elastic stress intensity factor with a plastically corrected crack length. The square of the stress intensity factor for the combined thermomechanical loading, divided by the effective elastic modulus (E'), is then added to a deformation plasticity solution term (based on only the mechanical loading). The result is a total J-integral expression for the full range of material/specimen behavior, from linearly elastic to fully plastic. The total J-integral expression is then divided by the elastic J-integral for combined thermomechanical loading to give $J/\bar{J}^e$, which is identical to $1/K_r^2$. This then forms the basis of the thermomechanical failure assessment diagram.

This proposed procedure was validated using test cases run on the ADINA finite element computer program for the following models:

• Single edge cracked plate with a crack length-to-plate width of 0.25, subjected to thermomechanical loading.
• Axisymmetric internally circumferentially flawed cylinder with a crack depth-to-thickness of 0.25, subjected to thermomechanical loading.

The comparisons made in terms of (S_r, K_r) coordinates were found to be excellent for $0.05 \leq S_r \leq 1.5$. Details of the proposed thermomechanical failure assessment diagram procedure are presented in the following section; validation test cases are detailed in a subsequent section.

A Failure Assessment Diagram Procedure for Thermal Stresses

A method was developed to include thermal stresses in the deformation plasticity failure assessment procedure. It can be shown that the thermal stresses do not contribute to plastic collapse [5,6] but act like mechanical stresses in the LEFM regime. Therefore a treatment of thermal stresses as mechanical ones can lead to overly conservative assessments if thermal stresses acting over the crack are tensile, and nonconservative assessments if these thermal stresses are compressive. A simple solution for this is to include these thermal stresses in the calculation of K_r but exclude them from the calculation of S_r. However, it is known [5] that these thermal stresses do contribute in the intermediate region of elastic-plastic behavior at the knee of the failure assessment curve.

Starting with the formulation of the crack-driving force in terms of the J-integral for deformation plasticity, the total J-integral expression can be written as

$$J = J^e(a_{\text{eff}}, P) + J^p(a, P, n) \tag{1}$$

where Shih and Hutchinson [7] showed that this equation is a good estimate for J over the complete range of applied stress or load (P) and strain. The term $J^p(a, P, n)$ is the fully plastic (p) contribution, which can be found in Ref 1 and is a function of the mechanical stress only. In the case of thermal stresses, $J^e(a_{eff}, P)$, the elastic contribution (e) is a function of both the thermal plus the mechanical stresses. The adjusted crack length (a_{eff}) due to Irwin accounts for small-scale plasticity in J^e. Equation 1 can then be rewritten as

$$J = \frac{[\bar{K}_e(a_{eff}, P)]^2}{E'} + J^p(a, P, n) \tag{2}$$

where $\bar{K}_e(a_{eff}, P)$ is the plastically adjusted elastic stress intensity factor for the combined thermomechanical loading (mechanical-plus-thermal stresses) denoted by the overbar symbol, and E' is the effective Young's modulus where $E' = E/(1 - \nu^2)$ for plane strain and $E' = E$ for plane stress, and ν is Poisson's ratio. In terms of the failure assessment diagram, Eq 2 can be rewritten as

$$1/K_r^2 = J/\bar{J}^e = \left[\frac{\bar{K}_e(a_{eff}, P)}{\bar{K}_e(a, P)} \right]^2 + \frac{E'J^p(a, P, n)}{[\bar{K}_e(a, P)]^2} \tag{3}$$

At this point, it must be remembered that the definitions of K_r and S_r for the thermomechanical loading conditions are different from those for mechanical loading only. The coordinates S_r, K_r of the failure assessment curve are now defined by

$$S_r = \frac{\sigma_{applied}}{\sigma_{limit\ load}} \quad \text{(mechanical loading only)} \tag{4}$$

and

$$K_r = [\bar{J}^e/J]^{1/2} \tag{5}$$

where J is defined by Eq 2 and $\bar{J}^e$ is the thermomechanical elastic stress intensity factor in terms of J.

Once the failure assessment curve is established as described, the coordinates of the assessment point (S_r', K_r') must be determined. Coordinates of an assessment point are designated by primed quantities, while the failure assessment line itself is designated by unprimed quantities. If for a given thermomechanical loading scenario and flaw size, the assessment point lies inside the failure assessment curve, the structure is safe (Fig. 1). For combined thermomechanical loading, S_r' and K_r' are defined by

$$K'_r = K'^m_r + K'^{th}_r$$

and   (6)

$$S'_r = S'^m_r$$

where the superscript refers to the mechanical/pressure (m) or thermal (th) load distribution.

For crack initiation, $K'_r(a, P)$ can be interpreted in terms of the stress intensity factor (K_I) or the J-integral as

$$K'_r(a, P) = K_I/K_{Ic} = \sqrt{J_{IE}(a, P)/J_{Ic}}$$

and   (7)

$$S'_r = \sigma/\sigma_1$$

where K_{Ic} (J_{Ic}) is the fracture toughness of the material, σ is the applied stress on the structure, σ_1 is the plastic collapse stress, and $J_{IE}(a, P)$ is the J-value of the elastically calculated stress intensity factor given by

$$J_{IE}(a, P) = \frac{K_I^2(a, P)(1 - \nu^2)}{E}$$ (8)

where E and ν are Young's modulus and Poisson's ratio respectively.

For ductile tearing, K'_r and S'_r are defined as

$$K'_r(a_0 + \Delta a) = \sqrt{J_{IE}(a_0 + \Delta a)/J_R(\Delta a)}$$

and   (9)

$$S'_r(a_0 + \Delta a) = \sigma/\sigma_1(a_0 + \Delta a)$$

where J_{IE} and σ_1 are functions of the amount of slow stable crack growth (Δa), J_R is the experimentally measured J-resistance curve plotted as a function of slow stable crack growth, and J_{IE} is the elastically calculated crack-driving force $(J_{applied})$ calculated from the stress intensity factor for the current crack length $(a_0 + \Delta a)$.

Finite-Element Incremental Plasticity Validation

To validate the failure assessment approach for the handling of thermomechanical loading, several modifications to the 1977 version of ADINA were required. These modifications are briefly discussed below.

A new material model was added to the ADINA program. The model represents a two-dimensional thermo-elastic-plastic material with its stress-strain relationship defined by the Ramberg-Osgood law

$$\epsilon_e = \frac{\sigma_e}{E} \qquad \text{for} \quad \sigma_e \leq \sigma_y \tag{10}$$

and

$$\epsilon_e = \frac{\sigma_e}{E} + (\sigma_e/B)^n \qquad \text{for} \quad \sigma_e > \sigma_y \tag{11}$$

where σ_e and ϵ_e are the effective stress and effective strain respectively, σ_y is the yield stress, E is Young's modulus, and B, n are constants. This model is applicable to an incremental small strain analysis with isotropic hardening.

A postprocessor program was developed to calculate the thermal J-integral for two-dimensional finite element models. In the presence of a thermal strain gradient, the two-dimensional J-integral has the form [6,8]

$$J_\theta = \int_\Gamma \left(W dy - \sigma_{ij} \frac{\partial u_i}{\partial x} n_j \, dS \right) + \iint_A \sigma_{ij} \frac{\partial \theta_{ij}}{\partial x} \, da \tag{12}$$

where Γ is an arbitrary curve starting on the lower crack surface and following a counterclockwise path, ending on the upper crack surface; W is the strain energy density; σ_{ij} is the stress tensor; u_i is the displacement; n_j is the normal vector; x and y are the coordinate axes along and perpendicular to the crack line; θ_{ij} is the thermal strain; A is the area enclosed within the contour Γ; and S is an arc length along the contour Γ. W is defined as $W_e + W_p$, where W_e is the elastic strain energy density and W_p is the plastic work, given by

$$W_p = \int_0^{\epsilon_p} \sigma_e d\epsilon_p \tag{13}$$

where ϵ_p is the effective plastic strain.

The required input data for the thermal J-integral consists of the path along the element integration points and the element numbers for the area integral. The postprocessor rearranges the input data in an ascending order of element numbers. It then reads and stores the required nodal and element information. From the nodal displacements, it interpolates the displacements and their derivatives at the integration points. The Gaussian nine-point integration formulas were subsequently used for evaluating both the contour and the area integrals. Several test cases for a center-cracked panel were run to check both the ADINA modifications and the postprocessor program.

In addition to the thermal J-integral capabilities, an axisymmetric J-integral postprocessor was added to the existing ADINA computer program capabilities. The postprocessor J-integral program was modified to compute various quantities needed for J-values in axisymmetric cylindrical configurations. These modifications include elastic-plastic mechanical as well as thermal stress contributions. The thermomechanical J-integral expression, based on the analysis presented in Ref 6, is

$$J_\theta = \left\{ \oint_\Gamma \left[W dz - \sigma_{ij} \frac{\partial u_i}{\partial r} n_j dS \right] \right.$$

$$\left. - \iint_A \frac{1}{r} \left[\sigma_r \epsilon_r - \sigma_\theta \epsilon_\theta - \tau_{rz} \frac{\partial u_z}{\partial r} \right] dA - \iint_A \sigma_{ij} \frac{\partial \theta_{ij}}{\partial r} dA \right\} \quad (14)$$

where

r and θ = radial and tangential coordinates,
$\epsilon_r = \partial u_r / \partial_r$ = radial strain,
$\epsilon_\theta = u_r / r$ = tangential strain for axisymmetric case,
u_r and u_z = radial and axial displacement, respectively,
Γ = an arbitrary curve starting on the lower crack surface and following a counterclockwise path, ending on the upper crack surface,
A = area enclosed by the path Γ in the r-z plane,
$W = (W_e + W_p)$ = mechanical strain energy containing elastic and plastic contributions,
σ_r, σ_θ, τ_{rz} = radial, tangential, and shear stress at a point,
θ_{ij} = thermal strain tensor,
z = axial coordinate normal to the crack surface, and
n_j = unit normal vector to path Γ with dS as arc length.

These modifications were tested for a circumferentially flawed cylinder ($a/t = 0.25$) under pure mechanical load. Details are provided in Ref 4.

To validate the proposed assessment approach, two incremental finite element test cases were run using the modified ADINA computer program:

• A single edge cracked plate with a crack length-to-plate width (a/b) of 0.25, subjected to thermomechanical loading.
• An axisymmetric internally circumferentially flawed cylinder with a crack depth-to-thickness (a/t) of 0.25, subjected to thermomechanical loading.

The first test case was that of a single edge cracked plate with $a/b = 0.25$ subjected to thermomechanical loads. Figure 2 illustrates the prescribed temperature gradient through the plate width. The plate edge with the postulated flaw is at the lower temperature of 93°C (200°F), while the opposite edge is at a higher temperature of 288°C (550°F). This temperature distribution is typical of a temperature profile found a few minutes after loss-of-coolant accident (LOCA) conditions occur in nuclear reactor pressure vessels when an emergency core cooling system (ECCS) is activated. Distribution of the thermoelas-

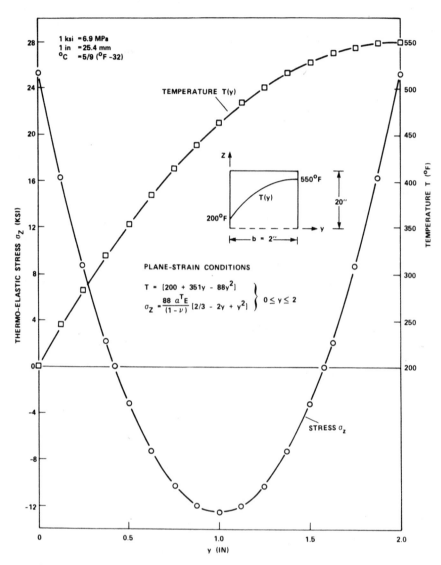

FIG. 2—*Prescribed temperature and resulting thermoelastic stress distribution in a plate.*

tic stress (σ_z) [9] normal to the plane of the postulated flaw is also shown in Fig. 2. For such large thermoelastic stresses near the plate edges, the actual stresses are elastic-plastic with the plastic zone spreading inward from the plate edges, as indicated by the ADINA solution for lower values of applied mechanical loads.

The stress intensity factor (K_I) due to the thermal stress, $(\sigma_z)_{th}$, was obtained from the BIGIF (*B*oundary *I*ntegral equation *G*enerated *I*nfluence *F*unctions) solution method [9] and is plotted in Fig. 3 as a function of fractional crack length a/b.

If a uniform tensile (mechanical) load is superimposed over the thermal stress, $(\sigma_z)_{th}$, the thermomechanical stress intensity factor $(\bar{K}_e)$ can be obtained as the algebraic sum

$$\bar{K}_e = [(K_I)_{thermal} + (K_I)_{mechanical}] \tag{15}$$

$(K_I)_{mechanical}$ is denoted by K_e in the remainder of this paper.

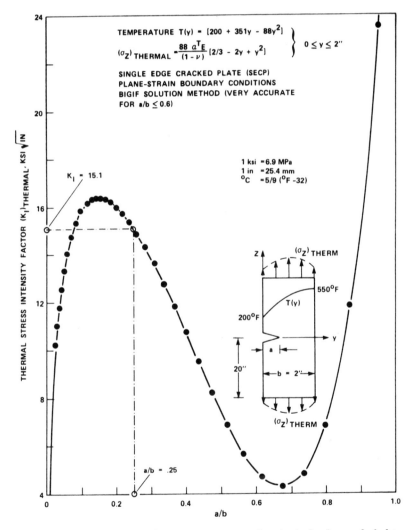

FIG. 3—*Stress intensity factor due to temperature gradient in single edge cracked plate.*

Under plane strain conditions, the thermomechanical elastic J-integral is defined by

$$\bar{J}^e = \frac{(1 - \nu^2)}{E} (\bar{K}_e)^2 \tag{16}$$

Figure 4 shows the total J-integral values obtained from the ADINA postprocessor computer program for purely mechanical and thermomechanical load cases. The J-axis in Fig. 4 includes both the thermal and mechanical loads for the thermomechanical test case; while for the pure mechanical load case, the J-axis is a function solely of the mechanical loading. For lower values of applied stresses, the J-integral values for the thermomechanical case are somewhat greater than the corresponding purely mechanical load. But, as the applied stress increases, this difference vanishes. The limit load (P_{LL}) for a single edge cracked plate [1] is

$$P_{LL} = 1.455 \eta c \sigma_{flow} \tag{17}$$

where

$c = (b - a)$, and
$\eta = \{[1 + (a/c)^2]^{1/2} - a/c\}$.

The S_r and K_r parameters of the failure assessment diagram (Fig. 5) for thermomechanical loads were determined from

$$S_r = \frac{\sigma_{appl}}{P_{LL}/(bt)} \tag{18}$$

$$K_r = \left[\frac{J_{ADINA}}{\bar{J}^e}\right]^{1/2} \tag{19}$$

where J_{ADINA} values are as shown in Fig. 4.

Analytically, the thermomechanical stress intensity factor is given by Eq 15. The effective crack length (corrected for small-scale plasticity) [4] required for the calculation of the proposed failure assessment diagram with secondary stresses is given by

$$a_{eff} = a + \frac{1}{6\pi} \left[\frac{\bar{K}_e(a, P)}{\sigma_y}\right]^2 \tag{20}$$

where both the strain hardening $(n - 1)/(n + 1)$ and ϕ correction of Shih [1] were neglected, and where σ_y is the yield strength of the material. The ϕ correction proposed by Shih can be expressed in terms of S_r as

$$\phi = \frac{1}{1 + S_r^2} \tag{21}$$

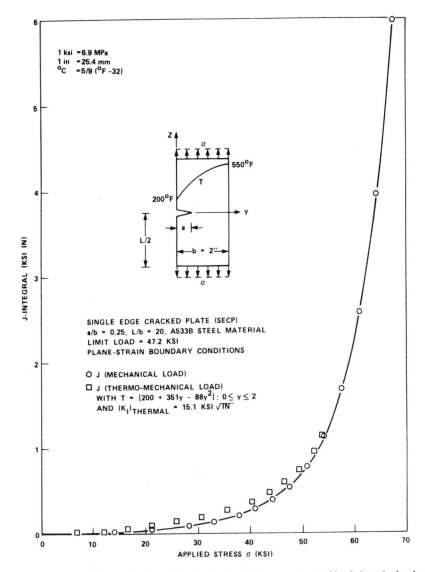

FIG. 4—*ADINA J-integral values of mechanical and thermomechanical loads in a single edge cracked plate.*

Figure 5 shows the comparisons between the finite element results and the analytically derived failure assessment curve for both the mechanical and thermomechanical cases. The solid line shown in Fig. 5 is based on Eq 20, where $\bar{K}_e(a, P)$ is replaced by $K_e(a, P)$. The agreement with the finite element results for mechanical loading only in terms of (S_r, K_r) is excellent for this

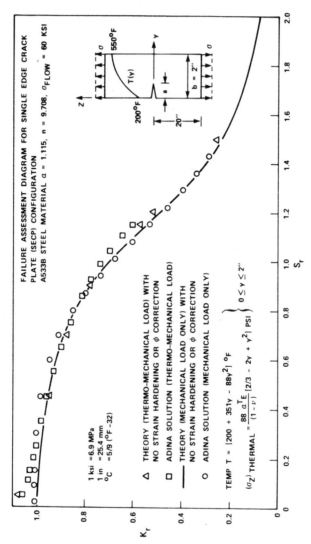

FIG. 5—*Comparison of ADINA finite element solution in terms of* (S_r, K_r) *and the deformation plasticity derived failure assessment diagram for mechanical and thermomechanical loads—single edge cracked plate.*

case. The final form of the expression for the failure assessment curve for this thermomechanical test case [4] can be written as

$$1/K_r^2 = J/\bar{J}^e = \left[\frac{\bar{K}_e(a_{\text{eff}}, P)}{\bar{K}_e(a, P)}\right]^2 + \frac{1}{(0.170/S_r + 1)^2}\left[\frac{J^p}{[K_e(a, P)]^2}\right] \quad (22)$$

which is based upon Eq 3.

The expression $J^p/[K_e(a, P)]^2$ is obtained from the expression for the failure assessment curve of the mechanical loading of a single edge cracked plate with $a/b = 0.25$ [4]. Table 1 gives the numerical results of the calculations made for this test case.

The results of these calculations in terms of (S_r, K_r) are shown by the triangular symbols in Fig. 5. These compare quite well with the square symbols determined from the corresponding finite element results.

The second test case was that of an axisymmetric internally circumferentially flawed cylinder with $a/t = 0.25$ subjected to thermomechanical stresses. A temperature gradient (ΔT) of 117°C (210°F) through the wall was prescribed. This is representative of thermal shock conditions found in a nuclear reactor pressure vessel a few minutes following a loss-of-coolant accident when an emergency core cooling system is activated. The temperature gradient and the resulting thermo-elastic stress distribution (σ_z) normal to the plane of the postulated crack extension is shown in Fig. 6.

The stress intensity factor due to this thermal stress (σ_z) was obtained from a comparison of the solution methods of BIGIF [9], Buchalet and Bamford [10], and Labbens et al [11] and is plotted in Fig. 7 as a function of a/t. The BIGIF solution method is very accurate for $a/t \leq 0.6$. It can be seen that for $a/t \leq 0.35$, all three solution methods yield almost the same values of $(K_I)_{\text{thermal}}$.

TABLE 1—Numerical results based on thermomechanical loading failure assessment approach for single edge cracked plate.[a]

$\left[\dfrac{\bar{K}_e(a_{\text{eff}}, P)}{\bar{K}_e(a, P)}\right]^2$	S_r	K_r
0.862	0.05	1.007
1.117	0.45	0.946
1.288	0.70	0.877
1.513	0.90	0.780
1.666	1.0	0.708
1.927	1.145	0.572
2.056	1.20	0.516
3.014	1.50	0.254

[a]$a/b = 0.25, \alpha = 1.115, n = 9.708, \sigma_y = 414$ MPa (60 ksi).

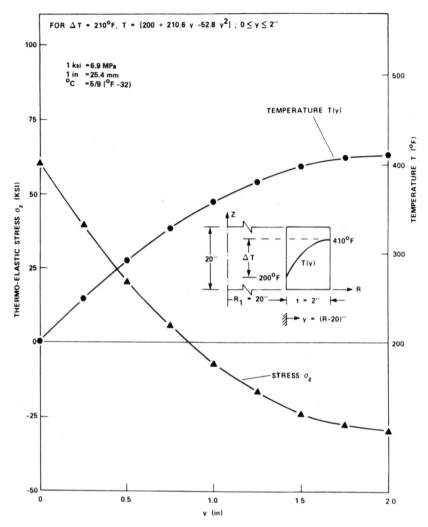

FIG. 6—*Prescribed temperature gradient and resulting thermoelastic stress distribution in axisymmetric hollow cylinder.*

The elastic-plastic J-integral values obtained from the ADINA post-processor computer program for purely mechanical and thermomechanical loads are shown in Fig. 8. For lower values of applied loads, the thermomechanical J-integral values are higher than those for mechanical load only. For higher values of mechanical loads, however, the difference in the J-integral values between the cases of mechanical and thermomechanical loads decreases gradually. Again, the J-axis in Fig. 8 includes both the thermal and mechanical loads for the thermomechanical test case. For the pure mechanical load case, the J-axis is based only on the mechanical loading.

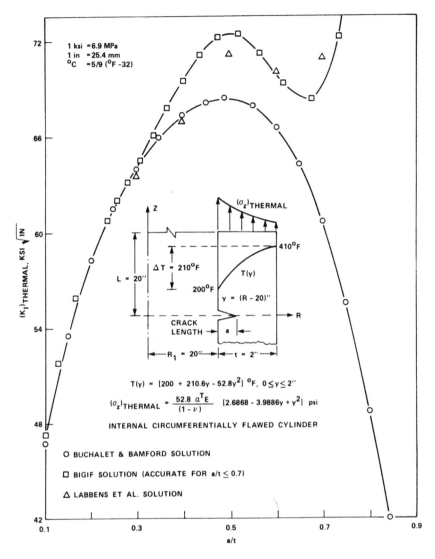

FIG. 7—*Stress intensity factors due to temperature gradient of* $\Delta T = 117°C$ *(210°F) in circumferentially flawed cylinder.*

Figure 9 shows the comparisons between the ADINA finite element results and the theoretically derived failure assessment diagram for both the mechanical and thermomechanical cases. The solid line is based upon

$$a_{\text{eff}} = a + \frac{1}{6\pi}\left(\frac{n-1}{n+1}\right)\left[\frac{K_e(a, P)}{\sigma_y}\right]^2 \phi \qquad (23)$$

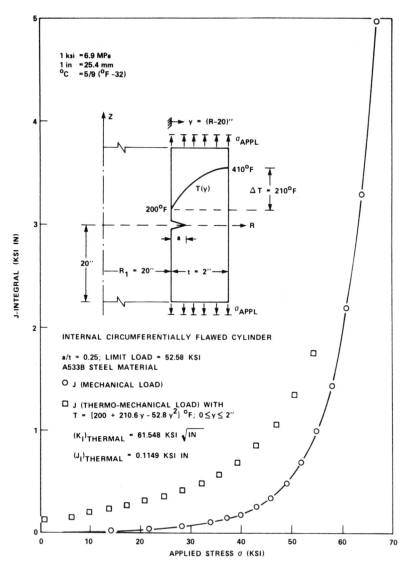

FIG. 8—*ADINA J-integral values for mechanical and thermomechanical loads on internally circumferentially flawed cylinder.*

The calculations for the thermomechanical case also are based on Eq 23 with $\bar{K}_e(a, P)$ replacing $K_e(a, P)$. The expression for the failure assessment diagram for this thermomechanical test case can be written as

$$1/K_r^2 = J/\bar{J}^e = \left[\frac{\bar{K}_e(a_{eff}, P)}{\bar{K}_e(a, P)}\right]^2 + \frac{1}{(0.714/S_r + 1)^2}\left[\frac{J^p}{[K_e(a, P)]^2}\right] \quad (24)$$

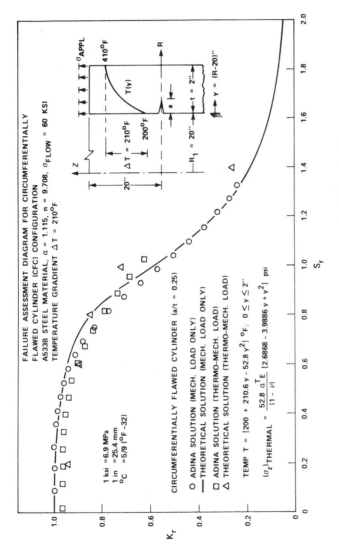

FIG. 9—*Comparison of ADINA finite element solution in terms of* (S_r, K_r) *and the deformation plasticity derived failure assessment diagram for mechanical and thermomechanical loads—circumferentially flawed cylinder.*

where the term $J^p/[K_e(a, P)]^2$ is obtained from the expression for the failure assessment curve of the mechanical loading of an internally circumferentially flawed cylinder with $a/t = 0.25$. Table 2 gives the numerical results of the calculations for this test case.

The results of these calculations in terms of (S_r, K_r) are shown by the triangular symbols in Fig. 9. The triangular symbols compare quite well with the square symbols determined from the finite element results in terms of (S_r, K_r). For both the mechanical and thermomechanical loading cases, the agreement between the ADINA results and the theoretically derived failure assessment diagram is good, except in the transition region (the knee of the failure assessment curve). In this comparison, the analytical solutions retained the original strain hardening $(n - 1)/(n + 1)$ and ϕ corrections for the effective crack length term given by Eq 23. This trend in the transition region has been observed earlier with various other configurations under pure mechanical loads [4]. The comparisons of the thermomechanical loading results to those of the pure mechanical loading results appear to be consistent. One would expect that the addition of a tensile thermal load to the mechanical load is more detrimental to a structure. This would produce a failure assessment curve for thermomechanical loading which would fall inside the failure assessment curve for mechanical loading.

Sample Problem

A sample problem of a longitudinal flaw in a nuclear reactor pressure vessel under combined pressure/temperature loading was chosen to illustrate the procedure discussed. The pressure vessel is assumed to be under a postulated accident scenario.

This postulated accident condition is based on the occurrence of an overcooling transient (thermal loading) combined with an uncontrolled repressurization (primary loading) of the reactor vessel. This transient occurs at a par-

TABLE 2—*Numerical results based on proposed thermomechanical loading failure assessment diagram approach for internally circumferentially flawed cylinder.[a]*

$\left[\dfrac{\bar{K}_e(a_{\text{eff}}, P)}{\bar{K}_e(a, P)} \right]^2$	S_r	K_r
1.103	0.20	0.952
1.246	0.60	0.895
1.294	0.80	0.858
1.336	1.00	0.726
1.386	1.40	0.259

[a] $a/t = 0.25$, $\Delta T = 117°C$ (210°F), $\alpha = 1.115$, $n = 9.708$, $\sigma_y = 414$ MPa (60 ksi).

ticular time during the accident. The assumed flaw is a continuous (full-length) longitudinal crack of $a/t = 0.25$. The thermal load is due to the inner wall of the vessel being cooled to 177°C (350°F) from 274°C (525°F). This temperature difference (ΔT) of 97°C (175°F) produces a hoop stress. The resulting hoop stress due to this thermal gradient in the presence of a defect was input into an expression for K_I for a longitudinal flaw in a cylinder developed by Buchalet and Bamford [10]. The resulting thermal stress intensity factor as a function of a/t is shown in Fig. 10. The combined thermomechanical stress intensity factor ($\bar{K}_e$) for this problem is given by

$$\bar{K}_e = (K_I)_{thermal} + (K_I)_{pressure}$$

or   (25)

$$\bar{K}_e = (K_I)_{thermal} + 10P \sqrt{\pi a} \, F(a/t)$$

where $(K_I)_{thermal}$ is as shown in Fig. 10 and

$$F(a/t) = \frac{1.165 - 1.339a/t}{(1 - a/t)^{5/2}} \tag{26}$$

for the range of $0.125 \leq a/t \leq 0.75$.

For this sample problem, two pressures were assumed and the corresponding factors of safety were calculated.

The first step in the assessment procedure is the generation of the appropriate failure assessment curve for the longitudinal flaw in a pressurized cylinder subjected to thermal loading. The general expression for the failure assessment curve is given by Eq 3. For the thermal load of this sample problem, Eq 3 can be rewritten as

$$1/K_r^2 = J/\bar{J}^e = \left[\frac{\bar{K}_e(a_{eff}, P)}{\bar{K}_e(a, P)}\right]^2 + \frac{1}{(0.378/S_r + 1)^2} \left[\frac{J^p(a, P, n)}{[K_e(a, P)]^2}\right] \tag{27}$$

The expression $J^p/[K_e(a, P)]^2$ is obtained from the equation for the failure assessment curve of the mechanical loading of a longitudinal flaw in a pressurized cylinder from Ref 4 and is repeated for completeness:

$$J^p/[K_e(a, P)]^2 = \frac{0.0026\alpha(1 - a/t)h_1 S_r^{n-1}}{\left[0.1F(a/t)\dfrac{(1 - a/t)}{(1 + 0.1a/t)}\right]^2} \tag{28}$$

where $F(a/t)$ is given by Eq 26, and α, n are the Ramberg-Osgood strain hardening parameters. The function h_1 was taken from Fig. C-35 of Ref 1. Note that for the h_1 functions given by that figure, S_r must be defined as

$$S_r = P/P_0 \qquad (29)$$

where

$$P_0 = \frac{0.2}{\sqrt{3}} \, \sigma_0 \, \frac{(1 - a/t)}{(1 + 0.1a/t)} \qquad \text{for} \quad t/R_i = 0.10 \qquad (30)$$

and σ_0 is the yield strength of the material. The expression $\bar{K}_e(a, P)$ was determined from Eq 25. The expression $\bar{K}_e(a_{\text{eff}}, P)$ was also determined from this

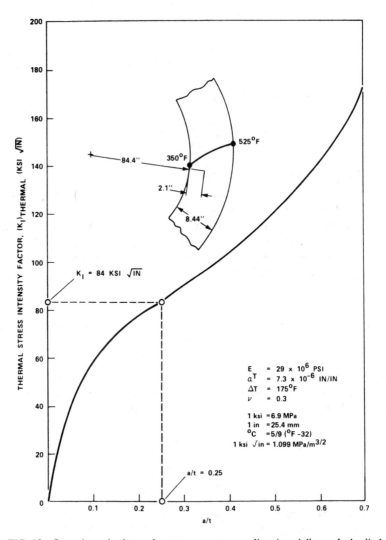

FIG. 10—*Stress intensity factor due to temperature gradient in axially cracked cylinder.*

same equation, but with the effective crack depth (a_{eff}) replacing the actual crack depth (a). The expression used to determine a_{eff} was

$$a_{eff} = a + \frac{1}{6\pi}\left(\frac{n-1}{n+1}\right)\left[\frac{K_e(a, P)}{\sigma_0}\right]^2 \phi \tag{31}$$

The unknown Ramberg-Osgood parameters (α, n) were determined by a least-square fit of the *true* stress/*true* plastic strain curve of the A533B material at the temperature of interest. The derived failure assessment diagram is illustrated in Fig. 11. Numerical values determined from Eq 27 are given in Table 3.

The next step in the assessment procedure is the determination of the locus of points (S_r', K_r') for each pressure loading. The equations are

$$K_r' = K_r'^P + K_r'^S$$

and $\tag{32}$

$$S_r' = S_r'^P$$

where

$$K_r' = \sqrt{J_{IE}(a_0 + \Delta a)/J_R(\Delta a)}$$

and $\tag{33}$

$$S_r'^P = P/P_0(a_0 + \Delta a)$$

The material resistance curve $J_R(\Delta a)$ used in this sample problem was taken from data on 4T compact A533B steel specimens tested at General Electric [12]. Table 4 presents the numerical results obtained from the assessment of the pressure vessel for the two assumed pressures. Figure 11 shows the assessment points for both the 6.9 MPa (1 ksi) pressure (unprimed numbered points) and the 13.8 MPa (2 ksi) pressure (primed numbered points). The double-primed numbered points and the points with superscript "o" are common to both pressure loading cases and are used in the factor of safety calculations. The factors of safety given in Table 4 were determined from the expression

$$\text{factor of safety} = \frac{n^o n''}{n^o n} \tag{34}$$

where n refers to the numbered points labelled 1, 2, 3, . . .; and Eq 34 refers to the case of 6.9 MPa (1 ksi) pressure. The safety factors for the case of the 13.8 MPa (2 ksi) pressure were determined by replacing n by n' in Eq 34. It can be seen readily that the factors of safety are directly proportional to the corresponding primary loadings, namely the applied pressures.

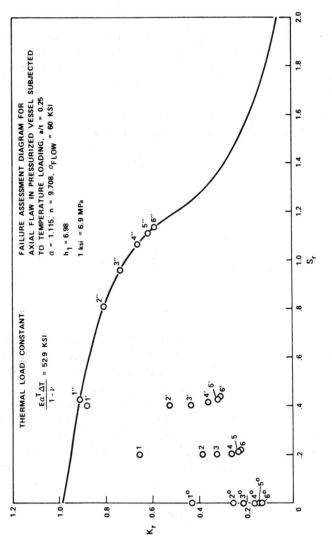

FIG. 11—*Assessment of an axial flaw in pressurized vessel subjected to temperature loading.*

TABLE 3—*Numerical values for derived failure assessment diagram for longitudinal crack in pressurized cylinder subjected to thermal loading, a/t = 0.25 (see Fig. 11).*

S_r	$\left[\dfrac{\bar{K}_e(a_{\mathrm{eff}}, P)}{\bar{K}_e(a, P)}\right]^2$	$\dfrac{E'J^P(a, P, n)}{[\bar{K}_e(a, P)]^2}$	K_r
0.01	1.030	0.000	0.985
0.20	1.105	0.000	0.952
0.60	1.318	0.005	0.870
1.00	1.486	0.519	0.706
1.20	1.548	2.779	0.481
1.40	1.599	11.381	0.278
2.00	1.670	293.404	0.058

[a] $\alpha = 1.115$; $n = 9.708$; $\sigma_{\mathrm{flow}} = 60$ ksi; $h_1 = 6.98$; 1 ksi = 6.9 MPa.

$$\frac{E \bar{\alpha} \Delta T}{(1 - \nu)} = 52.9 \text{ ksi}$$

$$1/K_r^2 = J/\bar{J}_e = \left[\frac{\bar{K}_e(a_{\mathrm{eff}}, P)}{\bar{K}_e(a, P)}\right]^2 + \frac{E'J^P(a, P, n)}{[\bar{K}_e(a, P)]^2}$$

Conclusions

A proposed procedure was presented that can handle the failure assessment of nuclear pressure vessels and piping subjected to combined thermomechanical loadings. Two finite element crack models simulating a single edge cracked plate and a circumferentially cracked cylinder subjected to combined mechanical-plus-thermal stresses were used to benchmark the proposed assessment procedure. Based on these cases run using the ADINA computer program, the proposed assessment approach seems to be adequate for structures subjected to combined thermomechanical stresses.

A sample problem of a postulated accident condition for a typical nuclear reactor vessel illustrated the applicability of the proposed assessment procedure.

Acknowledgments

The authors acknowledge the support of this work by the Electric Power Research Institute, Palo Alto, California (Contract RP 1237-2). Thanks is extended to Dr. D. Norris of EPRI, without whose support this work would not have been possible. The authors also thank Hwa Zien, Babcock & Wilcox Research and Development Division, for his work on the modification of the computer program ADINA; and Paul Sensmeier, Babcock & Wilcox Research and Development Division, and Kenneth Yoon, Babcock & Wilcox Nuclear Power Generation Division, for their encouragement, support, and helpful discussions.

TABLE 4—Numerical results for sample problem of pressurized cylinder with longitudinal crack (a/t = 0.25) subjected to thermal loading.[a]

Point	Δa, in.[a]	J_R, 10^3 lb/in.	$K_r'^S$	$K_r'^P$		$S_r'^P$		K_r'		Factor of Safety	
				A	B	A	B	A	B	A	B
1	0.005	1.20	0.431	0.225	0.450	0.197	0.394	0.656	0.881	1.07	2.14
2	0.053	3.52	0.254	0.134	0.268	0.199	0.398	0.388	0.522	2.04	4.08
3	0.094	5.16	0.212	0.113	0.226	0.200	0.400	0.325	0.438	2.35	4.70
4	0.245	8.44	0.170	0.094	0.188	0.206	0.412	0.264	0.358	2.54	5.08
5	0.412	11.56	0.148	0.086	0.172	0.212	0.424	0.234	0.320	2.63	5.26
6	0.531	13.02	0.143	0.085	0.170	0.217	0.434	0.228	0.313	2.61	5.22

[a]$J_r(\Delta a)$ curve obtained from Ref 13; 1 ksi = 6.9 MPa; 1 lb/in. = 175.1 N/m; 1 in. = 25.4 mm.

$$\frac{E \bar{\alpha} \Delta T}{(1 - \nu)} = 52.9 \text{ ksi}$$

Pressure A = 1000 psi; Pressure B = 2000 psi.

References

[1] Kumar, V., German, M. D., and Shih, C. F., "An Engineering Approach for Elastic-Plastic Fracture Analysis," EPRI Topical Report NP-1931, Research Project 1237-1, Electric Power Research Institute, July 1981.

[2] "Resolution of Reactor Vessel Materials Toughness Safety Issue," NUREG-0744, U.S. Nuclear Regulatory Commission, Sept. 1981.

[3] Harrison, R. P., Loosemore, K., and Milne, I., "Assessment of the Integrity of Structures Containing Defects," CEGB Report R/H/6, Central Electricity Generating Board, United Kingdom, 1976.

[4] Bloom, J. M. and Malik, S. N., "A Procedure for the Assessment of the Integrity of Nuclear Pressure Vessels and Piping Containing Defects," EPRI Topical Report NP-2431, Research Project 1237-2, Electric Power Research Institute, June 1982.

[5] Chell, G. G. and Ewing, D. J. F., "The Role of Thermal and Residual Stresses in Linear Elastic and Post Yield Fracture Mechanics," Central Electricity Generating Board Report RD/L/N216/76, 1976; *International Journal of Fracture*, Vol. 13, No. 4, Aug. 1977, pp. 467–479.

[6] Ainsworth, R. A., Neale, B. K., and Price, R. H., "Fracture Behavior in the Presence of Thermal Strains," Conference on Tolerance of Flaws in Pressurized Components, Paper C96/78, Institution of Mechanical Engineers, London, 1978.

[7] Shih, C. F. and Hutchinson, J. W., "Fully Plastic Solutions and Large-Scale Yielding Estimate for Plane Stress Crack Problems," *Journal of Engineering Materials and Technology, Transactions of ASME*, Series H, Vol. 98, No. 4, Oct. 1976, p. 289.

[8] Blackburn, W. S., *International Journal of Fracture Mechanics*, Vol. 8, 1972, pp. 343–346.

[9] "BIGIF—Fracture Mechanics Code for Structures," EPRI Topical Report NP-838, Research Project 700, Electric Power Research Institute, Dec. 1978.

[10] Buchalet, C. B. and Bamford, W. H., in *Mechanics of Crack Growth, ASTM STP 590*, American Society for Testing and Materials, 1976, pp. 385–402.

[11] Labbens, R., Pellissier-Tanon, A., and Heliot, J. in *Mechanics of Crack Growth, ASTM STP 590*, American Society for Testing and Materials, 1976, pp. 345–365.

[12] Shih, C. F. et al, "Methodology for Plastic Fracture," EPRI Topical Report NP-1735, Research Project 601-2, Electric Power Research Institute, March 1981.

[13] Timoshenko, S. P. and Goodier, J. N., *Theory of Elasticity*, McGraw-Hill, New York, 1970, pp. 433–436.

Fatigue Crack Growth

Adnan Sahli[1] and Pedro Albrecht[2]

Fatigue Life of Welded Stiffeners with Known Initial Cracks

REFERENCE: Sahli, A. and Albrecht, P., **"Fatigue Life of Welded Stiffeners with Known Initial Cracks,"** *Fracture Mechanics: Fifteenth Symposium, ASTM STP 833,* R. J. Sanford, Ed., American Society for Testing and Materials, Philadelphia, 1984, pp. 193–217.

ABSTRACT: The objective of this study was to determine the crack propagation life of a weldment and to assess the significance of the results with regard to the design and inspection of bridge girders. The experimental part consisted of marking initial crack sizes in 68 fillet-welded transverse stiffeners subjected to constant-amplitude loading or spectrum loading typical of highway bridges. The crack propagation lives were then calculated with fracture mechanics methods using average crack growth rates and a reasonably simple method of determining the stress intensity factor. In general, more cycles were needed to grow the crack than were calculated in this study. The predictions were conservative in the sense that they underestimated the observed lives. The results correlated better for the larger than for the smaller marked crack sizes. The propagation life of the specimens tested in this study corresponded to the stage of part-through crack growth in transverse stiffeners welded to the web and the flange of girders, or to the web alone. The generally low crack detection probabilities with ultrasonic and radiographic inspection and the results of this study suggest that it would be prudent to design structures with long service lives, such as highway bridges, to the fatigue limit so that in most cases cracks would not initiate.

KEY WORDS: fatigue, cracks, welding, steel, crack detection, bridges

Fracture mechanics calculations of fatigue crack propagation have been shown to predict well the number of cycles required to grow a fatigue crack from an initial size to a final size. The results are particularly good for standard specimens, such as the compact specimen. These ideal specimens exhibit the following advantages: two-dimensional problem, low applied nominal stresses, a single deep crack ($a/W > 0.4$), about constant stress intensity factor along the nearly straight crack front, and mostly uniform material properties through the specimen depth.

[1]Engineer, Blunt and Evans Consulting Engineers; formerly, Graduate Student, Department of Civil Engineering, University of Maryland, College Park, Md. 20742.

[2]Professor, Department of Civil Engineering, University of Maryland, College Park, Md. 20742.

In contrast, fatigue life calculations of welded structural details involve many uncertainties that arise from the following factors:

1. Cracks at fillet welds can either initiate at one dominant point along the weld toe and grow as a single part-through crack, or they can initiate at multiple adjacent points and eventually join to form a long and shallow crack. The number and location of flaws that serve as crack initiation points are unpredictable.

2. Cracks propagate through regions of varying microstructure such as the weld, the heat affected zone (HAZ), and the base metal.

3. The major part of the propagation life is spent growing the crack at low ΔK values while it is shallow.

4. Cracks tend to initiate at points of geometrical discontinuities where the stress concentration may cause local yielding, even when there is no crack.

5. Most details have complex geometries, and expressions for the stress intensity factor for a crack embedded in such geometries are necessarily numerical approximations.

6. The by-pass plastic zone may cause crack closure in the wake of the full length of crack extension, whereas in compact specimens the crack closes along the fatigue crack but not along the saw-cut notch.

7. The welding residual stresses, which affect the stress ratio (R) and hence the crack-growth rate, vary along the path of crack extension through the weld, HAZ, and base metal.

The aforementioned uncertainties make it difficult to accurately predict the crack propagation life of welded details.

Objective

The objective was to determine how well one can predict, with an engineering approach, the crack propagation life of a weldment under service loads typical of highway bridges. The fracture mechanics approach was utilized, average material properties were assumed, and simplifications guided by engineering judgement were made.

The objective was pursued by analyzing crack propagation at non-load-carrying fillet-welded transverse stiffeners subjected to either constant-amplitude or spectrum loading. The number of cycles needed to propagate the cracks from a marked size to failure were calculated and compared with the observed values.

Experimental Work

Test Specimens

The tension specimens consisted of a main plate, 10 by 25 by 330 mm (⅜ by 1 by 13 in.), and two transverse stiffener plates, 7 by 25 by 50 mm (¼ by 1 by 2 in.), attached with 7-mm (¼-in.) fillet welds (Fig. 1). The plate material from which the specimens were fabricated conformed to ASTM Specification for High-Strength Low-Alloy Structural Steel with 50 ksi (345 MPa) Minimum Yield Point to 4 in. Thick (A 588). One batch of specimens was welded automatically, the other batch was welded manually. Cross sections of both types of welds are shown in Fig. 2. Since the heat input is larger in automatic than in manual welding, the weld cross section was larger in the former case.

This detail simulates the stress condition at a transverse stiffener welded to the web or to the flange of bridge girders, as well as diaphragm gussets welded to the web. It is classified in the fatigue specifications [1] as a Category C detail.

Specimen Fabrication

All specimens were made by a bridge fabricator who used the same techniques, workmanship, and inspection as required by the State of Maryland for their highway bridges.

The main plates and the stiffener plates were flame cut from larger plates to widths of 330 and 50 mm (13 and 2 in.), respectively. The rolling direction of the main plate corresponded to the longitudinal axis of the specimens and to the direction of loading. The plates were shot blasted before welding in the manner specified by the Structural Steel Painting Council (SSPC) for Class SP6-63, commercial blast.

Two types of welding were used. One batch of specimens was welded by the automatic submerged-arc process with 2.4-mm (³⁄₃₂-in.)-diameter L61 wire and L761 flux, the other batch by the shielded metal arc process with 1.6-mm (¹⁄₁₆-in.) Lincoln E-8018-C3 electrodes. The stiffener plates were tack welded to both sides of the main plate. The two longitudinal 7-mm (¼-in.) fillet welds on one side of the main plate were laid first. The plate was then turned and the remaining two welds were laid. After welding, the specimens were saw cut from the assembly to a width slightly larger than 25 mm (1 in.). The saw cut

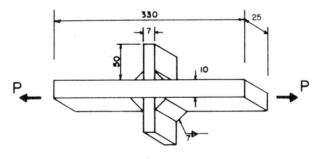

FIG. 1—*Test specimen.*

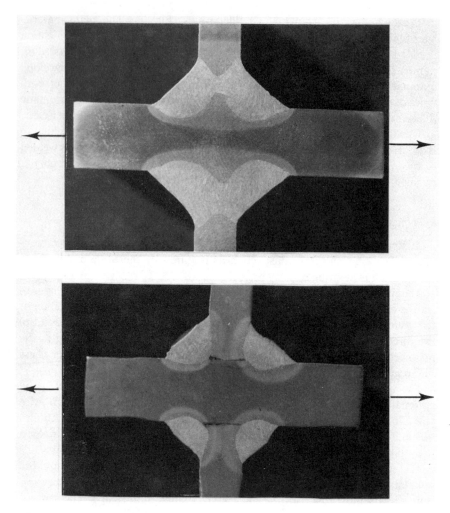

FIG. 2—*Polished and etched longitudinal section through automatically* (top) *and manually* (bottom) *welded specimens.*

surfaces were then wet ground to a surface roughness not exceeding 125 ASA micro.

Testing Procedures

All specimens were tested in the engineering laboratory at room temperature and humidity levels. The constant-amplitude tests were run with sinusoidal wave loading, the variable-amplitude tests with saw-tooth wave loading. The minimum stress was 3.5 MPa (0.5 ksi) in all tests.

Previous studies at the University of Maryland had examined the variable-amplitude fatigue life [2,3] and the effect of weathering on the fatigue behavior of transverse stiffeners [4,5]. All specimens in those studies were stress cycled from the as-fabricated condition to failure. For the purposes of the present study, the crack sizes in 68 specimens were marked at a predetermined number of cycles that varied from 17 to 98% of the total fatigue life of the specimens. They became the known initial crack sizes for the calculation of the crack propagation life. Two methods of marking were employed. In one, the fillet weld toes were brushed with Dykem Steel Blue dye that was sucked into the crack. Judging by the sharply delineated front of the marked cracks, the dye dried rapidly. In the other method, the specimens were left outdoors for six months during which corrosion marked the crack size [4,5]. No significant difference was found in the fatigue lives of the specimens marked by these two methods.

After the specimens had failed by breaking into two parts, the marked crack sizes were measured with a travelling microscope. The crack depth (a_m) was taken as the distance along a normal line from the weld toe to the deepest point of the crack front. The crack length ($2c_m$) was measured between the two points where the crack front intersected the weld toe line. In corner cracks, the half-length (c_m) was taken from the intersection point with the weld toe line to the measurement point for the crack depth. For this reason, the product $2c_m$ exceeded in a few cases the specimen width. Cracks could not be found in those specimens identified with arrows in Fig. 8.

Fatigue Test Data

Crack Initiation and Propagation

The cracks initiated at one or more points along the weld toe line and propagated through the thickness of the main plate in a plane normal to the applied load. The specimens eventually failed when the net ligament ruptured in a ductile manner at an average net section stress about equal to the ultimate tensile strength.

Figure 3 shows the fatigue crack surface of two manually welded specimens. The first specimen (E233) was subjected to 962 000 cycles of 138-MPa (20-ksi) stress range. The three waves in the leading edge of the front suggest three crack initiation points from which individual cracks grew until they joined each other. This crack was weather-marked at 720 000 cycles. It grew from the marked size to failure in an additional 242 000 cycles. The second specimen (E132) was subjected to 3 254 000 cycles of 90-MPa (13-ksi) stress range. After 2 640 000 cycles, the largest crack was 3.86 mm (0.152 in.) deep and had the approximate shape of a quarter ellipse. It grew to failure after an additional 614 000 cycles. Also shown are two smaller cracks growing adjacent to the quarter-elliptical crack.

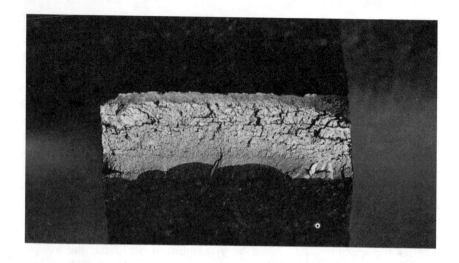

FIG. 3—*Fatigue crack surface in manually welded specimens E233* (top) *and E132* (bottom).

Automatically Welded Specimens

Table 1 summarizes the fatigue test data for the automatically welded specimens. Listed for each specimen are the designation, stress range (f_r), number of cycles at marking (N_m), fatigue life (N_f), and observed crack propagation life

$$N_0 = N_f - N_m \qquad (1)$$

The seven Series G specimens were subjected to constant-amplitude stress ranges of 176, 229, and 262 MPa (25.5, 33.2, and 38 ksi). The cracks in these specimens were weather-marked at $N_m/N_f = 62$ to 93% of their fatigue lives. The observed crack propagation lives varied from 10 000 to 270 000 cycles, depending on time of marking and stress range.

Listed next in Table 1 are the 31 Series C and D automatically welded specimens cycled under variable-amplitude loading. The load history corresponded to the mean of 106 individual stress-range histograms for truck traffic on 29 bridges in eight states [2,6]. In the Series C tests, the normalized stress ranges of the mean histogram between $0.5 \leq f_r/f_{r,max} \leq 1.0$ were retained, and the specimens were tested under the randomly ordered 10-block spectrum shown in Fig. 4. The number of cycles per spectrum were 100, 1000, and 10 000 for the specimens designated CX0X, CX1X, and CX2X, respectively. The full histogram, $0.25 \leq f_r/f_{r,max} \leq 1.0$, was used for the Series D specimens, which were tested under the randomly ordered 15-block spectrum of 1000 cycles shown in Fig. 5. In both Series C and D, the normalized stress ranges were factored by 207,304, and 386 MPa (30, 44, and 56 ksi). Adding to that the 3.5 MPa (0.5 ksi) minimum stress still left the maximum stress below the measured 425-MPa (61.7-ksi) yield strength of the 10-mm-thick (⅜ in.) main plate. The cracks were dye-marked at 27 to 98% of their fatigue lives. The observed crack propagation lives varied from 33 000 to 1 921 000 cycles.

Manually Welded Specimens

Table 2 summarizes the fatigue test data for the manually welded specimens. The 20 Series E specimens were cycled under constant-amplitude stress ranges of 90, 138, 207, and 290 MPa (13, 20, 30, and 42 ksi). The cracks in these specimens were marked at 24 to 97% of their fatigue lives. The observed crack propagation lives varied from 13 000 to 5 881 000 cycles. The 15 specimens with a number three in the second-to-last digit of the specimen designation were weather-marked. The others were dye-marked.

Also listed in Table 2 are the fatigue test data for the 10 Series D manually welded specimens which were stress cycled under the 15-block load spectrum shown in Fig. 5. The normalized stress ranges were factored by 152, 207, 303, and 386 MPa (22, 30, 44, and 56 ksi). The cracks in these specimens were dye-marked at 17 to 47% of their fatigue lives. The observed crack propagation lives varied from 406 000 to 7 005 000 cycles.

Analysis of Crack Propagation

Crack Propagation Model

The crack propagation lives were calculated with the Paris equation:

$$\frac{da}{dN} = C\Delta K^n \tag{2}$$

TABLE 1—Fatigue test data and analysis of automatically welded specimens.[a]

Specimen No.	Stress Range (f_r), MPa	Marked Crack Size a_m, mm	Marked Crack Size $2c_m$, mm	ΔK_{min}, MPa$\sqrt{m}$	No. of Cycles at Marking (N_m), kilocycles	Fatigue Life (N_f), kilocycles	Crack Propagation Life Observed ($N_0 = N_f - N_m$), kilocycles	Crack Propagation Life Calculated (N_c), kilocycles
Constant-Amplitude Loading								
G231	176	0.79	4.85	11.7	441	711	270	204
G232	176	2.69	11.3	16.3	441	586	145	96
G332	229	1.52	17.04	20.0	203	246	43	50
G334	229	1.07	14.0	18.2	203	265	62	63
G432	262	2.36	37.1	27.5	133	143	10	19
G433	262	0.69	tunnel	19.1	133	205	72	46
G434	262	2.11	20.3	25.1	133	157	24	24
Variable-Amplitude Loading								
C101	103–207	3.68	20.3	12.2	1 910	2 028	118	110
C102	103–207	0.15	tunnel	5.3	1 622	2 798	1 176	575
C103	103–207	1.27	9.96	8.6	1 520	1 782	262	333
C111	103–207	0.2	1.50	5.2	1 500	2 717	1 217	677
C112	103–207	...	...	...	1 079	2 528	1 449	...
C113	103–207	0.51	3.15	6.6	1 752	2 545	793	566
C122	103–207	0.28	3.35	6.0	2 000	3 921	1 921	558

C123	103–207	0.69	tunnel	7.9	2 000	2 676	676	370
C201	152–304	0.96	6.99	11.6	787	937	150	117
C202	152–304	1.65	14.5	13.9	390	504	114	74
C203	152–304	0.38	tunnel	9.9	290	485	195	137
C211	152–304	1.91	tunnel	15.4	400	445	45	53
C212	152–304	..	..	..	167	556	389	..
C213	152–304	0.38	5.59	9.6	300	530	230	148
C221	152–304	0.23	1.27	7.7	229	840	611	227
C222	152–304	..	..	..	301	849	548	..
C223	152–304	1.91	tunnel	15.5	501	560	59	53
C302	193–386	1.71	tunnel	17.0	150	183	33	33
C303	193–386	0.25	tunnel	21.8	101	242	141	69
C311	193–386	0.23	tunnel	11.0	150	239	89	71
C312	193–386	0.33	tunnel	12.1	150	247	97	63
C313	193–386	0.36	tunnel	12.4	139	278	139	62
C321	193–386	..	..	..	100	326	226	..
C322	193–386	0.33	tunnel	12.1	143	267	124	64
C323	193–386	0.38	tunnel	12.6	168	286	118	60
D111	52–207	5.54	18.9	9.9	16 551	16 912	361	106
D211	76–304	0.96	5.49	8.0	1 878	2 373	495	361
D212	76–304	3.1	11.3	10.6	2 527	2 760	233	136
D311	97–386	0.36	tunnel	8.8	370	592	222	172
D312	97–386	0.25	tunnel	8.1	461	794	333	191
D313	97–386	0.51	tunnel	9.7	501	790	289	150

^{a} 1 ksi $\sqrt{\text{in.}}$ = 1.0989 MPa $\sqrt{\text{m}}$; 1 in. = 25.4 mm.

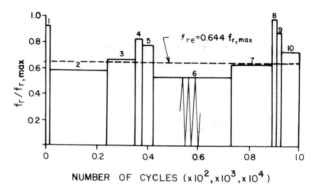

FIG. 4—*Normalized stress range histogram applied to Series C specimens.*

where

da/dN = crack growth rate,
ΔK = range of stress intensity factor, and
C and n = material constants.

Equation 2 was assumed to be log-log linear between the threshold value of the stress intensity range (ΔK_{th}) below which a crack would not propagate and the final crack size.

For constant-amplitude loading, f_r = constant, Eq 2 was solved for the number of cycles required to propagate the crack from the initial size (in this case the marked crack size, a_m) to the final size, a_f. Separation of the variables and integration yielded the calculated crack propagation life (N_c):

$$N_c = \frac{1}{C} \int_{a_m}^{a_f} \frac{1}{\Delta K^n} \, da \tag{3}$$

Equation 3 was solved by 32-point Gaussian quadrature.

For variable-amplitude loading, f_r not constant, Eq 2 was solved by the Runge-Kutta method for the crack increment (Δa) after each block of each spectrum. Summing the number of cycles in succeeding blocks, as the crack grew in increments (Δa) from a_m to a_f, yielded the calculated crack propagation life.

It is worth noting an alternative procedure of directly calculating the variable-amplitude fatigue life with Eq 3. It consists of replacing the cycles in a load spectrum by an equal number of equivalent constant-amplitude stress range cycles

$$f_{re} = (\sum_i \gamma_i f_{ri}^n)^{1/n} \tag{4}$$

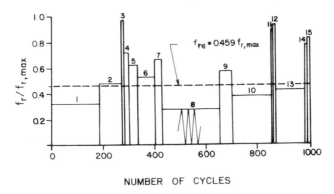

FIG. 5—*Normalized stress range histogram applied to Series D specimens.*

where γ_i and f_{ri} are the frequency of occurrence and the corresponding value of stress range for each block of the spectrum, and n is the exponent in Eq 3. The equivalent stress ranges were 64.4 and 45.9% of the maximum stress range, respectively, for the 10-block and 15-block spectra shown in Figs. 4 and 5. The explicit calculation of the variable-amplitude crack propagation life, using the equivalent stress range, gives the same result as the crack increment procedure provided that:

1. All values of ΔK, during crack growth from a_m to a_f, are larger than ΔK_{th}.

2. The number of cycles in a low stress range block, following a high stress range block, is small enough to preclude crack growth delay.

3. The number of load spectra applied during the crack propagation life is large enough so that, when the end of life is reached at some point in the spectrum, the cumulative $(1/n)$th power average of all cycles is not significantly different from the $(1/n)$th power average of the cycles in one spectrum.

The validity of the first condition will be examined later. It was shown experimentally that the second condition is satisfied for the block spectra of up to 10 000 cycles used in this study [2]. With regard to the third condition, Table 1 shows that 59 000/10 000 = 5.9 block spectra were applied during the crack propagation life of Specimen C223. At least twice as many load spectra were applied to all other specimens. Although Specimen C223 had seen a rather small number of block spectra during crack growth from a_m to a_f, termination of the test at the 9/10th fraction of the last spectrum reduced the equivalent stress range for all 59 000 cycles by only 0.5% of the value one would obtain for six complete block spectra.

To check the numerical accuracy of the calculations, the variable-amplitude fatigue lives of some specimens were calculated by the Runge-

TABLE 2—Fatigue test data and analysis of manually welded specimens. [a]

Specimen No.	Stress Range (f_r), MPa	Marked Crack Size		ΔK_{mm}, MPa√m	No. of Cycles at Marking (N_m), kilocycles	Fatigue Life (N_f), kilocycles	Crack Propagation Life	
		a_m, mm	$2c_m$, mm				Observed ($N_f - N_m$), kilocycles	Calculated (N_c), kilocycles
Constant-Amplitude Loading								
E131	90	0.35	19.8	5.4	2 640	8 521	5 881	1 570
E132	90	3.86	25.4	10.7	2 640	3 254	614	339
E134	90	0.31	9.14	5.2	2 640	8 266	5 626	1 668
E434	90	1.17	26.9	24.0	82	134	52	26
E142	90	0.94	4.08	5.9	2 000	3 475	1 475	1 860
E145	90	0.60	2.22	5.1	1 000	3 000	2 000	2 488
E212	138	0.25	6.45	9.7	536	993	457	367
E231	138	1.50	6.45	10.5	720	1 419	699	369
E232	138	1.76	25.1	12.7	720	930	210	212
E233	138	1.65	17.9	12.3	720	962	242	233
E234	138	1.30	10.8	11.6	720	878	158	196
E311	207	0.25	2.24	11.0	63	259	196	163
E312	207	0.56	15.1	14.3	122	315	193	107
E331	207	1.63	21.3	18.6	216	269	53	57
E332	207	1.26	27.4	17.5	216	319	103	74
E333	207	1.00	28.4	16.4	216	302	86	84
E334	207	3.07	tunnel	24.0	216	246	30	28
E431	290	0.84	tunnel	22.1	82	168	86	28
E432	290	2.49	28.2	29.7	82	95	13	14
E433	290	0.94	27.9	22.0	82	133	51	30
Variable-Amplitude Loading								
D221	38–152	0.63	2.72	2.5	3 254	9 863	6 609	4 890
D223	38–152	0.66	5.0	2.8	6 308	13 313	7 005	3 455
D321	52–207	0.48	1.73	3.1	817	4 104	3 287	2 270
D322	52–207	0.66	3.66	3.7	1 713	4 325	3 612	1 472
D323	52–207	0.23	15.6	2.8	1 037	4 308	3 271	1 925
D421	76–304	0.13	0.61	3.3	292	1 120	828	804
D422	76–304	...	...	...	292	1 361	1 069	...
D423	76–304	0.23	0.53	3.1	292	1 221	929	1 320
D522	97–386	0.51	2.65	6.3	140	803	663	223
D523	97–386	0.46	1.07	4.9	197	603	406	470

[a] 1 ksi √in. = 1.0989 MPa√m; 1 in. = 25.4 mm.

Kutta method and by direct integration using the equivalent stress range. The difference was found to be at most 0.1%. The latter method is well suited for structures, such as highway bridges, whose load histories do not normally exhibit load interaction effects.

In six specimens (D221, D223, D321, D323, D421, and D423) the ΔK values for some variable-amplitude cycles during the early stages of crack growth were smaller than ΔK_{th}. These cycles were counted in the calculation of the crack propagation life but were assumed to produce no crack extension. Eventually, as the lowest ΔK value exceeded ΔK_{th}, all cycles contributed to crack extension.

Crack Growth Rates

The crack growth equation used was

$$\frac{da}{dN} = 4.8 \times 10^{-12} \Delta K^3 \tag{5}$$

with ΔK in units of MPa$\sqrt{m}$ and da/dN in units of m/cycle. Equation 5 falls in the middle of a data band for four ferritic steels A36, ABS-C, A302-B, and A537-A above $\Delta K = 16.5$ MPa$\sqrt{m}$ (15 ksi $\sqrt{in.}$) [7]. It follows the mean of the data for A533 and A508 steels [8], and the mean of A36 steel data [9] down to ΔK values near the threshold.

Threshold values of ΔK reported in the literature were summarized in Refs 9 and 10. It is well known that ΔK_{th} decreases with increasing minimum-to-maximum stress ratio (R). For example, ΔK_{th} for A533 and A508 steels was found to vary from 8 MPa$\sqrt{m}$ (7.3 ksi $\sqrt{in.}$) at $R = 0.1$ to 3.0 MPa$\sqrt{m}$ (2.75 ksi $\sqrt{in.}$) at $R = 0.8$ [8].

The 3.5-MPa (0.5-ksi) minimum stress, to which all specimens in this study were subjected, suggests a small minimum R-ratio and consequently a high ΔK_{th}. In reality, residual tensile stresses near the weld toe elevate the R-ratio above nominal values, thus effectively decreasing the threshold. The reduction depends on the depth of the marked crack size, because the residual tensile stresses decay with increasing distance from the weld toe. Since a growing crack spends most of its propagation life while it is shallow, however, one must assume a large R-ratio effect. In this study it was assumed that $\Delta K_{th} = 3.5$ MPa$\sqrt{m}$ (3.2 ksi $\sqrt{in.}$).

Stress Intensity Factor

Stress intensity factors for ideal specimen configurations and different types of loading can be found in handbooks, but none applies to structural details typical of highway bridge construction. One must, therefore, resort to numerical solutions. The three-dimensional finite-element method, aided by special crack tip elements, for example, is quite powerful. But the need to repeat the calculations for many crack sizes and aspect ratios does not justify this costly approach. Furthermore, any potential gain in accuracy over more approximate methods would have to be weighed against unavoidable losses in accuracy from variations in weld profile and number of crack initiation sites along the weldment. These cannot be adequately modeled.

An alternative approach is used herein [11]. It consists of determining the stress distribution in a plane strain strip of the uncracked body and creating a stress-free crack surface by integrating away the normal stresses perpendicular to the line where the crack is to be inserted. The solution is modified to account for the effect of a part-through crack of semielliptical shape in a plate of finite thickness. The stress intensity factor at the deepest point of the crack front is given by

$$K = F_S F_W F_E F_G f \sqrt{\pi a} \qquad (6)$$

It consists of $K = f\sqrt{\pi a}$, the solution for the central crack in an infinite plate subjected to remote applied stresses (f), and the correction factors for the free surface:

$$F_S = 1.12 \qquad (7)$$

the finite thickness of the plate:

$$F_W = \sqrt{\frac{2t}{\pi a} \tan \frac{\pi a}{2t}} \qquad (8)$$

the elliptical shape of the crack front:

$$F_E = \frac{1}{E_k} \qquad (9)$$

with the complete elliptical integral of the second kind given by

$$E_k = \int_0^{\pi/2} \left[1 - \left(1 - \frac{a^2}{c^2} \sin^2 \theta \right) \right]^{1/2} d\theta \qquad (10)$$

where a/c is the semi-axis ratio. Finally, the local stress gradient caused by the stress concentration at the weldment is accounted for by the geometry correction factor [11]:

$$F_G = \frac{2}{\pi} \sum_i \frac{f_{bi}}{f} \left(\text{arc sin} \frac{b_{i+1}}{a} - \text{arc sin} \frac{b_i}{a} \right) \qquad (11)$$

where f_{bi} is the piece-wise constant stress along the path of the crack, between b_i and b_{i+1}. The factor F_G represents the ratio of the stress intensity factor for a nonuniform stress distribution along the line of the crack to the stress intensity factor for an uniform stress applied remotely. It conveniently models the stress concentration produced by a structural detail. The stress distribution in the plane through the weld toe (the crack path) was obtained by a finite-element analysis of the uncracked transverse stiffener specimen shown in Fig. 1. Only one quarter of the double symmetrical specimen was modeled. The size of the smallest finite element near the weld toe was $b_2 - b_1 = 0.05$ mm (0.002 in.). The stress distribution along the line of the crack was calculated by linearly extrapolating the stresses from the first two elements on either side and averaging the values.

The stress intensity factor varies along the front of an elliptical crack in accordance with the factor

$$F = \left[\sin^2 \beta + \frac{a^2}{c^2} \cos^2 \beta \right]^{1/4} \tag{12}$$

in which the parametric angle β is measured from the major axis of the ellipse. This factor increases from the point where the crack front intersects the free surface at the weld toe to the deepest point of the crack front. Conversely, F_G decreases as one moves in the same direction along the crack front. In this study, it was assumed that the product of the two factors remains constant. Consequently, the aspect (semi-axis) ratio of the elliptical crack would also remain constant during the entire life. Therefore the crack propagation needs to be computed at the deepest point of the crack front alone, utilizing K given by Eq 6.

Crack Sizes

The depth (a_m) and the surface length ($2c_m$) of all marked crack sizes are listed in Tables 1 and 2 for the automatically and manually welded specimens, respectively. The corresponding aspect ratios (a_m/c_m) are plotted in Figs. 6 and 7. The 45-deg reference line, at $a/c = 1$, corresponds to a circular crack front. For each type of welding, constant and variable amplitude loading generated cracks of comparable aspect ratios, as one would expect.

In general, manual welds are more prone to defects and weld bead irregularities which act as starting points for dominant single cracks. Automatic welds, on the other hand, have more regular weld contours which enhance multiple crack initiation and crack joining. Indeed, observed tunnel cracks were common in the automatically welded specimens, regardless of crack depth. In contrast, only a few manually welded specimens developed tunnel cracks, and then only when $a_m > 0.8$ mm. Typical weld contours for both types of welding are shown in Fig. 2.

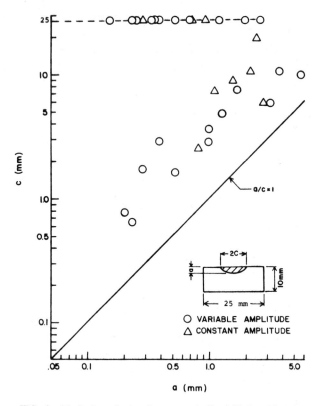

FIG. 6—*Marked crack sizes in automatically welded specimens.*

Figure 8 shows the marked crack depth (a_m) in percent of the fatigue life. The earliest times were 27% for automatically welded specimen C221 and 17% for manually welded specimen D522. When expressed in percent of the fatigue life, the automatically welded specimens were marked later than the manually welded specimens because they had a longer crack initiation life and a longer fatigue life. In five specimens, which were dyed between 21% (D422) and 43% (C112), no cracks were visible on the fracture surface after the specimens had failed. The smallest crack sizes marked were $a_m = 0.15$ mm (0.006 in.) for automatically welded specimen C102 and $a_m = 0.13$ mm (0.005 in.) for manually welded specimen D421.

An analysis of the fracture surface indicated that the specimens failed when the average stress on the net ligament reached the ultimate tensile strength. Assuming, for simplicity, a straight crack front over the entire specimen width, the net ligament at failure is given by

$$A_{net} = \frac{P_{max}}{F_u} = (t - a_f)w \qquad (13)$$

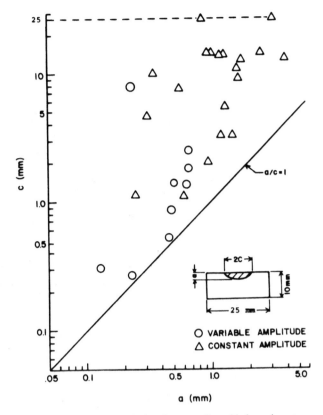

FIG. 7—*Marked crack sizes in manually welded specimens.*

where

P_{max} = maximum applied load,
F_u = ultimate tensile strength,
t = 10 mm (0.39 in.), and
w = 25 mm (1 in.).

The final crack size for each specimen is therefore given by

$$a_f = t - \frac{P_{max}}{wF_u} \qquad (14)$$

Calculations of fatigue life are not sensitive to final crack size. Hence, assuming a straight crack front versus the actual crack front has no significant effect on the result.

Summary of Results

The results for automatically welded specimens are shown in Fig. 9. The

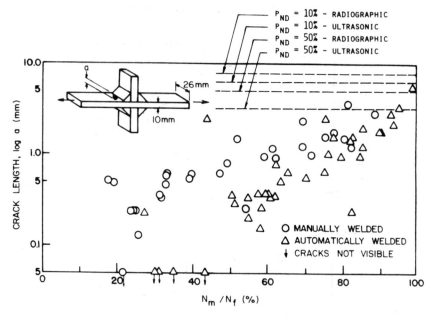

FIG. 8—*Crack depth at time of marking in precent of fatigue life.*

observed crack propagation lives of those specimens varied from 10 000 cycles (G432) to 1 921 000 cycles (C122), depending on time of marking and stress range. The results for manually welded specimens are shown in Fig. 10. The observed crack propagation lives varied in the latter case from 13 000 cycles (E432) to 7 005 000 cycles (D223).

For both types of welding, most of the calculated lives fell within a factor of two of the observed lives, as shown by the dashed lines in Figs. 9 and 10. The average ratios of observed-to-calculated fatigue life were 1.09 for automatic welds/constant-amplitude loading, 1.70 for automatic welds/variable-amplitude loading, 1.57 for manual welds/constant-amplitude loading, and 1.56 for manual welds/variable-amplitude loading.

In general, more cycles were needed to grow the crack than were calculated in this study. The predictions were conservative in the sense that they underestimated the observed lives. The correlation tended to improve the larger the marked crack size. This finding supports the belief held by some investigators that calculations of crack propagation life are less reliable for shallow cracks than deep cracks.

For comparison, the fatigue lives of Specimens D221, D223, D321, D323, D421, and D423 were also analyzed assuming that ΔK has no threshold value. The lives predicted in this manner were shorter than those based on the assumption made herein that a threshold of 3.5 MPa$\sqrt{m}$ exists by amounts varying from 5% (D221) to 0.6% (D423).

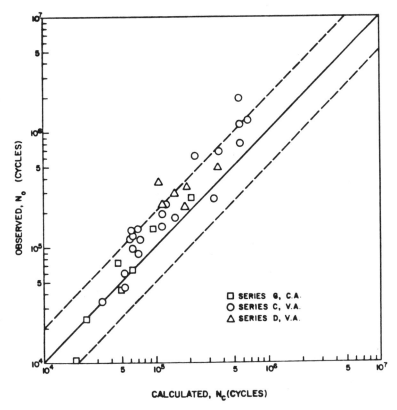

FIG. 9—*Comparison between observed and computed crack propagation lives of automatically welded specimens.*

Application to Design and Inspection

A method was presented for predicting the part-through crack propagation life of a non-load-carrying fillet weld detail that was subjected to constant-amplitude loading and to variable-amplitude service loading. This type of detail is found in highway bridges and crane girders, for example, in which stiffeners welded to the web and flange of plate girders resist web buckling.

Fatigue cracks at stiffeners welded to girder webs grow in three stages: (1) part-through crack in web, (2) two-ended through crack in web, and (3) three-ended crack in web and flange [12]. These three stages of growth consume, on average, 80, 15, and 5% of the fatigue life, respectively. They are illustrated in Fig. 11 for a 965-mm (38-in.)-deep girder subjected to 4 433 000 cycles of 106 MPa (15.4 ksi) nominal bending stress range in the web at the end of the stiffener. Figure 12 shows one beam elevation with a crack growing from the stiffener end and three cross sections of the web with various part-through cracks.

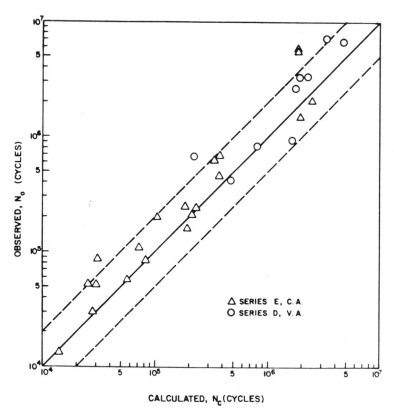

FIG. 10—*Comparison between observed and computed crack propagation life of manually welded specimens.*

The bending stress range in the web at the end of the stiffener and the number of applied cycles are listed in the figure caption. Note in Fig. 12*b*, for example, the 0.89-mm (0.035-in.) deep crack which initiated from a slag inclusion at the weld toe.

Fatigue cracks at stiffeners welded to girder web and flanges grow in two stages: (1) part-through crack in flange, and (2) two-ended through crack in flange [*12*]. The two stages consume about 95% and 5% of the fatigue life, respectively. They are illustrated in Fig. 13 for a 965-mm (38-in.)-deep girder subjected to 2 012 000 cycles of 95 MPa (13.8 ksi) nominal bending stress range in the flange at the weld toe. Figure 14 shows a fatigue crack at failure. Figure 15 illustrates the stage of part-through crack growth from the toe of the stiffener-to-flange weld, with (1) a single 0.71-mm (0.028-in.)-deep crack, and (2) multiple crack growth.

The fatigue life of the specimens examined in the present study is about the same as the number of cycles needed to grow part-through cracks at stiffeners

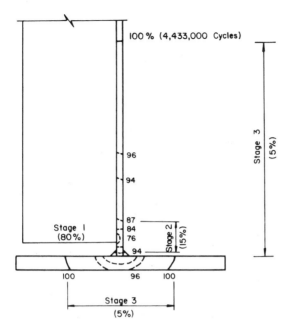

FIG. 11—*Stages of crack growth at stiffener welded to web* [12].

welded to the web or flanges or both. As Figs. 11 and 13 show, that stage consumes most of the fatigue life. It can be said, therefore, that the cruciform specimens of Fig. 1 model well the transverse stiffeners in bridge girders.

The results of this study have important applications to bridge inspection. Figure 16 shows the probability of not detecting part-through cracks with ultrasonic and radiographic inspection methods [13]. Assuming that a 90% crack detection probability is desired for the purpose of determining if bridges could remain in service, only cracks deeper than the following would be found: 6.3 mm (0.25 in.) for ultrasonic inspection and 7.8 mm (0.31 in.) for radiographic inspection. Comparing these values with the crack sizes marked in the present study (see Fig. 8) indicates that the part-through crack growth stage is practically over by the time cracks could be found with a 90% reliability. Note that even if a 50% crack detection probability were acceptable, over 80% of the part-through crack growth stage would have elapsed.

Reliable crack detection requires through cracks which have typically a larger opening than part-through cracks. They would break the paint and form a rusted line trace on the paint film. But the stage of through crack growth is, in general, only 20% and 5% of the total number of cycles to failure of web and flange weldments, respectively. Hence, to have a resonable chance of finding cracks before the moment carrying capacity of a girder is lost, the inspection intervals towards the end of the 50-year bridge design life should

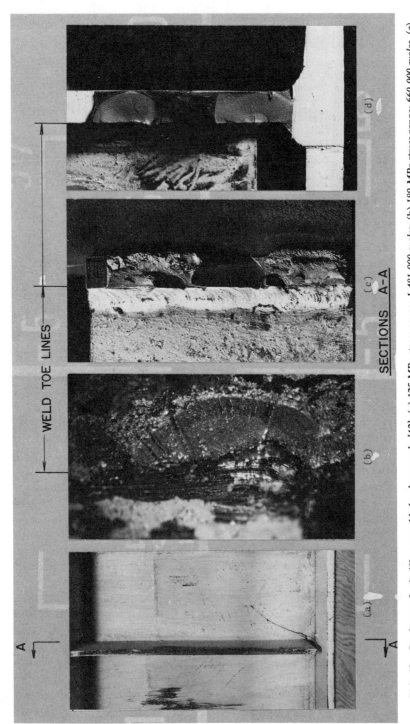

FIG. 12—*Crack growth of stiffeners welded to beam webs [12]. (a) 125-MPa stress range; 1 481 000 cycles. (b) 189-MPa stress range; 669 000 cycles. (c) 164-MPa stress range; 1 017 000 cycles. (d) 157-MPa stress range; 1 119 000 cycles.*

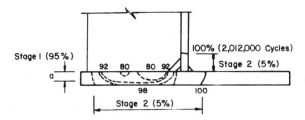

FIG. 13—*Stages of crack growth at stiffener welded to flange* [12].

not exceed 20% of 50 years = 10 years for web weldments or 5% of 50 years = 2.5 years for flange weldments. Clearly, such short inspection intervals would make fatigue rating of the over 600 000 bridges in the nation very expensive. Considering that, on average, every year some 12 000 highway bridges in the nation reach their intended service life, it is not economically feasible to inspect them at short intervals.

The currently low crack detection probabilities, the difficulty of developing more reliable field methods of crack detection, the large number of bridges

FIG. 14—*Crack growth at stiffener welded to flange* [12].

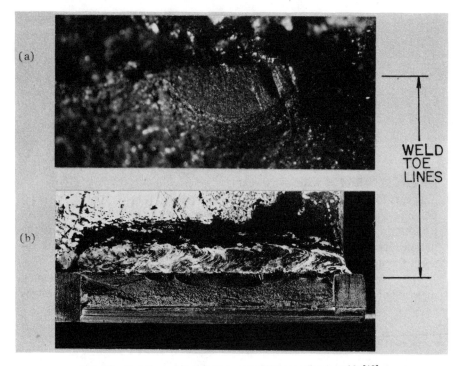

FIG. 15—*Part-through cracks at toe of stiffener-to-flange welds* [12].

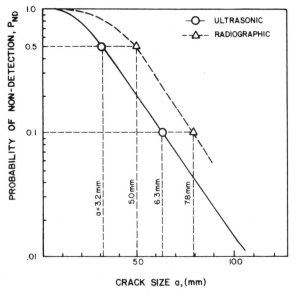

FIG. 16—*Probability of not detecting a crack by ultrasonic and radiographic inspection as a function of crack depth (adapted from Ref 13).*

reaching their service life every year, and the high cost of inspection, repair, and traffic rerouting are all compelling reasons for good detailing and for designing bridges with high safety factors against fatigue. Any bridge on a primary system, which usually has an average daily truck traffic of 2500 or more (2500 × 365 days × 50 years = 45 625 000 load cycles), should be fatigue proof. The benefits would far exceed the cost, if any, of improving the details and the cost of the additional steel needed to lower the stress range below the fatigue limit.

References

[1] "Standard Specifications for Highway Bridges," 12th ed., American Association of State Highway and Transportation Officials, Washington, D.C., 1977.
[2] Albrecht, P. and Yamada, K. in *Service Fatigue Loads Monitoring, Simulation, and Analysis, ASTM STP 671*, American Society for Testing and Materials, 1979, pp. 255-277.
[3] Albrecht, P. and Friedland, I. M., *Journal of the Structural Division*, American Society of Civil Engineers, Vol. 105, No. ST12, Dec. 1979, pp. 2657-2675.
[4] Albrecht, P., "Fatigue Behavior of Weathered Steel Bridge Components," Report FHWA/MD-81/02, University of Maryland, College Park, Dec. 1982.
[5] Albrecht, P. and Cheng, J. G., *Journal of Structural Engineering*, Vol. 109, No. 9, Sept. 1983, pp. 2048-2065.
[6] Yamada, K. and Albrecht, P. A., "A Collection of Live Load Stress Histograms of U.S. Highway Bridges," Civil Engineering Report, University of Maryland, College Park, 1975.
[7] Barsom, J. M., "Fatigue-Crack Propagation in Steel of Various Yield Strength," U.S. Steel Corp., Applied Research Laboratory, Monroeville, Pa., 1971.
[8] Paris, P. C., Bucci, R. J., Wessel, E. T., Clark, W. G., and Mager, T. R. in *Stress Analysis and Growth of Cracks, Proceedings of the 1971 National Symposium on Fracture Mechanics, Part I, ASTM STP 513*, American Society for Testing and Materials, 1972, pp. 141-176.
[9] Klingerman, D. J. and Fisher, J. W., "Threshold Crack Growth in A36 Steel," Fritz Engineering Report 386.2, Lehigh University, Bethlehem, Pa., 1971.
[10] Sahli, A., "Calculation of Fatigue Life of Transverse Stiffeners with Known Initial Crack Sizes," Master of Science thesis, Department of Civil Engineering, University of Maryland, College Park, Dec. 1981.
[11] Albrecht, P. and Yamada, K., *Journal of the Structural Division*, American Society of Civil Engineers, Vol. 103, No. ST2, Feb. 1977, pp. 377-389.
[12] Albrecht, P., "Fatigue Strength of Welded Beams with Stiffeners," Ph.D. dissertation, Lehigh University, Bethlehem, Pa., June 1972.
[13] Harris, D. O., "An Analysis of the Probability of Pipe Rupture at Various Locations in the Primary Cooling Loop at a Babcock and Wilcox 177 Fuel Assembly Pressurized Water Reactor—Including the Effects of a Periodic Inspection," Report SAI.050.77-PA, Science Applications, La Jolla, Calif., Sept. 1977.

Brian J. Schwartz,[1] Nancy G. Engsberg,[1] and Dale A. Wilson[1]

Development of a Fatigue Crack Propagation Model of Incoloy 901

REFERENCE: Schwartz, B. J., Engsberg, N. G., and Wilson, D. A., **"Development of a Fatigue Crack Propagation Model of Incoloy 901,"** *Fracture Mechanics: Fifteenth Symposium, ASTM STP 833*, R. J. Sanford, Ed., American Society for Testing and Materials, Philadelphia, 1984, pp. 218–241.

ABSTRACT: An interpolative crack propagation model was developed for Incoloy 901 using a hyperbolic sine function. The equation

$$\log(da/dN) = C_1 \sinh[C_2(\log(\Delta K) + C_3)] + C_4$$

is used to describe crack growth rates at various operating conditions. Crack propagation rate data for varying stress ratios were examined over a broad range of temperatures and frequencies with this model. Stress ratios (R) investigated included both positive and negative values ranging from -1.0 to a maximum of 0.7. The influence of increasing temperature on crack growth was examined over a temperature range of 260 to 620°C (500 to 1150°F). Frequencies from 0.00833 to 30 Hz were investigated to simulate conditions experienced by a gas turbine engine disk of this material. The resulting model can be used as an effective tool to predict residual life of a gas turbine engine disk, because growth rates can be obtained even where data are unavailable.

KEY WORDS: crack propagation, fracture mechanics, Incoloy 901, modeling, fatigue, elevated temperature

Describing crack growth rates of materials at high temperatures is a complex problem. Changing operating conditions (for example, frequency) will alter the material behavior. To make accurate life predictions, an extensive data base that characterizes the behavior at all operating conditions is required. An interpolative model, used to describe crack growth behavior at conditions between test result conditions, is a useful tool for predicting life when available test results are limited. Furthermore, with careful planning of a test matrix for characterizing material behavior, very accurate predictions can be made for a wide range of operating conditions.

Pratt & Whitney Aircraft has an empirical model that describes crack

---

[1]Pratt & Whitney Aircraft, West Palm Beach, Fla. 33402.

growth rates at all conditions within the bounds of the data [1-4]. A form of the hyperbolic sine (see Eq 1) is used for this purpose. A curve fit is calculated using the least (minimum) summed squared error between calculated and observed values for the dependent variable. Each hyperbolic sine curve representation of fatigue crack propagation is related to the others, the relationship depending on differences in frequency, stress ratio (minimum/maximum load), and temperature. The finished model can then be interpolated to describe crack growth rates at any temperature, stress ratio, or frequency within the bounds of the data base.

This model was applied to Incoloy 901, and an interpolative model was developed for positive and negative stress ratios as well as frequencies at various temperatures. The effects of these changing operating conditions were investigated and reported. Predictions were then made and compared with actual cyclic lives of laboratory test specimens.

Experimental Program

Test Specimens and Procedures

ASTM Test for Constant-Load-Amplitude Fatigue Crack Growth Rates Above 10^{-8}m/Cycle (E 647) through-thickness center crack tension specimens (CCT) and compact type specimens (CT) were used to obtain crack propagation data. The material composition and mechanical properties of Incoloy 901 are listed in Tables 1 and 2 respectively. Testing was conducted on servohydraulic, closed-loop, load-controlled testing machines. Specimens were precracked using procedures outlined in ASTM E 647. Precracking was performed at room temperature at a cyclic frequency of 10 or 20 Hz; anomalous precracking effects were easily recognizable. The crack propagation tests were conducted with a triangular loading waveform.

Specimen heating was provided by resistance clamshell furnaces having windows to allow observation of crack growth at the test temperature. Crack lengths were measured on both surfaces of the propagation specimen using a traveling microscope. This was facilitated by interrupting the cyclic loading and applying the mean test load. This procedure held the specimen rigid while increasing crack tip visibility. A high-intensity light was used to provide oblique illumination to the crack and further increase crack visibility. In general, crack length measurements were taken at increments no larger than 0.50 mm (0.020 in.). Crack length measurements are considered to be accu-

TABLE 1—*Composition (weight percent) of Incoloy 901.*

Cr	Ni & Co	Mo	Ti	B	C	Fe
11.00	40.00	5.00	2.60	0.01	0.02	remainder

TABLE 2—*Mechanical properties of Incoloy 901.*

Temperature					
°C	20	260	427	538	621
°F	70	500	800	1000	1150
Tensile strength					
MPa	1210	1105	1080	1036	1008
psi	175	160	156	150	146
Yield strength					
MPa	900	795	760	753	746
psi	130	115	110	109	108
Elongation, %	17	17	18	19	17
Reduction of Area, %	20	20	23	26	23

rate to within ±0.025 mm (0.001 in.). Table 3 lists all crack propagation tests used in the model development.

Mathematical Model Development

Model Description—The interpolative hyperbolic sine (SINH) model is in the form of computer software capable of describing crack propagation at various stress ratios, temperatures, and frequencies representative of turbine disk operation. The model is based on the hyperbolic sine equation

$$\log(da/dN) = C_1 \sinh[C_2(\log(\Delta K) + C_3)] + C_4 \tag{1}$$

where the coefficients have been shown [1-4] to be functions of test frequency (ν), stress ratio (R), and temperature (T); that is, C_1 = material constant, $C_2 = f_2(\nu, R, T)$, $C_3 = f_3(\nu, R, T)$, and $C_4 = f_4(\nu, R, T)$.

The hyperbolic sine model is easily adapted to describe fundamental parametric effects of operating conditions such as stress ratio, frequency, and temperature on crack growth rate.

Figure 1 schematically depicts the qualitative effects on crack propagation rate of frequency (Fig. 1*a*), stress ratio (Fig. 1*b*), and temperature (Fig. 1*c*). Because of the simple relationships observed between the coefficients of the SINH model and the fundamental propagation controlling parameters, interpolations are straightforward. It is here that the model demonstrates its great usefulness: the SINH model provides descriptions of crack propagation characteristics where data are unavailable.

Advanced Regression Considerations—Interpolative modeling of crack propagation as a function of operating parameters, such as frequency, stress ratio, and temperature (Fig. 1), requires a multiple regression capability which allows simultaneous consideration of several different collections of data, each differing from the others by only one fatigue crack propagation (FCP) controlling parameter. Separate regression of the individual data sets contributing to an interpolative model does not allow data at one condition to

TABLE 3—*Crack propagation test specimens for Incoloy 901.*

Specimen No.	Temperature °C	Temperature °F	Cyclic Frequency	Stress Ratio	Thickness mm	Thickness in.
1178	538	1000	0.17 Hz (10 cpm)	0.05	7.62	0.300
1180	538	1000	120-s dwell	0.05	7.67	0.302
1181	538	1000	20 Hz	0.05	7.62	0.300
1191	538	1000	30 Hz	0.5	6.53	0.257
1192	427	800	20 Hz	0.05	7.65	0.301
1193	427	800	30 Hz	0.5	7.80	0.307
1194	538	1000	30 Hz	0.05	7.70	0.303
1195	538	1000	30 Hz	0.3	7.70	0.303
1227	427	800	0.0083 Hz (0.5 cpm)	0.05	7.65	0.301
1228	427	800	0.17 Hz (10 cpm)	0.05	7.59	0.299
1229	427	800	0.17 Hz (10 cpm)	0.05	7.65	0.301
1230	427	800	0.17 Hz (10 cpm)	0.5	7.29	0.287
1412	538	1000	0.17 Hz (10 cpm)	−1.0	7.59	0.299
1421	538	1000	0.17 Hz (10 cpm)	0.05	12.73	0.501
1422	538	1000	0.17 Hz (10 cpm)	0.3	12.83	0.505
1424	621	1150	20 Hz	0.5	12.78	0.503
1538	621	1150	120-s dwell	0.05	7.59	0.299
1539	538	1000	30 Hz	0.7	7.65	0.301
1540[a]	538	1000	0.17 Hz (10 cpm)	−1.0	7.57	0.298
1543	621	1150	20 Hz	0.7	7.42	0.292
1544	621	1150	0.0083 Hz (0.5 cpm)	0.05	7.59	0.299
1571	621	1150	0.17 Hz (10 cpm)	0.05	7.54	0.297
1587[b]	260	500	0.17 Hz (10 cpm)	0.05	7.62	0.300
1597	260	500	0.0083 Hz (0.5 cpm)	0.05	12.70	0.500
1600	260	500	20 Hz	0.05	6.35	0.250
1601	260	500	0.17 Hz (10 cpm)	0.05	6.38	0.251
1222	260	500	0.17 Hz (10 cpm)	0.05	7.59	0.299
0463	538	1000	0.17 Hz (10 cpm)	0.5	6.32	0.249
0462	538	1000	0.17 Hz (10 cpm)	0.05	6.30	0.248

[a]Uneven crack front.
[b]Prestrain.

influence the model at another condition. However, the final model was to have broad interpolative capability, with behavior at one condition used to describe FCP at another condition. Therefore it was desirable to permit the data to exhibit their mutual influence during the modeling process.

Pratt & Whitney Aircraft has developed a mathematical technique to accomplish this. Individual sets of data are treated independently relative to some of the SINH coefficients, while the entire collection is treated as an entity with respect to the interpolative coefficients (functions of v, R, T, etc.). This improved method was used to develop the propagation models presented in this report.

Interpolation Algorithm—The fundamental strength of the hyperbolic sine model is its interpolative capacity. The procedure known as the interpolation algorithm for calculating the SINH coefficients, describing FCP under repre-

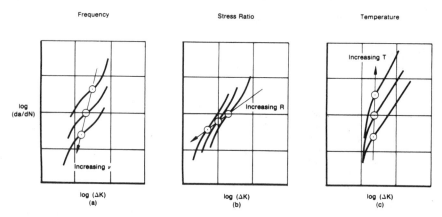

FIG. 1—*Crack propagation as influenced by* (a) *frequency,* (b) *stress ratio, and* (c) *temperature.*

sentative engine operating conditions, is illustrated in the following paragraphs.

The coefficients (for example, C_2, C_3, and C_4) at any intermediate value of an element life controlling parameter can be determined from

$$C_j = C_{j\,\text{base}} + \Delta C_j, \quad j = 2, 3, 4 \tag{2}$$

where

$$C_j = \begin{bmatrix} C_2 \\ C_3 \\ C^4 \end{bmatrix} = \text{interpolated value of coefficients}$$

and

$$\Delta C_j = \begin{bmatrix} \Delta C_2 \\ \Delta C_3 \\ \Delta C_4 \end{bmatrix} = \text{differences from baseline values}$$

Assuming the SINH coefficients are linear functions of the controlling parameters, it is evident that

$$\begin{matrix} \begin{bmatrix} \Delta C_2 \\ \Delta C_3 \\ \Delta C_4 \end{bmatrix} \\ N \times 1 \end{matrix} = \begin{matrix} \begin{bmatrix} \partial C_2/\partial \nu, & \partial C_2/\partial R, & \partial C_2/\partial T \\ \partial C_3/\partial \nu, & \partial C_3/\partial R, & \partial C_3/\partial T \\ \partial C_4/\partial \nu, & \partial C_4/\partial R, & \partial C_4/\partial T \end{bmatrix} \\ N \times N \end{matrix} \times \begin{matrix} \begin{bmatrix} \Delta \nu \\ \Delta R \\ \Delta T \end{bmatrix} \\ N \times 1 \end{matrix}$$

where

$$\begin{bmatrix} \Delta \nu \\ \Delta R \\ \Delta T \end{bmatrix} = \text{differences from baseline values}$$

and the $N \times N$ partial derivative matrix is easily determined from the slopes of the lines relating each coefficient with each rate controlling parameter. The computation of the intermediate coefficients using Eq 2 is then straightforward. Actually, the coefficients are nonlinear functions of ν, R, and T; however, they are linear functions of other functions [for example, $\log(1 - R)$] and therefore the foregoing argument applies.

The model is now in a form which is influenced by all operating conditions available, making it possible to predict crack growth rates anywhere within the bounds of 260 to 620°C (500 to 1150°F), frequencies of 0.00833 to 30 Hz, and stress ratios from -1.0 to 0.7.

Results and Conclusions

Interpolative crack propagation models of Incoloy 901 developed from the data generated under Navy contract (J-52 Product Support N00019-79-C-0368) are presented in the following sections. Each of the individual models describes an influence of a single test parameter (stress ratio, frequency, and temperature) on crack propagation. The combination of these models forms a unified description of the crack growth behavior over the range of operating conditions tested.

All crack length versus cycles (a,N) data were regressed using the seven-point incremental polynomial technique (Ref 5) to produce crack growth rate versus applied stress intensity range $(da/dN, \Delta K)$ data.

Effect of Loading Rate (Frequency)

The influences of loading rate on crack propagation in Incoloy 901 at 260°C (500°F) and 538°C (1000°F), $R = 0.05$, are illustrated in Figs. 2 and 5 and the conditions are shown in Table 4. The data were gathered at frequencies ranging from 0.0083 Hz (0.5 cpm) to 30 Hz and show no appreciable difference in crack growth.

The effect of frequency at a higher temperature and stress ratio [538°C (1000°F), $R = 0.5$] displays a much greater influence (Fig. 3). There are two possible explanations for this phenomenon. The increase of temperature allows oxidation at the crack tip as well as increasing the creep effects. The greater stress ratio ($R = 0.5$) keeps the crack opened for a longer percentage of time. At lower frequencies (10 cpm) the crack will be open for a long time

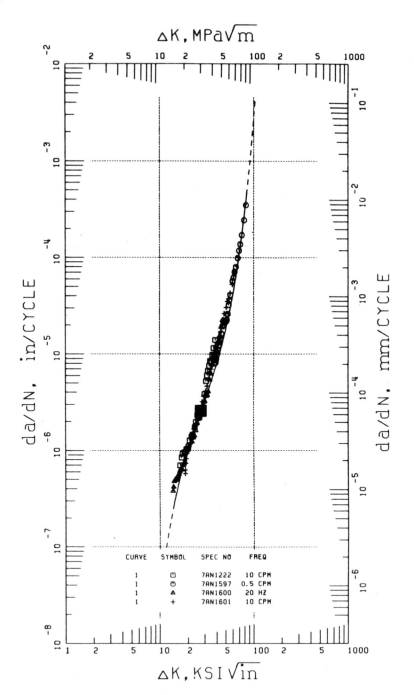

FIG. 2—*Frequency effect of Incoloy 901 (PWA 1003) on crack propagation rate, R = 0.05, 260°C (500°F).*

compared with higher frequencies (30 Hz); this will allow the oxidation and creep effects to become more damaging.

A plot of the correlative parameters is presented in Fig. 4. Note the coefficients are linear functions of frequencies; that is, $C = A + B \log$ (frequency). The model can be interpolated by using the equations in Fig. 4 to describe crack propagation at any frequency within the limits of the tested area.

Effect of Positive Stress Ratio

Crack propagation at varying positive stress ratios is displayed in Fig. 5. Tests were run at 538°C (1000°F), 0.1667 Hz (10 cpm), and stress ratios of R = 0.3 and 0.5. Data displayed at $R = 0.05$ were tested at various frequencies (as shown in Fig. 5, frequency has no effect at these conditions). A plot of the correlative parameters is shown in Fig. 6.

Effect of Negative Stress Ratio

An interpolative model of the effect of stress ratio was developed from data over the range of $-1.0 \leq R \leq 0.0$. The cyclic frequency was 0.1667 Hz (10 cpm) and the test temperature was 538°C (1000°F). The base condition of $R = 0.0$ was extrapolated from the positive stress ratio model; the resulting model is shown in Fig. 7. A plot of the correlative parameters with the corresponding equations is shown in Fig. 8.

Note that ΔK is defined as $K_{max} - K_{min}$ for $R > 0$ and as $K_{max} - 0$ for $R \leq 0$.

Effect of Temperature

The effect of operating temperature on crack propagation is illustrated in Fig. 9 and described in Table 5. As expected, the crack growth rate increases as temperature increases. The correlative parameter equations and plots are shown in Fig. 10. The relationship between the coefficients did not appear to be linear functions for the full temperature range; therefore a model which is only linear for each segment (Figs. 9 and 10) was developed.

Prior Plastic Deformation

Component crack propagation life predictions are made using da/dN data obtained from virgin (unstrained) material. However, cracks occurring in real hardware often initiate in regions of high stress concentration such as boltholes or blade attachment areas. Material in these regions may be subjected to many strain cycles in excess of 1.0%. It is assumed that FCP behavior (da/dN versus ΔK) in this material is essentially the same as that for virgin material and that data generated on virgin specimens can be used to describe

TABLE 4—*Operating conditions for Incoloy 901 (PWA 1003) crack propagation, effect of frequency.*

HYPERBOLIC SINE MODEL COEFFICIENTS

$Y = C1 \times SINH(C2 \times (X + C3)) + C4$

WHERE $Y = LOG (da/dN)$ AND $X = LOG (\Delta K)$

CURVE	SYMBOL	C1	C2	C3	C4	K RANGE	R²	SEE
1	▣ ◉ ▲ +	0.7000	3.8840	-1.4400	-5.5890	(13.77, 83.02)	0.9841	0.1194

R² = Coefficient of Determination
SEE = Standard Error of the Estimate

METRIC CONVERSIONS

C1 METRIC = C1 ENGLISH   C2 METRIC = C2 ENGLISH
C3 METRIC = C3 ENGLISH − 0.040935   C4 METRIC = C4 ENGLISH + 1.40483

Curve	Symbol	Spec No.	Temperature °C	(°F)	Freq	R	Type	Thickness mm	(in.)
1	□	7AN1222	260	(500)	10 cpm	0.05	CCT	7.59	(0.299)
1	⊖	7AN1597	260	(500)	0.5 cpm	0.05	CT	12.70	(0.500)
1	◄	7AN1600	260	(500)	20 Hz	0.05	CT	6.35	(0.250)
1	+	7AN1601	260	(500)	10 cpm	0.05	CT	6.38	(0.251)

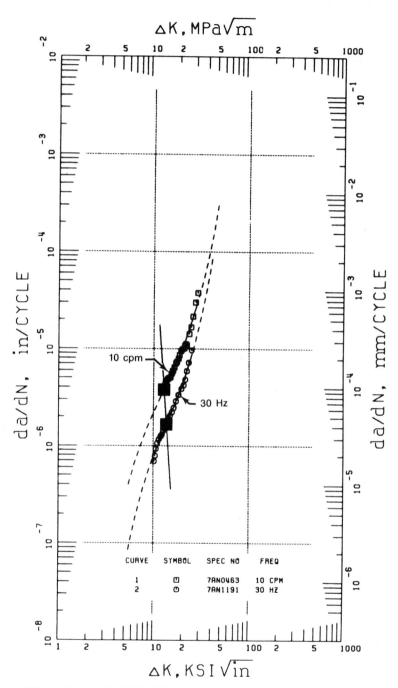

FIG. 3—*Incoloy 901 (PWA 1003) frequency model, R = 0.5, 538°C (1000°F).*

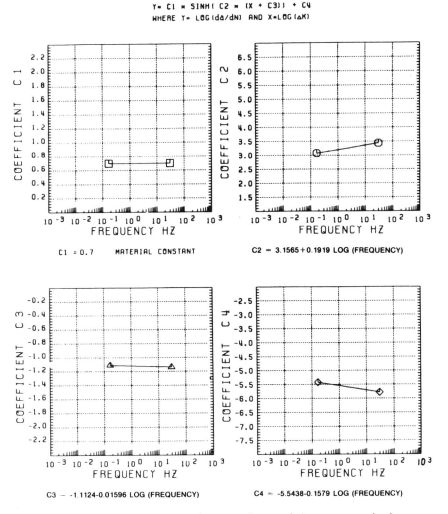

FIG. 4—*Incoloy 901 (PWA 1003) crack propagation correlative parameters for frequency model, R = 0.5, 538°C (1000°F).*

FCP in prestrained material. The purpose of this auxiliary investigation was to test the above hypothesis using Incoloy 901 (PWA 1003).

In order to simulate the cyclic loading experienced in the bore of a J-52 Incoloy 901 turbine disk, 100 strain cycles having a $\epsilon_{max} = 0.017$ and $\epsilon_{min} = 0.0088$ were applied to an unnotched center crack tension specimen at 0.1667 Hz (10 cpm) and room temperature.

The unnotched specimens were removed from the test machine, and a through-thickness center flaw was cut using electrical discharge machining

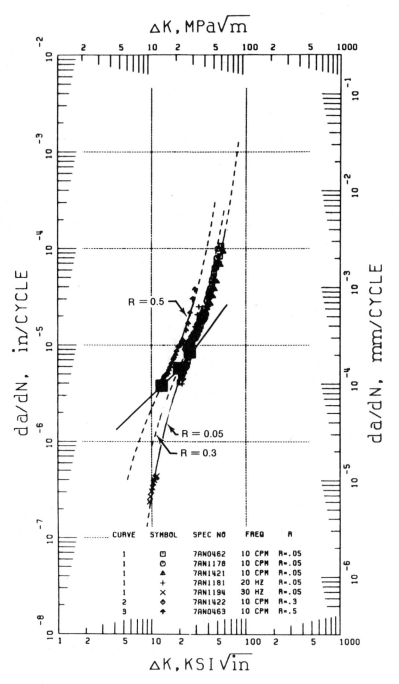

FIG. 5—*Incoloy 901 (PWA 1003) stress ratio model, varying frequencies, 538°C (1000°F).*

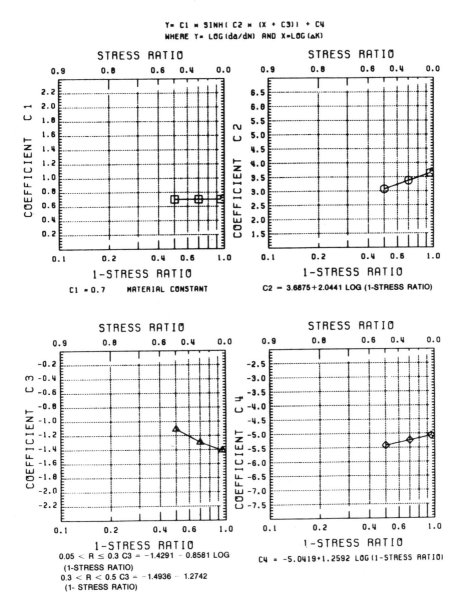

FIG. 6—*Incoloy 901 (PWA 1003) crack propagation, correlative parameters, positive stress ratio model, varying frequencies, 538°C (1000°F).*

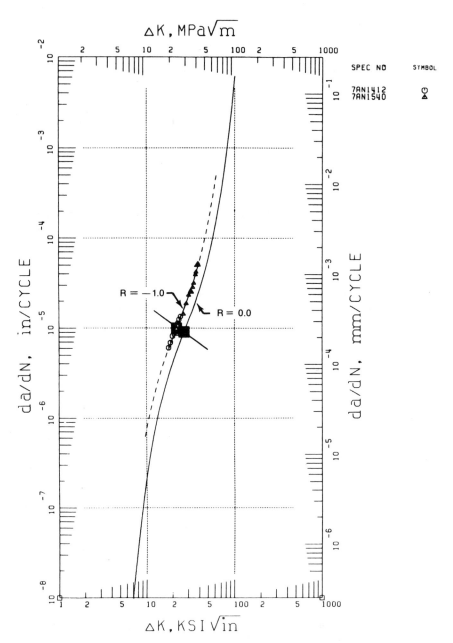

FIG. 7—*Incoloy 901 (PWA 1003) negative stress ratio model, 0.1667 Hz (10 cpm), 538°C (1000°F).*

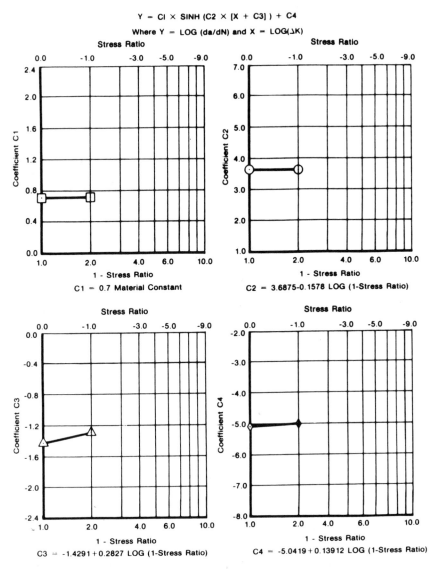

FIG. 8—*Incoloy 901 (PWA 1003) crack propagation correlative parameters, negative stress ratio model, 0.1667 Hz (10 cpm), 538°C (1000°F).*

(EDM). From this point the specimen was treated the same as virgin specimens. The notched specimen was returned to the testing machine, precracked at room temperature, and the crack was propagated to failure under constant load amplitude fatigue at a temperature of 260°C (500°F), 0.1667 Hz (10 cpm), and $R = 0.05$. The crack propagation rate data (da/dN versus ΔK) of virgin (unstrained) and prestrained materials are compared in Fig. 11.

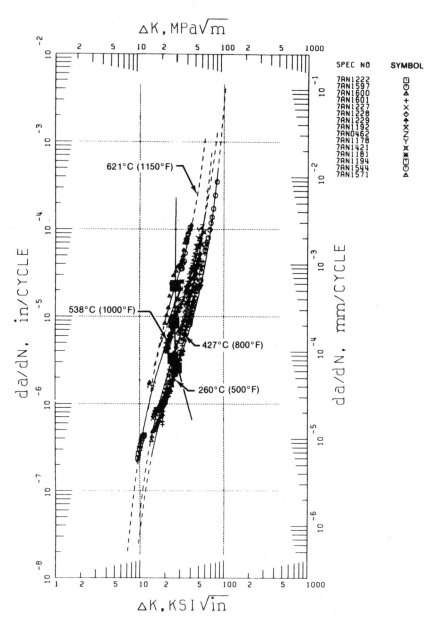

FIG. 9—*Incoloy 901 (PWA 1003) temperature model,* R = 0.05, *varying frequencies.*

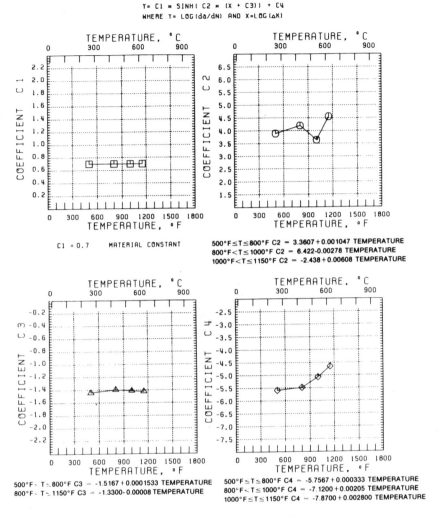

FIG. 10—*Incoloy 901 (PWA 1003) crack propagation correlative parameters, temperature model, R = 0.05, varying frequencies.*

Crack growth in the prestrained specimen is approximately 35% faster than observed from the virgin material. However, the prestrained specimen data differed in crack length from side to side in excess of 17.8 mm (70 mils). This deviation may have had a substantial effect on crack growth rates.

Statistics of the Model

The interpolation algorithm was used to predict crack growth lives of specimens listed in Table 3. Hyperbolic sine coefficients were defined for each of

TABLE 5—*Operating conditions for Incoloy 901 (PWA 1003).*

HYPERBOLIC SINE MODEL COEFFICIENTS

$Y = C1 \times SINH(C2 \times (X + C3)) + C4$

WHERE $Y = LOG (da/dN)$ AND $X = LOG (\Delta K)$

CURVE	SYMBOL	C1	C2	C3	C4	K RANGE	R^2	SEE
1	⊟ ⊙ ◢ +	0.7000	3.8840	-1.4400	-5.5890	(13.77, 83.02)	0.9841	0.1194
2	✕ ◇ ✦ ✘	0.7000	4.1980	-1.3940	-5.4800	(14.40, 53.80)	0.9868	0.0751
3	Ⓩ Ⓨ ✘ ✻ ⊟	0.7000	3.6420	-1.4100	-5.0700	(9.48 , 58.08)	0.9876	0.0765
4	⊙ ◢	0.7000	4.5540	-1.4220	-4.6500	(19.89, 40.94)	0.9503	0.0904

R^2 = Coefficient of Determination
SEE = Standard Error of the Estimate

METRIC CONVERSIONS

C1 METRIC = C1 ENGLISH
C3 METRIC = C3 ENGLISH - 0.040935
C2 METRIC = C2 ENGLISH
C4 METRIC = C4 ENGLISH + 1.40483

Curve	Symbol	Spec No.	Temperature °C (°F)	Freq	R	Type	Thickness mm (in.)
1	⊟	7AN1222	260 (500)	10 cpm	0.05	CCT	7.59 (0.299)
1	⊖	7AN1597	260 (500)	0.5 cpm	0.05	CT	12.70 (0.500)
1	◄	7AN1600	260 (500)	20 Hz	0.05	CT	6.35 (0.250)
1	+	7AN1601	260 (500)	10 cpm	0.05	CT	6.38 (0.251)
2	X	7AN1227	427 (800)	0.5 cpm	0.05	CCT	7.65 (0.301)
2	◇	7AN1228	427 (800)	10 cpm	0.05	CCT	7.59 (0.299)
2	✦	7AN1229	427 (800)	10 cpm	0.05	CCT	7.65 (0.301)
2	⊠	7AN1192	427 (800)	20 Hz	0.05	CT	7.65 (0.301)
3	N	7AN0462	538 (1000)	10 cpm	0.05	CT	6.30 (0.248)
3	Y	7AN1178	538 (1000)	10 cpm	0.05	CCT	7.62 (0.300)
3	✖	7AN1421	538 (1000)	20 Hz	0.05	CT	12.73 (0.501)
3	✳	7AN1181	538 (1000)	30 Hz	0.05	CCT	7.62 (0.300)
3	⊟	7AN1194	538 (1000)	10 cpm	0.05	CT	7.70 (0.303)
4	⊖	7AN1544	621 (1150)	0.5 cpm	0.05	CCT	7.59 (0.299)
4	◄	7AN1571	621 (1150)	10 cpm	0.05	CCT	7.54 (0.297)

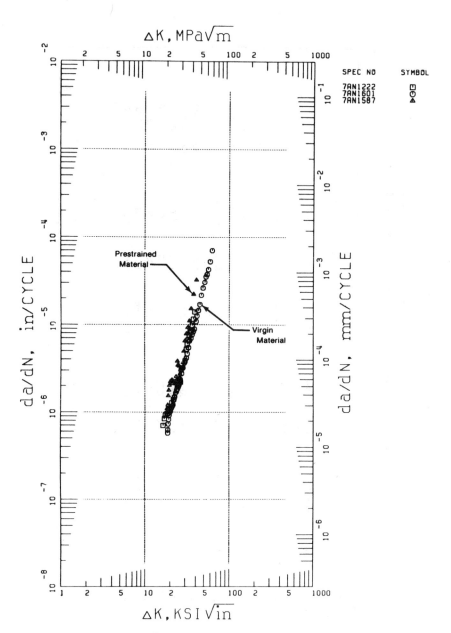

FIG. 11—*Incoloy 901 (PWA 1003) crack propagation rate, effect of prior plastic deformation,* R = 0.05, 0.1667 Hz (10 cpm), 260°C (500°F).

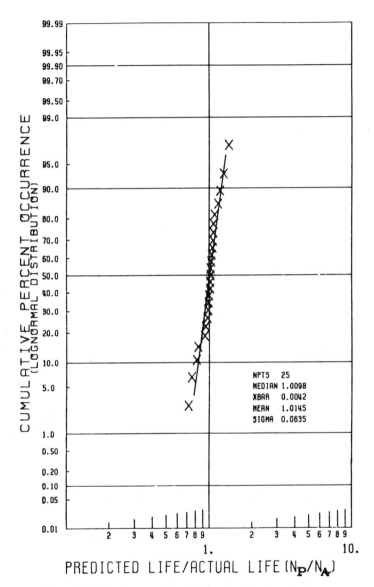

FIG. 12—*Lognormal probability plot of test results.*

TABLE 6—*Incoloy 901 (PWA 1003) interpolative life predictions.*

Specimen No.	N_p/N_A
1230	0.7172
1193	0.7530
1421	0.8147
1539	0.8373
1544	0.9298
0462	0.9463
1191	0.9745
1601	0.9858
1178	0.9919
1422	1.003
1424	1.022
0463	1.022
1181	1.027
1192	1.028
1227	1.038
1571	1.049
1195	1.056
1229	1.069
1228	1.070
1194	1.070
1412	1.097
1222	1.158
1600	1.199
1597	1.266
1543	1.373

log standard deviation = 0.0622; mean = 1.00975

the specimen test conditions. The hyperbolic sine equation was then integrated over the tested crack lengths and the resulting cycles to failure were compared with actual test results. These predicted cycles/actual cycles (N_p/N_a) are shown in Table 6.

The mean of the N_p/N_a's plotted in Fig. 12 is 1.0145. That is, the average life prediction is 1.5% more than the actual life. The log standard deviation is 0.0635. The model is considered a highly accurate tool for describing the crack propagation rates in Incoloy 901 (PWA 1003) under constant peak amplitude loading.

References

[1] Annis, C. G., Jr., Wallace, R. M., and Sims, D. L., "An Interpolative Model for Elevated Temperature Fatigue Crack Propagation," AFML-TR-76-176, Part I, Air Force Materials Laboratory, Wright-Patterson Air Force Base, Ohio, Nov. 1976.
[2] Wallace, R. M., Annis, C. G., Jr., and Sims, D. L., "Application of Fracture Mechanics at Elevated Temperature," AFML-TR-76-176, Part II, Air Force Materials Laboratory, Wright-Patterson Air Force Base, Ohio, Nov. 1976.

[3] Sims, D. L., Annis, C. G., Jr., and Wallace, R. M., "Cumulative Damage Fracture Mechanics at Elevated Temperature," AFML-TR-76-176, Part III, Air Force Materials Laboratory, Wright-Patterson Air Force Base, Ohio, Nov. 1976.

[4] Larsen, J. M., Schwartz, B. J., and Annis, C. G., Jr., "Cumulative Damage Fracture Mechanics under Engine Spectra," AFML-TR-79-4159, Air Force Materials Laboratory, Wright-Patterson Air Force Base, Ohio, Jan. 1980.

[5] Hudak, S. J., Jr., Bucci, R. J., and Malcolm, R. C., "Development of Standard Methods of Testing and Analysis Fatigue Crack Growth Rate Data," Technical Report AFML-TR-78-40, Air Force Materials Laboratory, Wright-Patterson Air Force Base, Ohio, May 1978.

P. E. Bretz, [1] *A. K. Vasudévan,* [1] *R. J. Bucci,* [1] *and R. C. Malcolm* [1]

Fatigue Crack Growth Behavior of 7XXX Aluminum Alloys under Simple Variable Amplitude Loading

REFERENCE: Bretz, P. E., Vasudévan, A. K., Bucci, R. J., and Malcolm, R. C., **"Fatigue Crack Growth Behavior of 7XXX Aluminum Alloys under Simple Variable Amplitude Loading,"** *Fracture Mechanics: Fifteenth Symposium, ASTM STP 833,* R. J. Sanford, Ed., American Society for Testing and Materials, Philadelphia, 1984, pp. 242-265.

ABSTRACT: This investigation examines the influence of temper and purity on the fatigue crack growth (FCG) behavior of four 7XXX aluminum alloys (7075-T6, 7075-T7, 7050-T6, 7050-T7) tested using a periodic single overload spectrum at low stresses. The loading sequence consists of a periodic single overload spike occurring once every 8000 constant amplitude cycles; the overload spike is 1.8 times the constant amplitude load peak. All tests have been performed in high humidity (>90% relative humidity) air at a frequency of 20 Hz and a constant amplitude load ratio of $R = 0.33$. The observed microstructural effects on transient (that is, spectrum) FCG behavior are contrasted with those for steady-state (constant load amplitude) fatigue performance.

KEY WORDS: fracture mechanics, metal fatigue, crack propagation, microstructure, retardation, variable amplitude, spectrum loading, aluminum alloy

The fatigue crack growth (FCG) life of structural members often is dominated by the propagation of flaws at very low stress intensity factor ranges (ΔK). For this reason, FCG rate information in the low ΔK regime can aid in designing structures for long fatigue life. Unfortunately, for aluminum alloys this near-threshold fatigue information has only recently become available [1-3].

Extensive studies [1-9] of constant load amplitude FCG behavior in aluminum alloys have shown that crack growth rates are influenced by alloy microstructure and particularly by the size, nature, and distribution of various second-phase particles: coarse constituents, intermediate size dispersoids, and fine precipitates. Recent work [1,2] has illustrated that the relative effect of various microstructural features on FCG behavior varies with ΔK level and

[1]Senior Engineer, Staff Scientist, Technical Specialist, and Senior Materials Analyst, respectively, Alloy Technology Division, Aluminum Company of America, Alcoa Center, Pa. 15069.

that the greatest differences in fatigue resistance among aluminum alloys occur in the low ΔK (near-threshold) regime. Thus the greatest opportunity to improve FCG resistance in aluminum alloys appears to be in the structure-sensitive low ΔK region.

Constant load amplitude fatigue behavior can be regarded as the "steady-state" response of a material to cyclic loading. However, cyclic loading in service usually is comprised of a spectrum of variable amplitude load excursions. Fatigue crack growth behavior under variable amplitude loading conditions includes transient response characteristics not addressed in constant amplitude tests; in particular, the ability of a material to retard crack growth following overloads. Because transient behavior is not measured in constant amplitude tests, the relative rankings of alloy FCG resistance under spectrum loading can be opposite to rankings obtained from constant load amplitude tests [10,11].

This study evaluates the effect of 7XXX aluminum alloy microstructure on transient FCG response using simple tensile overloads and is intended to link the understanding of steady-state and transient FCG response at relatively low ΔK levels. This study complements earlier work [7-9] which evaluated overload-microstructure interactions in 7XXX alloys tested at moderate-to-high stress intensity factor levels. Since FCG performance depends upon details of the loading history, the clarification of transient alloy response to elements of variable cyclic load histories represents an additional step towards knowing which alloy, component design, or test procedure is optimum for a particular class of application (for example, fighters as opposed to transport aircraft).

The specific objectives of this investigation are:

1. Clarify the role of 7XXX aluminum aircraft alloy microstructure, as influenced by composition and temper, on crack growth retardation characteristics for single periodic overloads superimposed on low ΔK constant load amplitude cycles.

2. Consolidate results of this investigation with previous work on identical alloys and tempers [7-9] to clarify the role of microstructure on steady-state and transient FCG response over a broad range of applied stress intensity factors.

Experimental Procedure

Materials

The alloys chosen for study are the same laboratory-fabricated samples of 7075 and 7050 type high-strength aluminum alloys which have been used previously in two Navy contracts [1,7] to characterize both constant load amplitude and simple overload FCG behavior. Complete descriptions of the fabrication practice and microstructural characterization are available in Ref 7.

The compositions of the two alloys are given in Table 1; results of tension tests are listed in Table 2.

Simple Overload FCG Tests

Test specimens were loaded in a sequence consisting of a periodic overload spike occurring once every 8000 constant amplitude cycles (OCR = 1:8000); the magnitude of the overload spike was 1.8 times the maximum peak load for the constant load amplitude cycles (OLR = 1.8). The nomenclature for overload tests is shown in Fig. 1. The stress ratio (R) for the constant amplitude cycles was $\frac{1}{3}$, and the applied frequency was 20 Hz. All testing was conducted in high humidity air ($> 90\%$ relative humidity). The targeted crack growth rates correspond to about 1.5 to 2 orders of magnitude decrease over those rates obtained in Ref 7 under comparable loading conditions (OLR = 1.8, OCR = 1:8000) at a higher baseline ΔK.

Center-crack-tension (CCT) specimens of identical width (102 mm), thickness (6.35 mm), and orientation (L-T) to those of Ref 7 were used in the overload tests. Precracking was accomplished without load shedding, using the same load range as that for the constant load amplitude portion of the simple overload spectrum. Programmed loads were provided by an MTS load profiler interfaced with the electrohydraulic test system.

Crack growth in the overload tests was monitored electronically using crack propagation gages [7]. In addition, crack length was measured visually using grids applied to one side of each CCT specimen.

Fractographic Examination

The scanning electron microscope (SEM) was selected as the primary tool for characterization of fracture appearance, with the object of identifying variations in fracture topography which correspond to observed differences in mechanical behavior of the alloys.

Results and Comparison with Previous Work

Simple Overload FCG Tests

Test results are presented in Fig. 2 in the form of total crack length versus total number of cycles (constant amplitude plus overload) for each alloy and

TABLE 1—*Alloy compositions (weight percent).*

	Si	Fe	Cu	Zr	Cr	Mg	Zn	Ti
7075	0.23	0.28	1.51	0.00	0.23	2.26	6.34	0.02
7050	0.07	0.11	2.10	0.13	0.00	2.16	6.16	0.02

TABLE 2—*Longitudinal tensile properties.*

	Tensile Strength, MPa (ksi)	Yield Strength, MPa (ksi)	Elongation, %
7075-T6	558 (81)	530 (77)	9.5
7075-T7	505 (73)	454 (66)	11
7050-T6	612 (89)	570 (83)	12
7050-T7	511 (74)	440 (64)	14

temper. Data for both alloys in the T7 temper represent the average response of duplicate tests, while the data for each alloy in the T6 temper represent the results of single tests.

Two observations are apparent when examining the data in Fig. 2. Most striking is the three-fold increase in life of the T6 over the T7 temper for both 7075 and 7050. Less dramatic, but also significant, is the longer life of 7075

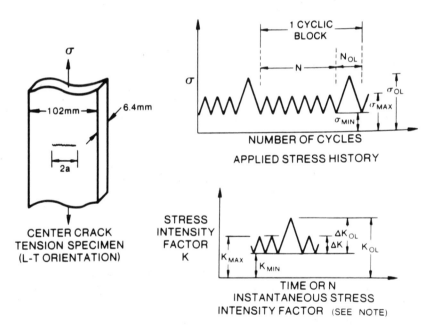

R = CONSTANT AMPLITUDE STRESS RATIO = $\sigma_{MIN}/\sigma_{MAX}$ = K_{MIN}/K_{MAX} = 1/3

OLR = OVERLOAD RATIO = σ_{OL}/σ_{MAX} = K_{OL}/K_{MAX} = 1.8

N = NUMBER OF CONSTANT AMPLITUDE CYCLES = 8000

N_{OL} = NUMBER OF OVERLOAD CYCLES = 1

OCR = OCCURRENCE RATIO = N_{OL}/N = 1/8000

FIG. 1—*Description of periodic single overload tests.*

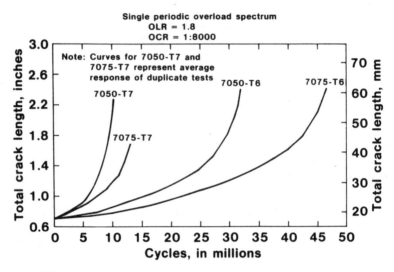

FIG. 2—*Results of periodic single overload tests on four aluminum alloys.*

versus 7050 for equivalent tempers, an increase of approximately 50%. The greater FCG resistance of both alloys in the T6 temper over the T7 temper for an overload (OL) history of OLR = 1.8, OCR = 1:8000 is consistent with constant load amplitude (CA) data for these alloys at near-threshold rates below about 1×10^{-9} m/cycle (4×10^{-8} in./cycle) [1,2]. It is not consistent, however, with either CA or OL rankings previously reported for these same materials tested at higher growth rates [7,9]. In these earlier studies, the T7 temper of both alloys had longer FCG lives than T6 for both CA and OLR = 1.8, OCR = 1:8000 (OL) histories. In contrast, though, FCG behavior in these earlier studies was examined at ΔK levels above approximately 6.6 MPa $\sqrt{m}$ (6 ksi $\sqrt{in}$.). Since CA FCG rates for the T6 temper do not fall below those for T7 until ΔK decreases to about 3 MPa $\sqrt{m}$ (2.7 ksi $\sqrt{in}$.) [1,2], the inverse ranking observed in the current study could not be predicted from the previous work.

The increase in FCG life for 7075 over 7050 of comparable temper has been noted before for an identical OLR = 1.8, OCR = 1:8000 OL sequence at higher applied stress levels; the opposite ranking is seen at high growth rates when an OLR = 1.4, OCR = 1:4000 spectrum is employed [7,9]. Furthermore, the superior FCG resistance of 7075 over 7050 at low ΔK levels for the OLR = 1.8, OCR = 1:8000 load history does not agree with data in the literature on spectrum FCG in which 7050 generally is superior to 7075 [10,12,13]. The referenced studies used load spectra more complex than a single periodic OL sequence, and often were conducted at significantly higher applied stress than the present work. In one case, the FCG resistance of 7075 was better than 7050 in comparable temper for an aircraft fighter spectrum, while 7050 FCG resistance was superior for CA loading. Thus it should not be expected that CA and spectrum FCG alloy rankings will be identical, nor will all spectra result in the same performance rankings.

Fractography

All OL crack growth experiments were conducted by first precracking each CCT specimen under constant amplitude loading using the same load range as for baseline CA cycles of the spectrum. After a well-established fatigue crack was developed, the periodic overload spectrum was applied. The application of this OLR = 1.8, OCR = 1:8000 spectrum caused an immediate and dramatic change in fracture surface topography, as shown in Fig. 3 for 7050-T7. The texture of the CA fracture surfaces in these figures is crystallographic, much like that noted previously for low ΔK constant amplitude loading [1,2]. This is to be expected, since the stress intensity range at the end of CA precracking was approximately 3.3 MPa $\sqrt{m}$ (3 ksi $\sqrt{in}$.). The distinctly different fracture topography after beginning the OL sequence, however, suggests that crystallographic fracture has been suppressed in favor of some other crack growth mechanism. This, in turn, suggests that different microstructural features may control crack growth under a simple overload spectrum in comparison to growth under constant amplitude loading.

Inspection of this initial OL region (Fig. 3b) reveals regularly spaced lines oriented perpendicular to the crack growth direction. Examination of test specimen fatigue surfaces at various crack lengths reveals that these fracture surface lines dominate the topography of both 7075 and 7050 in the T6 and T7 tempers regardless of ΔK levels and that spacings of these lines increase with ΔK magnitude. Figure 4 illustrates this fracture mechanism at various ΔK levels for 7075-T6. To confirm the origin of these markings, measurements of line spacings were made at a series of crack lengths (that is, various ΔK levels) for each alloy and temper. The fact that these measurements are much larger than macroscopic growth rates per cycle for the OL FCG tests confirms that this fracture mechanism is not classical fatigue striation formation. When the line spacing measurements are divided by 8001 cycles/block and plotted along with simple OL crack growth rate data (Fig. 5), however, a close correspondence is seen for 7075-T6 and also for the other alloys. (The calculation of ΔK for simple OL FCG data is described in the next section.) This confirms that these markings represent the increment of crack advance in each load block. Unfortunately, it is not possible to distinguish the fraction of each block band produced during an OL cycle from that part created by a segment of 8000 CA cycles.

No evidence of classical fatigue striations was found within each block band at any crack length. Indeed, no discernible features within a band could be seen for macroscopic growth rates less than 2.5×10^{-9} m/cycle (1×10^{-7} in./cycle). By contrast, the distinctive feathery appearance of crystallographic fatigue fracture was evident where measured OL growth rates exceeded 2.5 $\times$

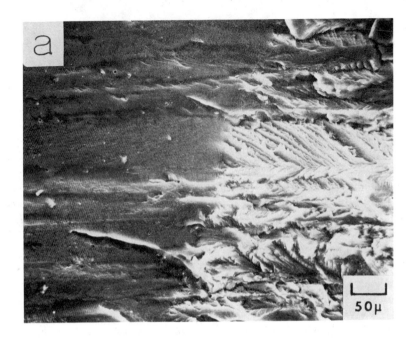

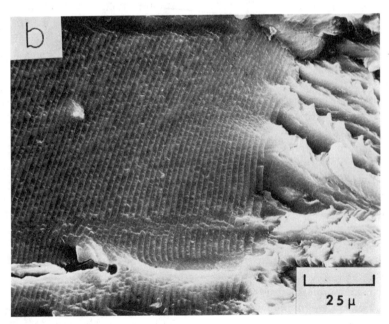

FIG. 3—*SEM fractograph of 7050-T7 showing transition from CA precrack to OL spectrum.* $\Delta K = 3.1$ *MPa* $\sqrt{m}$ *(2.8 ksi* $\sqrt{in.}$*). (a) General view. (b) Close-up view showing fine lines in OL region.*

10^{-9} m/cycle (Fig. 6). It is important to point out, however, that this crystallographic fracture has occurred at $\Delta K = 8.1$ MPa $\sqrt{m}$ (7.4 ksi $\sqrt{in.}$). At an equivalent ΔK in CA specimens, crystallographic fracture already has been superseded by void coalescence [1,2]. Thus, although crystallographic fatigue fracture is a mechanism common to both CA and OL tests, it occurs at significantly higher ΔK levels for the OL spectrum than for CA loading.

Another interesting contrast in fracture topographies between CA and OL FCG specimens relates to the appearance of second-phase particles on the fracture surfaces. The presence of second-phase particles on CA fracture surfaces is limited to higher ΔK where void coalescence is the dominant fracture mechanism [1,2] Figure 7a, however, shows an abundance of constituent particles on the OL fracture surface of 7075-T6 at applied baseline CA stress intensities for which little or no constituent particle fracture is seen in CA specimens. In fact, most of the constituents appear to be fractured, which certainly must be a result of the periodic OL spikes. This evidence suggests that second-phase particles play a different role in the fatigue fracture process during this OL spectrum than during CA loading.

The density of fractured constituent particles varies among the four alloy/temper combinations. Figure 7, which compares all four alloys, shows a greater density of fractured constituent particles on the OL fracture surface of 7075 than 7050 in the same temper. This is not surprising, since 7050 is a higher purity alloy (that is, contains a lower volume fraction of constituents) than 7075. Temper also appears to influence constituent particle fracture, since particle density is much higher for 7075-T6 (Fig. 7a) than for 7075-T7 (Fig. 7c). These observations will be discussed in the next section. Figure 7 also shows that the band formation is not entirely regular; however, the reason for the occasional coarse bands (Figs. 7b and 7c) is not clear.

Discussion

Analytical Model for FCG Performance

It is not evident from Fig. 2 to what extent crack growth was retarded in alloys 7075 and 7050 as a result of the simple OL sequence. To examine this question, a modified version of the computer program EFFGRO was used to separate the linear damage and retardation components of FCG life for the variable load history. The EFFGRO program is based on the Vroman retardation model [14],[2] and has previously been used to predict experimental lives for these alloys under spectrum loading conditions at higher ΔK levels [7,9].

For each alloy and temper, the fatigue life of a CCT test specimen was pre-

---

[2]The Vroman model reflects "state-of-the-art" FCG life prediction methodology and accounts for retardation through the use of effective stress intensity factor concepts based on crack tip plasticity. See *Part-Through Crack Fatigue Life Prediction, ASTM STP 687*, American Society for Testing and Materials, 1979.

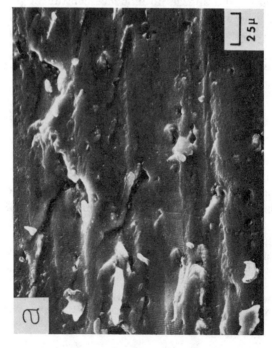

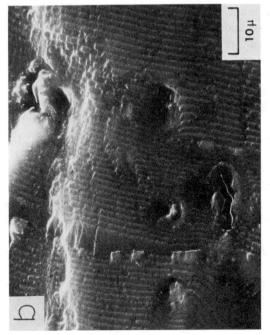

FIG. 4—*SEM fractograph of 7075-T6 showing increasing band spacing with* ΔK. (a) $\Delta K = 3.4\ MPa\ \sqrt{m}$ (3.1 ksi $\sqrt{in}$.). (b) Close-up view of (a). (c) $\Delta K = 3.7\ MPa\ \sqrt{m}$ (3.4 ksi $\sqrt{in}$.). (d) $\Delta K = 7.4\ MPa\ \sqrt{m}$ (6.7 ksi $\sqrt{in}$.).

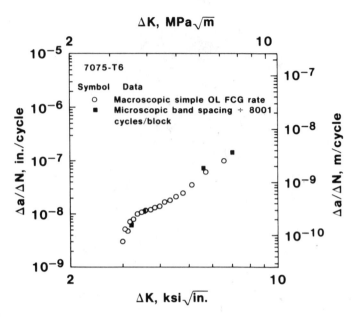

FIG. 5—*Correlation between macroscopic FCG rates and measured fracture surface band spacing for 7075-T6.*

dicted for three conditions: (1) the actual OLR = 1.8, OCR = 1:8000 OL spectrum incorporating OL-induced retardation; (2) the identical loading sequence as (1), but assuming no retardation; and (3) the OL history with a constant amplitude sequence consisting solely of the overloads. The second case assumes that crack extension is a linear damage summation of the independent contributions of OL and CA cycles in the spectrum without load interaction effects. Conversely, the third case removes the baseline CA cycles and assumes that crack extension takes place only during the OL cycle in each load block.

Figure 8 shows predictions for the growth of a crack from a half-crack length of 8.9 to 27.9 mm (0.35 to 1.1 in.) in a 102-mm (4-in.)-wide CCT panel. In these predictions, one block represents 8000 CA cycles + 1 OL cycle, as defined in Fig. 1. Several observations can be made with regard to the data in Fig. 8. The actual experimental lives of these alloys are much longer than predicted by the linear damage assumption (no retardation), with actual FCG lives greater than those predicted by a factor of 5 for 7050-T7 to a factor of 46 for 7075-T6. Clearly, the periodic OL (OLR = 1.8, OCR = 1:8000) causes a significant retardation of crack growth during baseline CA cycles. In contrast, actual lives of all four alloys are orders of magnitude shorter than predicted for FCG during only the OL cycles. This suggests that crack extension occurring during the CA portion of each loading block comprises a significant fraction of total spectrum life.

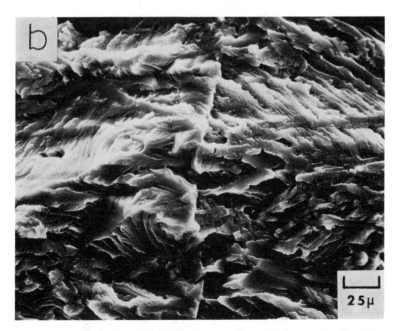

FIG. 6—*SEM fractograph of 7050-T7 showing crystallographic fracture details within coarse bands at* $\Delta K = 8.1$ *MPa* $\sqrt{m}$ *(7.4 ksi* $\sqrt{in.}$*). (a) General view. (b) Close-up view showing fine details.*

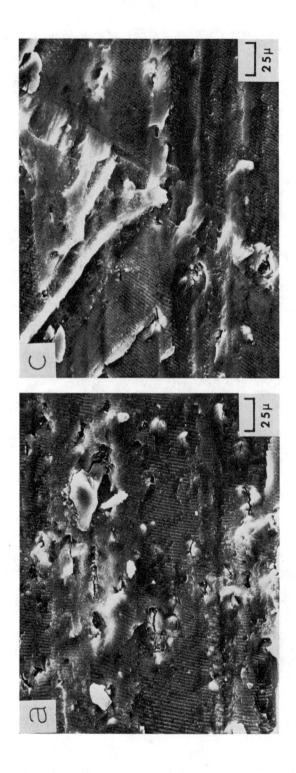

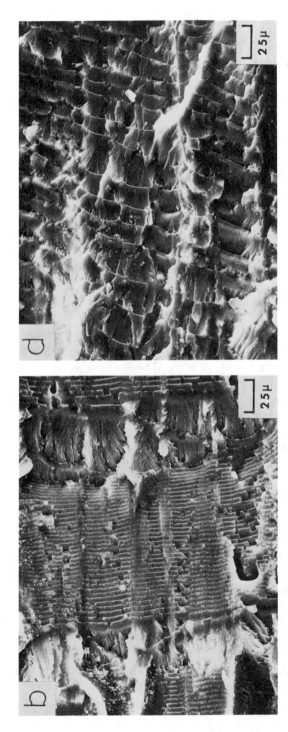

FIG. 7—SEM fractograph comparing fracture surface features and constituent particle density for all four alloys at da/dN = 2.5 × 10⁻¹⁰ m/cycle (1 × 10⁻⁸ in./cycle). (a) 7075-T6. (b) 7050-T6. (c) 7075-T7. (d) 7050-T7.

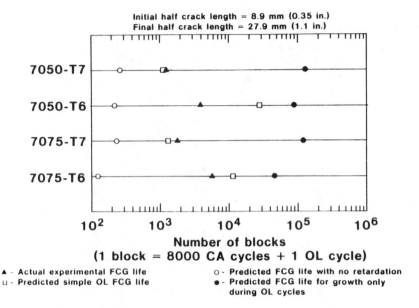

FIG. 8—*Life predictions using the Vroman retardation model.*

When retardation is considered, the accuracy of the EFFGRO predictions varies with alloy and temper (Fig. 8). Whereas the predictions for both alloys in the T7 temper are conservative and within 75% of actual lives, the predictions for T6 temper alloys are nonconservative and as much as 700% greater than actual lives. Upon first examination, this inaccuracy seems surprising. As will be discussed later, there are multiple crack growth retardation mechanisms operating during variable amplitude loading. However, one mechanism is expected to dominate spectrum FCG behavior at low stress intensities: overload-induced localized plastic deformation, which reduces the effective stress intensity factor (ΔK_{eff}) during subsequent baseline CA cycles. Since this is the retardation mechanism on which EFFGRO is based, these life predictions should have been reasonably accurate. This was not the case for alloys in the T6 temper.

One clue to the inaccuracy of EFFGRO for the T6 temper alloys appears when a detailed examination of the Vroman model calculations is made. The model determines a value of ΔK_{eff} for CA cycles following each overload excursion, and then uses CA FCG data to determine a growth rate (and, thus, a crack growth increment) for each cycle. For CA cycles immediately following the first OL, ΔK_{eff} is calculated to be 2.0 MPa $\sqrt{m}$ (1.8 ksi $\sqrt{in.}$), whereas the applied stress intensity (ΔK_{app}) is 3.3 MPa $\sqrt{m}$ (3 ksi $\sqrt{in.}$). This value for ΔK_{eff} is below the range of near-threshold FCG data for both T6 alloys but not outside the measured range for either T7 alloy [1,2]. Thus EFFGRO is forced to extrapolate CA FCG data for 7075-T6 and 7050-T6 to make life predictions.

The nonconservative life predictions for both alloys in the T6 temper imply that extrapolated growth rates are themselves nonconservative. On the other hand, no extrapolation of FCG data was necessary for either T7 temper alloy, so the life estimations are relatively accurate.

Crack Growth Retardation

While the data in Fig. 8 indicate the degree to which FCG lives in 7XXX alloys were increased by an OLR = 1.8, OCR = 1:8000 OL spectrum, the actual retardation of crack growth rates relative to CA loading cannot be ascertained from this information. To do so, we must compare da/dN from CA and OL tests for equivalent loading conditions. For each OL test this is accomplished by calculating ΔK corresponding to the baseline CA cycles in the OL spectrum. Figure 9 presents CA and OL crack growth rate data versus baseline ΔK for 7075-T6 and -T7. From this figure it is clear that the periodic 1.8 P_{max} OL has caused substantial retardation of crack growth rates in subsequent baseline CA cycles.

The data in Fig. 9 also suggest that the degree of retardation is not the same for all aluminum alloys and tempers. The amount of retardation can be measured conveniently by using a retardation ratio (RR) which is defined as the ratio of CA da/dN to OL da/dN at equivalent base ΔK levels. Thus a larger RR indicates greater retardation at comparable applied ΔK. Values of RR at several ΔK levels are presented in Fig. 10; they show that the T6 temper alloys

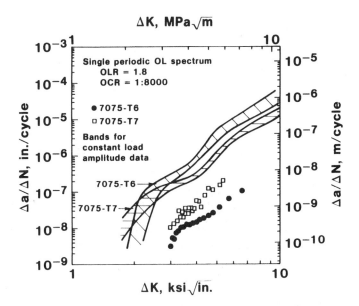

FIG. 9—Comparison of CA and OL crack growth data for 7075-T6 and -T7.

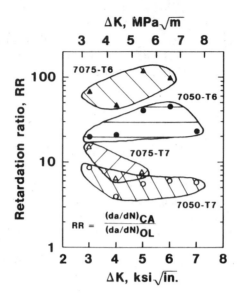

FIG. 10—*Comparison of calculated retardation rates for all four alloys as a function of* ΔK.

exhibit greater retardation than T7, while alloy 7075 has greater retardation capability than 7050 for equivalent temper. Keep in mind, though, that high values of *RR* do not guarantee superior spectrum FCG resistance if, for example, CA FCG performance is extremely poor.

In earlier work at higher K levels, Sanders, Bucci, and co-workers [7,9] considered how the ratio of crack growth rate per load block (da/dB) to overload crack tip plastic zone size ($r_{y\,OL}$) influences retardation. They concluded that as the value of $da/dB/r_{y\,OL}$ decreases, the degree of interaction between successive overloads and, therefore, the amount of retardation increases. In particular, if $da/dB/r_{y\,OL} < 0.18$, the next OL is applied before the crack has grown out of the so-called reverse plastic zone, and retardation is maximized [15]. Figure 11 shows that $da/dB/r_{y\,OL}$ versus ΔK for all four materials is less than 0.18 for at least the early part of the OL tests. Thus differences in degree of retardation between alloys and tempers cannot be explained solely by plastic zone size arguments at these low K magnitudes.

Microstructural Effects on OL Fatigue Crack Growth

The previous discussion of the fracture mechanics approach to retardation illustrates how greater attenuation of FCG rates following an overload is assumed to be related to an increase in the plastic zone size at the crack tip [15-17]. Because plastic zone size varies inversely with yield strength, greater retardation is expected in low strength materials. This clearly is not the case for the data shown in Fig. 10, where 7075 and 7050 in the low strength T7

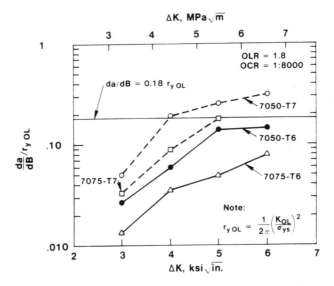

FIG. 11—*Crack growth per block normalized by plastic zone size for the periodic single overload spectrum. A value below 0.18 indicates crack tip always with reverse plastic zone of 1.8 overload.*

temper exhibited much less retardation than these alloys in the high strength T6 temper.

The key to this apparent inconsistency lies in the fact that, at low stresses such as were used in this study, ΔK_{eff} within the OL plastic zone is suppressed to near-threshold magnitudes. For example, if a 1.8 OL is superimposed on a baseline ΔK_{app} of 3.3 MPa $\sqrt{m}$ (3 ksi $\sqrt{in.}$), ΔK_{eff} for the first CA cycle after the OL will be less than 2.2 MPa $\sqrt{m}$ (2 ksi $\sqrt{in.}$) according to the Vroman model. At these ΔK_{eff} values, growth rates will be much lower for the T6 temper than for T7. Thus, even though ΔK_{eff} may be slightly smaller for a T7 temper than for T6 because of the decreased yield strength in an overaged temper, the much lower crack growth rates for a peak-aged alloy result in greater retardation of crack growth in comparison to the T7 temper when ΔK_{eff} is reduced to near-threshold values.

Fatigue crack growth retardation is also expected when OL excursions cause local fracture of coarse second-phase particles within the OL plastic zone [7,9,18]. These secondary fractures relieve some of the crack tip strain energy, thereby lowering ΔK_{eff} and reducing crack growth rates. Since the degree to which crack tip strain energy is relieved should be related to the number of fractured particles, greater retardation is expected in alloys containing a large amount of coarse second-phase particles. This was observed in previous studies when low purity 7075 exhibited greater retardation than higher purity 7050 for the OLR = 1.8, OCR = 1:8000 OL spectrum at high stress levels [7,9] and when greater retardation was observed in 7075 when compared to higher purity 7475 in both aircraft spectra and simple high-low block tests [18].

In the present results, 7075 again exhibits longer FCG lives (Fig. 2) and greater retardation (Fig. 10) than 7050 in the same temper. If this ranking were caused by differences in strength level, then 7075 should have a lower yield strength than 7050 in both T6 and T7 tempers. Table 2 shows that this is not the case for the T7 temper; while in the T6 temper, the strength of 7050 is only 10% higher than that of 7075. This FCG performance ranking could, however, be caused by the higher constituent volume fraction in 7075, and is reflected in the higher constituent particle density on the fracture surface of 7075 as compared to 7050 (Fig. 7).

There also is reason to believe that this fracture of constituents, and thus the degree of retardation, is influenced by temper. As mentioned previously, the density of broken particles on the fracture surface is much higher for 7075-T6 than for 7075-T7 (Fig. 7). This greater propensity for fracture of constituents in the T6 temper may be due to the greater degree of slip planarity in this temper relative to the T7 condition. In the crack tip plastic zone, intense slip bands characteristic of the T6 temper [8] can impinge on constituents; the resulting stress concentration should fracture these particles. In contrast, the more diffuse (homogeneous) slip in an overaged condition will provide neither the same degree of stress concentration nor, therefore, the same tendency towards constituent fracture. Thus the metallurgical change in the nature of slip which occurs with aging practice can influence the degree of retardation.

Knott and Pickard [19] recently have discussed the effect of cyclic yield strength on OL fatigue crack growth retardation in an Al-Zn-Mg alloy. These authors observed that delay periods during which crack growth is retarded following overloads are greatest for the underaged condition and decrease progressively with further aging to T6 and T7 conditions. This shortening of the delay period was related to a decrease in cyclic work-hardening capability with increased aging. In a separate investigation of the fatigue properties of various aluminum alloys [20], measurements of cyclic properties show that 7075-T651 cyclically hardens to a small degree, while 7075-T7351 is cyclically stable. According to Knott and Pickard's arguments, this would be expected to result in greater post-OL retardation for the peak-aged alloy relative to an overaged condition.

The influence of cyclic yield strength on FCG retardation can contribute to some fraction of the increase in fatigue life for alloys in the T6 temper over T7 in the present investigation. We believe, however, that the observed temper effect on simple OL FCG retardation primarily is due to the higher ΔK_{th} and to the greater tendency towards constituent fracture in the T6 temper. As previously discussed, CA crack growth rates are ten times lower for alloys in the T6 temper than for those in the T7 temper at the stress intensity levels to which ΔK_{eff} is suppressed initially by the 1.8 OL excursions. Surely this difference in near-threshold CA crack growth rates has a greater effect on simple OL FCG lives than does the modest difference in cyclic tensile properties between T6 and T7 tempers. The evidence for a temper effect on constituent fracture is clear, while the resulting influence on retardation seems to be straightforward.

There are two other metallurgical differences between 7075 and 7050 not previously discussed. Table 1 shows that the copper content in 7050 is higher (2.10% versus 1.51% for 7075) and that zirconium replaces chromium as the primary dispersoid-forming element. Although an increased copper content reduces CA FCG rates modestly in the intermediate ΔK region [7-9], this increase from 1.5 to 2.10% has no effect on near-threshold fatigue resistance [1,2]. Similarly, the substitution of zirconium for chromium has little or no effect on constant amplitude growth rates at intermediate and low ΔK levels [1,2], despite increasing toughness [8]. For these reasons, it seems unlikely that either copper content or the chromium versus zirconium difference can have an appreciable effect on OL fatigue behavior, particularly at the lower K magnitudes examined in this study.

Summary

The retardation process involves a reduction of the effective crack tip stress intensity factor (ΔK_{eff}) during low amplitude stress cycles following an overload. These simple overload experiments and a previous study [7,9] have identified two retardation mechanisms which can reduce ΔK_{eff}. The first is a result of overload plasticity, which increases crack closure forces [21] and promotes clamping forces developed by elastic constraints surrounding the oversized plastic zone of the overload. This retardation mechanism is yield strength dependent; that is, lower strength leads to larger overload plasticity and greater retardation. The second mechanism involves local cracking at constituent particles, which occurs during high tensile overloads. The rate of fatigue crack growth during subsequent lower stress cycles decreases because part of the crack tip stress intensity is distributed among the local fractures, reducing the driving force for crack extension. Since this mechanism is controlled by constituent volume fraction, a greater second-phase particle density increases FCG retardation. Of course, if the OL ΔK approaches alloy toughness, crack growth can actually be accelerated by constituent fracture during overloads.

Comparison of results from CA and OL studies indicates that the same alloy microstructural characteristics may not optimize both CA FCG resistance and spectrum retardation behavior. For this reason, we have chosen to consider variable load amplitude crack growth as a two-stage process (Fig. 12). In the first stage, a combination of load history and metallurgical variables such as yield strength and constituent volume fraction control retardation and define ΔK_{eff}. Crack extension for each load cycle is then specified in Stage 2 by the CA da/dN versus ΔK relationship, where ΔK now equals ΔK_{eff} for each load excursion. Because retardation can reduce ΔK_{eff} to very small values, near-threshold FCG behavior may contribute significantly to spectrum fatigue per-

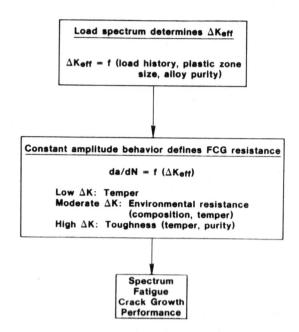

FIG. 12—*Schematic diagram showing the influence of metallurgical variants on both retardation and crack growth during spectrum fatigue loading.*

formance. Therefore the sensitivity of near-threshold fatigue resistance to alloy microstructure can carry over to FCG under variable amplitude load spectra.

General statements about the variable amplitude FCG resistance of a certain alloy microstructure are difficult to formulate, since metallurgical effects on both retardation and crack growth characteristics must be considered simultaneously. Nevertheless, the understanding of fatigue mechanisms acquired from simple overload history experiments does permit certain generalizations to be made. Such statements describe microstructural characteristics which enhance FCG resistance for classes of simple load sequences typical of many applications. Tables 3 and 4 summarize the statements listed below, which are the result of this study and prior work [7,9] on the Al-Zn-Mg-Cu alloy system:

1. When prior overload cycles force effective ΔK values for numerous baseline cycles to near-threshold levels, longer fatigue crack growth lives may be expected for alloys in the T6 temper than those in a T7 temper. This is due to the greater constant load amplitude FCG resistance of the T6 temper at nearthreshold crack growth rates. Load spectra for which a T6 temper should exhibit better FCG performance than T7 includes those histories where intermediate ΔK overloads are imposed at moderate-to-low frequencies on low to moderately low ΔK base cycles.

TABLE 3—*Effect of temper on Al-Zn-Mg-Cu alloy FCG resistance.*

Type of Loading		Dominant Metallurgical Factor
T6 favored	Low ΔK CA High OL on low ΔK base	Higher ΔK_{th} for T6 increases FCG life
	High OL on low/moderate ΔK base	Higher matrix strength enhances secondary particle fracture which lowers ΔK_{eff} after overloading
T7 favored	Low level overloads	Lower matrix strength enhances plastic zone retardation mechanisms
	Moderate ΔK CA Low/moderate OL on intermediate ΔK base	Greater resistance to degradation by environment
	High ΔK CA Extremely high OL	Higher toughness

2. When frequently applied overloads or baseline cycles at intermediate ΔK levels are the cycles in the load history during which most crack growth occurs, maximum spectrum fatigue life is achieved through optimizing intermediate ΔK FCG resistance. In this case, a T7 temper is favored over T6 because of the greater resistance to environmental degradation of FCG performance in the overaged condition [7,8].

3. One primary mechanism of overload-induced FCG retardation is controlled by the scale of overload plasticity. Crack growth retardation increases with crack tip plastic zone size, which itself varies inversely with yield strength. Therefore, overaging to a lower strength (T7) condition generally provides increased FCG resistance over the T6 temper when retardation is both significant (actual life appreciably greater than predicted by linear damage models) and controlled by crack tip plasticity. This retardation mechanism dominates when overload plastic strains are too small to cause appreciable secondary cracking at constituents (see Statement 4 below).

4. When applied ΔK is high (approaching alloy K_{Ic}), secondary cracking at coarse constituents contributes to reduced toughness and lower FCG resistance. Therefore greater alloy purity (lower Fe, Si content) improves FCG re-

TABLE 4—*Effect of purity (Fe, Si content) on Al-Zn-Mg-Cu alloy FCG resistance.*

	Type of Loading	Dominant Metallurgical Factor
Low purity favored	High OL on low-to-moderate ΔK base	Second-phase particle fracture after overloading reduces ΔK_{eff}
High purity favored	High ΔK CA Very high OL	Particle fracture contributes to significant crack extension during overloading

sistance when baseline cycles are at high ΔK or when extremely high overloads cause appreciable crack extension. By contrast, secondary cracking at constituents can reduce ΔK_{eff} by lowering strain energy at the main crack tip. This in turn reduces crack growth rates during baseline load cycles following an overload excursion. As long as no appreciable growth occurs during an overload, the net effect can be to lower overall FCG rates. When this occurs (for example, high overloads applied with moderate frequency on low-to-medium ΔK base cycles), increasing alloy constituent volume fraction can increase fatigue life. This retardation mechanism is more effective in increasing FCG life when matrix yield strength is high, which favors the T6 temper over T7 for greater FCG resistance.

Acknowledgments

We would like to thank the Department of the Navy, Naval Air Systems Command, who sponsored this work under Contract N00019-79-C-0258.

References

[1] Bretz, P. E., Vasudévan, A. K., Bucci, R. J., and Malcolm, R. C., "Effect of Microstructure on 7XXX Aluminum Alloy Fatigue Crack Growth Behavior Down to Near-Threshold Rates," Final Report, Naval Air Systems Command Contract N00019-79-C-2058, 9 Oct. 1981.
[2] Bretz, P. E., Bucci, R. J., Malcolm, R. C., and Vasudévan, A. K. in *Fracture Mechanics: Fourteenth Symposium—Volume II: Testing and Applications*, ASTM STP 791, J. C. Lewis and G. Sines, Eds., American Society for Testing and Materials, 1983, pp. II-67–II-86.
[3] Kirby, B. R. and Beevers, C. J., *Fatigue of Engineering Materials and Structures*, Vol. 1, 1979, pp. 203–215.
[4] Staley, J. T. in *Properties Related to Fracture Toughness, ASTM STP 605*, American Society for Testing and Materials, 1976, pp. 71–103.
[5] Truckner, W. G., Vasudévan, A. K., Bucci, R. J., and Thakker, A. B., "Effects of Microstructure on Fatigue Crack Growth of High-Strength Aluminum Alloys," USAF Technical Report AFML-TR-76-169, Wright-Patterson AFB, Ohio, Oct. 1976.
[6] Staley, J. T., Truckner, W. G., Bucci, R. J., and Thakker, A. B., *Aluminum*, Vol. 53, No. 11, Nov. 1977, pp. 667–669.
[7] Sanders, T. H., Jr., Sawtell, R. R., Staley, J. T., Bucci, R. J., and Thakker, A. B., "Effect of Microstructure on Fatigue Crack Growth of 7XXX Aluminum Alloys under Constant Amplitude and Spectrum Loading," Final Report, Naval Air Development Center, Contract N00019-C-0482, 14 April 1978.
[8] Sanders, T. H., Jr., and Staley, J. T., "Review of Fatigue and Fracture Research on High-Strength Aluminum Alloys," in *Fatigue and Microstructure*, American Society for Metals. Metals Park, Ohio, 1979, p. 467.
[9] Bucci, R. J., Thakker, A. B., Sanders, T. H., Jr., Sawtell, R. R., and Staley, J. T. in *Effect of Load Spectrum Variables on Fatigue Crack Initiation and Propagation, ASTM STP 714*, American Society for Testing and Materials, 1980, pp. 41–78.
[10] Bucci, R. J. in *Flaw Growth and Fracture, ASTM STP 631*, American Society for Testing and Materials, 1977, pp. 388–401.
[11] Bucci, R. J., *Engineering Fracture Mechanics*, Vol. 12, 1979, pp. 407–441.
[12] Jones, R. L. and Coyle, T. E., "The Mechanical, Stress-Corrosion Fracture Mechanics, and Fatigue Properties of 7050, 7475, and Ti-8 Mo-8 V-2 Fe-3 Al Plate and Sheet Alloys," Report FGT-5791, General Dynamics, Fort Worth Division, Tex., 1976.

[13] Schra, L., "Engineering Property Comparisons for Four Al-Zn-Mg-Cu Type Forging Alloys," Report NLR TR 79022U, National Aerospace Laboratory, The Netherlands, 1979.
[14] Vroman, G., "Analytical Prediction of Crack Growth Retardation Using a Critical Stress Intensity Concept," Technical Report TFR 71-701, North American Rockwell, Los Angeles Division, Calif., 1971.
[15] Trebules, V. W., Roberts, R., and Hertzberg, R. W. in *Progress in Flaw Growth and Fracture Toughness Testing, ASTM STP 536*, American Society for Testing and Materials, 1973, pp. 115–146.
[16] Wheeler, O. E., *Journal of Basic Engineering Transactions*, Series D, Vol. 94, 1972, pp. 181–186.
[17] Mills, W. J. and Hertzberg, R. W., *Engineering Fracture Mechanics*, Vol. 8, No. 4, 1976, p. 657.
[18] Schulte, K., Trautmann, K. H., and Nowack, H., "Influence of the Microstructure of High-Strength Aluminum Alloys on Fatigue Crack Propagation under Variable Amplitude Loading," presented at International Conference on Analytical and Experimental Fracture Mechanics, Rome, 1980.
[19] Knott, J. F. and Pickard, A. C., *Metal Science Journal*, Aug./Sept. 1979, pp. 399–404.
[20] Sanders, T. H., Jr., Staley, J. T., and Mauney, D. A., "Strain Control Fatigue as a Tool to Interpret Fatigue Initiation of Aluminum Alloys," presented at 10th Annual International Symposium on Materials Science, Seattle, 1975.
[21] Elber, W. in *Damage Tolerance in Aircraft Structures, ASTM STP 486*, American Society for Testing and Materials, 1971, p. 236.

*R. W. Lang,[1] M. T. Hahn,[1] R. W. Hertzberg,[2]
and J. A. Manson[2]*

Effects of Specimen Configuration and Frequency on Fatigue Crack Propagation in Nylon 66

REFERENCE: Lang, R. W., Hahn, M. T., Hertzberg, R. W., and Manson, J. A., "**Effects of Specimen Configuration and Frequency on Fatigue Crack Propagation in Nylon 66,**" *Fracture Mechanics: Fifteenth Symposium, ASTM STP 833*, R. J. Sanford, Ed., American Society for Testing and Materials, Philadelphia, 1984, pp. 266–283.

ABSTRACT: Fatigue crack propagation (FCP) rates were measured in nylon 66 as a function of test frequency using center-cracked-tension (CCT) and wedge-open loading (WOL) specimens. In order to enhance any effects of stress- and frequency-induced changes in the dynamic modulus and to maximize the capability for hysteretic heat generation, the nylon was equilibrated to a water content of 4.7% by weight. For this water content the glass transition region of the material occurs at around the fatigue test temperature. FCP rates determined using CCT specimens were found to be consistently higher at all frequencies than those using WOL specimens. In addition, hysteretic heat-up measured at the crack tip and across the whole unbroken ligament was higher in the CCT specimens. These results are related to the differences in the far field stress profiles and their effect on the relative specimen stiffness and on the heat generation capability. Effects of test frequency are discussed in terms of changes in strain rate and the beneficial and detrimental effects of local crack tip and general specimen heating.

KEY WORDS: fatigue crack propagation, fracture mechanics, engineering plastics, nylon 66, specimen configuration, frequency, hysteretic heating, nonlinear viscoelasticity

The increasing interest in engineering plastics for lightweight structural components which are exposed to cyclic loads has directed the attention of many researchers to the investigation of the fatigue behavior of these materials. Since initiation and growth of fatigue cracks are being increasingly recognized as a potential source of failure, many of the studies conducted over the past decade have focused on evaluating fatigue crack propagation (FCP) be-

[1] Research Assistant, Materials Research Center, Lehigh University, Bethlehem, Pa. 18015.
[2] New Jersey Zinc Professor and Professor of Chemistry, respectively, Materials Research Center, Lehigh University, Bethlehem, Pa. 18015.

havior in terms of fracture mechanics concepts [1,2]. Analogous to the case of metals, crack growth rates per cycle (da/dN) are most commonly related to the stress intensity factor range (ΔK) according to a relationship proposed by Paris [3]:

$$\frac{da}{dN} = A \cdot \Delta K^m \tag{1}$$

where A and m are constants depending on the material, material parameters, and test conditions.

The dependence of A and m in Eq 1 on various test parameters is in part related to the viscoelastic nature of polymeric materials, that is, to the time and temperature dependence of mechanical properties. The most widely studied effects of any test parameter at present involved variations in frequency [1,2,4-14]. While crack growth rates were found to decrease with increasing frequency for some polymers, FCP rates remained essentially unchanged in others. In addition, increasing growth rates with increasing frequency have been reported for at least one group of polymers.

Several attempts have been made to explain these contradicting trends. For example, it has been argued [8,9] that frequency sensitivity is related to the frequency dependence of the dynamic modulus. While it is certainly true that a strain rate or for that matter a frequency-induced increase in modulus can be expected to decrease growth rates by decreasing the local cyclic strains at the crack tip, it has been shown [11] that such changes in specimen stiffness cannot in general account for the large frequency sensitivity observed in some polymers. Alternatively it has been proposed [1] that the beneficial effect of increasing frequency on FCP resistance is a result of a localized temperature rise at the crack tip which blunts the crack and results in a lower effective ΔK. Furthermore, it has been shown [12-14] that increasing growth rates with increasing frequency are due to hysteretic heating on a large scale in which a large part of the specimen heats up significantly. This decrease in FCP resistance has been rationalized in terms of a temperature-induced decrease in the overall elastic modulus.

Since the actual temperature rise depends on the balance between the heat generation rate and the heat transfer rate, it has been pointed out previously [13] that specimen heating will also depend upon such geometrical factors as specimen thickness, specimen planar dimensions, and net section stress. Therefore the relative fatigue performance can be expected to depend on specimen geometry when extensive heating occurs.

General Considerations

The object of the present study was to investigate the influence of specimen configuration and frequency on FCP resistance in nylon 66. According to the

principles of linear elastic fracture mechanics, no such dependence on specimen configuration is expected since the propagation of a fatigue crack is assumed to be controlled entirely by the stress field near the crack tip and therefore to be a function of ΔK only. However, in engineering plastics that exhibit
pronounced viscoelastic behavior at the prevailing test conditions, variations
in specimen configuration may alter FCP resistance as a result of changes in
the overall stress range ($\Delta \sigma$) acting on the specimen. This is readily understood in terms of the general representation of ΔK as

$$\Delta K = \frac{\Delta P}{B \cdot \sqrt{w}} \cdot Y$$

where ΔP is the applied load range, and the geometrical calibration factor Y
is a function of specimen configuration, crack length (a), and specimen
width, (W). Hence for given values of ΔK, a, and W the overall loading range
varies inversely with the Y calibration factor. Since Y is much smaller for a
center-cracked-tension (CCT) specimen than for a wedge-open loading
(WOL) specimen, it follows that ΔP_{CCT} is much larger than ΔP_{WOL}. This indicates that the cyclic stresses away from the crack tip are also larger in the
CCT specimen.

A qualitative representation comparing the distributions of stresses acting
in the load direction and in the crack plane across the unbroken ligament in
CCT and WOL specimens is illustrated in Fig. 1 for a given value of K. As the
specimen is loaded, a plastic zone develops at the crack tip, thereby shifting
the purely elastic stress profile towards the plasticity-corrected (elastic-plastic) profile. While the stresses in the near field where near-crack-tip field
equations apply are independent of specimen configuration, variations occur
at some distance away from the crack tip. Thus the steeper gradient of the far
field stresses in the WOL specimen results in lower ligament stresses than in
the CCT specimen. Furthermore, due to the bending moment acting on the
WOL specimen, stresses for this configuration become negative near the back
face.

In explaining effects of specimen configuration on fatigue crack growth, it
is important to recognize that the dynamic modulus (E^*) of polymers may
depend on the applied stress amplitude as well as test frequency and temperature. In this regard, several investigations have shown noticeable decreases in
modulus with increasing tensile stress amplitude [15–19]. In addition, moduli
determined under compression are generally found to be higher than their
counterparts determined under tension [15]. Consequently, the overall effective modulus of the unbroken ligament that determines the specimen compliance can be expected to be higher for the WOL configuration than for the
CCT configuration if ligament stresses are sufficient to introduce nonlinear
behavior. Of course, any change in the effective specimen modulus will in
turn affect the cyclic stresses and strains locally at the crack tip and therefore

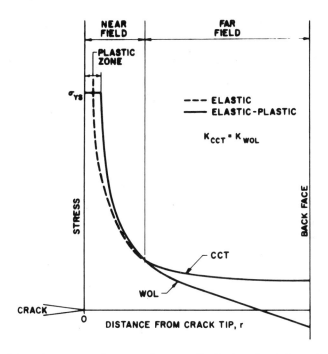

FIG. 1—*Schematic representation comparing stress distributions ahead of a crack tip for CCT and WOL specimens.*

also the plastic zone size. Specifically, the lower effective modulus in the CCT specimen will result in larger crack tip strains and hence tend to increase crack growth rates at a given value of ΔK.

Another important point to be considered in the case of polymer fatigue is the potential for hysteretic heat generation. The magnitude and extent of cyclic-load-induced specimen heating depends on both mechanical and material factors.

Assuming a sinusoidal load application in the linear viscoelastic range, several authors have shown [20,21] that the temperature rise per unit time (neglecting heat losses) ($\Delta \dot{T}$) becomes

$$\Delta \dot{T} = \frac{f \cdot \pi \cdot D'' \ (f, T) \cdot \Delta \sigma^2}{4 \cdot c_p \cdot \rho} \tag{2}$$

where

$f =$ frequency,
$D'' =$ dynamic loss compliance,
$\Delta \sigma =$ applied stress range,
$T =$ temperature,

c_p = specific heat, and
ρ = density.

Taking the triaxiality of the stress field at the crack tip into account by using a vibrational stress tensor, Barenblatt et al [22] derived the following expression for the temperature distribution close to the crack tip:

$$\Delta T (r, \theta) = \frac{f \cdot D'' \cdot \Delta K^2}{8 \cdot r \cdot a_h} \cdot \psi (\theta, \nu) \tag{3}$$

where

r, θ = polar coordinates measured from the crack tip,
f = frequency,
D'' = dynamic loss compliance,
a_h = coefficient of heat exchange, and
$\psi(\theta, \nu)$ = polynomial function of θ and Poisson's ratio ν.

Equations 2 and 3 are consistent in that they describe specimen temperature elevations with a relationship of the general form

$$\Delta T \, \alpha \, \frac{\Delta S^2 \cdot f \cdot D''}{h} \tag{4}$$

where h refers to the heat transfer characteristics. The parameters ΔT and ΔS can be thought of as representing the crack tip temperature rise and the applied stress intensity range in the case of crack tip heating; in terms of the heat-up in the unbroken ligament they refer to the temperature rise away from the crack tip and the applied stress range of the far field stresses. Again, due to the variations in the stress profiles shown in Fig. 1, we can expect significant differences in the heat-up characteristics of different specimen configurations.

Although it seems unrealistic to expect Eq 2 to 4 to hold in an absolute sense in the case of FCP due to the many assumptions made in the derivations, they have been used successfully for several practical applications at least in a qualitative sense [23,24]. This is not surprising since in any case the temperature rise will intuitively increase with frequency, damping capacity D'', and cyclic stress range or ΔK. Hence the difference in the magnitude of the cyclic far field stresses for different specimen configurations also suggests that variations in overall specimen heating may occur and may alter FCP resistance, at least at higher frequencies. While preliminary results supporting these speculations have been reported recently [25], it is the purpose of this study to examine and to discuss more closely the underlying mechanisms of the combined effects of frequency and specimen configuration on FCP in nylon 66.

Experimental Procedure

Materials and Specimen Preparation

The material used for the present investigation was commercially available nylon 66 (Zytel 101, number average molecular weight, $\overline{M}_n$ = 17 000) and was received in the form of injection-molded plaques (125 by 75 by 8.9 mm) in the dry, as-molded (water content ≈ 0.2%) condition. The plaques were equilibrated with water by immersion in a boiling, saturated aqueous solution of sodium acetate for 120 h. Specimens cut from these plaques were then stored in a closed container above a saturated aqueous solution of calcium nitrate until immediately before testing. The final water content of 4.7 (± 0.3)% by weight was determined by drying samples to constant weight under vacuum at about 80 to 90°C. This water content was chosen on the basis of previous experience [24] to ensure that significant hysteretic heating would occur in FCP tests at higher frequencies.

Center-cracked-tension and wedge-open loading specimens were machined with the notch oriented transverse to the original processing direction. Figure 2 shows these specimen configurations along with specimen dimensions. The CCT specimens were pin-loaded by means of aluminum tabs bolted onto each side in order to obtain more uniform loading and to prevent failure at the loading pins.

Fatigue Crack Propagation Tests

Fatigue crack propagation testing was performed on an electrohydraulic closed-loop testing machine. Environmental conditions were laboratory air at

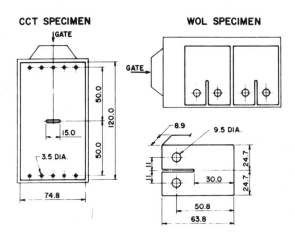

FIG. 2—*Geometry, orientation, and position of FCP specimens in injection-molded plaques (dimensions in mm).*

24 ($\pm$ 1)°C. The applied waveform was sinusoidal with constant load amplitude and a minimum/maximum load ratio (R) of 0.1. Crack growth data were obtained at frequencies from 1 to 70 Hz. The machined slot in all specimens was first sharpened with a razor blade, and a fatigue crack then initiated by cycling at frequencies above 10 Hz or at somewhat higher loads than used in the actual test procedure. Frequency and loads were then gradually changed to the actual test conditions. To eliminate any effects of precracking, the crack was then allowed to grow for about 0.3 mm before FCP data were taken. Crack length measurements were conducted after typical increments of 0.2 to 0.4 mm using a traveling microscope. Since the test machine had to be stopped briefly for such readings, the distance for these increments was extended to about 0.5 mm whenever significant crack tip heating occurred in order to minimize the effects of specimen cooling during the time period for crack tip readings.

Incremental crack growth rates were calculated using the modified secant formula

$$\left(\frac{da}{dN}\right)_n = \frac{a_{n+1} - a_{n-1}}{N_{n+1} - N_{n-1}}$$

where a is the crack length and N is the total number of cycles at the time of each crack tip reading. The stress intensity range ΔK for the two specimen configurations used was computed using the equations [26]

$$\Delta K = \frac{\Delta P}{B \cdot \sqrt{W}} \cdot \sqrt{\pi \left(\frac{a}{W}\right) \sec \pi \left(\frac{a}{W}\right)}$$

for CCT specimens, and

$$\Delta K = \frac{\Delta P}{B \cdot \sqrt{W}} \cdot \frac{2 + a/W}{(1 - a/W)^{3/2}} \cdot \left[0.8072 + 8.858 \left(\frac{a}{W}\right) - 30.23 \left(\frac{a}{W}\right)^2 \right.$$

$$\left. + 41.088 \left(\frac{a}{W}\right)^3 - 24.15 \left(\frac{a}{W}\right)^4 + 4.951 \left(\frac{a}{W}\right)^5 \right]$$

for WOL specimens, where

ΔP = applied load range,
B = specimen thickness,
W = specimen width, and
a = crack length in WOL specimens and the half-crack length in CCT specimens.

Fatigue crack propagation data were then plotted in the conventional way as log da/dN versus log ΔK.

Crack tip temperatures and temperature profiles across the specimen width were monitored using an infrared microscope (Model RM-2B, Barnes Engineering Company) which was mounted on an XYZ-positioner. Temperature measurements were made only while the specimen was being cycled, immediately before cycling was interrupted to record the crack tip position. A more detailed description of the measurement procedure is given elsewhere [27].

Dynamic Mechanical Testing

Small-strain dynamic mechanical spectra at 3.5 and 110 Hz were recorded using an Autovibron apparatus (Model DDV-III-C, IMASS Corporation). In order to avoid any effects of orientation, specimens for these tests were cut from the plaques in such a way that the loading direction was identical with the loading direction in the FCP specimens. Details of the test procedure are discussed by Webler et al [28]. To calculate the tensile loss compliance (D'') the following relationship was used [20]:

$$D'' = \frac{E''}{E'^2 + E''^2} = \frac{1/E''}{1 + (\tan^2\delta)^{-1}}$$

where

E' = dynamic storage modulus,
E'' = dynamic loss modulus, and
$\tan \delta$ = dynamic loss factor ($= E''/E'$).

Experimental Results

Fatigue crack growth rates versus ΔK are shown in Fig. 3 comparing data generated from both specimen configurations at frequencies ranging from 1 to 50 Hz. It is clear that FCP rates are indeed affected by specimen configuration in that crack growth rates determined with WOL specimens are consistently lower than those obtained from CCT specimens at all frequencies. In addition, there are differences in the FCP frequency sensitivity for the two configurations.

While growth rates for WOL specimens at 10 and 50 Hz are somewhat lower than at 1 Hz, there is a transition in the relative ranking of CCT test data with regard to test frequency as one proceeds from low to high ΔK levels. For example, FCP rates in CCT specimens at $\Delta K < 1.1$ MPa $\sqrt{m}$ decrease continuously as the test frequency increases from 1 to 50 Hz. At ΔK levels above 1.1 MPa$\sqrt{m}$ growth rates are lowest at a frequency of 10 Hz. An addi-

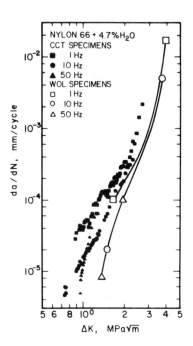

FIG. 3—*Fatigue crack propagation behavior at various frequencies for CCT and WOL specimens.*

tional test (not shown) performed on a CCT specimen at 70 Hz yielded even higher crack growth rates than were obtained in the 50-Hz experiment. Reproducibility tests performed on CCT specimens at 1 and 50 Hz showed good agreement and revealed the same tendencies as those illustrated in Fig. 3.

While essentially no hysteretic heat-up could be observed in the 1-Hz experiments, at least in the lower ΔK range, significant specimen heating occurred at higher frequencies. As indicated by Eqs 2 to 4, the tendency for hysteretic heat generation depends not only on various test parameters but also on the viscoelastic state of the material, specifically on the magnitude of D''.

Dynamic mechanical spectra recorded at two different frequencies are shown in Fig. 4. It should be noted that the test temperature of the FCP experiments (24°C) is within the temperature range over which the glass transition (T_g-region) of nylon 66 with 4.7% water occurs. Hence even a minor increase in temperature due to hysteretic heating in a fatigue test will result in a noticeable decrease in modulus. While there is essentially no strain rate effect on E' at low and high temperatures, some frequency dependence in the T_g-region is apparent. The high values of D'' in this temperature range are of

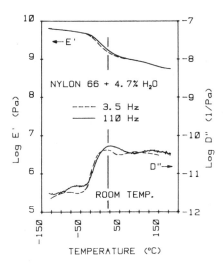

FIG. 4—*Storage modulus* (E') *and loss compliance* (D") *obtained from small-strain dynamic mechanical spectroscopy.*

course in agreement with the significant amount of heating that took place in FCP tests conducted at higher frequencies.

The measured variations in maximum temperature at the crack tip are shown in Fig. 5 as a function of ΔK for various frequencies and for both specimen configurations. Consistent with Eq 3, it is seen that crack tip temperatures increase with increasing frequency and applied ΔK range, reaching temperatures as high as 83°C in the CCT specimens at 50 Hz. It should be noted, however, that crack tip temperatures in the WOL specimens are invariably lower at any given frequency and ΔK level (except at low ΔK levels at 1 Hz where no significant heating was observed in either configuration).

The tendencies for local crack tip heating were paralleled by the amount of heat-up across the entire unbroken ligament. Figures 6 and 7 show temperature profiles recorded at various ΔK levels for CCT and WOL specimens, respectively, and reveal the extent of specimen heating along the crack plane as a function of the distance from the back face of the specimen. In accordance with Eqs 2 to 4 and the stress profiles shown in Fig. 1, hysteretic heat-up is maximized at or near the crack tip where cyclic stresses are highest, and then decreases monotonically with distance. As expected, overall specimen heating is seen to increase with increasing frequency, with ligament temperatures in the CCT specimens being significantly higher than in the WOL specimens, at least at 10 and 50 Hz. While heat-up away from the crack tip was insignificant at 1 Hz for the WOL specimen over the entire ΔK range and for the CCT specimen over most of the ΔK range, ligament temperatures in the

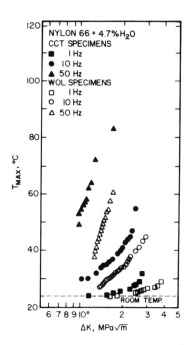

FIG. 5—*Comparison of the maximum temperature rise* (T_{max}) *at the crack tip for CCT and WOL specimens at various frequencies.*

latter configuration increased to values ranging from 30°C near the back face to 32°C at the crack tip just before failure.

Discussion

Temperature Measurements

The variations in specimen temperature elevations with different configurations apparently reflect the differences in the contribution of the cyclic far field stresses to overall specimen heating. Although a dependence of D'' on the magnitude of the cyclic stress range cannot be excluded, the higher temperatures away from the crack tip observed in the CCT specimens are certainly due at least in part to the higher cyclic stresses in this region according to Fig. 1 and Eqs 2 and 4. In turn, a higher ligament temperature restrains the extent of heat transfer from the principal heat source at the crack tip towards the remaining part of the specimen, thereby leading to higher crack tip temperatures in the CCT configuration despite nominally equivalent ΔK levels. Thus the discrepancies in crack tip heating shown in Fig. 5 for both configurations are believed to be caused mainly by differences in heat transfer

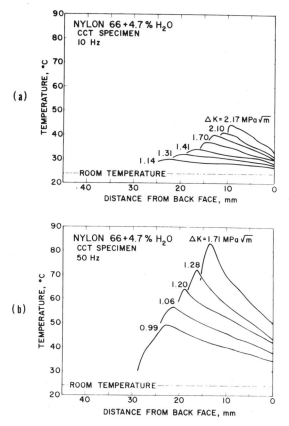

FIG. 6—*Temperature profiles recorded at various* ΔK *levels in the CCT specimen.* (a) *10 Hz.* (b) *50 Hz.*

characteristics and not by differences in local heat generation capability at the crack tip.

While the foregoing results clearly indicate the importance of the far field stresses on specimen heat-up for the CCT specimens, their role on the temperature rise in the WOL specimens is less clear. In fact, for the hypothetical case in which no heat exchange occurs at all, temperature elevations in any given volume element of the specimen should be proportional to the square of the local stress range according to Eq 2. For this scenario the temperature should reach a maximum at either end of the unbroken ligament (that is, a higher and a lower maximum at the crack tip and the back face, respectively) and reveal a minimum with no heating at the point of rotation where the local stress range is zero (Fig. 1). Since there is absolutely no indication for such a minimum in the temperature profiles in Fig. 7, it appears that overall specimen heating in the WOL specimens is controlled by heat transfer effects.

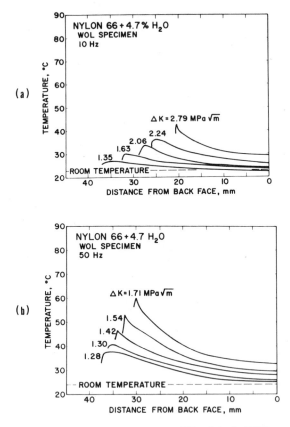

FIG. 7—*Temperature profiles recorded at various* ΔK *levels in the WOL specimen.* (a) *10 Hz.* (b) *50 Hz.*

The effect of test frequency and specimen configuration on the FCP resistance of the investigated material can now be re-examined in light of the specimen temperature measurements and the data from dynamic mechanical tests.

Fatigue Crack Propagation Behavior

As pointed out above, several attempts have been made to explain frequency sensitivity of FCP rates in polymers in terms of the effects of changes in strain rate and on the basis of the effects of localized versus generalized heating. Since all these factors (that is, frequency-dependent dynamic modulus, local crack tip heating as well as generalized specimen heating) have been found to occur to various degrees throughout this investigation, it is obvious that the effect of frequency on FCP behavior also must reflect the competition between these conflicting trends. In this context a model has been proposed recently [29,30] relating frequency sensitivity of FCP rates in polymers to such fracture mechanical parameters as plastic zone size (r_y) and critical crack opening displacement (δ_c). Thus it has been suggested that the net effect of increasing frequency on FCP resistance (in the range where dynamic creep crack extension can be neglected) will be determined by a balance among:

1. The effect of increasing strain rate resulting in a decrease of r_y and δ_c (detrimental).

2. The effect of increasing strain rate resulting in an increase in dynamic modulus (beneficial).

3. The hysteretic temperature rise at the crack tip resulting in an increase in r_y and δ_c (beneficial).

4. The hysteretic temperature rise of the bulk material in the far field resulting in a modulus decrease (detrimental).

Since the crack tip temperature elevations found in the FCP tests at 10 and 50 Hz were quite substantial, it can be assumed that Condition 1 is being overcompensated by Condition 3, at least in this case, and can therefore be neglected in the following discussion.

Effect of Frequency in the CCT Specimens

When examining the FCP data for CCT specimens in Fig. 3 it becomes apparent that there is a noticeable increase in the slopes of the curves as the frequency increases. It has been proposed [29] that such a converging tendency along with crossovers of FCP curves at various frequencies can be interpreted in terms of the increasing importance of a modulus drop due to gross specimen heating. Since high cyclic stresses occur at the crack tip even at low ΔK levels, the effects of local crack tip heating can be expected to dominate the behavior at low stress intensity ranges. With increasing crack length and hence increasing ΔK in a constant load-range experiment, the local stress amplitudes in the unbroken ligament rise. Consequently, the tendency for overall specimen heat-up increases continuously due to both (1) the increasing heat generation capability caused by the increasing cyclic stresses of the far field stress field, and (2) the heat transfer from the primary heat source at the crack tip. The fact that the glass transition region associated with a major modulus drop occurs right around the test temperature (Fig. 4) is without doubt a factor that favors the conditions for a temperature-induced decrease in specimen stiffness at higher ΔK levels.

Therefore the decrease in growth rates when increasing the frequency from 1 to 10 Hz clearly must be a result of the beneficial influence of the local crack tip temperature rise. Based on the results from the dynamic mechanical experiments (Fig. 4) one might expect that an interactive effect of the higher frequency towards an increase in modulus is probably limited by the overall ligament heat-up at 10 Hz even at low ΔK values (Fig. 6a). In fact, the convergence of the curves for 1 and 10 Hz at higher ΔK levels is an indication of the increasingly adverse influence of the decrease in modulus due to overall specimen heat-up at the higher frequency. A further increase in frequency to 50 Hz yields again lower growth rates at $\Delta K < 1.1$ MPa$\sqrt{m}$ due to beneficial crack tip heating. With increasing stress intensity range, however, the deleterious effect of the ligament temperature rises (Fig. 6b) becomes progressively more important. This leads to a crossover with the 10-Hz data, and growth rates approach the results for 1 Hz just before fast fracture. The reproducibly lower values for K_{cf} (maximum value of K associated with stable FCP) obtained in the 50-Hz experiments provide further confirmation for the adverse influence of gross specimen heating.

Effect of Frequency in the WOL Specimens

Compared to the variations in the CCT specimens, much less of an effect of frequency on relative fatigue performance was revealed in the WOL specimens, at least over the ΔK range investigated. Although crack growth rates at 10 and 50 Hz are somewhat lower than at 1 Hz for ΔK values below ~ 2.2 MPa$\sqrt{m}$, indicating a slightly dominant role of the consequences of beneficial crack tip heating, the overall FCP response is rather frequency insensitive. The quite significant differences observed in the hysteretically induced temperature elevations at 10 and 50 Hz (Figs. 5 and 7), on the other hand, suggest that an equilibrium exists between the beneficial and detrimental effects of the changes in strain rate, crack tip heating, and overall specimen heating.

Effect of Specimen Configuration

The effects of specimen configuration on FCP rates depicted in Fig. 3 are clearly in disagreement with expectations based on the concepts of linear elastic fracture mechanics. Some of the discrepancies found at higher frequencies certainly reflect at least in part the differences in the observed heat-up characteristics (Figs. 5 to 7). However, the higher growth rates in the CCT specimens tested at 1 Hz in a ΔK region where essentially no heat-up occurred must be caused by a different mechanism. As mentioned before, it is possible that the higher stresses in the unbroken ligament of the CCT specimen lead to a nonlinearity in the deformation process and hence to a lower effective dynamic modulus. Such a stress-induced modulus decrease has been reported for several polymers [15–19] and would obviously have a deleterious effect on FCP resistance similar to that of a stiffness decrease due to gross specimen heating. Also consistent with this argument is the reported increase in growth rates with increasing mean stress level (at a given frequency and ΔK) for nylon 66 containing various amounts of water [24,31]. Although it is true that under these circumstances the amount of crack extension should increase with increasing mean stress reflecting the more severe local crack tip stress conditions associated with the higher values for K_{min} and K_{max}, a possible contribution of a mean stress-induced modulus decrease is also conceivable.

It should be recognized, however, that such a stress-dependence of the dynamic modulus is a consequence of the viscoelastic nature of polymeric materials and may in fact occur at stresses far below the point at which irreversible plastic deformation (in the conventional sense of yielding) takes place. For example, it is well known that the stress-strain response of polymers can be curved in a region where the deformation is still fully recoverable. In this context it is interesting to compare the applied maximum net section stress with the reported static yield strength for this material, $\sigma_{ys} \approx 45$ MPa [32]. The net section stresses in the CCT specimen tested at 1 Hz, which varied from about 7 to 27 MPa over the entire test range for ΔK, indeed represent a significant fraction of the yield strength, at least at high stress intensities. While these numbers certainly tend to support the speculation on the stress dependence of the modulus, a more quantitative evaluation must await dynamic mechanical testing in the nonlinear range. The slight heat-up of the unbroken ligament in the CCT specimen tested at 1 Hz which was observed when recording the last few crack growth data also may have contributed towards a further decrease in modulus at the highest ΔK values.

Superimposed on the influence of the larger gross stress acting on the CCT specimens is the effect of hysteretic heating on growth rates at higher frequencies. Since both the local temperatures at the crack tip and the temperatures experienced by the remaining part of the specimen were found to be higher for the CCT configuration (Figs. 5 to 7), it appears that the net effect of the complex competition among the stress-dependent specimen stiffness, localized crack tip, and generalized specimen heating also results in higher FCP resistance of the WOL specimens at 10 and 50 Hz.

Summary and Conclusions

The effects of specimen configuration and test frequency on the FCP behavior of water-equilibrated nylon 66 (4.7% water) having a glass transition region at around room temperature were examined. In general it was found that crack growth rates determined using CCT specimens exceeded those using WOL specimens. In addition, specimen temperature measurements revealed more hysteretic crack tip heating and more gross specimen heating in the CCT specimens.

The results have been interpreted in terms of the differences in the far field stress distributions and their effect on the relative specimen stiffness and on the hysteretic heat generation capability. It has been shown that the dependence of growth rates on test frequency reflects a complex competition between the beneficial and detrimental effects associated with the changes in strain rate, and the occurrence of local crack tip and overall specimen heating. Thus the combination of heat-up measurements and dynamic mechanical spectroscopy represents a powerful approach for studying FCP phenomena in polymers.

Nonlinear viscoelasticity (as a result of the magnitude of stresses in the stress profiles) or temperature variations across the unbroken ligament (as shown in the measured temperature profiles) or both result in corresponding variations in modulus across the specimen width. These conditions represent a breakdown of the linear relationship between stress and strain and are therefore a violation of the basic assumptions of linear elastic fracture mechanics. The use of ΔK to describe growth rate data in these cases is therefore justifiable only in that the stress intensity factor range describes the combined conditions of loading and specimen geometry without having an accurate implication on the stress field ahead of the crack tip.

Although the effect of the overall stress on the effective specimen modulus is only tentative at this point, the results clearly indicate that one must proceed cautiously when attempting to apply ΔK-based FCP data for this material (and probably for other polymers in a similar viscoelastic state as well) in component design, especially in cases where cracks are physically short and applied stress levels are high. Whether or not alternative crack tip field parameters such as J or C^* can be used to model viscoelastic nonlinearity is currently being considered. Further studies of this point are indicated.

Acknowledgments

This work was supported in part by the Chemistry Division of the Office of Naval Research. The material tested was supplied by E. I. Du Pont de Nemours & Company, through the courtesy of Mr. E. Flexman.

References

[1] Hertzberg, R. W. and Manson, J. A., *Fatigue of Engineering Plastics*, Academic Press, New York, 1980.
[2] Manson, J. A. and Hertzberg, R. W., *Critical Review of Macromolecular Science*, Vol. 1, 1973, pp. 433–500.
[3] Paris, P. C. in *Proceedings*, 10th Sagamore Conference, Fatigue—An Interdisciplinary Approach, Syracuse, N.Y., 1964, p. 107.
[4] Hertzberg, R. W., Manson, J. A., and Skibo, M. D., *Polymer Engineering and Science*, Vol. 15, No. 4, 1975, pp. 252–260.
[5] Arad, S., Radon, J. C., and Culver, L. E., *Journal of Mechanical Engineering Science*, Vol. 13, 1971, p. 75.
[6] Mukherjee, B. and Burns, D. J., *Experimental Mechanics*, Vol. 2, 1971, pp. 433–439.
[7] Skibo, M. D., Hertzberg, R. W., and Manson, J. A., *Journal of Materials Science*, Vol. 11, 1976, pp. 479–490.
[8] Wnuk, M. P., *Journal of Applied Mechanics*, Vol. 41, No. 1, 1974, p. 243.
[9] Williams, J. G., *Journal of Materials Science*, Vol. 12, 1977, p. 2525.
[10] Hertzberg, R. W., Manson, J. A., and Skibo, M. D., *Polymer*, Vol. 19, 1978, p. 358.
[11] Hertzberg, R. W., Skibo, M. D., Manson, J. A., and Donald, J. K., *Journal of Materials Science*, Vol. 14, 1979, pp. 1754–1758.
[12] Skibo, M. D., Hertzberg, R. W., and Manson, J. A., *Deformation, Yield and Fracture of Polymers*, Plastics and Rubber Institute, Cambridge, England, 1979, pp. 4.1–4.5.

[13] Hertzberg, R. W., Skibo, M. D., and Manson, J. A. in *Fracture Mechanics, ASTM STP 700*, American Society for Testing and Materials, 1979, pp. 49-64.
[14] Hahn, M. T., Hertzberg, R. W., Lang, R. W., Manson, J. A., Michel, J. C., Ramirez, A., Rimnac, C. M., and Webler, S. M., *Deformation, Yield and Fracture of Polymers*, Churchill College, Cambridge, England, 1982, pp. 19.1-19.5.
[15] Koppelmann, J., Hirnböck, R., Leder, H., and Royer, F., *Colloid and Polymer Science*, Vol. 258, No. 1, 1980, pp. 9-23.
[16] Royer, F., *Colloid and Polymer Science*, Vol. 259, No. 2, 1981, pp. 202-208.
[17] Davis, W. M. and Macosko, C. W., *Journal of Rheology*, Vol. 22, No. 1, 1978, pp. 53-71.
[18] Isayev, A. I. and Katz, D., *International Journal of Polymeric Materials*, Vol. 8, 1980, pp. 25-43.
[19] Rahaman, M. N. and Scanlan, J., *Polymer*, Vol. 22, 1981, pp. 673-681.
[20] Ferry, J. D., *Viscoelastic Properties of Polymers*, Wiley, New York, 1961.
[21] Oberbach, K., *Kunststoffe*, Vol. 59, No. 1, 1969, pp. 37-39.
[22] Barenblatt, G. I., Eutov, V. M., and Salganik, R. L., IUTAM Symposium on Thermoinelastics, International Union of Theoretical and Applied Mechanics, East Kilbridge, England, 1968, pp. 33-46.
[23] Kramer, O. and Ferry, J. D. in *Science and Technology of Rubber*, Chapter 5, F. R. Eirich, Ed., Academic Press, New York, 1978.
[24] Hahn, M. T., Hertzberg, R. W., Manson, J. A., Lang, R. W., and Bretz, P. E., *Polymer*. Vol. 23, 1982, pp. 1675-1679.
[25] Lang, R. W., Hahn, M. T., Hertzberg, R. W., and Manson, J. A., *Journal of Materials Science Letters*, Vol. 3, 1984, pp. 224-228.
[26] Saxena, A. and Hudak, S. J., *International Journal of Fracture*, Vol. 14, No. 5, 1978, pp. 453-467.
[27] Hahn, M. T., Hertzberg, R. W., and Manson, J. A., *Review of Scientific Instruments*, Vol. 54, No. 5, 1983, pp. 604-606.
[28] Webler, S. M., Manson, J. A., and Lang, R. W. in *Polymer Characterization, Advances in Chemistry Series 203*, American Chemical Society, 1983, pp. 109-122.
[29] Lang, R. W., "Applicability and Limitations of the Linear Elastic Fracture Mechanics Approach to Fatigue in Polymers and Short-Fiber Composites," Ph.D. dissertation, Lehigh University, Bethlehem, Pa., 1984.
[30] Lang, R. W., Manson, J. A., and Hertzberg, R. W. in *The Role of the Polymeric Matrix in the Processing and Structural Properties of Composite Materials*, L. Nicholais and J. C. Seferis, Eds., Plenum, New York, 1983, pp. 377-396.
[31] El-Hakeem, H. A. and Culver, L. E., *International Journal of Fatigue*, Vol. 1, 1979, pp. 133-140.
[32] "Zytel Design Handbook," technical literature, E. I. Du Pont de Nemours & Company, Wilmington, Del., 1972.

D. F. Socie,[1] N. E. Dowling,[2] and P. Kurath[3]

Fatigue Life Estimation of Notched Members

REFERENCE: Socie, D. F., Dowling, N. E., and Kurath, P., **"Fatigue Life Estimation of Notched Members,"** *Fracture Mechanics: Fifteenth Symposium, ASTM STP 833,* R. J. Sanford, Ed., American Society for Testing and Materials, Philadelphia, 1984, pp. 284–299.

ABSTRACT: From an engineering view, it is convenient to separate the total fatigue life of notched members into two portions: crack initiation, which is spent nucleating and growing small cracks, and the crack propagation life, which is spent growing these cracks to final fracture. The difficulty in applying this concept has been in defining the size of an initiated crack in a smooth specimen and dealing with small crack growth in the plastic zone near the notch root.

These difficulties may be overcome by considering a simple model where the total fatigue life is the summation of the portion of life controlled by notch plasticity and the portion controlled by nominal stress and crack length. The local strain approach is used to compute the initiation life. Growth of small cracks in the notch plastic zone is assumed to be part of the initiation life. Fracture mechanics concepts are employed to estimate crack propagation lives assuming an initial crack size equal to the notch depth.

Twelve sets of fatigue data reported in the literature were analyzed to assess the validity of the model. These include variations in specimen type, notch size, notch acquity, and material properties. Good correlation between the analytical estimate and experimental data was observed.

KEY WORDS: fatigue, notches, crack growth, crack initiation, life prediction

During the early stages of the fatigue life of a notched member, the initiation and growth of cracks is controlled by the cyclic plastic strain at the notch root. Nominal stress controls the growth of crack during the later stages of the fatigue life. The transition from one control mode to the other is not well understood and is the subject of much research. From an engineering view, it is con-

[1] Associate Professor, Department of Mechanical and Industrial Engineering, University of Illinois at Urbana-Champaign, Urbana, Ill. 61801.

[2] Associate Professor, Engineering Science and Mechanics, Virginia Polytechnic Institute and State University, Blacksburg, Va. 24061.

[3] Research Assistant, Department of Theoretical and Applied Mechanics, University of Illinois at Urbana-Champaign, Urbana, Ill. 61801.

venient to separate the total fatigue life (N_T) into two portions: the crack initiation life (N_I), which is spent in developing and growing small cracks, and the crack propagation life (N_p), which is spent in growing cracks to failure.

Strain life concepts may be used to estimate the fatigue life when notch root plasticity is the controlling parameter. In this approach [1,2], a given surface strain and mean stress is assumed to always result in the same fatigue life. Fatigue data are in the form of a strain versus cyclic life curve. Usually, total cycles to failure of a small unnotched axial specimen are used. When combined with a relationship between nominal stresses and notch root strains, the strain life curve provides an estimate of the initiation life of the notched member. Neuber's rule is often used to estimate the relationship between the notch root stresses and strains and the nominal stresses and strains. Elastic-plastic finite element models are also used. Practical implementation of these techniques often requires knowledge of the fatigue notch factor (K_f) to account for notch size and acquity; however, this approach is not employed in the present paper.

Linear elastic fracture mechanics (LEFM) techniques are usually employed to estimate the crack propagation lives of components and structures [3,4] when nominal stress and crack length are the controlling parameters. Fatigue data employed in the analysis are in the form of cyclic stress intensity versus crack growth rate.

Stress intensity factors for cracks eminating from notches have been obtained [5] or can be estimated from approximate techniques [6]. The fatigue life is computed by integrating the crack growth rate for the material from some initial crack size to the final crack size. The results are not sensitive to the choice of final crack size, but are very sensitive to the selection of initial crack size. If the initial crack size is small, the crack will be growing in the plastic field of the notch and LEFM will not describe its growth. Conservative life estimates will result if the chosen initial crack size is too large.

In this paper, a simple approach is adopted for estimating crack propagation life. This approach avoids the issues of (1) the special behavior of K near a notch, (2) plasticity corrections for small cracks, and (3) choosing a specific initial crack size. In particular, LEFM will be used with the stress intensity ignoring the notch stress field, and the initial condition for crack propagation will simply be a crack length equal to the notch size. Crack propagation lives estimated on this basis will simply be added to initiation lives based on total life to failure of unnotched axial specimens.

Crack Initiation Approach

In addition to the materials data, a critical piece of information for making fatigue life predictions for notches is a relationship between load and the maximum stress and strain (σ and ϵ respectively) which occurs locally at the notch. This relationship may be determined from finite elements or other sophisticated analyses, or it may be estimated from an approximate method such as Neuber's rule.

In dealing with notched members, a nominal stress S is usually employed. The exact geometry of the notch, of course, determines its severity, which is characterized by an elastic stress concentration factor K_t. Note that K_t and S must be defined consistently so that the product $K_t S$ is the maximum stress in the notch; values of $K_t S$ are fictitious stresses. Analysis that specifically considers plastic deformation must then be employed to obtain the actual notch stresses and strains.

There are three regions of behavior for notched members: (1) simple elastic behavior, (2) local notch yielding under gross deformations that are elastic, and (3) general yielding. The transition between the first and second regions occurs at the point where $K_t S$ exceeds the yield strength of the material (σ_y). At the transition marking the beginning of general yielding, the strain begins to increase rapidly with $K_t S$, with the $K_t S$ value where this occurs being geometry dependent.

Approximate methods of estimating load versus notch strain curves are needed for the many situations where time or funds do not permit more detailed analysis. The most widely used approximate method is Neuber's rule [7]. This rule postulates that beyond yielding the geometric mean of the stress and strain concentration factors remains equal to K_t.

Load versus notch strain curves from Neuber's rule have the desirable property of exhibiting the three regions discussed earlier: elastic, local yielding, and general yielding. The transitions between these occur, respectively, at $K_t S = \sigma_y$ and at $S = \sigma_y$. For various cases of notched members under axial or bending loads or both, Neuber's rule has been compared to more exact analysis or to notch strain measurements [8,9]. In general, Neuber's rule appears to predict strains which are somewhat larger than the correct values; this results in conservative life predictions.

Some care is needed in using Neuber's rule. The transition to general yielding behavior is predicted by Neuber's rule to occur at $S = \sigma_y$, resulting in the ridiculous situation of predicting two different load-strain curves for two different definitions of S, such as definitions based on net section versus gross section area. Also, for the most commonly used definitions of S in many situations, the transition to general yielding is predicted to occur at the wrong load level.

This problem is discussed in Ref 8, where it is proposed that it be eliminated by using Neuber's rule in the modified form

$$\sigma\epsilon = K_p^2 \, S^* \, e^* \qquad (1)$$

where

$$K_p = \frac{S \text{ at fully plastic limit load}}{S \text{ at first notch yielding}} = \frac{S_p}{\sigma_y / K_t} \qquad (2)$$

in which K_t and S are defined in any desired mutually consistent manner, and S_p is the particular value of S corresponding to fully plastic behavior. Specifically, S_p is calculated for an ideal elastic, perfectly plastic, material having the same yield strength (σ_y) as the real material. The quantity S^* in Eq 1 is defined by

$$S^* = (K_t/K_p)S \tag{3}$$

The nominal strain quantity e^* in Eq 1 is obtained by entering the material's σ–ϵ curve with S^*. Where $S^* < \sigma_y$, Eq 2 reduces to the following form analogous to Eq 1:

$$\sigma - \epsilon = (K_p S^*)^2/E \tag{4}$$

where E is Young's modulus.

The result of Eqs 1 to 4 is that the onset of general yielding is predicted to occur at a level which is based on a fully plastic limit load calculation, specifically at $S^* = \sigma_y$. The predicted level is independent of the manner of defining S.

This modified form of Neuber's rule is employed here so that better estimates of the crack initiation life can be obtained when the nominal stresses approach yielding. In its original form, conservative estimates will be made for short lives. Once the notch strains have been determined, it is simply a matter of entering the strain life curve of the material to determine the fatigue life.

The strain life curve for completely reversed cycling is represented by the equation

$$\epsilon_a = (\sigma_f'/E)(2N)^b + \epsilon_f'(2N)^c \tag{5}$$

where

E = modulus of elasticity,
σ_f' = fatigue strength coefficient,
ϵ_f' = fatigue ductility coefficient,
b = fatigue strength exponent, and
c = fatigue ductility exponent.

To include mean stress effects in the life prediction, strain life curves from tests at various mean stresses, as in Ref 10, could be employed directly. Otherwise, an estimate of the effect is needed.

An approach similar to the Goodman diagram, with the assumption that the relationship between stress amplitude and plastic strain amplitude is independent of mean stress, implies that Eq 5 could be modified as follows to include the mean stress effect:

$$\epsilon_a = (\sigma_f'/E)[1 - (\sigma_0/\sigma_f')](2N)^b + \epsilon_f'[1 - (\sigma_0/\sigma_f')]^{c/b}(2N)^c \tag{6}$$

Equation 6 provides a value of life $(2N)$ for any combination of values of ϵ_a and σ_0, with iterative solution being required.

For most materials, particularly ferrous metals, a fatigue limit (S_{FL}) is observed at long lives. These effects are not included in Eqs 5 or 6 and should be considered separately. In this life regime the notch stresses remain nearly elastic. The threshold stress (S_{TH}) below which cracks will not initiate can be computed from the elastic stress concentration factor

$$S_{TH} = S_{FL}/K_t \tag{7}$$

The fatigue limit for ductile steels is often approximated by half of the ultimate strength.

Crack Propagation Approach

Crack growth calculations require a description of the crack growth rate of the material and the stress intensity factor for the geometry and loading of interest. Linear Elastic Fracture Mechanics is only applicable where the extent of plasticity is small compared with other geometric dimensions such as crack length. Since notch plasticity is usually present in finite life fatigue, LEFM is not generally applicable for short cracks where the behavior is controlled by the local stress-strain field of the notch.

For long cracks, that is, for cracks which have grown out of the local stress field of the notch, the bulk stress controls and the behavior is the same as if the notch were collapsed to become part of the crack:

$$K = FS\,(\pi a)^{1/2} \tag{8}$$

where $a = c + \ell$, c being the notch depth and ℓ the crack length measured from the notch root. The quantity F is a finite width correction factor which is a dimensionless function of the geometry and loading configuration.

In the region of dominance of Eq 8, the plastic zone size may be compared with the extended crack length, $a = c + \ell$, and the limitations of LEFM are much more easily satisfied. Since the geometry term F is often near unity for typical notch geometries, Eq 7 implies that the plastic zone size is small compared with crack length a, except for nominal stresses approaching general yielding. Hence the plasticity corrections are not generally needed. For example, the detailed elastic-plastic analysis of Sumpter and Turner [11] indicates that for a crack growing from a circular hole, the plasticity effect on K is 10% at $S/\sigma_y \simeq 0.73$ and 20% at $S/\sigma_y \simeq 0.85$.

Perhaps the most widely accepted correlation between constant-amplitude fatigue crack growth and applied loads has been proposed by Paris [12]. The rate of crack propagation per cycle is directly related to cyclic stress intensity in the following form:

$$da/dN = C(\Delta K)^m \qquad (9)$$

where C and m are material constants. In the simplest form, crack propagation lives are obtained by substituting an effective stress intensity and integrating Eq 9 from the initial crack size (a_0) to the final crack size (a_f):

$$N_P = \int_{a_0}^{a_f} da/[C(\Delta K_{eff})^m] \qquad (10)$$

Several models have been proposed for determining effective stress intensity factors that account for load ratio, sequence, and crack closure effects. A simple definition of effective stress intensity will be used here. Cracks are assumed to grow under tensile loads only. The compressive portion of the loading cycle is neglected.

The initial crack size is assumed to be equal to the notch depth. This approach may encounter difficulty for very brittle materials where the critical crack size is much less than the notch depth.

No plasticity corrections are made to the stress intensity. For ductile materials, propagation is predicted to terminate when the uncracked ligament reaches a net section stress equal to the yield strength. This assumption results in an estimated crack propagation life curve that approaches an upper limiting stress corresponding to net section yielding of an uncracked specimen. Fracture toughness is used to compute the final crack length and upper limiting stress for materials that meet plane strain requirements. This crack length was used to terminate the propagation calculations if it was smaller than the crack length calculated from the limit load.

Analysis of a wide varity of notched fatigue data [13–15] reveals that there is a limiting stress below which failure does not occur. This is consistent with the fracture mechanics concept of a threshold stress intensity factor (K_{th}) below which cracks will not grow even if they are initiated. Considering the notch depth c as the crack length, these limiting stresses are given by

$$S_{th} = K_{th}/F(\pi c)^{1/2} \qquad (11)$$

Again, F is typically the finite width correction and has a value near unity.

Results and Discussion

The procedure for estimating the fatigue lives of notched members may be divided into three life regimes:

1. *Short lives ($< 10^2$ cycles)*—Limit load concepts are used to determine the overall load-carrying capacity of the notched member. For situations that meet plane strain requirements, fracture toughness would be used to obtain

an estimate of the maximum load. Calculations show that the difference in allowable stress for fatigue lives from 1 to 100 cycles is usually less than 10%. This life regime can be treated as a simple strength problem because the behavior is dominated by the limit load.

2. *Intermediate lives (10^2 to 10^6 cycles)*—Both initiation and propagation lives are computed. Total life is taken as the summation of the two portions of the life. Fatigue notch factors are not used to compute the initiation life. The notch depth is used as the initial crack size for the propagation calculations.

3. *Long lives ($>10^6$ cycles)*—The threshold stress for both initiation and propagation are computed. The behavior of the notched member will be determined by the larger of the two threshold stresses.

For the model to be useful, it must be able to assess the fatigue life for a wide variety of experimental conditions. These include variations in specimen type, notch size, notch acquity, and material properties. Twelve sets of fatigue data found in the literature were analyzed to determine the validity of the model. Experimental data (symbols) and analytical results (lines) are shown in Figs. 1 to 12.

These data include three specimen types: double side notch, center notched, and blunt notch compact tension. Materials included 4340 and A-36 steels as well as 2024-T351, 7075-T6, and 7075-T651 aluminum. Stress concentration factors ranged from 2.1 to 10.7. Notch depths ranged from 2.5 to 31 mm. Stress ratios include values near $R = 0$ and $R = -1$.

Strain life material properties were obtained from the same references as the notch fatigue data. For steels, the fatigue limit was estimated from the ultimate strength. Long-life fatigue data were available for the aluminum alloys. In some instances, crack growth data were not published with the notch fatigue data and were obtained from other literature sources. Threshold stress intensity factors were not available for any of the tests. Best estimates were made from published data, generally for $R = 0$.

Effects of the stress concentration factor are shown in Figs. 1 and 2. The specimens were of the same geometry except that one had a sharper notch. At lives lower than 10^3 cycles, the limit load of the specimen governs the fatigue life. The simple limit load analysis is applicable to elastic–perfectly plastic materials. To account for the strain hardening of this material, an effective yield strength was taken as the average of the yield and tensile strength. These computations are described in more detail in Ref *16*. If the yield strength were used, the predictions would be lower and provide a better fit to the data. Between 10^3 and 10^5 cycles, crack growth consumes the majority of the fatigue life for both high and low K_t. At long lives, propagation dominates at high K_t and initiation at low K_t. At high values of K_t, crack initiation can be neglected. Correlation at long lives is affected by the simple assumptions made to obtain values for the fatigue limit and threshold stress intensity.

Figures 3 and 4 show the effect of specimen thickness. Both analytical and

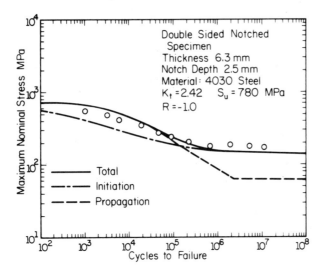

FIG. 1—*Experimental data* [17] *and analytic estimates for a blunt notched specimen with* $K_T = 2.42$.

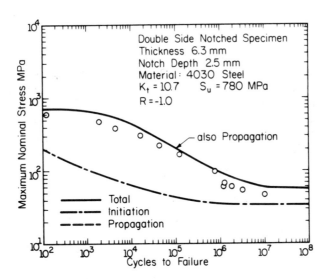

FIG. 2—*Experimental data* [17] *and analytic estimates for a sharp notched specimen with* $K_T = 10.7$.

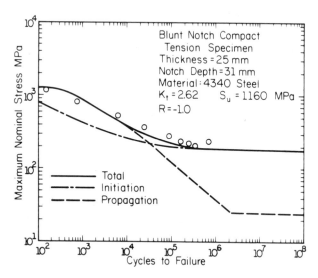

FIG. 3—*Experimental data and analytic estimates for a 25-mm-thick specimen* [18].

experimental results show no effect of thickness for the 5 and 25 mm thick specimens. Strain gage measurements show that the thinner specimen was in plane stress. The thicker specimen was between plane stress and plane strain. Effects of notch depth are shown in Figs. 4 and 5. The observed differences are beyond those that can be explained by the strength of the material. Calculated crack initiation lives will always be independent of specimen size and thickness, since they only depend on K_t and S. Crack growth will be influenced by notch depth and width of specimen. These tests show the same trends observed in the double side notch specimen: limit load failure at short lives, propagation dominance at intermediate lives, and initiation at long lives because of the low K_t.

Similar trends are also seen in the center notched data shown in Figs. 6 and 7 for two notch sizes. Note that as the notch size becomes larger, initiation contributes a larger fraction of the total life.

The preceding data were all determined on a high strength steel. Similar trends are observed on a lower strength steel. Results are shown in Figs. 8 and 9. Figure 10 gives similar results for a high strength aluminum alloy.

Effect of stress ratio change from 0.1 to −1.0 for an aluminum alloy is shown in Fig. 11. Decreasing stress ratio increases the propagation portion of the life because only the tensile portion of the load cycle is employed for crack growth while the entire stress range is effective in crack initiation.

The crack propagation life was a significant fraction of the total fatigue life at some or all stress levels in all the tests and analyses with the exception of that shown in Fig. 12 for a high strength aluminum alloy. This was observed

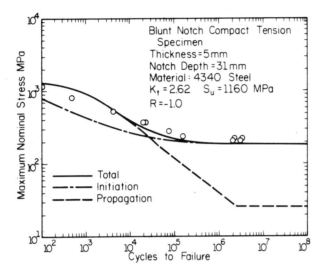

FIG. 4—*Experimental data* [18] *and analytic estimates for a 5-mm-thick specimen.*

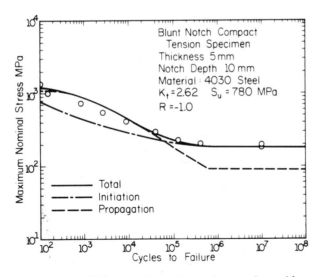

FIG. 5—*Experimental data* [19] *and analytic estimates for a specimen with a notch depth of* 10 mm.

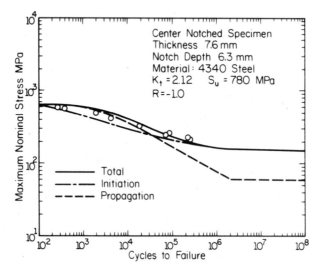

FIG. 6—*Experimental data [19] and analytic estimates for a center notched specimen with a 6.3 mm notch depth.*

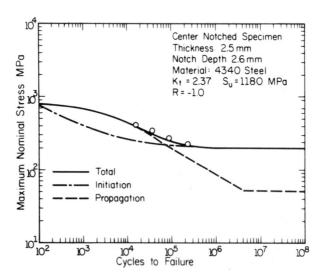

FIG. 7—*Experimental data [20] and analytical results for a center notched specimen with a 2.6 mm notch depth.*

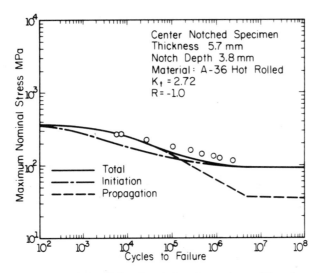

FIG. 8—*Experimental data* [21] *and analytic estimates for a mild steel specimen.*

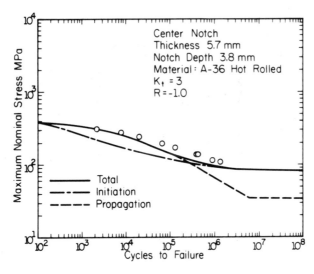

FIG. 9—*Experimental data* [21] *and analytic estimates for a mild steel specimen with higher* K_T.

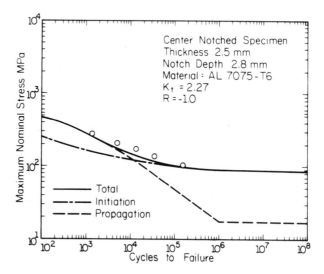

FIG. 10—*Experimental data* [20] *and analytic estimates for a center notched aluminum specimen.*

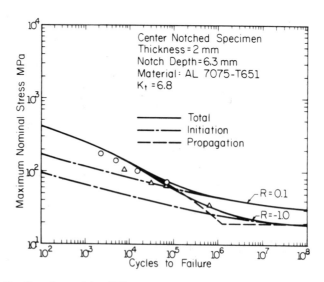

FIG. 11—*Experimental data* [22] *and analytic estimates showing stress ratio effects.*

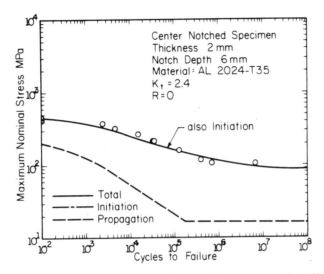

FIG. 12—*Experimental data [23] and analytic estimates for a center notched 2024 specimen.*

for 2024 aluminum with a K_t of 2.4 and stress ratio of 0.1 where crack initiation dominates at all stress levels and lives.

Recall the comments made earlier concerning the use of a crack length equal to the notch depth, $a_0 = c$, as the initial condition for calculating the crack propagation life. The smallest c included in the data presented is 2.5 mm, so that some caution is advised in dealing with very shallow notches. For example, the model used in this paper may not be very accurate where it is desired to treat very small, relatively smooth, corrosion pits as geometric notches.

Consider the case of a 90-deg fillet, such as in the "T-head" connection, or a step change in a shaft diameter. In such situations, there is obviously a stress raiser, but a notch size dimension analogous to c in a notched plate is not readily defined. Hence the model of this paper is not sufficiently general to include such cases, and additional work is needed.

For deeply notched members, the model of this paper is applicable, provided appropriate modifications are made as suggested in Ref 17. In particular, the notch dimension c is replaced by the width of the material remaining after the notch is cut out. See Ref 17 for details.

Summary

A simple model for estimating fatigue lives of notched plates has been presented. The following conclusions can be drawn for low and medium strength steels:

1. Reliable fatigue life estimates can be obtained without detailed consideration of the growth of small cracks in the notch plastic zone.

2. At high stress levels, net section plasticity has the dominant effect on fatigue lives.

3. Crack growth consumes the majority of the fatigue life at intermediate stress levels for high and low values of K_t.

4. At long lives, crack nucleation dominates the behavior for values of K_t between 2 and 3. Nonpropagating cracks dominate the behavior for higher values of K_t.

For the high strength aluminum alloys considered, the following conclusions can be drawn:

1. Crack growth consumes the majority of the fatigue life for intermediate lives and high values of K_t.

2. At lower stress levels, crack initiation consumes the majority of the fatigue life even for higher values of K_t.

3. Crack initiation consumes all of the fatigue life for lower values of K_t for all stress levels for $R = 0$ loading.

A similar model is expected to be applicable to deeply notched members, but the model does not apply in its present form to stress raisers similar to 90-deg fillets. The model has not been examined for very shallow notches, such as tiny corrosion pits, and the possibility exists that it may be inaccurate in such cases.

References

[1] Wetzel, R. M., Ed., *Fatigue under Complex Loading: Analysis and Experiments*, Society of Automotive Engineers, Warrendale, Pa., 1977.

[2] Morrow, J. and Socie, D. F. "The Evolution of Fatigue Crack Initiation Life Prediction Methods," in *Fatigue 81*, Society of Environmental Engineers, England, 1981, pp. 3-21.

[3] *Design of Fatigue and Fracture Resistant Structures, ASTM STP 761*, P. R. Abelkis and C. M. Hudson, Eds., American Society for Testing and Materials, 1982.

[4] *Fracture and Fatigue Control in Structures*, S. T. Rolfe and J. M. Barsom, Prentice-Hall, Englewood Cliffs, N.J., 1977.

[5] Newman, J. C., "An Improved Method of Collocation for the Stress Analysis of Cracked Plates with Various Shaped Boundaries," NASA TN D-6376, Washington, D.C., Aug. 1971.

[6] Emery, A. F. and Walker, G. E., "Stress Intensity Factor for Edge Cracks in Rectangular Plates with Arbitrary Loadings," ASME Publication 68-WA/MET-18, presented at ASME Winter Annual Meeting and Energy Systems Exposition, New York, Dec. 1968.

[7] Neuber, H., *Journal of Applied Mechanics, Transactions of ASME*, Vol. 28, No. 4, 1961, pp. 544-560.

[8] Seeger, T. and Heuler, P., *Journal of Testing and Evaluation*, Vol. 8, No. 4, 1980, pp. 199-204.

[9] Wilson, W. K., *Journal of Pressure Vessel Technology, Transactions of ASME*, Vol. 96, No. 4, Nov. 1974, pp. 293-298.

[10] Topper, T. H. and Sandor, B. I. in *Effects of Environmental and Complex Load History on Fatigue Life, ASTM STP 462*, American Society for Testing and Materials, 1970, pp. 93-104.

[11] Sumpter, J. D. G. and Turner, C. E., "Fracture Analysis in Areas of High Nominal Strain," presented at 2nd International Conference on Pressure Vessel Technology, San Antonio, Tex., 1973.

[12] Paris, P. C., "The Fracture Mechanics Approach to Fatigue," in *Proceedings*, 10th Sagamore Conference, Syracuse Univ. Press, N.Y., 1963.

[13] Frost, N. E., *Journal of Mechanical Engineering Science*, Vol. 2, No. 2, 1960, pp. 109–119.

[14] Frost, N. E. and Dugdale, D. S., *Journal of the Mechanics and Physics of Solids*, London, Vol. 5, 1957, pp. 182–192.

[15] Frost, N. E., *Proceedings of Institution of Mechanical Engineers*, London, Vol. 173, No. 35, 1959, pp. 811–827.

[16] Cox, E. P. and Lawrence, F. V. *ASTM STP 700*, American Society for Testing and Materials, 1980, pp. 529–551.

[17] Dowling, N. E., *Fatigue of Engineering Materials and Structures*, Vol. 2, No. 2, 1979, pp. 129–138.

[18] Dowling, N. E. and Wilson, W. K. "Analysis of Notch Strain for Cyclic Loading," presented at 5th International Conference or Structural Mechanic, in *Reactor Technology*, West Berlin, Germany, Aug. 1979, Vol. L, Paper L13/4.

[19] Dowling, N. E. and Wilson, W. K., "Geometry and Size Requirements for Fatigue Life Similitude among Notched Members," presented at Fifth International Conference on Fracture, Cannes, France, March 1981, in *Advances in Fracture Research*, Vol. 2, E. Francois, Ed., Pergamon Press, New York, 1982, pp. 581–588.

[20] Raske, D. T., "Section and Notch Size Effects in Fatigue," TAM Report 360, College of Engineering, University of Illinois at Urbana-Champaign, 1972.

[21] Sehitoglu, H., "Fatigue of Low Carbon Steels as Influenced by Repeated Strain Aging," Fracture Control Program Report 40, College of Engineering, University of Illinois at Urbana-Champaign, 1981.

[22] Kurath, P., Socie, D. F., and Morrow, JoDean, "A Nonarbitrary Fatigue Crack Size Concept to Predict Total Fatigue Lives," AFFDL-TR-79-3144, Wright-Patterson AFB, Ohio, 1979.

[23] Topper, T. H., Wetzel, R. M., and Morrow, JoDean, *Journal of Materials*, Vol. 4, No. 1, 1969, pp. 200–209.

M. Jolles[1] and V. Tortoriello[2]

Effects of Constraint Variation on the Fatigue Growth of Surface Flaws

REFERENCE: Jolles, M. and Tortoriello, V., **"Effects of Constraint Variation on the Fatigue Growth of Surface Flaws,"** *Fracture Mechanics: Fifteenth Symposium, ASTM STP 833*, R. J. Sanford, Ed., American Society for Testing and Materials, Philadelphia, 1984, pp. 300–311.

ABSTRACT: Studies have shown that fatigue growth of semielliptic surface flaws in plates is not adequately predicted solely by stress intensity factor analysis. The variation in stress field triaxiality along the flaw border must be an important factor contributing to differences between predicted and observed fatigue crack growth behavior. An analysis is formulated to examine surface flaw fatigue growth accounting for constraint variation using the concept of crack closure. The extent of crack closure is quantified for plane stress versus plane strain conditions. Changes in the geometric parameters describing flaw size and shape are studied. Applications of the analysis are compared with experiments. The use of the concept of crack closure to account for constraint variation along the border of a surface flaw yields excellent results.

KEY WORDS: fracture mechanics, fatigue (materials), surface flaw, fatigue crack growth

Accurate analyses of fatigue crack growth and fracture for three-dimensional geometries such as a surface flawed plate have developed slowly due to complications such as the lack of a closed-form solution for the stress intensity factor, multidirectional and nonselfsimilar crack growth, and variations in the stress intensity factor and stress field triaxiality along the flaw border. Although reviews of stress intensity factor solutions [1–2] and fatigue life predictions [3] indicate that great progress has been made, it has become apparent that some of the complicating factors, such as variation in the stress field triaxiality along the flaw border, may have significant effects on fatigue crack growth analysis.

Studies have shown that fatigue growth of semielliptic surface flaws in plates is not adequately predicted solely by stress intensity factor analysis [4–5]. The variation in constraint along the flaw border, from that of plane strain at the point of maximum flaw penetration to plane stress at the point of flaw intersec-

[1]Head, Fracture Mechanics Section, Naval Research Laboratory, Washington, D.C. 20375.
[2]Captain, Escuela Superior de la Fuerza Aerea Venzzolana, Caracas 1010, Venezuela.

tion with the plate surface, must be an important factor contributing to differences between predicted and observed fatigue crack growth behavior.

The variation in constraint along the flaw border results in a variation in the
extent of near crack tip yielding [6]. Since crack closure is caused by plastic
deformation left in the wake of the advancing crack front [7], it follows that the
variation in constraint can be accounted for by varying the amount of crack
closure occuring along the flaw border. Such an approach has been used by the
authors [8] to study semielliptic surface flaw geometry variations during low
stress ratio fatigue growth. Here, the analysis of constraint effects on crack
closure and surface flaw fatigue growth is further developed and compared
with results of experiments.

Analysis

Surface Flaw Geometry

Consider the semielliptic surface flaw in a plate as shown in Fig. 1. The dimension a is the crack depth, c is the crack half-length along the plate surface,
w is the plate width, t is the plate thickness, and ϕ is the parametric angle defining a location on the flaw border.

Since observations of such cracks are that they remain essentially semielliptic
[9], the growth of the surface flaw can be studied by analyzing the growth at the
points of maximum flaw penetration (Point a) and intersection of the flaw and
free surface (Point c).

Crack Closure

Implicit in an approach relating the fatigue crack growth rate to the stress
intensity factor range (ΔK) are the concepts that only the tensile portion of the
loading cycle contributes to the growth of the crack and that the crack surfaces

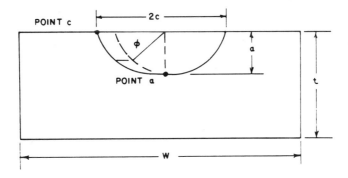

FIG. 1—Surface flaw geometry.

are fully open during the positive loads. Elber [7] first reported observing that fatigue cracks close at positive loads during constant-amplitude loading. It follows that fatigue crack growth will occur only during the portion of the loading cycle when the crack surfaces are fully open. Thus it is appropriate to use the crack opening stress level (σ_{op} in Fig. 2) as the reference stress level and to introduce an effective stress intensity factor range (ΔK_{eff}) defined over the portion of the loading cycle when the crack is fully open.

The extent of crack closure can then be conveniently quantified by use of the crack closure parameter

$$U = \frac{\sigma_{max} - \sigma_{op}}{\sigma_{max} - \sigma_{min}} = \frac{\Delta K_{eff}}{\Delta K} \tag{1}$$

which must fall within the limits $0 \le U \le 1$. For a given material, the crack closure parameter is found to be only a function of the stress ratio (R) for constant-amplitude loading [10].

The crack opening stress at the intersection of a flaw and a free surface can be determined experimentally from a plot of applied stress versus displacement across the crack faces (Fig. 3). The crack opening stress corresponds to the value of the stress at which a linear response is attained. Then, Eq 1 can be used to determine the value of the crack closure parameter under plane stress conditions at Point c, U_c.

Results from such experiments using compact tension specimens of 2024-

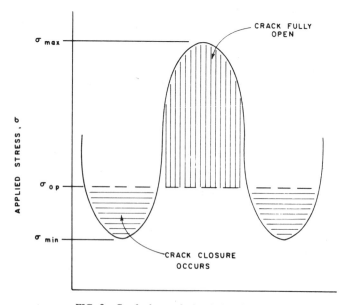

FIG. 2—*Crack closure during fatigue loading.*

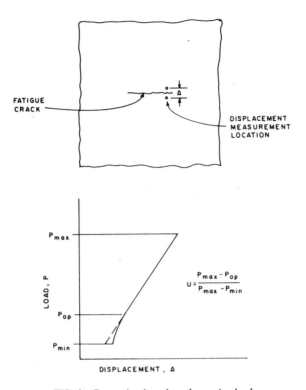

FIG. 3—*Determination of crack opening load.*

T351 aluminum are shown in Fig. 4. A linear least squares analysis of measurements of the crack closure parameter at different stress ratios results in

$$U_c = 0.707 + 0.408R \qquad (2)$$

The determination of the crack closure parameter under plane strain conditions at the point of maximum flaw penetration (U_a) would be difficult due to lack of physical access to the interior of the specimen. However, previous work by the authors [8] have determined

$$\frac{U_c}{U_a} = 0.911 \qquad (3)$$

Thus U_a can be deduced from knowledge of U_c and imposing the limit $U \leq 1$. This result can be expected to be valid, independent of material, provided that the flaw and plate geometry are such that plane strain conditions exist at the crack tip region in the interior of the plate.

A plot of U_c and the deduced value of U_a versus the stress ratio is shown in

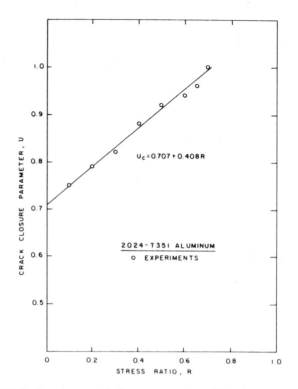

FIG. 4—*Plane stress crack closure parameter for 2024-T351 aluminum.*

Fig. 5. It can be noted that no crack closure will occur under plane stress conditions ($U_c = 1$) when $R \geq 0.718$; no crack closure will occur under plane strain conditions ($U_a = 1$) when $R \geq 0.500$. This is in excellent agreement with finite element analyses of crack closure by Newman [11].

Surface Flaw Growth

For the case of isotropic fatigue crack growth properties, within the range where a linear relationship is observed on a logarithmic plot of the fatigue crack growth rate versus effective stress intensity factor range, the crack growth rates at Points a and c can be written as

$$\frac{da}{dN} = C(\Delta K_{\text{eff}_a})^n \qquad (4)$$

$$\frac{dc}{dN} = C(\Delta K_{\text{eff}_c})^n \qquad (5)$$

where C and n are material properties.

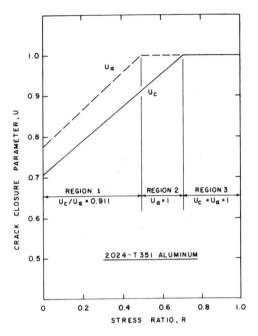

FIG. 5—*Crack closure parameters for 2024-T351 aluminum.*

The stress intensity factor for remote tension (σ) is given by Newman and Raju [12] in the form

$$K = \sigma \left(\frac{\pi a}{Q} \right)^{1/2} F\left(\frac{a}{t}, \frac{a}{c}, \frac{c}{w}, \phi \right) \qquad (6)$$

Combining Eqs 1, 4, 5, and 6 and writing in incremental form yields

$$\Delta a = \left(\frac{U_a F_a}{U_c F_c} \right)^n \Delta c \qquad (7)$$

It is interesting to note that this relation is independent of the magnitude of loading.

Thus, for any given geometry, Ref 12 allows the determination of the boundary correction factors at the points of interest, F_a and F_c. Fatigue crack growth rate testing permits the determination of n. Then, knowledge of the crack closure parameters under plane stress and plane strain conditions allows the solution for the increase of flaw depth (Δa) given an arbitrary small extension of flaw length (Δc). Repeated application of Eq 7, incrementing the flaw dimensions after each calculation, allows geometry variations during fatigue growth to be studied.

Flaw Growth Behavior

Equation 7 implies that given a flaw geometry which determines the boundary correction factors, the change in the size and shape of a flaw in a given material is governed solely by the ratio of the crack closure parameters U_c/U_a. It is apparent, from observation of Fig. 5, that there are three distinct regions of flaw growth behavior. For $0 \leq R \leq 0.500$, crack closure will occur along the entire flaw border. Here, $U_c/U_a = 0.911$ and the flaw geometry variation will be independent of the stress ratio. For $0.500 < R < 0.718$, crack closure will not occur along the interior of the flaw border where plane strain conditions exist, but will occur near the free surface. Here, the ratio U_c/U_a and the flaw geometry variation are a function of the stress ratio. For $R \geq 0.718$, no crack closure occurs along the entire flaw border. Here, $U_c = U_a = 1$ and the flaw geometry variation is independent of the stress ratio.

Analyses were performed to study flaw geometry changes in view of the foregoing observations. Fatigue growth of initially semicircular flaws of relative flaw depth $a/t = 0.1$ were predicted for loading by various stress ratios through

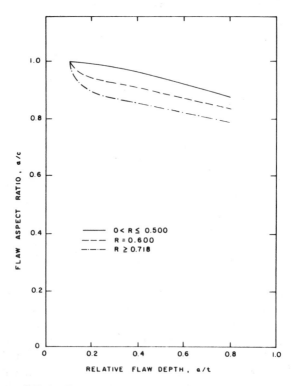

FIG. 6—*Geometry histories for initially semicircular flaws.*

application of Eq 7. Results of the analyses are presented in Fig. 6. A wide variation in the subsequent flaw geometries is observed.

Additional analyses were performed to study changes in geometry of cracks of initial relative flaw depth $a/t = 0.2$ and various initial aspect ratios for fatigue loading with a range of stress ratios. The results, shown in Fig. 7, indicate that the growth pattern of the initially shallow semicircular flaw always serves as an asymptote for the growth patterns of all other initial geometries.

Experiments

Experiments were conducted to assess the accuracy of the analyses. Eight fatigue crack growth rate tests were conducted in accordance with ASTM Test for Constant-Load-Amplitude Fatigue Crack Growth Rates Above 10^{-8} m/Cycle (E 647) on compact tension specimens of 2024-T351 aluminum to determine the range of ΔK_{eff} for which Eq 7 is valid as well as the crack growth rate exponent n. The results are presented in Fig. 8. If $2 \leq \Delta K_{eff} \leq 20$ MPa

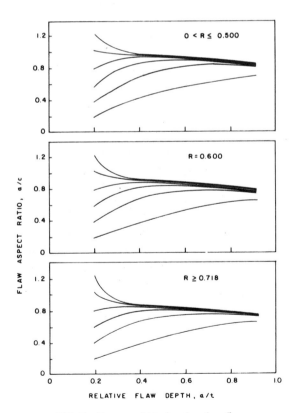

FIG. 7—*Geometry histories of surface flaws.*

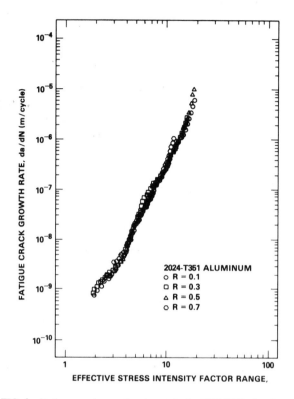

FIG. 8—*Fatigue crack growth rate results for 2024-T351 aluminum.*

$m^{1/2}$ a nearly linear relation is observed and the analytic procedure will be valid. At least squares analysis results in $n = 3.76$. Note that when an effective stress intensity factor range is used, the measured fatigue crack growth rates are independent of the stress ratio.

Eight surface flaw specimens were fabricated of 2024-T351 aluminum with a width $w = 2.54$ cm and thickness $t = 2.54$. Groups of four specimens were prepared with similar initial flaw geometry obtained by electrode-discharge machining (EDM). Within each group, one specimen was fatigued in a servo-hydraulic test system at each stress ratio $R = 0.1$, 0.3, 0.6, and 0.75, a maximum load of 124 MPa, and frequency of 10 Hz until the flaw reached a predetermined length along the surface. The specimen was then loaded to failure and the initial and final flaw dimensions measured.

The initial and final flaw geometry and loading parameters for each experiment are summarized in Table 1. The range of ΔK_{eff} achieved during all experiments is within the limits qualified above.

Although the crack length to plate width ratios ($2c/w$) obtained during the experiments are as large as 0.96, the stress intensity factor solution utilized in

TABLE 1—*Flaw geometry and loading parameters.*

| Experiment Number | Initial Geometry | | | Final Geometry | | | | Range of ΔK_{eff}, MPa m$^{1/2}$ |
| | | | | Experiment | | Analysis | | |
	a/t	a/c	Stress Ratio	a/t	a/c	a/t	a/c	
1	0.12	1.02	0.10	0.22	0.99	0.22	0.99	6.07 to 8.45
2	0.16	0.99	0.30	0.43	0.93	0.43	0.95	6.01 to 13.80
3	0.15	0.99	0.60	0.42	0.90	0.42	0.90	3.66 to 9.00
4	0.15	1.00	0.75	0.40	0.84	0.40	0.86	2.28 to 4.81
5	0.12	0.68	0.10	0.45	0.94	0.45	0.94	6.21 to 17.98
6	0.14	0.70	0.30	0.44	0.92	0.44	0.94	5.71 to 14.61
7	0.15	0.79	0.60	0.41	0.89	0.41	0.90	3.90 to 8.89
8	0.17	0.80	0.75	0.39	0.83	0.39	0.85	2.74 to 5.57

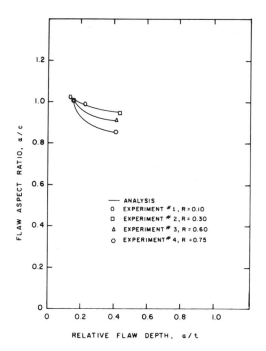

FIG. 9—*Comparison of geometry histories of experiments with analyses for Group 1.*

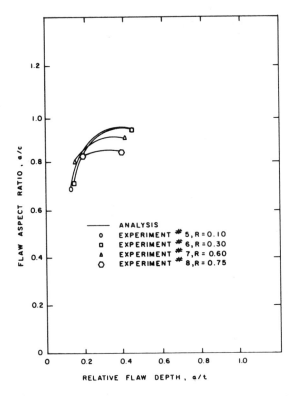

FIG. 10—*Comparison of geometry histories of experiments with analyses for Group 2.*

the analysis has a stated limit $2c/w < 0.5$ [12]. Previous work by the authors [8], however, has applied the stress intensity factor solution to surface flaws as large as $2c/w = 0.94$ without any apparent loss in accuracy. In addition, the term in the boundary correction factor of Eq 6 which is a function of c/w was developed utilizing experimental stress intensity factors for surface flaws as large as $2c/w = 0.82$ [13]. Thus the extrapolation of the stress intensity factor solution is expected to have little significance on the predicted results.

The results from the experiments and application of the analysis to predict the flaw geometry variations are presented in Figs. 9 and 10. Excellent agreement is obtained between the experiments and analyses. It is noted that the flaw geometries observed when $R = 0.1$ and 0.3 are essentially the same. The flaw geometries observed when $R = 0.6$ and 0.75 are distinctly different from each other as well as the low stress ratio flaw growth. These observations are in agreement with the aforementioned arguments and analyses.

Summary and Conclusions

Fatigue growth of surface flaws cannot be predicted solely by a stress intensity factor analysis. Studies of geometry variations indicate that the effects of variation of the stress field triaxiality are significant. The concept of crack closure is used to account for constraint variation.

The extent of crack closure, as quantified by the crack closure parameter U, is significantly different under plane stress versus plane strain conditions for low stress ratios.

There exists a distinct value of the stress ratio above which crack closure will not occur. This is at a moderate stress ratio for plane strain conditions and at a high stress ratio for plane stress conditions.

The relative amount a crack closure along the flaw border determines the geometric growth patterns of a semielliptic surface flaw. Thus the geometry history of a given flaw is only a function of the stress ratio.

The use of the concept of crack closure to account for constraint variation along the border of a surface flaw yields excellent results.

Acknowledgments

Portions of this work were conducted by the authors at the University of Missouri.

References

[1] Newman, J. C., Jr., "A Review and Assessment of the Stress Intensity Factors for Surface Cracks," NASA TM 78805, Washington, D.C., 1978.

[2] McGowan, J. J., Ed., *Experimental Mechanics*, Vol. 20, No. 8, 1980, pp. 253-264.

[3] Chang, J. B., Ed., *Part Through Crack Fatigue Life Prediction, ASTM STP 687*, American Society for Testing and Materials, 1979.

[4] Hodulak, L., Kordisch, H., Kunzelmann, S., and Sommer, E. in *Fracture Mechanics, ASTM STP 677*, American Society for Testing and Materials, 1979, pp. 399-410.

[5] Newman, J. C., Jr., and Raju, I. S., "Analyses of Surface Cracks in Plates under Tension or Bending Loads," NASA TP 1578, Washington, D.C., 1979.

[6] Aurich, D., *Engineering Fracture Mechanics*, Vol. 7, No. 4, 1975, pp. 761-765.

[7] Elber, W., *Engineering Fracture Mechanics*, Vol. 2, No. 1, 1970, pp. 37-46.

[8] Jolles, M. and Tortoriello, V., in *Fracture Mechanics: Fourteenth Symposium—Volume I: Theory and Analysis, ASTM STP 791*, American Society for Testing and Materials, 1983, pp. I-297-I-307.

[9] Corn, D. L., *Engineering Fracture Mechanics*, Vol. 3, No. 1, 1971, pp. 45-52.

[10] Elber, W. in *Damage Tolerance in Aircraft Structures, ASTM STP 486*, American Society for Testing and Materials, 1971, pp. 230-242.

[11] Newman, J. C., Jr., "A Crack Closure Model for Predicting Fatigue Crack Growth under Aircraft Spectrum Loading," NASA TM 81941, Washington, D.C., 1981.

[12] Newman, J. C., Jr., *Engineering Fracture Mechanics*, Vol. 15, No. 1-2, 1981, pp. 185-192.

[13] Jolles, M., McGowan, J. J., and Smith, C. W., *Journal of Engineering Materials and Technology*, Vol. 97, No. 1, 1975, pp. 45-51.

Material Influences on Fracture

V. P. Swaminathan[1] and J. D. Landes[2]

Temperature Dependence of Fracture Toughness of Large Steam Turbine Forgings Produced by Advanced Steel Melting Processes

REFERENCE: Swaminathan, V. P. and Landes, J. D., **"Temperature Dependence of Fracture Toughness of Large Steam Turbine Forgings Produced by Advanced Steel Melting Processes,"** *Fracture Mechanics: Fifteenth Symposium, ASTM STP 833*, R. J. Sanford, Ed., American Society for Testing and Materials, Philadelphia, 1984, pp. 315–332.

ABSTRACT: Three advanced steel melting processes—low sulfur vacuum silicon deoxidation, electroslag remelting, and vacuum carbon deoxidation—were applied to produce three Cr-Mo-V ASTM A470, Class 8 steel forgings for steam turbine application. Plane strain fracture toughness (K_{Ic}) values were obtained using 2T and 3T compact specimens between 24 and 93°C (75 and 200°F). The elastic-plastic fracture toughness (J_{Ic}) was obtained using 1T-CT and round compact specimens at 149, 260, and 427°C (300, 500, and 800°F). Both multiple specimen "heat-tint" techniques and single specimen "unloading compliance" techniques were employed at 149°C (300°F) to generate *J-R* curves. Both of these techniques yielded results with excellent agreement. Round CT specimen results agree well with standard CT specimen results.

These advanced technology forgings show significant improvement (factors of two to three higher) in fracture toughness over conventionally produced forgings. This increase in toughness is attributed mainly to the ability of these processes in producing cleaner steel, especially with very low sulfur content, and the associated reduction in the amount of nonmetallic inclusions. Minimums in toughness and the tearing modulus at 260°C (500°F) were observed. These minimums correspond to the occurrence of minimum in the tensile ductility at the same temperature which may be due to the dynamic strain-aging phenomenon.

KEY WORDS: advance melting, Cr-Mo-V steel, forgings, steam turbines, fracture toughness, J_{Ic}, *J-R* curve, K_{Ic}, tearing modulus, mechanical properties

Metallurgically cleaner and higher quality steel rotor forgings can make an important contribution to both the performance and the reliability of steam turbines used in the power generation industry. Because of some rotor failures

---

[1] Senior Engineer, Westinghouse Steam Turbine Generator Division, Orlando, Fla. 32817.
[2] Manager, Mechanics of Materials, Westinghouse R&D Center, Pittsburgh, Pa. 15235.

315

in the 1950s, considerable effort was directed toward improving the quality of large rotor forgings used in the manufacturing of these steam turbines. Significant improvements have been achieved in the areas of steel refining, vacuum casting technology, ingot making, and forging methods. To study the effects of three of these new steel melting and ingot making processes, three forgings were produced using these processes. These processes were low sulfur vacuum silicon deoxidation (low S), vacuum carbon deoxidation (VCD), and electroslag remelting (ESR). The material was 1Cr-1Mo-0.25V (ASTM A470, Class 8) steel, which is widely used for high-temperature steam turbine rotor forgings. A description of these processes can be found in Ref 1. The three forgings were to be used for manufacturing turbine rotors to be placed in service in three operating power plants.

Fracture mechanics principles and related material properties are increasingly used in flaw tolerant design and structural integrity evaluation of large structures. Fracture toughness determination has been an important task in the material evaluation program. Since the steam turbines experience a wide range of temperatures, it is important to obtain the fracture toughness values through the temperature range of operation. At low temperatures valid plane strain fracture toughness (K_{Ic}) values for the Cr-Mo-V steel can be obtained using 2T and 3T compact tension specimens. However, at and above upper-shelf temperature the elastic-plastic fracture toughness (J_{Ic}) is measured using relatively small specimens [2–4]. The purpose of this investigation was to determine fracture toughness K_{Ic} at lower temperatures (24 to 85°C) [75 to 185°F] and J_{Ic} toughness at the upper-shelf temperature (149°C) [300°F] and at two higher temperatures (260 and 427°C) [500 and 800°F] for the three advanced-technology forgings.

Materials, Specimens, and Test Techniques

The chemistry of the forgings is in accordance with ASTM A470, Class 8 specification. Since the three forgings were manufactured by three different steel mills there are some variations in the chemical composition (Table 1). Each of the forgings was forged from an ingot weighing approximately 100 metric tons. Two test ends (1070 mm diameter by 915 mm long) were provided

TABLE 1—*Chemical composition (weight percent) of the Cr-Mo-V rotor forgings.*

Process	C	Mn	P	S	Si	Ni	Cr	Mo	V	Sn	Al
Low S	0.31	0.78	0.007	0.001	0.23	0.33	1.13	1.15	0.23	0.002	0.004
VCD	0.28	0.76	0.004	0.001	0.05	0.40	1.18	1.21	0.26	0.010	0.005
ESR	0.31	0.78	0.009	0.002	0.19	0.27	1.18	1.18	0.26	0.003	0.009
Conventional											
Cr-Mo-V [6]	0.30	0.82	0.006	0.010	0.32	0.15	1.02	1.12	0.25	...	...
Joppa 3 [12]	0.26	0.75	0.017	0.020	0.30	...	0.99	1.1	0.25	...	...

at the top and bottom of the main body of the rotor forgings. Heat treatment for properties consisted of austenitizing at 950°C (1742°F), followed by forced air cooling and then tempering at 680°C (1256°F). The room temperature tensile properties and the Charpy impact properties are summarized in Table 2 for the test end and bore bar locations.

From each rotor, specimens were machined from three locations: top end, bottom end, and bore bar. The bore bars were obtained from the axial centerline of the rotor. For K_{Ic} testing, compact tension specimens of 2T and 3T geometries were machined from the top and bottom test ends. Two types of geometries were used for the J_{Ic} tests. Standard 1T-CT specimens were machined from the test ends, and round compact specimens were obtained from the round bore core bars. Sketches of the specimen geometries are shown in Fig. 1, and a photograph of the specimens used for this investigation is shown in Fig. 2.

The K_{Ic} tests were conducted on the 2T and 3T-CT specimens in accordance with ASTM E 399. Specimen orientation was such that the crack was oriented along the axial-radial plane and the crack propagated in the radial direction. For tests above room temperature, a three-zone resistance furnace was used. The temperature control was within ±2°C (3.6°F). Crack opening was monitored on the front face by a clip-on gage.

Two test techniques were used to obtain the J-R curves. The first was the multiple specimen "heat-tint" technique [4], and the second was the unloading compliance technique using a single specimen to obtain the entire J-R curve [5]. The standard and the round CT specimens had a thickness of 25.4 mm (1 in.). Single specimen and multiple specimen curves were generated for comparison at 149°C (300°F) only. At 260 and 427°C (500 and 800°F) only one specimen per curve was used. The specimens were heated using electrical heat-

TABLE 2—*Tension (25°C) [75°F] and Charpy impact properties.*[a]

Property	Test Ends				Bore Bar		
	Low S	VCD	ESR	Ref 6	Low S	VCD	ESR
Tension							
0.2% yield strength, MPa	631	641	634	626	627	644	662
Tensile strength, MPa	786	779	772	784	786	785	814
Elongation, %	22	21	21	17	21	22	21
Reduction of area, %	60	61	60	47	62	62	60
Charpy Impact							
Energy at −18°C, J	7.9	12.2	12.2		4.7	9.2	9.5
Energy at 24°C, J	14	41	23.3		10.8	18	16.7
Upper-shelf energy, J	136	142	136		134	136	109
FATT, °C	96	60	74		99	65	95

[a] 6.895 MPa = 1 ksi; 1.356 J = 1 ft · lb.

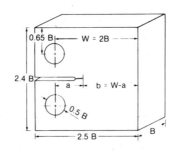

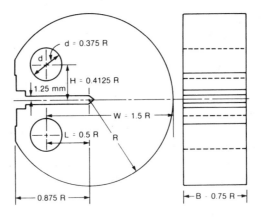

FIG. 1—*Standard and round compact tension specimens.*

ing tapes. The temperature was maintained within ±3°C (5.5°F) during the tests using thermocouples mounted in the specimen notch. All tests were conducted on servohydraulic test frames. A computer with appropriate interface system and X-Y recorder provided acquisition for load displacement and unloading compliance data. The procedure for calculation of J-values from load versus load-line displacement curves is covered in ASTM E 813.

Results and Discussion

K_{Ic} Testing

The results of the plane strain fracture toughness tests for the three forgings are summarized and presented in Table 3 and Fig. 3. Test temperature range was from room temperature (24°C) to 93°C (75 to 200°F). It was attempted to obtain valid K_{Ic} results at the maximum temperature possible without violating the size criterion. The maximum temperature where valid K_{Ic}-values could be obtained from 3T-CT specimens was 70°C (158°F). Above this temperature, shear lips developed on the fracture surfaces and the thickness criterion

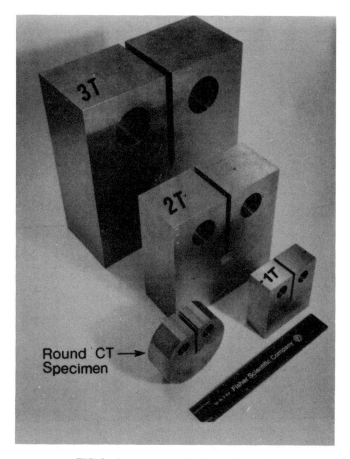

FIG. 2—*Specimens used for* K_{Ic} *and* J_{Ic} *testing.*

was also violated. All the 2T specimens were tested at room temperature. Two of the 3T specimens, one from the low-sulfur top end and one from VCD bottom end, were also tested at room temperature to see if any effect of thickness on the K_{Ic}-values existed. As seen in Fig. 3 (solid points), 3T specimen results fall in the scatter band for the corresponding rotor.

Fracture toughness results of a conventionally produced Cr-Mo-V forging from Ref 6 are also plotted in Fig. 3. The three-advanced technology rotors show a significant improvement in the fracture toughness values in this temperature range. The mean K_{Ic} values at 24°C (75°F) show improvement of about 25% for the low S, 50% for the ESR, and 75% for the VCD rotor when compared with conventional rotor toughness. Comparison of the room temperature K_{Ic}-values for the top and bottom ends of the ESR and VCD rotors reveals that the bottom end toughness is slightly but consistently higher. Since

TABLE 3—*Plane strain fracture toughness* (K_{Ic}) *values for the three advanced-technology forgings.*

Material	Specimen Type	Temperature		K_{Ic}		Valid
		°C	°F	MPa$\sqrt{m}$	ksi$\sqrt{in.}$	
Low Sulfur						
Top end	2T-CT	24	75	56.6	51.5	yes
	2T-CT	24	75	50.1	45.6	yes
	3T-CT	24	75	58.7	53.4	yes
	3T-CT	93	200	143.2	130.3	no
Bottom end	3T-CT	70	158	82.0	74.6	yes
	3T-CT	72	162	84.7	77.1	yes
VCD						
Top end	2T-CT	24	75	79.7	72.5	yes
	2T-CT	24	75	110.4	100.5	no
	2T-CT	24	75	78.5	71.4	yes
	3T-CT	38	100	82.5	75.1	yes
	3T-CT	66	151	110.6	100.6	yes
Bottom end	2T-CT	24	75	113.2	103	no
	2T-CT	24	75	85.8	78.1	yes
	2T-CT	24	75	91.3	83.1	yes
	3T-CT	24	75	92.1	83.8	yes
	3T-CT	52	125	121.3	110.4	no
ESR						
Top end	2T-CT	24	75	67.8	61.7	yes
	2T-CT	24	75	63.5	57.8	yes
	3T-CT	68	155	110.1	100.2	yes
	3T-CT	70	158	109.9	100.0	yes
Bottom end	2T-CT	24	75	70.2	63.9	yes
	2T-CT	24	75	76.3	69.4	yes
	3T-CT	70	158	139.3	126.8	no
	3T-CT	60	140	103.6	94.3	yes

only two or three specimens were tested from each end at 24°C (75°F), it can be said that for practical purposes the top and bottom end toughness is the same. The toughness increased as the test temperature was raised, ESR forging showing more rapid increase than the low S and VCD forgings. Although the VCD toughness is higher than that of the ESR at 24°C (75°F), at higher temperatures it converges towards the ESR toughness.

The improvement in the fracture toughness of these advanced-technology forgings is mainly attributed to the metallurgical cleanliness and reduction in segregation streaks that may contain nonmetallic inclusions such as manganese sulfide. The presence of inclusions can be detrimental to the tension, impact, and fracture toughness properties. Wilson reported that by reducing the sulfur in a plate steel, significant improvement in the fracture toughness can

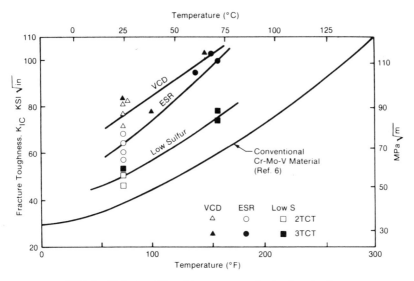

FIG. 3—*Variation of* K_{Ic} *with temperature for the three forgings.*

be obtained [7]. For the three subject forgings the sulfur level was reduced by about an order of magnitude when compared with the conventional steel (Table 1). As a result, a significant reduction in the sulfide inclusions was achieved as revealed by sulfur prints [8]. Hebsur et al reported that significant improvements in the fracture toughness and fatigue crack growth behavior of AISI 4340 were achieved by electroslag refining process [9]. Again, this was attributed to the removal of nonmetallic inclusions and the reduction of sulfur content.

The K_{Ic} results show that the toughness of the ESR is higher than that of the low S forging, and the VCD toughness is higher than that of the others. The improved cleanliness of the steel as a result of these advanced processes indirectly contributes to improved toughness. However, the material toughness and the mechanical properties are very much determined by the heat treatment variables such as austenitizing temperature, soak time, cooling rates, and tempering temperatures. Since these forgings were received from three different procedures, there may be variations in the aforementioned factors.

Charpy impact properties are very sensitive to the heat treatment conditions that determine the final microstructure. Typical curves of impact energy or percent brittle fracture versus temperature for the three forgings are shown in Fig. 4. For the VCD forging the curves were shifted towards the left, a result of higher impact energy values in the transition region and lower fracture appearance transition temperature (FATT). These curves for the ESR and low S forgings are positioned towards the right, resulting in higher FATT values.

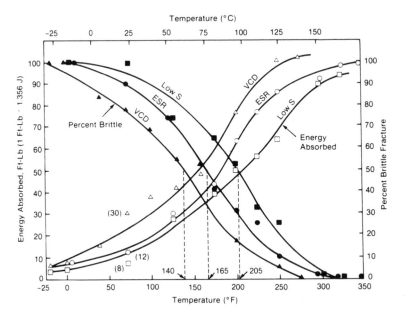

FIG. 4—*Typical Charpy impact curves for the three forgings.*

J_{Ic} Testing

The three rotor forgings are identified as Rotor A, produced by the low S process; Rotor B, produced by the VCD process; and Rotor C, produced by the ESR process. The specimens were numbered to identify the forging and location from which they were taken. For example, AB refers to specimens from the low S forging bottom end. Bore bar specimens are identified by EL. CEL refers to ESR forging bore bar location.

The multiple specimen "heat-tint" method contributes one point per specimen on the J versus crack extension plot (J-R curve). The J_{Ic}-value is obtained by fitting a straight line through points that fall within a range of crack extension (Δa) values bounded by two exclusion lines [4]. The exclusion lines are taken parallel to the blunting line at 0.15 and 1.5 mm along the Δa axis. The J_{Ic}-value is determined at the intersection of the best fit line and the blunting line given by

$$J = 2\sigma_0 \Delta a \tag{1}$$

where σ_0 is the flow stress obtained from the average of the 0.2% yield stress and the ultimate tensile stress.

The unloading compliance technique was used on the standard 1T-CT specimens from the top and bottom ends of each rotor. Comparison of the unloading compliance and the heat-tint data points is shown in Fig. 5. Excellent agreement is found between the data obtained by both of these techniques. This is typical for all other test locations. All the specimens satisfied the size requirement specified by

$$B, b > 25J/\sigma_0$$

where B is the thickness and b is the remaining ligament size.

The J_{Ic}-values for the various test locations and the slope of the R-curve (dJ/da) are summarized and presented in Table 4. Only the standard CT specimens from the test ends were tested at 260 and 427°C (500 and 800°F). In addition, K_{Ic}-values from the J_{Ic}-values were also calculated from the relationship [11]

$$K_{Ic}^2 = \frac{EJ_{Ic}}{1 - \nu^2} \tag{2}$$

where $\nu = 0.3$ (Poisson's ratio) and the values of E for the different temperatures are as given in Table 4. A summary plot of all the J-R curves at 149°C (300°F) is presented in Fig. 6. The scatter in the J_{Ic}-values for different test locations for the low S rotor is about 35 kJ/m² (200 in. · lb/in.²), which is not significant; that is, the different sections of the rotor had similar toughness. Similarly, the top and bottom ends of the VCD and ESR forgings have similar toughness within the same rotor. When the three rotors are compared, how-

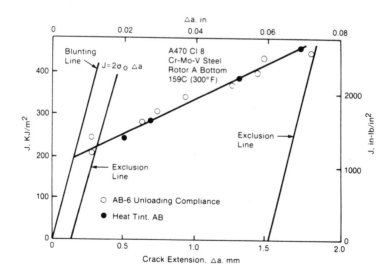

FIG. 5—*A typical plot of J versus crack extension, comparing unloading compliance with "heat-tint" technique.*

TABLE 4—*Summary of* J_{Ic} *test results for advanced rotor forgings and conventional forging data from the literature.*

Material	Temper-ature[a]	J_{Ic}		$K_{Ic}(J_{Ic})$		dJ/da		T^b
		KJ/m^2	in. · lb/in.^{2}	MPa√m	ksi√in.	MPa	ksi	
Low Sulfur (Rotor A)								
Top end	T1	149	852	181	165	132	19.1	61
	T2	105	598	151	137	93	13.5	44
	T3^c	287	1645	245	223	242	35.2	132
Bottom end	T1	188	1075	203	185	169	24.5	78
	T2	137	786	174	158	103	15	49
	T3	174	996	195	177	221	32	120
Bore bar	T1	151	861	181	165	166	24.1	77
VCD (Rotor B)								
Top end	T1	256	1462	236	215	136	19.8	63
	T2	214	1222	214	195	123	17.8	62
	T3	273	1563	228	207	193	28	105
Bottom end	T1	204	1164	212	193	152	22	70
	T2	193	1104	201	183	88	12.8	44
	T3	197	1125	202	184	143	20.7	84
Bore bar	T1	146	834	179	163	177	25.7	79
ESR (Rotor C)								
Top end	T1	145	826	178	162	209	30.3	96
	T2	73	416	125	114	162	23.5	77
Bottom end	T1	132	756	170	155	123	17.9	57
	T2	179	1024	197	179	111	16.1	52
	T3	231	1320	220	200	184	26.7	100
Bore bar	T1	104	593	151	137	206	29.9	95
Conventional [6]	T1	82	470	134	122	52	7.5	24
	T2	73	420	125	114	41	6.0	20
	T3	79	450	129	117	62	9.0	34
Joppa 3 [12]	T4	53	300	107	98	30	4.4	14

[a]T1 = 149°C (300°F); T2 = 260°C (500°F); T3 = 427°C (800°F); T4 = 204°C (400°F).
[b]To calculate T and K_{Ic} (J_{Ic}), the elastic modulus E is 200×10^3 MPa (29×10^3 ksi) at T1; 197×10^3 MPa (28.5×10^3 ksi) at T2; and 190×10^3 MPa (27.5×10^3 ksi) at T3.
[c]Invalid test.

ever, VCD top and bottom test ends show higher toughness. ESR bore bar shows somewhat lower toughness than the other locations.

The improvement in toughness of these advanced rotors over the conventional [6] and the vintage rotor toughness [12] is very significant (Fig. 7 and Table 4). The J_{Ic}-values are a factor of two to three higher and dJ/da slopes a factor of three to eight higher. Certainly, the goal of improving toughness by applying advanced steel melting techniques has been achieved. For the Joppa 3 rotor [12], the results from 204°C (400°F) tests are considered here, since this temperature was the upper-shelf temperature for that material.

Since the turbine rotors experience much higher temperatures in service, J_{Ic}-

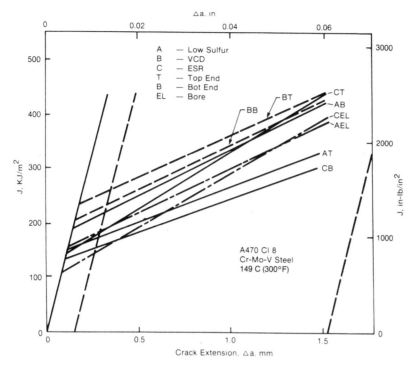

FIG. 6—*Summary of* J-R *curves obtained at 149°C (300°F).*

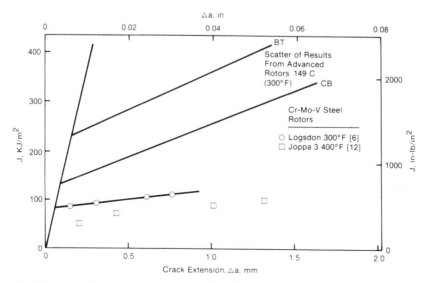

FIG. 7—*Comparison of* J-R *curve results at 149°C (300°F) of advanced rotors with the conventional and vintage rotors.*

tests were conducted to obtain toughness properties at 260°C (500°F) and 427°C (800°F). The *J-R* curves obtained from single specimen tests are summarized in Table 4 and in Fig. 8. It is seen that the *R*-curves at 427°C (800°F) are consistently positioned higher compared with 260°C (500°F) results. The *J-R* maximum and minimum curve bands at the three test temperatures are compared in Fig. 9. The 149°C (300°F) band partly overlaps the 260°C and 427°C (500°F and 800°F) bands. Variation of J_{Ic} with temperature is illustrated in Fig. 10. Even though the scatter is quite large, there is a distinct drop of toughness at 260°C (500°F). This result is quite surprising and contrary to the general belief that there is an upper-shelf toughness or drop in toughness at higher temperatures for many steels [6]. The two test locations that are exceptions to this rule are VCD and ESR bottom ends. Also given in Table 4 are the values of T, the tearing modulus, developed by Paris et al [13]. It is given by the expression

$$T = \frac{E}{\sigma_0^2} \, (dJ/da) \qquad (3)$$

where dJ/da is the slope of the best fit straight line as described earlier to obtain J_{Ic}-values. The higher the value of T, the higher the resistance of the material for slow stable crack growth after the crack initiation (J_{Ic} point). The T-values for the subject forgings are much higher compared with the conventional and vintage rotors. Variation of T with temperature is shown in Fig. 11.

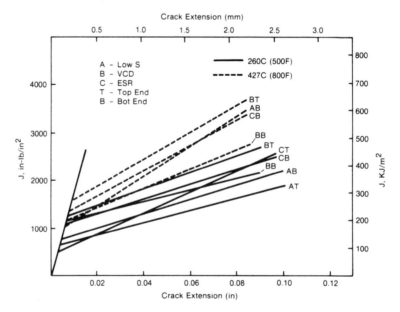

FIG. 8—*Summary of* J-R *curves from single specimens at 260°C (500°F) and 427°C (800°F).*

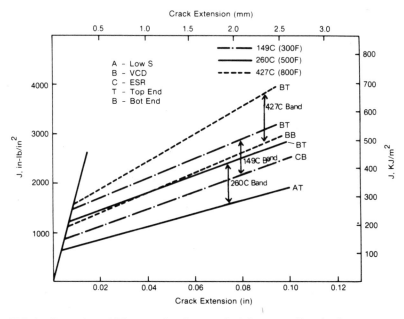

FIG. 9—*Comparison of J-R curves (maximum and minimum trend) at the three test temperatures for all test locations.*

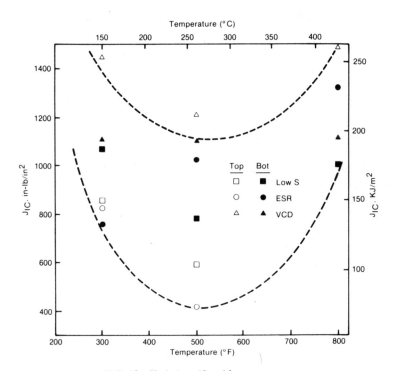

FIG. 10—*Variation of J_{Ic} with temperature.*

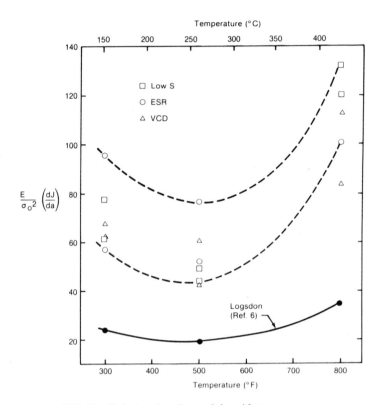

FIG. 11—*Variation of tearing modulus with temperature.*

The data at 427°C (800°F) from low S top end are not included in Fig. 10 and 11 since it was not a valid test. Occurrence of minimum at 260°C (500°F) is quite apparent in this plot. The rapid rise of the curve after 260°C (500°F) is due to a more rapid decrease in the value of the flow stress (σ_0) at the higher temperatures (Fig. 12). The behavior of T is more consistent than the J_{Ic} variation. Even though the VCD material shows somewhat higher J_{Ic}-values at all three temperatures (Fig. 10), when T-values are compared (Fig. 11) the VCD data points fall at the lower end of the scatter band. This may indicate that the mechanism(s) controlling the crack initiation (J_{Ic} point) do not necessarily control the crack propagation under stable crack growth conditions, which means J_{Ic} and T may be two independent material properties.

The possibility that the observed minimums in J_{Ic} and T at 260°C (500°F) could be due to specimen geometry or thickness is excluded. Specimen thickness was the same (25.4 mm) in all cases. The round specimen geometry and the standard CT specimens show essentially similar behavior in these tests. One possible explanation for this drop in toughness is a temperature and strain

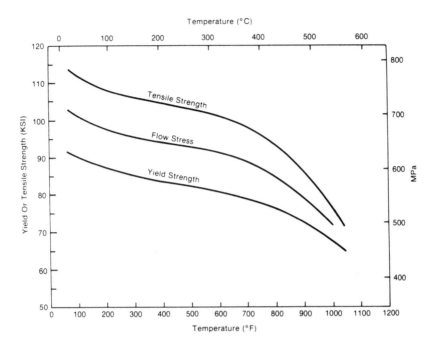

FIG. 12—*Typical plot of yield and tensile strengths or flow stress versus temperature.*

rate sensitive mechanism such as dynamic strain-aging (less frequently known as blue brittleness). This phenomenon results in reduced tensile ductility (both elongation and reduction of area) when the tests are performed in the temperature range 150 to 350°C (300 to 660°F) for carbon steels [*14*]. Typical variations of elongation and reduction of area with temperature are summarized in Fig. 13 for the forgings in this study. The curves go through a minimum at about 260°C (500°F). It should be pointed out that this behavior is in no way related to the advanced steel melting processes but is a general trend in many materials. The point of minimum toughness corresponds to the minimum in tensile ductility, an observation which supports the foregoing theory.

Evidence is also found in the literature that dynamic strain-aging during the development of the plastic zone in front of the crack may decrease fracture toughness. Oestensson reported [*15*] that the fracture toughness of an A533 pressure vessel steel exhibited a minimum at about 250°C (500°F) which was attributed to a dynamic strain-aging phenomenon caused by interstitial nitrogen. The temperature at which the minimum toughness occurred was influenced by the loading rates. The higher the loading rate, the higher this temperature [*15*]. However, more work is needed on the Cr-Mo-V steels to unequivocally prove that the dynamic strain-aging does play a role in influencing the fracture toughness.

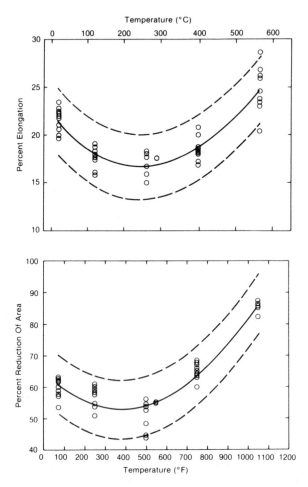

FIG. 13—*Typical variation of elongation and reduction of area with temperature.*

Summary

A summary plot of the fracture toughness variation with temperature is presented in Fig. 14. The K_{Ic}-values at and above 149°C (300°F) were calculated from J_{Ic}-values using Eq 2. VCD forging toughness is at the upper end of the scatter band at all test temperatures; however, a similar trend is not observed for the low S and ESR forgings. Even though the low S has lower toughness below 149°C (300°F) (valid K_{Ic} tests), it has somewhat higher toughness than the ESR at 149°C (300°F). At 260°C (500°F), low S data points fall within the scatter of the ESR data points. This indicates that there seems to be no direct correlation between lower temperature brittle fracture behavior and higher temperature ductile behavior; that is, if a rotor has higher toughness at lower temperatures (transition region), it may not have the same tendency at higher

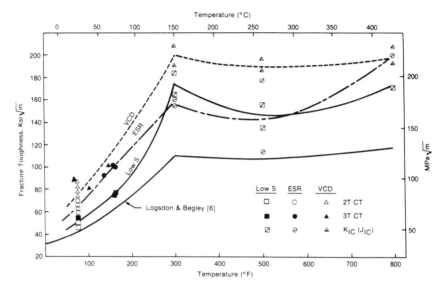

FIG. 14—*Summary plot of fracture toughness variation between 24°C (75°F) and 427°C (800°F) for the three advanced-technology forgings and the conventional forging from Ref 6.*

temperatures (ductile fracture regime) and a cross-over is possible. The important point is that when compared with conventionally produced forgings, the advanced technology forgings show superior toughness at all temperatures. This improvement can be attributed to the very low sulfur content in these forgings and reduction in the amount of nonmetallic inclusions. Improvements in the solidification patterns in the original ingots as well as currently applied forging technology may have also contributed to improved toughness. The differences in toughness among the three forgings are probably related to the variations in the heat-treatment conditions rather than to the processes, as indicated by the relative positions of the Charpy impact property curves.

Conclusions

1. The three advanced steel melting processes—low S, VCD, and ESR—significantly improved the fracture toughness (K_{Ic} and J_{Ic}) of Cr-Mo-V turbine rotor forgings. Improvement in K_{Ic} is about 25 to 75% and in J_{Ic} is about 200 to 300%. These improvements are attributed to low sulfur levels, cleaner steel, and improved forging techniques.

2. The *J-R* curves from single specimen (unloading compliance) techniques are in good agreement with multiple specimen results.

3. Round compact tension specimen results agree well with standard CT specimen results.

4. Occurrence of minimums in J_{Ic} and T at 260°C (500°F) appear to be related to the drop in tensile ductility due to a dynamic strain-aging phenomenon.

Acknowledgments

The subject work was partly funded under Electric Power Research Institute Contract RP 1343-1, R. I. Jaffee, Program Manager. The authors acknowledge and thank the EPRI and the Westinghouse Electric Corporation for permission to publish this work. The authors acknowledge the assistance of P. J. Barsotti, E. J. Helm, and A. R. Petrush of the Westinghouse R&D Center and J. K. Donald, M. R. Morgan, and G. Miller of the DEL Research Laboratory with the testing phase of the J_{Ic} evaluation. R. E. Lewis of the Steam Turbine Generator Division is thanked for performing the K_{Ic} tests. Japan Casting and Forging Corporation produced the low S forging, ARBED Saarstahl produced the ESR forging, and the Japan Steel Works produced the VCD forging.

References

[1] Swaminathan, V. P., Steiner, J. E., and Jaffee, R. I., "High Temperature Turbine Rotor Forgings by Advanced Steel Melting Technology," presented at the ASME Joint Power Generation Conference, 82-JPGC-Pwr-24, Denver, Oct. 1982.

[2] Begley, J. A. and Landes, J. D. in *Fracture Toughness, Proceedings of the 1971 National Symposium on Fracture Mechanics, Part II, ASTM STP 514*, American Society for Testing and Materials, 1972, pp. 24–39.

[3] Landes, J. D. and Begley, J. A. in *Developments in Fracture Mechanics Test Methods Standardization, ASTM STP 632*, W. F. Brown and J. G. Kaufman, Eds., American Society for Testing and Materials, 1977, pp. 57–81.

[4] Clarke, G. A. et al, *Journal of Testing and Evaluation*, Vol. 7, No. 1, Jan. 1979, pp. 49–56.

[5] Clarke, G. A., Andrews, W. R., Paris, P. C., and Schmidt, D. W. in *Mechanics of Crack Growth, ASTM STP 590*, American Society for Testing and Materials, 1976, pp. 27–42.

[6] Logsdon, W. A. and Begley, J. A., *Engineering Fracture Mechanics*, Vol. 9, 1977, pp. 461–47ᴄ.

[7] Wilson, A. D., *Journal of Engineering Materials and Technology, Transactions of ASME*, Vol. 101, July 1979, pp. 265–274.

[8] Swaminathan, V. P. and Argo, H. C., "Advanced Steel Melting Technology and Turbine-Generator Rotor Quality," presented at the Steam Turbine-Generator Technology Symposium, Westinghouse Electric Corporation, Charlotte, N.C., Oct. 1980.

[9] Hebsur, M. G., Abraham, K. P., and Prasad, Y. V. R. K., *Engineering Fracture Mechanics*, Vol. 13, No. 4, 1980, pp. 851–864.

[10] Barsom, J. M. and Rolfe, S. T., *Impact Testing of Metals, ASTM STP 466*, American Society for Testing and Materials, 1970, pp. 281–302.

[11] Begley, J. A. and Landes, J. D. in *Fracture Toughness, Proceedings of the 1971 National Symposium on Fracture Mechanics, Part II, ASTM STP 514*, American Society for Testing and Materials, 1972, pp. 1–23.

[12] Clarke, G. A., Shih, T. T., and Kramer, L. D., "Evaluation of the Fracture Properties of Two 1950 Vintage CrMoV Steam Turbine Rotors," EPRI Final Report on RP 502, Electric Power Research Institute, March 1978.

[13] Paris, P. C., Tada, H., Zahoor, A., and Ernst, H. in *Elastic-Plastic Fracture, ASTM STP 668*, American Society for Testing and Materials, 1979, pp. 124–159.

[14] Reed-Hill, R. E., *Physical Metallurgy Principles*, Van Nostrand, New York, 1973, pp. 347–351 and 796–797.

[15] Oestensson, B. in *Reliability Problems of Reactor Pressure Components*, Vol. 1, International Atomic Energy Agency, 1978, pp. 303–313.

S. J. Garwood[1]

Fracture Toughness of Stainless Steel Weldments at Elevated Temperatures

REFERENCE: Garwood, S. J., **"Fracture Toughness of Stainless Steel Weldments at Elevated Temperatures,"** *Fracture Mechanics: Fifteenth Symposium. ASTM STP 833,* R. J. Sanford, Ed., American Society for Testing and Materials, Philadelphia, 1984, pp. 333–359.

ABSTRACT: This paper comprises a study of the resistance to the propagation of a ductile tear in 316 stainless steel parent plate in all six orientations, and submerged-arc and manual metal arc weld metal and heat-affected zone (HAZ) regions in stainless weldments at 370 and 540°C. Tests were carried out in bending and tension in both load and displacement control at 370°C. The propagation resistances of the various regions are assessed from the measurement of resistance curves and maximum load toughness using both crack tip opening displacement (CTOD) and *J*-integral parameters.

The T-L orientation appears to have the poorest resistance to tearing of the four normal cracking orientations for parent plate. The S-L orientation, however, has the worst toughness overall. Heat-affected zone toughness of both manual and submerged-arc welds is equivalent to parent plate. Submerged-arc weld metal toughness is comparable with the T-L orientation in plate material. Manual metal arc weld metal toughness is inferior to the toughness exhibited by submerged-arc and the four normal cracking orientations in parent plate. Increasing the test temperature from 370 to 540°C for manual metal arc weld metal appears to produce a distinct increase in resistance to ductile crack propagation.

Provided adequate thickness constraint is ensured, maximum load toughness determinations from the standard laboratory three-point bend geometry would appear to give conservative estimates of tensile instabilities under load control. The relevance of this behavior is discussed in terms of the use of single-parameter design curve methods for the estimate of tolerable flaw sizes in structural applications

KEY WORDS: stainless steel, welded joints, plate, bend tests, crack propagation, ductile fracture, fracture toughness

Following preliminary studies, the first experimental fast breeder reactor, the Dounreay Fast Reactor (DFR), was constructed and became operational in 1963. The next step towards a commercial fast reactor (CFR) in the United Kingdom was the prototype fast reactor (PFR) which was approved in 1966

[1]Head, Fracture Section, Engineering Department, The Welding Institute, Research Laboratory, Abington, Cambridge, England.

and commissioned in 1975. The design of the first CFR is currently under consideration by the National Nuclear Corporation (NNC).

At the time of the instigation of this research program the design of the CFR was essentially based on the PFR, employing a pool system using sodium as the coolant, the core inlet temperature being 370°C with the outlet at 540°C. As with the PFR the core was to be supported by a separate removable diagrid which is in turn mounted on a circular stainless steel strongback. In the latest reference design the strongback arrangement has been eliminated and the integrity of the reactor depends on the primary vessel, which will be manufactured from 25 to 60 mm thick 316 stainless steel.

The core support structure of the CFR is obviously of extreme importance to the structural integrity of the reactor, since movement of the core away from the control rods could result in a whole core accident (WCA). Initial analyses indicated that the critical flaw size of this stainless steel application is apparently very small. There are a number of reasons behind the low numbers resulting from the analyses initially employed, not the least of which is the inapplicability of the fracture mechanics method adopted. Most currently accepted critical flaw analysis procedures operate from a linear elastic premise. The local stress is determined by multiplying the applied stress by the sum of the stress concentration factors operating. Additive to this stress are the local residual stresses, usually assumed to be of yield magnitude. This procedure can result in the determination of a local stress of seven to eight times yield. This value of local stress is then employed in the analysis with the initiation of tearing value of material toughness. The latter parameter is often determined on weldments which have experienced some degree of prestraining to simulate the effects of residual plastic strains.

This type of analysis is, of course, very conservative and may be criticized on a number of counts:

1. The use of linear elastic fracture mechanics (LEFM) is certainly inappropriate, since local yielding will tend to alleviate the high stresses in regions of stress concentration and hence invalidate the assumptions of LEFM.

2. The assumption of yield magnitude residual stresses may be inappropriate in certain situations, and it is almost certainly incorrect to apply the stress concentration factor to the residual stress component.

3. The use of prestrained material to determine toughness values relating to highly restrained welds or irradiated materials may be unduly conservative, since the effect of residual stresses is already included in the conservative assumptions of Points 1 and 2. Residual plastic strains must be allowed for, however, and this can be achieved by using simulated restraint conditions in the welded test blocks used for toughness determinations.

4. Toughness values determined at the point of initiation of tearing on small-scale laboratory specimens are employed because of the uncertainty of the stability of subsequent crack extension. In stainless steels, however, very

large increases of toughness are exhibited with quite small amounts of ductile crack extension [1].

Over the last few years, extensive efforts have been applied into research on yielding fracture mechanics in the United States and Europe. Of high importance is the description of the stability of ductile crack extension, and a number of analyses have been developed which enable greater insight into this problem [2-5].

The purpose of the work program contained in this report is to consider a stainless steel weldment appropriate to the welds in the primary vessel of CFR in order to:

1. Provide conservative estimates of the initiation and propagation toughness of the plate, weld metal (WM), and heat-affected zone (HAZ) by testing small-scale laboratory specimens in bending.

2. Determine the relationship between small-scale bend tests and larger scale, more structurally relevant tension tests in order to assess the relevance of design curve procedures in providing conservative estimates of structural ductile instabilities.

Test Program

The test program, which is detailed in Table 1, consisted of:

Phase I – Part 1:   Small-scale bend tests to investigate orientation effects in parent plate.

Phase I – Part 2:   Measurement of resistance curves for parent plate, weld metal, and HAZ regions at 370 and 540°C using three-point bend specimens.

Phase II: Large-scale tests at 370°C on three material conditions:
    i. Parent plate.
    ii. Weld metal.
    iii. Heat-affected zone.
    Three test types, one specimen for each material condition:
    (a) Surface notched bend.
    (b) Surface notched tension.
    (c) Center cracked tension.

Plate Material

To provide sufficient material for the parent plate test program and to manufacture the weld metal test blocks, two 50-mm-thick plates (2438 by

TABLE 1—Summary of testing program.

Number	a/W	W, mm	B, mm	Material	Orientation/Notch Position	Specimen Type	Test Temperature, °C	Comments
Phase I - Part 1								
(a) 3 off	0.5	50	50	parent plate	L-S	bend	370	
(b) 3 off	0.5	50	50	parent plate	L-T	bend	370	maximum
(c) 3 off	0.5	50	50	parent plate	T-S	bend	370	load
(d) 3 off	0.5	50	50	parent plate	T-L	bend	370	values
(e) 3 off	0.5	50	50	parent plate	S-T	bend	370	
(f) 3 off	0.5	50	50	parent plate	S-L	bend	370	
Phase I - Part 2								
(a) 6 off	0.5	100	50	parent plate	T-L	bend	370	
(b) 6 off	0.5	100	50	parent plate	S-L	bend	370	
(c) 6 off	0.5	100	50	submerged-arc	HAZ	bend	370	R-curve
(d) 6 off	0.5	100	50	submerged-arc	WM	bend	370	evaluation
(e) 6 off	0.5	100	50	MMA	HAZ	bend	370	
(f) 6 off	0.5	100	50	MMA	WM	bend	370	
(g) 6 off	0.5	100	50	MMA	WM	bend	540	
Phase II								
(a) 1 off	0.5	50	150	parent plate	L-S	bend	370	
(b) 1 off	0.5	50	150	MMA	WM	bend	370	
(c) 1 off	0.5	50	150	MMA	HAZ	bend	370	maximum
(d) 1 off	0.5	50	150	parent plate	L-S	SENT	370	load
(e) 1 off	0.5	50	150	MMA	WM	SENT	370	values
(f) 1 off	0.5	50	150	MMA	HAZ	SENT	370	
(g) 1 off	0.5	150	50	parent plate	L-T	CCT	370	
(h) 1 off	0.5	150	50	MMA	WM	CCT	370	
(i) 1 off	0.5	150	50	MMA	HAZ	CCT	370	

1219 mm) were employed. The chemical compositions of these plates, labelled A 7781 and P 7378, are shown in Table 2, which also gives their longitudinal tensile properties at room temperature quoted by the material supplier.

These plates (although both comply to 316 specifications) have entirely different carbon contents and are hereafter labelled Plate A (0.06 wt% C) and Plate P (0.03 wt% C).

Manufacture of Welded Test Blocks

Six welded test blocks were manufactured using manual metal arc (MMA) (Armex GT electrodes) and submerged-arc (BX 600 flux and 316 filler). Six 150 mm high by 1219 mm long strips were cut from each 50-mm thick plate.

TABLE 2—*Mechanical and chemical properties of 316 stainless steel plates.*

	Chemical Analysis, wt%							
Plate	C	Si	Mn	P	S	Cr	Mo	Ni
A7781	0.06	0.42	1.62	0.031	0.022	17.18	2.51	11.21
P7217	0.029	0.31	1.70	0.034	0.012	17.00	2.40	12.30

	Mechanical Properties at 20°C (Longitudinal Orientation)[a]		
Plate	σ_Y (2% Proof Stress), N/mm^2	σ_u (UTS), N/mm^2	Elongation, %
A7781	249	593	61
P7217	250	585	58

	Mechanical Properties at 370°C		
Material Condition	σ_Y (0.2% Proof Stress), N/mm^2	σ_u (UTS), N/mm^2	Elongation, %
Plate A			
Longitudinal	171	469	42
Transverse	225	487	41
Plate P			
Longitudinal	192	454	44
Transverse	267	466	40
Submerged-arc weld metal	325	473	22
MMA weld metal	386	471	28

[a]Quoted by material supplier.

These strips were labelled A1-6 and P1-6 respectively, and three MMA and three submerged-arc test blocks were manufactured using the weld procedures shown in Fig. 1a and Table 3. Macrosections of the welds produced are shown in Fig. 1b.

Each set of three test blocks for MMA and submerged-arc used the combination of Plates A to P, A to A, and P to P to ensure that variability in dilution was adequately covered. As shown in Fig. 1, the weld profile used was fairly unusual, being of double U configuration with steep (10 deg) sides. This procedure was not intended exactly to reproduce the welds likely to be used in CFR (these have not been established to date), but instead was designed to complement the profiles used in a residual stress study being carried out at The Welding Institute. The profile employed also has the advantage of providing an almost straight HAZ, which enables fairly easy positioning of the notches for the HAZ tests.

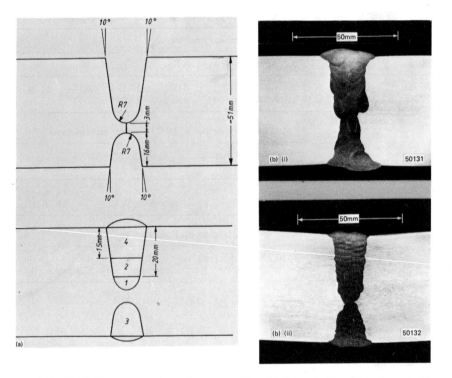

FIG. 1—(a) *Welding preparation and sequence:* (1) *weld clamped;* (2) *weld unclamped;* (3) *backgrind-unclamped to fill;* (4) *weld unclamped.* (b) *Macrosections of welds:* (i) *submerged-arc;* (ii) *manual metal arc.*

TABLE 3—*Weld specifications.*

Submerged-Arc Weld	
Wire	Type 316
Flux	BX 600
Root run specification	2 runs, Armex GT MMA electrode,
	4 mm diameter, 24 V, 150 A
Procedure:	
Current	500 A
Voltage	3 V
Travel speed	5 to 10 mm/s
Energy input	1.5 to 3.0 J/mm
Interpass temperature	100°C max
MMA Weld	
Electrode	Armex GT, 4 mm diameter
Procedure:	
Position	downhand
Run-out length	225 mm
Root runs (2)	
Current	150 A
Voltage	24 V
Filler	
Current	145 A
Voltage	22 V

The tensile properties of the plates and weld metal were determined at 370°C. These data are set out in Table 2.

Specimen Extraction Details

1. All parent plate samples for Phases I and II were notched in material extracted from Plate P. Plate A was only used for the end pieces of the S-L tests.

2. Two HAZ bend specimens for Phase I – Part 2 were extracted from each of the six welded test blocks and notched in alternate sides of the weld. Thus three P notch and three A notch HAZ specimens were obtained for both MMA and submerged-arc processes.

3. Two weld metal notched bend specimens for Phase 1 – Part 2 were extracted from each of the six welded test blocks. An additional three specimens were extracted from MMA test blocks for the 540°C test series.

4. Two weld metal surface notch specimens for the bend and tension tests of Phase II, a center-through-notched HAZ specimen (notched in A), a surface-notched HAZ bend and tension specimen (notched in P), and a center-through-notch weld metal specimen were also manufactured from the MMA test blocks.

Experimentation

Specimen Design

All six parent plate orientations were investigated in Phase I - Part 1 as defined in Fig. 2. Three-point bend specimens complying with the subsidiary testpiece of British Standard BS 5762 (1979), Methods for Crack Opening Displacement (COD) Testing, were employed with $B = W = 50$ mm and $a/W \simeq 0.5$ (Fig. 3a). Three specimens of each test orientation were extracted from Plate P. For the S-L and S-T orientation, end pieces were electron beam welded to 50-mm cubes to enable the required 200-mm span to be achieved.

For Phase I - Part 2, three-point bend specimens complying with the preferred geometry of BS 5762 were used; that is, $W = 2B = 100$ mm (Fig. 3a) with $a/W \simeq 0.5$. Parent plate testpieces were extracted from Plate P. To provide the required span of 400 mm for the S-L tests, end pieces were provided from both Plates P and A. Weld metal and HAZ bend testpieces were taken from the test blocks as described previously.

The following tests were employed in Phase II (Fig. 3b): oversquare bend tests ($B = 150$ mm, $W = 50$ mm, and $a/W \simeq 0.5$), single edge notch tension (SENT) ($B = 150$ mm, $W = 50$ mm, $a/W \simeq 0.5$), and center cracked tension (CCT) ($B = 50$ mm, $W = 150$ mm, $2a/W \simeq 0.5$). Clip gage positions for these specimens are also depicted in Fig. 3b.

All specimens were fatigue precracked before testing.

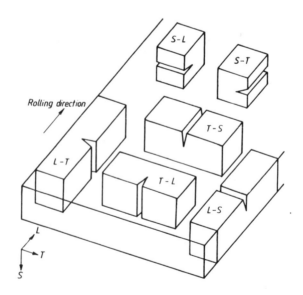

FIG. 2—*Specimen orientations.*

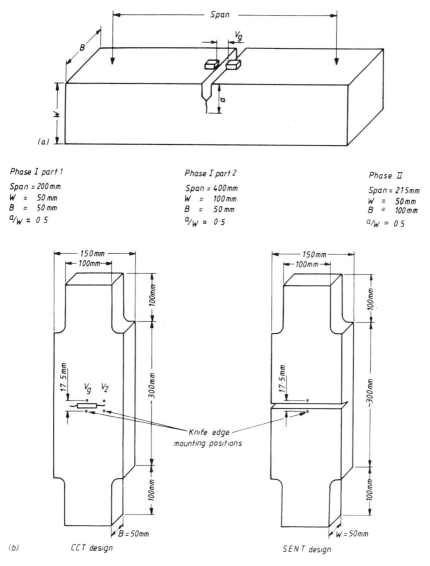

Phase I part 1
Span = 200mm
W = 50 mm
B = 50 mm
$a/w \approx 0.5$

Phase I part 2
Span = 400mm
W = 100mm
B = 50 mm
$a/w \approx 0.5$

Phase II
Span = 215mm
W = 50mm
B = 100mm
$a/w \approx 0.5$

CCT design

SENT design

FIG. 3—*Test specimen design. (a) Bend test design. (b) Tension tests.*

Experimental Details

Phase I - Part 1—All tests were carried out at 370°C using heating pads clamped to the surface of the specimen; the temperature was monitored and controlled throughout each test from thermocouples on the specimen. The tests were conducted in load control using a 60-tonne servohydraulic Instron at a loading rate of 0.5 kN/s.

Autographic records of load, crosshead displacement, and clip gage displacements were taken. Owing to the high temperature, insulated knife edges, shielding, and air cooling were employed to minimize clip gage drift. A correction to the crosshead reading for the extraneous displacements was determined by indenting the rollers into the broken halves of the specimens. Owing to the applied loading conditions all specimens failed by ductile instability at maximum load.

Phase I - Part 2—An identical test procedure to that employed for Part 1 was used for the Part 2 specimens. To enable the development of resistance curves, however, these specimens were tested effectively in displacement control using an 180-tonne Baldwin hydraulic machine at a crosshead speed of ~2 mm/min.

A multiple-specimen method was employed for the determination of resistance curves (six specimens per curve). Specimens were tested to various levels of displacement, unloaded, allowed to cool to room temperature and broken open to reveal the amount of stable crack extension which had been heat tinted by the test temperature of 370°C (one series at 540°C).

Phase II—All Phase II tests were carried out in the Baldwin testing machine. The bend specimens were tested using an identical procedure to the Phase I - Part 1 tests. In this situation, however, as in the Phase I - Part 2 tests, instability did not occur at maximum load. Thus after the attainment of maximum load the bend specimens were unloaded, allowed to cool, and broken open.

The tension specimens had to be reduced in width at each end to allow them to fit in the grips (Fig. 3*b*). The center cracked specimens were monitored using two clip gages, one (V_g) positioned in the center of the notch, the other (V_2) at one end of the fatigue crack (Fig. 3*b*).

All tension tests achieved a maximum load with stable crack extension and then failed in an unstable ductile manner under decreasing load. For the tension tests the crosshead displacements included a large component of extraneous displacement, and therefore the load versus clip gage measurements were used for *J* determinations.

General Testing Details

Although careful precautions were taken to minimize clip gage drift, in certain cases some drift was apparent in the clip gage outputs. This was a particular problem on the larger Phase II tension specimens, where heat output from the specimen was greater and air cooling of the clip gage was made difficult by the moving crossheads. All load-clip gage records were inspected carefully for evidence of drift.

The clip gage measurements from any test in Phase I experiencing drift were disregarded; thus the CTOD values for these tests were not calculated.

The corresponding J-values, which are based on crosshead displacements, were unaffected.

For the Phase II tests the clip gage results have to be taken to achieve any output data from the test. Thus, for the specimens experiencing clip gage drift, corrections to the load-displacement record were attempted to allow for the drift in order to provide a comparison with test results from Phase I.

For the bend geometries the crosshead displacements were corrected for extraneous displacement in all cases to ensure that overestimates of J were prevented.

Initial and final crack lengths were taken as an average of measurements taken along the specimen thickness, one measurement being taken per 10 mm of thickness. The crack lengths were determined with a travelling microscope, and measurements of stable crack extension were made excluding the stretch zone width.

Analysis

J-Integral Determinations

Bend Tests—Landes and Begley [6] demonstrated the use of estimation methods based on the energy (U) under the load-corrected crosshead displacement record to evaluate J for the growing crack:

$$J = \frac{\eta U}{B(W - a)} \tag{1}$$

where η is a geometric term which is approximately two for bend geometries. By ignoring crack extension and taking the total energy under the curve, a simple estimation method is obtained which forms the basis of the ASTM recommendations for J_{Ic} determinations (E 813); that is, for the three-point bend geometry,

$$J_0 = \frac{2 U_{total}}{B(W - a_0)} \tag{2}$$

where U_{total} is the total area under the load-corrected crosshead displacement record, B is specimen thickness, W is specimen width, and a_0 is initial fatigue crack length.

Tension Tests—Estimation formulae relating the area U under the load displacement record to J have also been derived for tension geometries [7,8]. Sumpter [8] quotes for CCT:

$$J_0 = \frac{\eta_e U_e + U_p}{B(W - 2a_0)} \tag{3}$$

and for SENT

$$J_0 = \frac{\eta_e U_e + U_p}{B(W - a_0)} \qquad (4)$$

where U_e and U_p are the elastic and plastic components of the energy under the load displacement record, and η_e is a geometric factor dependent on the gage length employed for the displacement measurement. In these tests the clip gage was placed across the notch mouth (a gage length of 17.5 mm was employed). The appropriate values of η_e are given in Fig. 4 for CCT and SENT geometries [9].

CTOD Determinations

Bend Tests—BS 5762 procedures were employed for all the crack tip opening displacement (CTOD) evaluations for the bend geometries. The symbol δ_0 refers to the "test stopped" CTOD value determined using the final value of clip gage (when the displacement-controlled test was stopped) and the initial fatigue crack length. Thus this CTOD value refers to the displacement at the original crack tip position once stable crack growth has occurred.

Values of δ_m refer to the value of CTOD calculated from the clip gage displacement at (1) instability in the load-controlled tests, and (2) the first attainment of the maximum load plateau for displacement control. The initial

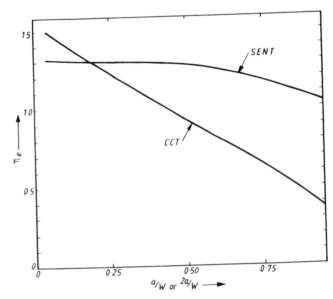

FIG. 4—*Elastic η-factors for tension geometries (with displacements measured at notch mouth).*

fatigue crack length was employed for all δ_m evaluations. The parent material 2% proof stress at room temperature was used for all CTOD calculations. For the weldments, where local σ_Y values may be higher than the parent material value, this would result in a very slight underestimate of CTOD. Owing to the plasticity encountered by all specimens tested in this program, however, this effect is negligible.

Tension Tests—The definition of CTOD for the tension geometries is rather ambiguous prior to crack extension and undefined subsequent to initiation of tearing. Crack infiltration methods prior to initiation [10] and studies of the behavior of the plate surface of tension specimens subsequent to stable crack extension carried out at The Welding Institute have indicated that the clip gage reading at the center of a CCT and at the mouth of an SENT specimen is approximately twice the value of the CTOD at the initial fatigue crack tip position and is relatively insensitive to crack length, specimen dimensions, and material. For example,

$$\delta_0 = (1/2)V_g \tag{5}$$

Therefore this formula is used for comparative purposes in the following section.

Results

Phase I - Part 1

The maximum load values of J and δ interpreted using the aforementioned formulae are set out in Table 4 for the six orientations from parent plate P. These tests were carried out under load control, and the maximum load value in fact refers to the instability point and not necessarily first attainment of maximum load.

Phase I - Part 2

J and CTOD values were derived for the seven sets of six specimens used for the R-curve determinations in this part of the test program. Where the stable crack extension induced in the test was sufficient to cause the attainment of maximum load in these displacement-controlled tests, δ_m and J_m values were also obtained referring to the position of first attainment of maximum load.

Although every precaution was taken to insulate the clip gage monitoring the mouth opening displacement from the hot specimen, clip gage drift was experienced on three specimens. Thus the CTOD values for these specimens have not been calculated.

The crosshead displacements are unaffected by temperature, and therefore the J-values for these specimens can be used for comparison purposes. The

TABLE 4—*Maximum load toughness results.*

Description	δ_m, mm	J_m, MJ/m^2
Parent Plate P – 370°C		
L-S Orientation		
Bend : $W = 50$ mm; $B = 50$ mm	>3.09 ⎱ 3.61 ⎰3.42 3.56 ⎰	1.81 ⎱ 1.93 ⎰1.90 1.96 ⎰
Bend : $W = 50$ mm; $B = 150$ mm	3.13	1.59
SENT : $W = 50$ mm; $B = 150$ mm	≡3.79	1.98
L-T Orientation		
Bend : $W = 50$ mm; $B = 50$ mm	>2.19 ⎱ 2.40 ⎰2.35 2.46 ⎰	1.21 ⎱ 1.28 ⎰1.21 1.13 ⎰
CCT : $W = 150$ mm; $B = 50$ mm	≡2.89	1.68
T-S Orientation		
Bend : $W = 50$ mm; $B = 50$ mm	2.07 ⎱ 1.75 ⎰1.75 1.51 ⎰	0.93 ⎱ 0.80 ⎰0.80 0.67 ⎰
T-L Orientation		
Bend : $W = 50$ mm; $B = 50$ mm	1.18 ⎱ 1.58 ⎰1.38 1.39 ⎰	0.58 ⎱ 0.73 ⎰0.66 0.68 ⎰
Bend : $W = 100$ mm; $B = 50$ mm	1.95 ⎱ 1.64 ⎰1.83 1.90 ⎰	0.66 ⎱ 0.64 ⎰0.66 0.68 ⎰
S-T Orientation		
Bend : $W = 50$ mm; $B = 50$ mm	1.12 ⎱ 0.68 ⎰0.78 0.54 ⎰	0.41 ⎱ 0.27 ⎰0.30 0.22 ⎰
S-L Orientation		
Bend : $W = 50$ mm; $B = 50$ mm	0.69 ⎱ 0.80 ⎰0.70 0.62 ⎰	0.26 ⎱ 0.30 ⎰0.26 0.23 ⎰
Bend : $W = 100$ mm; $B = 50$ mm	0.68 ⎱ 0.70 ⎰0.73 0.81 ⎰	0.24 ⎱ 0.28 ⎰0.27 0.31 ⎰
HAZ – 370°C		
Submerged-Arc		
Bend : $W = 100$ mm; $B = 50$ mm (cf L-T, P plate)	>1.38	>0.76
Bend : $W = 100$ mm; $B = 50$ mm (cf L-T, A plate)	1.57	0.74

TABLE 4—*Continued.*

Description	δ_m, mm	J_m, MJ/m^2
MMA		
Bend : $W = 100$ mm; $B = 50$ mm		
(cf L-T, P plate)	>1.31	>0.66
Bend : $W = 100$ mm; $B = 50$ mm		
(cf L-T, A plate)	>1.71	>0.78
SENT : $W = 50$ mm; $B = 150$ mm		
(cf L-S, A plate)	$\equiv 3.22^a$	1.96^a
CCT : $W = 150$ mm; $B = 50$ mm		
(cf L-T, A plate)	$\equiv 2.24^a$	1.44^a
Bend : $W = 50$ mm; $B = 150$ mm		
(cf L-S, P plate)	0.99^a	0.57
Weld Metal		
Submerged-Arc – 370°C		
Bend : $W = 100$ mm; $B = 50$ mm	$\left.\begin{array}{c}0.91\\1.02\end{array}\right\}0.97$	$\left.\begin{array}{c}0.49\\0.56\end{array}\right\}0.52$
MMA – 370°C		
Bend : $W = 100$ mm; $B = 50$ mm	$\left.\begin{array}{c}1.60\\1.03\end{array}\right\}1.31$	$\left.\begin{array}{c}0.83\\0.44\end{array}\right\}0.63$
Bend : $W = 50$ mm; $B = 150$ mm	0.41	0.26
SENT : $W = 50$ mm; $B = 150$ mm	$\equiv 1.17$	0.68
CCT : $W = 150$ mm; $B = 50$ mm	$\equiv 2.09^a$	1.26^a
MMA – 540°C		
Bend : $W = 100$ mm; $B = 50$ mm	1.39	0.66

^{a}Value obtained from clip gage corrected for zero drift (see text).

resistance curves in terms of J_0 and δ_0 versus stable crack extension excluding stretch zone (Δa) are plotted in Figs. 5 to 9.

Although procedures have been developed to derive J_{Ic} (ASTM E 813) and δ_i (BS 5762) from resistance curves, the curves drawn here do not comply with the requirements set out in ASTM E 813. This is due to the requirement of this program to investigate the R-curves after substantial crack extension; this conflicts with the requirements of the J_{Ic} initiation determination procedure (ASTM E 813), which requires at least four specimens with crack extension less than 1.5 mm. For this reason J_i and δ_i estimates only are quoted in Table 5 based on the extrapolation of the J_0 and δ_0 resistance curves respectively to $\Delta a = 0$ (excluding the stretch zone).

Although it is felt that these estimates give a reasonable comparison of the initiation toughness of the various plate orientations and weld and HAZ notch

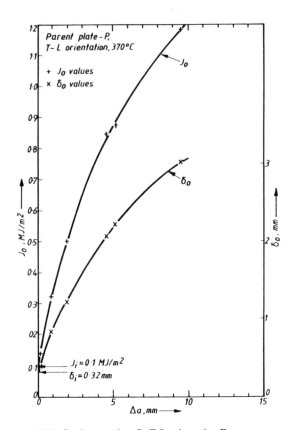

FIG. 5—*Parent plate P; T-L orientation R-curves.*

positions, their absolute values in terms of the parameters defined in the aforementioned standards are questionable. For the same reason, values of the T-parameter [2], which is normally based on the initial slope of the resistance curves, have not been quoted. It was felt that a more accurate assessment of the various propagation resistances could be made by direct comparison of the resistance curves as in Fig. 10. To a certain extent the maximum load toughness values quoted in Table 4 also reflect the relative propagation resistances.

For the HAZ tests the notch was positioned in either A or P plate (three specimens of each type for each set of specimens).

Phase II

The maximum load values of J and CTOD (interpreted from Eq 5 for the tension tests) are quoted in Table 4 together with the maximum load results from Phase I. Unfortunately, clip gage drift was experienced on four of the

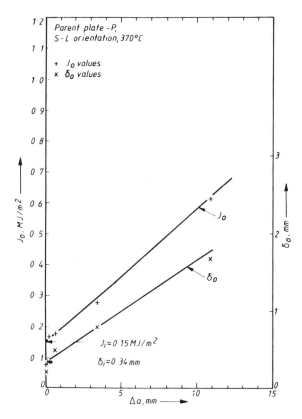

FIG. 6—*Parent plate P; S-L orientation R-curves.*

specimens in this series. As the only comparative measurements are based on this clip gage displacement, the drift experienced on these tests (evidenced by an incorrect unloading slope in the load displacement diagram) was allowed for and the displacements corrected accordingly. The results from these four tests must be treated with caution, however, and can only be used for comparative purposes within the overall program.

Discussion

Interpretation of Results

As stated in the previous section, the limited extent of the test program prevented an accurate evaluation of initiation toughness and the determination of the T-parameter. Because of the rapid increase in resistance with a small increment of crack extension exhibited by this material, the structural relevance of the initiation of tearing toughness as defined by the procedures of

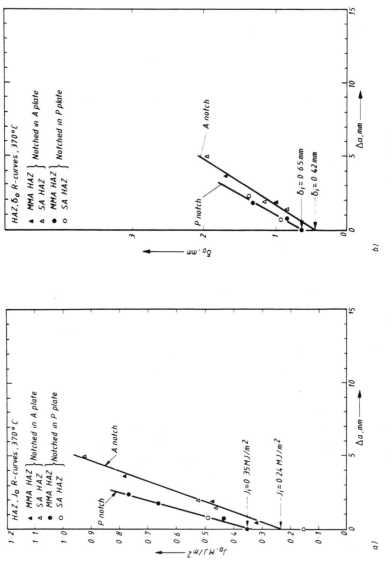

FIG. 7—(a) *Heat-affected zone* J_0 *R-curves.* (b) *Heat-affected zone* δ_0 *R-curves.*

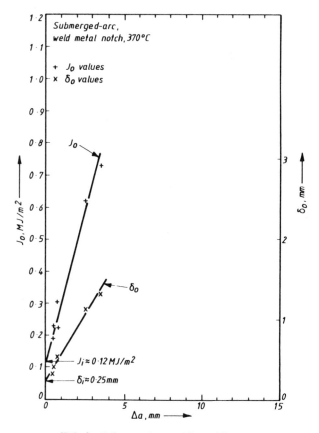

FIG. 8—*Submerged-arc weld metal R-curves.*

BS 5762 and ASTM E 813 is questionable. However, the evaluation of an elastic-plastic instability analysis based on appropriate resistance curves and estimates of driving force curves (for example, using the analysis of Kumar et al [11]) is complex and expensive.

It has recently been suggested that maximum load toughness estimates (δ_m and J_m) obtained from laboratory bend tests of specified dimensions [12] can be used safely in a single-parameter design curve equation to estimate allowable flaw sizes in ductile tearing situations [13].

The CTOD design curve [14] provides the most established single parameter elastic-plastic fracture assessment route, as set out in British Standard PD 6493 (1980), Guidance on Some Methods for the Derivation of Acceptance Levels for Defects in Fusion Welded Joints, which is used to determine the maximum *tolerable* length of a through-thickness crack ($2\bar{a}_m$).

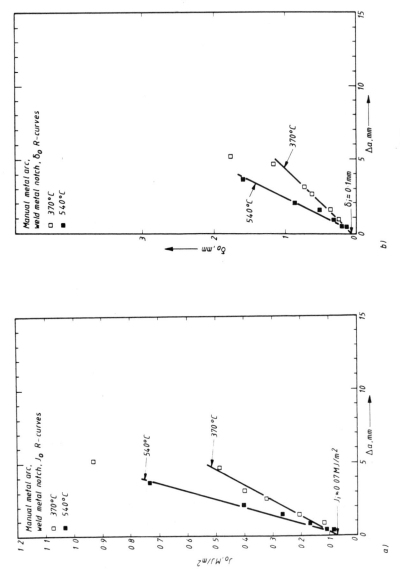

FIG. 9—*Manual metal arc.* (a) *Weld metal* J_0 *R-curves.* (b) *Weld metal* δ_0 *R-curves.*

TABLE 5—*Initiation of tearing estimates.*

Description	Temperature, °C	δ_i, mm	J_i, MJ/m^2
Parent Plate			
Type P			
T-L orientation	370	0.32	0.10
S-L orientation	370	0.34	0.15
HAZ (Submerged Arc + MMA)			
Notch position			
Plate P (L-T)	370	0.65	0.35
Plate A (L-T)	370	0.42	0.24
Weld Metal			
Submerged-arc	370	0.25	0.12
MMA	370 and 540	0.10	0.07

Thus, using a maximum load value of toughness (δ_m),

$$\bar{a}_m = \frac{\delta_m}{2\pi e_Y (e/e_Y)^2} \quad \text{for} \quad e/e_Y \leq 0.5$$

and   (6)

$$\bar{a}_m = \frac{\delta_m}{2\pi e_Y (e/e_Y - 0.25)} \quad \text{for} \quad e/e_Y \geq 0.5$$

where e/e_Y is the local applied strain ratio; that is,

$$\frac{e}{e_Y} = \frac{K_t\sigma + \sigma_R}{\sigma_Y}$$

where

K_t = geometric stress concentration factor,
 σ = applied stress,
 σ_R = residual stress, and
 σ_Y = yield stress.

Begley et al [15] and Turner [16] have proposed a J design curve; Burdekin et al [17] have recently suggested a three-tier definition based on Turner's proposals. This definition is similar to the simplification of Turner's curve defined in Ref 13, which states that, for maximum load toughness in terms of J_m,

$$\bar{a}_m = \frac{J_m}{2\pi E e_Y^2 (e/e_Y)^2} \quad \text{for} \quad e/e_Y \leq 1.0$$

and   (7)

$$\bar{a}_m = \frac{J_m}{2.5\,\pi E e_Y^2 (e/e_Y - 0.2)} \quad \text{for} \quad e/e_Y \geq 1.0$$

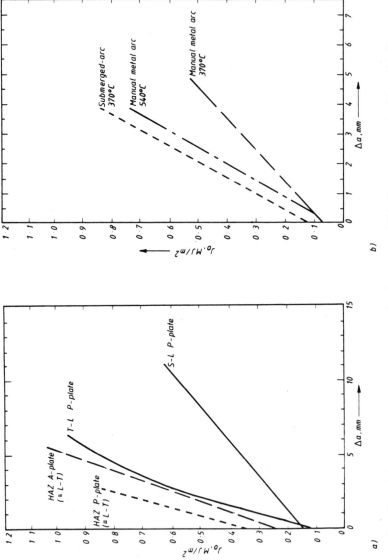

FIG. 10—(a) Comparison of parent plate and HAZ J_0 R-curves at 370°C. (b) Comparison of weld metal J_0 R-curves.

where E is Young's modulus and e_Y is the equivalent strain at yield, that is, σ_Y/E.

In order to substantiate the use of maximum load toughness values, as suggested in Refs *12* and *13*, in the foregoing assessment equations for austenitic applications, the toughnesses relating to various weld procedures, parent plate notch orientations, and specimen geometries are compared in the following subsections. Maximum load toughness and resistance curves are also employed in identical geometries to assess relative toughness of the plates and weldments under scrutiny.

Parent Plate Orientation Dependence

It is evident from inspection of the maximum load toughness results, which are summarized in Table 4, that a distinct orientation dependence is evident in 316 stainless steel plate at 370°C. Of the four "normal" cracking orientations, the L-S gave the highest δ_m (and J_m) values with 3.42 mm (1.90 MJ/m²), followed by the L-T orientation with 2.35 mm (1.21 MJ/m²), the T-S with 1.75 mm (0.80 MJ/m²), and the T-L with 1.38 mm (0.66 MJ/m²). As expected, the S-T and S-L orientations give the lowest values of 0.78 mm (0.3 MJ/m²) and 0.70 mm (0.26 MJ/m²).

Two parent plate orientations were selected for the R-curve evaluations of Part 2. The T-L which gave the lowest of the "normal" cracking directions and the S-L which gave the lowest overall were chosen on the basis of the maximum load results. The resulting R-curves are shown in Figs. 5 and 6 for the T-L and S-L orientations respectively. It is interesting to note that, although the propagation resistance and maximum load toughness for the T-L are much higher than the S-L, the estimated initiation toughnesses are very similar (Table 5). The maximum load values are compared in Table 4. From Ref *12* it might be expected that the 50 by 50 mm testpieces of Part 1 would give lower results than the 100 by 50 mm specimens of Part 2. In fact they are very similar, presumably owing to the effect of the load control used in Phase I – Part 1 compensating for the expected reduction in δ_m for B by B compared with $2B$ by B testpieces.

HAZ Tests

The resistance curves relating to the submerged-arc and MMA HAZ notched bend specimens tested in Phase I – Part 2 are shown in Fig. 7. There would appear to be no significant effect of weld procedure between the two sets of results. However, it does appear significant whether the notch was positioned in material from Plate P or Plate A.

All parent plate testpieces were extracted from Plate P. Thus the HAZ notched in P material should be compared with the L-T orientation in Phase I – Part 1. The HAZ regions tested in Part 2 were very tough, giving

$\delta_m = \geq 1.3$ mm for P plate. This compared with $\delta_m = 2.35$ mm for parent plate in Part 1 (Table 4). Unfortunately, since a δ_m-value was not obtained in the R-curve tests it is not possible to assess whether the HAZ regions compare favorably with the parent plate results in bending. For the SENT geometry tested in Phase II, however, very similar J_m-values were obtained: parent plate P, L-T orientation, recording 1.98 MJ/m^2, and the HAZ of the MMA weld, also from Plate P, giving 1.96 MJ/m^2 (Table 4).

The oversquare bend tests of Phase II can also be compared for the L-S orientation, with parent plate P giving $J_m = 1.59$ MJ/m^2 and the HAZ MMA test recording 0.57 MJ/m^2. In this latter case, however, the crack ran out of the HAZ region into the less tough MMA weld after a small amount of crack extension, which would effectively reduce the expected maximum load value.

Comparison of the resistance curves of the HAZ regions in Fig. 7 indicates that the HAZ region in Plate P has a substantially higher propagation resistance than the HAZ of Plate A. This is undoubtedly owing to the inherent toughness of Plate A being inferior to Plate P. As can be seen from Table 2, Plates P and A have quite different chemistries although both comply to 316 stainless steel specifications. Plate A has a much higher carbon and sulfur content than Plate P.

Weld Metal Toughness Tests

The resistance curves derived from the submerged-arc weld metal tests at 370°C of Phase I – Part 2 are shown in Fig. 8 and indicate good initiation and propagation toughness. The submerged-arc weld would appear to have equivalent toughness to the parent plate P (T-L orientation) shown in Fig. 5. In contrast, the MMA weldment at 370°C exhibited poor initiation toughness ($\delta_i = 0.1$ mm) and relatively low propagation resistance, although having a higher resistance curve than the Plate P (S-L orientation) results of Fig. 6.

A comparison of all the resistance curves derived from the Phase I – Part 2 tests can be made by reference to Fig. 10. Maximum load toughness determinations and initiation estimates are compared in Tables 4 and 5 respectively.

The MMA weld would appear to give the lowest initiation toughness of all tests. However, although the propagation resistance of the MMA welds is lower than Plate P in the "normal" cracking orientations, the HAZ regions (A and P notches), and the submerged-arc weld metal, the S-L orientation of Plate P is inferior after ~ 2 mm of stable crack growth. Since the S-L orientation is not normally regarded as a critical crack propagation direction, the MMA weld metal was chosen as the material exhibiting the poorest toughness and selected for the 540°C R-curve test in Phase I – Part 2 and the weld metal and HAZ tests of Phase II.

The 540°C resistance curve for MMA weld metal is compared with the 370°C results in Fig. 9. It is evident that the increase in test temperature induces an increase in propagation resistance.

Geometry Considerations

The relationship between the B by B and $2B$ by B maximum load test results for parent plate has been discussed previously. The Phase II tests incorporated an oversquare bend specimen with $W = 50$ mm and $B = 150$ mm for Plate P (L-S orientation). The maximum load results from this test can be compared with the Phase I - Part 1 tests in Table 4. The reduction of δ_m (J_m) from 3.42 mm (1.9 MJ/m^2) to 3.13 mm (1.59 MJ/m^2) that might be expected from an increase in thickness from 50 to 150 mm is compensated to a certain extent, since the Phase I - Part 1 tests were carried out in load control compared with displacement control for the Phase II tests.

In fact, the criterion suggested for adequate thickness constraint in Ref *12*, that is, $B > 25$ (J_m/σ_Y), would require a specimen 159 mm thick for a J_m value of 1.59 MJ/m^2. It was also suggested in Ref *12* that maximum load results from the bend geometries should give conservative estimates of their equivalent thickness tension counterparts. Table 4 would tend to substantiate this for austenitic plate, weld, and HAZ regions at 370°C.

In the L-S orientation, Plate P, the J_m-value for the SENT specimen is 1.98 MJ/m^2 compared with 1.59 MJ/m^2 in bending. For the L-T orientation, the B by B bend specimen has a J_m of 1.21 MJ/m^2 compared with 1.68 MJ/m^2 for the CCT geometry.

For MMA weld metal the 150-mm-thick SENT geometry gave $J_m = 0.68$ MJ/m^2 compared with $J_m = 0.26$ MJ/m^2 in bending, and for the 50-mm-thick CCT configuration $J_m = 1.26$ MJ/m^2 was obtained compared with 0.63 MJ/m^2 in bending.

The HAZ region of MMA weld (notched in Plate P) exhibited a $J_m = 1.96$ MJ/m^2 for SENT with $J_m = 0.57$ MJ/m^2 in bending with 150 mm section thickness (crack ran into weld metal). The value of J_m for parent plate P in the L-S orientation for a 150-mm-thick bend specimen was 1.59 MJ/m^2.

Future Work

If instability analyses are to be carried out for structural situations, then procedures for the determination of structurally relevant driving force and resistance curves must be established. The use of either initiation of tearing or maximum load toughness values in a single-parameter fracture assessment such as the CTOD or J design curve relies on the applicability of the design curve to the material under consideration. At present, design curve procedures have not been assessed for austenitic steels.

The weldments tested in this program were in the as-welded condition. The influence of residual stresses on crack growth resistance is not well understood at present, particularly for high work-hardening materials such as 316 stainless steel. Therefore a comparative program on stress-relieved joints would be very desirable.

The tests incorporated in this test program were on fully yielded bend and tension specimens. The relationship between these tests and contained yielding situations more applicable to structural situations has yet to be determined for austenitic materials.

Conclusions

1. A distinct orientation dependence is evident from maximum load bend test results in 50-mm-thick 316 stainless steel plate at 370°C. Of the four "normal" cracking orientations the L-S gave the highest maximum load toughness with $\delta_m = 3.42$ mm, followed by L-T with 2.35 mm, T-S with 1.75 mm, and T-L with 1.38 mm. As expected, the S-T and S-L orientations gave the lowest values, 0.78 and 0.70 mm respectively.

2. Load-controlled tests appear to give higher maximum load toughness estimates than equivalent displacement-controlled tests.

3. The effect of specimen thickness on the maximum load toughness of the bend tests caused a reduction of δ_m from 3.4 to 3.1 mm as thickness was increased from 50 to 150 mm. However, this effect is confused by the dependence of δ_m on the type of machine control.

4. For parent plate, the maximum load toughness determined from bend specimens is a good approximation (with a slight underestimate) of the load-controlled instability toughness of single edge notch tension tests (L-S orientation) and gives a conservative estimate of the center cracked tension (L-T orientation) instability toughness at 370°C in 316 stainless steel. The same relationship between geometries exists for MMA weld metal and HAZ regions.

5. No difference was evident between the propagation resistance of MMA and submerged-arc HAZs in 316 stainless steel at 370°C. A marked effect of parent plate chemistry was noticeable in these tests, however. The HAZ testpieces notched in the L-T orientation in Plate P were tougher than equivalent testpieces in Plate A, which has a higher carbon and sulfur content.

6. The propagation resistances of MMA HAZ in tension are in good agreement with parent plate tension results for the relevant orientations. In bending, the crack tends to propagate into MMA weld metal and the toughness is reduced accordingly.

7. Submerged-arc weld metal gave very good propagation toughness equivalent to parent plate in the T-L orientation.

8. Austenitic MMA weld metal showed inferior propagation toughness to that measured on 316 stainless steel submerged-arc weld metal and parent plate (normal cracking orientations) at 370°C. Although the propagation toughness of MMA weld metal was better than S-L orientated parent plate specimens, the initiation toughness was lower at $\delta_i = 0.1$ mm. Increasing the test temperature to 540°C caused an increase in propagation toughness of the MMA austenitic weld metal.

Acknowledgments

The views expressed in this paper are those of the author and do not necessarily represent the policy of the Nuclear Installations Inspectorate whose sponsorship of this work is gratefully acknowledged. The author would like to thank the staff of the Fracture Laboratory, who carried out the tests under the supervision of B. A. Wakefield. The assistance of D. A. Faux and M. Goble with the analyses of the results is also gratefully acknowledged.

References

[1] Tanaka, K. and Harrison, J. D., *Welding Research Abroad*, Vol. 24, No. 3, March 1978, pp. 2–21.

[2] Paris, P. C., Tada, H., Zahoor, A., and Ernst, H. in *Elastic-Plastic Fracture, ASTM STP 668*, American Society for Testing and Materials, 1979, pp. 5–36.

[3] Shih, C. F., de Lorenzi, H. G., and Andrews, W. R. in *Elastic-Plastic Fracture, ASTM STP 668*, American Society for Testing and Materials, 1979, pp. 65–120.

[4] Turner, C. E. in *Fracture Mechanics, ASTM STP 677*, C. W. Smith, Ed., American Society for Testing and Materials, 1979, pp. 614–628.

[5] Garwood, S. J. in *Fracture Mechanics, ASTM STP 677*, C. W. Smith, Ed., American Society for Testing and Materials, 1979, pp. 511–532.

[6] Landes, J. D. and Begley, J. A. in *Fracture Analysis, ASTM STP 560*, American Society for Testing and Materials, 1974, pp. 170–186.

[7] Rice, J. R., Paris, P. C., and Merkle, J. G. in *Progress in Flaw Growth and Fracture Toughness Testing, ASTM STP 536*, 1973, pp. 231–245.

[8] Sumpter, J. D. G., "Elastic-Plastic Fracture Analysis and Design Using the Finite Element Method," Ph.D. thesis, University of London, 1974.

[9] Tower, O. L., *Welding Institute Research Bulletin*, Vol. 22, Nov. 1981, pp. 319–323, and Dec. 1981, pp. 345–349.

[10] Robinson, J. N., *International Journal of Fracture*, Vol. 12, No. 5, 1976, pp. 723–738.

[11] Kumar, V., German, M. D., and Shih, C. F., "An Engineering Approach for Elastic-Plastic Fracture Analysis," EPRI Report NP 1931, Electric Power Research Institute, July 1981.

[12] Towers, O. L. and Garwood, S. J. in *Proceedings*, Third Colloquium on Fracture (ECF-3), London, Sept. 1980, Pergamon Press, pp. 57–68.

[13] Towers, O. L. and Garwood, S. J. in *Proceedings*, Fifth International Conference on Fracture (ICF-5), March 1981, Pergamon Press, pp. 1731–1740.

[14] Harrison, J. D., Dawes, M. G., Archer, G. L., and Kamath, M. S., in *Elastic-Plastic Fracture, ASTM STP 668*, American Society for Testing and Materials, 1979, pp. 606–631.

[15] Begley, J. A., Landes, J. D., and Wilson, W. K. in *Fracture Analysis, ASTM STP 560*, American Society for Testing and Materials, 1974, pp. 155–169.

[16] Turner, C. E., "A *J* Design Curve Based on Estimates for Some Two-Dimensional Shallow Notch Configurations," Organization for Economic Co-operation and Development (OECD)—Nuclear Energy Authority (NEA)—Committee on the Safety of Nuclear Installations (CSNI), Specialist Meeting on Elastoplastic Fracture Mechanics, Daresbury, U.K., CSNI Report 32, Vol. 2, 1978, pp. 18.1–20.

[17] Burdekin, F. M., Cowan, A., Milne, I., and Turner, C. E., "Comparisons of COD, R6 and *J*-Contour Integral Methods of Defect Assessment, Modified to Give Critical Flaw Sizes," Paper 41, presented at Conference on Fitness for Purpose Validation of Welded Constructions, The Welding Institute, Cambridge, England, Nov. 1981.

J. S. Huang[1] *and R. M. Pelloux*[2]

Application of High-Temperature Fracture Mechanics to the Prediction of Creep Crack Growth for a γ–γ′ Nickel-Base Superalloy

REFERENCE: Huang, J. S. and Pelloux, R. M., **"Application of High-Temperature Fracture Mechanics to the Prediction of Creep Crack Growth for a γ–γ′ Nickel-Base Superalloy,"** *Fracture Mechanics: Fifteenth Symposium, ASTM STP 833,* R. J. Sanford, Ed., American Society for Testing and Materials, Philadelphia, 1984, pp. 360–377.

ABSTRACT: Creep crack growth rates were measured in a γ–γ′ low-carbon Astroloy at 650 to 760°C. A theoretical model based upon the mechanism of cavity nucleation, growth, and coalescence on the grain boundaries near a crack tip was used to predict the creep crack growth rates in Astroloy. The model assumed that creep crack growth was an intermittent cracking process as observed in the experiments. A crack tip stress field of the HRR type in a creeping medium, derived previously by Riedel and Rice, was used for the analysis. The model gave a good prediction of the dependence of the crack growth rates on test temperature and stress intensity factor. The model does not take into account the crack tip stress triaxiality. However, by assuming that the crack tip stress triaxiality was intermediate between the plane strain and the plane stress, a good correlation was obtained between theory and experimental data.

KEY WORDS: creep crack growth, low-carbon Astroloy, fracture mechanics, creep cavity, stress intensity factor, C^*-integral, J-integral

Creep crack growth is a high-temperature crack propagation phenomenon under static load. Creep crack growth is usually the result of creep damage—that is, grain boundary cavity nucleation, growth, and coalescence. This type of crack growth mechanism is often observed during the creep fracture of a smooth-bar under a uniaxial stress.

With the growing application of fracture mechanics to engineering structure design, the study of creep crack growth of a single crack has recently become a topic of great interest. Several mechanical parameters have been

---

[1] Previously with Exxon Research and Engineering Company, Florham Park, N.J.; currently with Pfizer Incorporated, Wallingford, Conn. 06492.
[2] Massachusetts Institute of Technology, Cambridge, Mass. 02139.

used to correlate the crack growth rate data. These include the stress intensity factory (K) [1-4], the energy rate integral ($C*$) [5-7], and the net section stress (σ_n) [8]. The basis of using fracture mechanics parameters to characterize creep crack growth (CCG) has also been discussed by Riedel and Rice [9]. In their work they showed that the time-dependent stress field in a creeping solid could be related to one of three parameters (K, $C*$, or σ_n), depending upon the creep behavior of the material.

Different theoretical models predicting creep crack growth rates based on the micromechanisms of crack growth have also been published [10-12]. A theoretical approach to CCG is useful in predicting the creep crack growth rates (CCGR) outside the laboratory testing range and in understanding the relative effects of microstructure on CCGR.

This paper presents the results of a theoretical prediction of creep crack growth rate for a nickel-base superalloy. The predicted results are compared with experimental data. Future work required to improve the model is also discussed.

Experimental Procedure

Material Characterization

The material is a powder metallurgical hot isostatically-pressed Astroloy. Its chemical composition is given in Table 1. The 0.2% yield strength of the material is 932 MPa at 25°C and 794 MPa at 760°C.

Figure 1 shows a replica transmission electron micrograph (TEM) of the material microstructure. The average grain size of the material was 40 μm, and the average intercarbide particle spacing along the grain boundaries was 3.1 μm.

The relationships of the flow stress versus plastic strain and the flow stress versus secondary creep rate were also characterized. These relationships, fitted to power law equations, were

Plastic strain/stress relations (for $4 \times 10^{-4} \leq \epsilon_p \leq 4 \times 10^{-2}$):

$$\epsilon_p = 6 \times 10^{-39} \, [\sigma(MPa)]^{12.1} \quad \text{at } 650°C \tag{1}$$

$$\epsilon_p = 9.8 \times 10^{-39} \, [\sigma(MPa)]^{12.1} \quad \text{at } 700°C \tag{2}$$

$$\epsilon_p = 1.6 \times 10^{-38} \, [\sigma(MPa)]^{12.1} \quad \text{at } 760°C \tag{3}$$

Steady creep rate/stress relation:

$$\dot{\epsilon}_c(s^{-1}) = 3.94 \times 10^{-15} \, [\sigma(MPa)]^{12.31} \, \exp \frac{-511 \text{ kJ mole}^{-1}}{RT} \tag{4}$$

TABLE 1—*Chemical composition (weight percent) of test material.*

Cr	Co	Mo	Al	Ti	C	B	Ni
15.0	17.0	5.0	4.0	3.5	0.03	0.03	balance

where $650°C \leq T \leq 760°C$ and $400\ \text{MPa} \leq \sigma \leq 800\ \text{MPa}$ (R is universal gas constant and T is temperature).

Crack Growth Testing

The specimen used for the creep crack growth experiment was of single-edge-crack design (Fig. 2). It had a gage length of 25.4 mm and a gage width of 11.73 mm. The initial crack was prepared by cutting with a string saw of 0.1 mm diameter followed by fatigue precracking; the initial crack length was in the range of 1.0 to 1.5 mm. There were two semicircular side grooves machined on both flat sides of the specimen in order to minimize the crack tip curvature during the crack growth. The stress intensity factor was calculated with the equation given by Brown and Srawley [13]:

$$K = \sigma\sqrt{\pi a}\left[1.12 - 0.23\left(\frac{a}{w}\right) + 10.6\left(\frac{a}{w}\right)^2\right.$$
$$\left. - 21.7\left(\frac{a}{w}\right)^3 + 30.4\left(\frac{a}{w}\right)^4\right] \tag{5}$$

Creep crack growth tests were conducted under constant loading. For most of the tests, the nominal stress across the gage section was 414 MPa. Tests were conducted on an ATS level tester. A three-zone a-c resistance furnace was used to heat the specimen. The overall temperature uniformity within the gage section of the specimen was maintained to within $\pm 4°C$.

Some tests were also conducted in helium atmosphere of 99.999% purity in a study of the effect of oxidation on creep crack growth. For these tests, an Inconel vacuum retort was used to contain the specimen.

An electrical potential technique was used to measure the crack length. A constant direct current was passed through the gage section of the specimen, and the voltage across the crack faces was monitored. The relation between crack length and voltage output was calculated with the equation given by Johnson [14]:

$$\frac{V(a)}{V(a_o)} = \cosh^{-1}\left(\frac{\cosh\left(\pi Y/2W\right)}{\cos\left(\pi a/2W\right)}\right)\bigg/\cosh^{-1}\left(\frac{\cosh\left(\pi Y/2W\right)}{\cos\left(\pi a_o/2W\right)}\right) \tag{6}$$

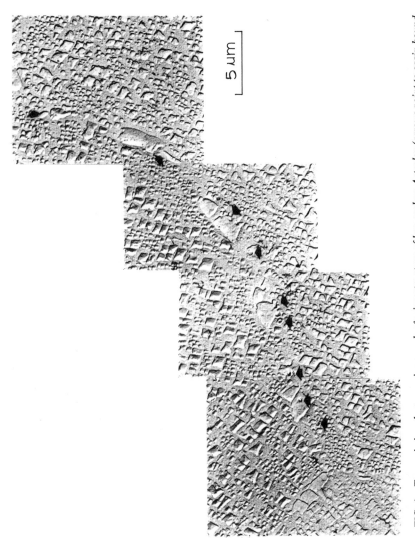

FIG. 1.—*Transmission electron micrograph of microstructure of low-carbon Astroloy (arrows point to grain boundary carbides).*

5 μm

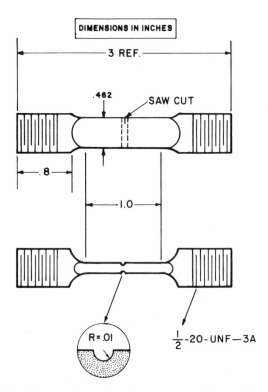

FIG. 2—*Creep crack growth specimen geometry (1 in. = 25.4 mm).*

where a is the crack length corresponding to the output voltage (V), a_0 is the initial crack length, W is the width of the specimen, and $2Y$ is the distance between the two locations where the electrical potential is measured. Equation 6 was also verified by fatigue cracking several specimens to different crack lengths. This verification showed that the maximum difference between prediction and actual crack lengths was 6%.

Experimental Results

Crack Growth Rate Data

Creep crack growth tests were conducted at 650, 700, and 760°C. Figure 3 shows typical data of crack length versus time. Note that the crack length increases in a discontinuous manner and the instantaneous crack growth rate is either zero or infinite. In this paper, however, the CCGR we refer to is the average CCGR. In order to obtain the average CCGR from the data shown in Fig. 3, we first drew a smooth curve through the data by inspection. The average CCGR was then obtained as the slope of the curve.

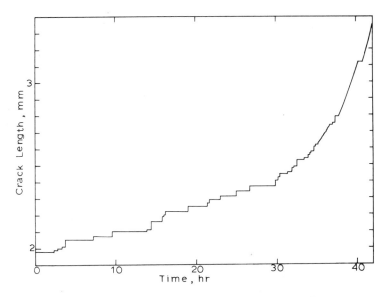

FIG. 3—*Typical crack growth data (crack length versus time).*

Figure 4 shows the creep crack growth rates in air and in helium at three different temperatures plotted versus the stress intensity factor (K). The data show that creep crack growth rate was strongly sensitive to temperature. At the same K-level, the creep crack growth rates cover three orders of magnitude within the temperature of 650 to 760°C. For K between 40 and 80 MPa$\sqrt{m}$, the activation energy for creep crack growth was estimated at 541 kJ/mole, which is close to the activation energy for steady-state creep of this material.

The data also show that at 760°C the effect of oxidation on the creep crack growth is negligible; at 650°C the creep crack growth rates are accelerated by oxidation by a factor of 2 to 3.

Crack Growth Mechanism

The crack growth mechanisms were investigated by scanning electron microscopy (SEM) of the fracture surfaces and by sectioning incompletely fractured samples. The typical fracture surface features (Fig. 5) show extensive grain boundary cavitation. There was no significant difference between the samples tested in air or in helium.

Figure 6 shows a micrograph taken on the longitudinal section of a sample. The cavities shown in the photograph were near a crack tip. There is no doubt from these results that creep crack growth in the low-carbon Astroloy is the consequence of cavity nucleation, growth, and coalescence. Further investi-

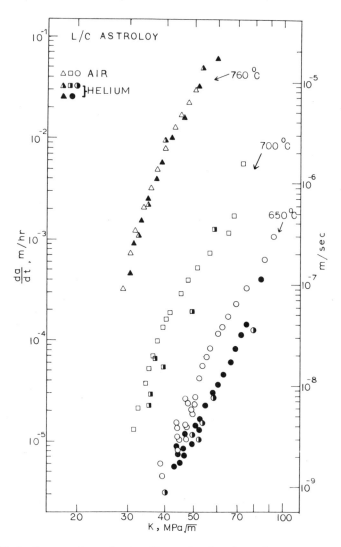

FIG. 4—*Creep crack growth rates of low-carbon Astroloy at three temperatures.*

gations of several samples showed that the average cavity spacing was 3.8 μm, which is approximately equal to the grain boundary carbide spacing. This observation will be used in the theoretical analysis.

Theoretical Prediction of Creep Crack Growth Rate

Based upon the observation that creep crack growth in the Astroloy was the consequence of cavity nucleation, growth, and coalescence ahead of a main

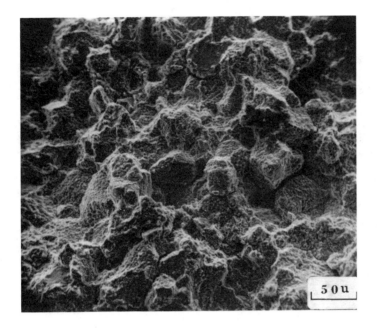

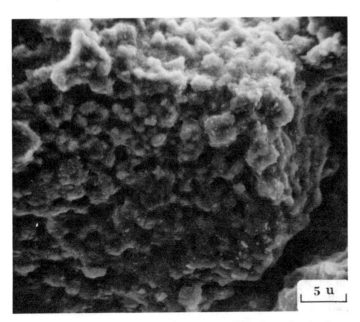

FIG. 5—*Fracture surface features of a creep crack growth specimen showing intergranular fracture mode.*

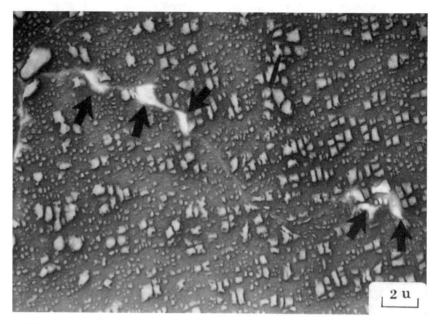

FIG. 6—*Typical features of cavities (pointed to by arrows) near the crack tip of a creep crack in low-carbon Astroloy.*

crack, the CCGR of Astroloy was predicted with the following assumptions: (1) the two-dimensional microstructure of the alloy consisted of a uniform hexagonal grain structure and the crack grew discontinuously from one triple point to the next triple point along the grain boundaries; (2) while the crack tip was stopped at one triple point, cavities ahead of the crack tip were nucleated and grew by power law creep; and (3) the crack would advance instantaneously to the next triple point when the cavities on the grain boundary facet ahead of the crack tip grew to a size equal to their mean spacing.

The analyses for cavity nucleation and growth and the stress distribution ahead of the crack tip are described in this section.

Cavity Nucleation

The cavity nucleation theory proposed by Raj and Ashby [15] was first applied to predict the stress necessary to nucleate a cavity. This stress is usually called the *threshold stress for cavity nucleation* (σ_{th}). The predicted σ_{th} was then modified with a factor derived from the work by Argon et al [16].

It was assumed that cavities were nucleated on the grain boundary carbide/matrix interfaces. The radius of a nucleated cavity was estimated as $2\gamma/\sigma_{th}$, where γ is the surface energy of the matrix. As proposed by Raj and Ashby [15], the cavity nucleation rate $(\dot{\rho})$ was given as

$$\dot{\rho} = \dot{\rho}_{max} \frac{4\pi\gamma\delta D_B}{\sigma\Omega^{4/3}} \exp\left(\frac{\sigma\Omega}{kT}\right) \exp\left(-\frac{4\gamma^3 F_v(\theta)}{\sigma^2 kT}\right) \qquad (7)$$

where

γ = surface free energy for matrix = 1.85 J/m^2 (Ni) [17],

Ω = atomic volume = 2.3 × 10^{-29} m^3 (for pure Ni),

δD_B = grain boundary diffusion coefficient multiplied by grain boundary thickness = 2.8 × 10^{-15} exp (−114.6 kJ mole^{-1}/RT) m^3 s^{-1} (for pure Ni) [18],

ρ_{max} = maximum number of nucleation sites per unit area of the boundary = $1/\pi\lambda^2$ = 8.8 × 10^{10}m^{-2}, where 2λ is the average spacing between cavities along grain boundaries (=3.8 × 10^{-6} m),

$F_v(\theta)$ = $(4\pi/3)(2 - 3\cos\theta + \cos^3\theta)$, a function which when multiplied by the cube of cavity radius gives the volume of cavity,

θ = a characteristic angle equal to $(\alpha + \beta - \mu)/2$,

α = $\cos^{-1}(\gamma_B/2\gamma)$,

β = $\cos^{-1}((\gamma_{IB} - \gamma_I)/\gamma)$,

μ = apex angle of an inclusion where the cavity is nucleated, assumed equal to 90 deg,

γ_B = grain boundary surface free energy = 0.87 J/m^2 (for pure Ni) [19],

γ_I = surface free energy of carbides where cavities are nucleated, assumed equal to 0.5 J/m^2 for low-carbon Astroloy,

γ_{IB} = free energy of the interface between a carbide particle and the matrix, assumed equal to 1.5 J/m^2 for low-carbon Astroloy,

T = absolute temperature, K, and

k = Boltzmann's constant = 1.38 × 10^{-23} J/K.

Using Eq 7 and the aforementioned values for its parameters, the cavity nucleation rate $\dot{\rho}$ was plotted in Fig. 7 versus the normalized stress (stress divided by Young's modulus E). For Astroloy, E = 1.517 × 10^5 MPa at 760°C; = 1.793 × 10^5 MPa at 650°C. As can be seen in Fig. 7, the cavity nucleation rate increases rapidly versus stress at the low stress end. It is common that the threshold stress for cavity nucleation (σ_{th}) is defined at $\dot{\rho}$ = 1 s^{-1}. It can then be seen from Fig. 7 that σ_{th} is approximately equal to 9 × 10^{-3}E for the Astroloy in the temperature range of 650 to 760°C.

Recent work by Argon et al [16], however, has shown that the foregoing cavity nucleation theory (Eq 7) overestimates σ_{th} by about a factor of 3, since it neglects considering the strain energy released by a nucleus during nucleation process. Therefore in our analysis we took σ_{th} for the low-carbon Astroloy as 3 × 10^{-3}E in the temperature range of 650 to 760°C.

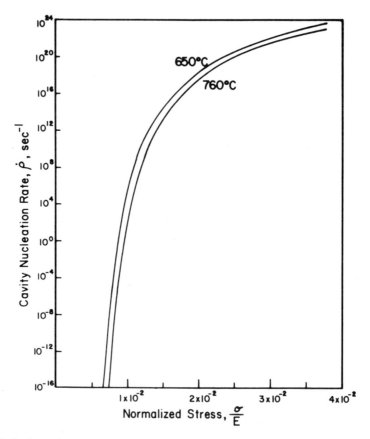

FIG. 7—*Dependence of the cavity nucleation rate on the normalized stress, at 650°C and 760°C, in low-carbon Astroloy.*

Cavity Growth

It was assumed that cavities grew as spheres; thus the size of a cavity was simply represented by its radius. It was further assumed that surface diffusion was so fast that it was not a rate-limiting step.

Cavities were assumed to grow by a power law creep mechanism as proposed by Hancock [20] or Beere and Speight [21]:

$$\frac{dR}{dt} = \frac{1}{2} R \dot{\epsilon}_c \left(\frac{\lambda^2}{\lambda^2 - R^2} \right)^{n_c} = \frac{1}{2} R B_c \sigma^{n_c} \left(\frac{\lambda^2}{\lambda^2 - R^2} \right)^{n_c} \qquad (8)$$

where R is the radius of a cavity, λ is one half of intercavity spacing, B_c and n_c are the constants in the power law creep constitutive relation, and σ and $\dot{\epsilon}_c$ are the stress and steady-state creep rate respectively.

Crack Tip Stress Field

The crack tip stress distribution used to calculate the cavity nucleation and growth rates is now described. Immediately following an instantaneous crack advance, the initial stress distribution ahead of a crack tip is described by a plastic singular field enclosed in an elastic singular field for small-scale yielding condition. The elastic stress distribution is given in terms of the stress intensity factor K and the angular function $f_{ij}(\theta)$ as

$$\sigma_{ij}(r, \theta) = \frac{K}{(2\pi r)^{1/2}} f_{ij}(\theta) \tag{9}$$

The plastic singular field is given by Hutchinson [22] and Rice and Rosengren [23] (HRR) as

$$\sigma_{ij}(r, \theta) = \left(\frac{J}{B_p I_{n_p} r}\right)^{1/(n_p+1)} \bar{\sigma}_{ij}(\theta) \tag{10}$$

where B, n_p are the exponents in the power law equation of hardening—that is, $\epsilon_p = B(\sigma)^{n_p}$, I_{n_p} is the constant depending on the exponent n_p, $\bar{\sigma}_{ij}(\theta)$ are functions of the polar angle θ and the exponent n_p, and J is a path-independent integral defined in Refs 22 and 23.

By combining Eqs 9 and 10, a simple expression for the crack tip stress at a point (r, θ) from the crack tip, under the initial loading condition, is

$$\sigma_{ij}(r, \theta, t = 0) = \text{MIN}\left(\frac{K}{\sqrt{2\pi r}} f_{ij}(\theta), \left[\frac{J}{I_{n_p} B_{n_p} r}\right]^{1/(n_p+1)} \bar{\sigma}_{ij}(\theta)\right) \tag{11}$$

As creep proceeds with time, the stress distribution ahead of the crack tip will relax. The stresses are given by Riedel and Rice [9] as

$$\sigma_{ij}(r, \theta, t) = \left[\frac{C_s(t)}{I_{n_c} B_c r}\right]^{1/(n_c+1)} \bar{\sigma}_{ij}(\theta) \tag{12}$$

where $C_s(t)$ is the amplitude of the HRR singularity with the forms

$$C_s(t) = \frac{K^2}{E(1 + n_c)t} \quad \text{for} \quad t < t_{tr} \tag{13}$$

and

$$C_s(t) = C^* = \int W^* \, dy - T_i \frac{\partial \dot{u}_i}{\partial x} \, ds \quad \text{for} \quad t \geq t_{tr} \tag{14}$$

where W^* is the strain energy rate density, T_i and $\dot{u}_i$ are traction and displacement rate along the path of the integral, B_c and n_c are the exponents in the power law creep equation, and t_{tr} is defined as a characteristic transition time as

$$t_{tr} = \frac{K^2}{E(1 + n_c)C^*} \tag{15}$$

Combining Eqs 11 and 12 gives the crack tip stress field after an instantaneous crack jump:

$$\sigma_{ij}(r, \theta, t) = \text{MIN} \left(\left[\frac{J}{B_p I_{n_p} r} \right]^{1/(n_p+1)} \tilde{\sigma}_{ij}(\theta), \quad \frac{K}{\sqrt{2\pi r}} f_{ij}(\theta), \right.$$

$$\left. \left[\frac{K^2}{E I_{n_c}(1 + n_c) r t B_c} \right]^{1/(n_c+1)} \tilde{\sigma}_{ij}(\theta) \right) \quad \text{for} \quad t < t_{tr} \tag{16}$$

and

$$\sigma_{ij}(r, \theta, t) = \text{MIN} \left(\left[\frac{J}{B_p I_{n_p} r} \right]^{1/(n_p+1)} \tilde{\sigma}_{ij}(\theta), \quad \frac{K}{\sqrt{2\pi r}} f_{ij}(\theta), \right.$$

$$\left. \left[\frac{C^*}{I_{n_c} B_c r} \right]^{1/(n_c+1)} \tilde{\sigma}_{ij}(\theta) \right) \quad \text{for} \quad t \geq t_{tr} \tag{17}$$

Thus Eq 11 describes the stress distribution immediately after a crack jump and Eqs 16 and 17 describe the same function at a given time after the crack jump. In our analysis, the parameters J and C^* were calculated as follows: J was calculated as K^2/E, since for all our tests linear elastic small-scale yielding conditions prevailed; C^*-integral was calculated by using the fully plastic J-integral solution for a single edge crack tension specimen derived by Kumar et al [24], with the creep strain rate replacing the plastic strain, because there is an analogy between the two.

Numerical Procedures

During the numerical calculations, the region ahead of the crack tip was divided into subregions with each subregion equal to one grain size. The cavities in each subregion were assumed to have the same size and to grow under the local stress described previously. The stress in each subregion was approximated as a constant equal to $\sigma_{22}(x, t)$, where $\sigma_{22}(x, t)$ is the stress component normal to the crack plane and x is the distance along the crack plane between the crack tip and the subregion.

To calculate $\sigma_{22}(x, t)$, the value of the angular function $\tilde{\sigma}_{22}$ for Eqs 11, 16, and 17 was taken as 2, an intermediate case between the plane stress condition and the plane strain condition ($\bar{\sigma}_{22} = 1.2$ for plane stress and $\bar{\sigma}_{22} = 2.6$ for plane strain, when the power law exponent, n_c or n_p, is equal to 12).

Figure 8 shows a schematic plot of the numerical calculation sequence. At Step 1, the crack length and specimen geometry are given. At Step 2, the stress intensity factor and the initial stress distribution are calculated and cavities in a subregion are nucleated if the local stress in the subregion is larger than the threshold stress for cavity nucleation. At Step 3, the cavity growth rate in each subregion is calculated based on the current stress distribution. At Step 4, a very small time step (Δt) is assigned and cavities in each subregion are allowed to grow for Δt. The next calculation sequence will depend on the cavity size distribution: if the cavities in some subregion have grown to the cavity spacing, the crack is assumed to advance to this subregion (Step 6) and then the calculation sequence goes back to Step 2 for calculating the new initial stress distribution ($t = 0$); on the other hand, if none of the cavities has grown to the cavity spacing, the calculation sequence goes to Step 5 for calculating the relaxed stress distribution $\sigma_{22}(x, t)$ and then to Step 3 for calculating new cavity growth rates.

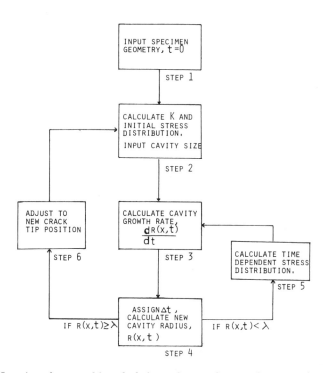

FIG. 8—*Iteration scheme used for calculating cavity growth rate and creep crack growth rate.*

Comparison Between Prediction and Experiment

The predicted creep crack growth rates at three temperatures are shown in Fig. 9 along with the experimental data. At low stress intensity factor levels (<40 MPa√m), the agreement between the observed and predicted creep crack growth rates is good. As the stress intensity factor increases, the difference becomes larger. The maximum difference is about a factor of five. At

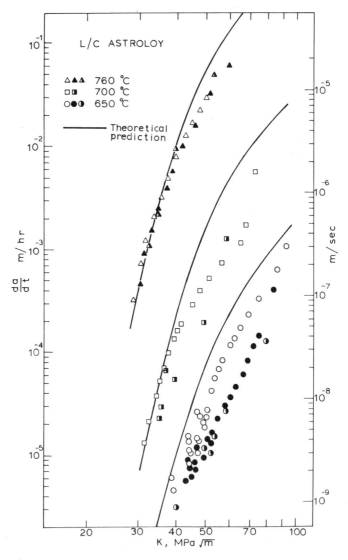

FIG. 9—*Predicted creep crack growth rates compared with experimental data for low-carbon Astroloy.*

low stress intensity factor levels, the predicted temperature dependence of the creep crack growth rates also compares well with that of the experimental data. The predicted creep crack growth rates show two apparent stages of creep crack growth. The experimentally observed third-stage creep crack growth at high stress intensity factor levels is not predicted. This difference may be due to the fact that crack growth in the high stress intensity factor regime is a combination of creep cavitation induced growth and fast fracture instability, the latter not being considered in the current creep crack growth model. Except for this difference, the model gives a good prediction of the dependence of crack growth rate on the stress intensity factor and temperature.

Discussion

In this paper the creep crack growth rate data are plotted versus the stress intensity factor (K) because K is the dominant parameter for the crack tip stress field for the test conditions in this work. The authors have shown elsewhere [25] that the parameter K did empirically correlate very well with the creep crack growth rates of Astroloy. It was also shown [25] that the time intervals between creep crack jumps for the experiments in this work were always less than the transition time (t_{tr}) as defined in Eq 15.

It is to be noted, however, that the present model of creep crack growth does allow for the prediction of creep crack growth rates for other alloys and loading conditions which may be creep-ductile, that is, with very short t_{tr}. For these materials and conditions, it would be misleading to plot the crack growth rates versus the K-parameter.

Several models of creep crack growth similar to the present model have been published by other researchers [10–12]. The major difference between the other models and the one in this paper is that the others do not allow for stress relaxation by plastic deformation and creep deformation in the crack tip region. This simplification results in the prediction of a relationship of the type $da/dt = A(K)^m$. This is apparently not consistent with the whole range of experimental data as shown here. Furthermore, when stress relaxation is neglected, the stresses in the crack tip region are overestimated, which lead to an overprediction of the creep crack growth rates.

The main difficulty associated with the application of the present model is with the selection of $\bar{\sigma}_{22}$ $(\theta = 0)$, which gives a measure of the stress triaxiality at the crack tip. It has been shown by the authors [25] and other researchers [2] that creep crack growth is so strongly sensitive to triaxiality that crack tunneling always occurred even though the specimens were side-grooved. Certainly there is a need to improve the present model to include triaxiality effects as a function of the specimen or component thickness.

Finally, more research remains to be done to account for the effect of a triaxial stress field on cavity growth. The cavity growth rate equation proposed by Beere and Speight [21] (Eq 8) does not consider the effect of triaxiality and may not be completely applicable.

Conclusion

A theoretical model predicting creep crack growth rate is proposed. The material parameters required for the prediction include grain size, intercavity spacing, threshold stress for cavity nucleation, and the stress/strain relationships for plastic deformation and creep. The model gives a good prediction of the K-dependence and temperature-dependence for the crack growth rate of a nickel-base superalloy. Further work remains to be done to improve its capability to account for crack tip stress triaxiality.

Acknowledgments

The authors wish to thank the Air Force Office of Scientific Research of the United States for the financial support of this work. The authors also thank Professors F. A. McClintock and A. S. Argon, Massachusetts Institute of Technology, for many interesting and helpful discussions on the paper.

References

[1] Floreen, S., *Metallurgical Transactions A*, Vol. 6A, 1975, p. 1741.
[2] Sadananda, K. and Shahinian, P., *Metallurgical Transactions A*, Vol. 6A, 1977, pp. 439–449.
[3] Sadananda, K., *Metallurgical Transactions A*, Vol. 7A, 1978, pp. 78–84.
[4] Koterazawa, R. and Iwata, Y., *Journal of Engineering and Materials Technology, Transactions of ASME*, Vol. 98, 1976, p. 296.
[5] Nibkin, K. M., Webster, G. A., and Turner, C. E. in *Cracks and Fracture, ASTM STP 601*, American Society for Testing and Materials, 1976, pp. 47–62.
[6] Harper, M. P. and Ellison, E. G., *Journal of Strain Analysis*, Vol. 12, 1977, p. 167.
[7] Landres, J. D. and Begley, J. A. in *Mechanics of Crack Growth, ASTM STP 590*, American Society for Testing and Materials, 1976, pp. 128–148.
[8] Taira, S. and Ohtani, R., Paper C234/173, International Conference on Creep and Fatigue in Elevated Temperature Applications, Institution of Mechanical Engineers, 1974.
[9] Riedel, H. and Rice, J. R. in *Fracture Mechanics: Twelfth Conference, ASTM STP 700*, American Society for Testing and Materials, 1980, pp. 112–130.
[10] Nix, W. D., Matlock, D., and Dimelfi, D., *Acta Metallurgica*, Vol. 25, 1977, p. 495.
[11] Vitek, V., *Acta Metallurgica*, Vol. 26, 1978, p. 1345.
[12] Pilkington, R. and Miller, D., *Metallurgical Transactions A*, Vol. 11A, 1980, p. 177.
[13] Brown, W. F., Jr., and Srawley, J. E. in *Current Status of Plane Strain Crack Toughness Testing of High Strength Metallic Materials, ASTM STP 410*, American Society for Testing and Materials, 1967.
[14] Johnson, H. H., *Materials Research and Standards*, Vol. 5, No. 9, 1965, p. 442.
[15] Raj, R. and Ashby, M. F., *Acta Metallurgica*, Vol. 23, 1975, p. 653.
[16] Argon, A. S., Chen, I. Y., and Law, C. W., "Intergranular Cavitation in Creep," in *Proceedings*, AIME Fall Meeting on Fatigue-Creep Environment Interactions, Milwaukee, Wisc., Oct. 1979.
[17] Blakely, J. M. and Maiya, P. S., *Surfaces and Interfaces*, Syracuse University Press, Syracuse, N.Y., 1967, p. 325.
[18] Wazzan, A. R., *Journal of Applied Physics*, Vol. 36, 1965, p. 3596.

[19] Murr, L., *Interface Phenomenon in Metals and Alloys*, Addison and Wesley, Reading, Mass., 1971, p. 132.
[20] Hancock, J. W., *Metal Science*, Vol. 10, No. 9, Sept. 1976, p. 319.
[21] Beere, W. and Speight, M. V., *Metal Science*, Vol. 12, No. 4, April 1978, p. 172.
[22] Hutchinson, J., *Journal of the Mechanics and Physics of Solids*, Vol. 16, 1968, p. 13.
[23] Rice, J. R. and Rosengren, G., *Journal of the Mechanics and Physics of Solids*, Vol. 16, 1968, p. 1.
[24] Kumar, V., German, M. D., and Shih, C. F., "An Engineering Approach for Elastic-Plastic Fracture Analysis," EPRI NP-1931, Electric Power Research Institute, Palo Alto, Calif., July 1981, pp. 3–12.
[25] Huang, J. S., "Fatigue Crack Growth and Creep Crack Growth of P/M HIP Low-Carbon Astroloy at High Temperature," Sc.D. thesis, Massachusetts Institute of Technology, Cambridge, Mass., Feb. 1981.

J. D. Landes[1] and D. E. McCabe[1]

Effect of Section Size on Transition Temperature Behavior of Structural Steels

REFERENCE: Landes, J. D. and McCabe, D. E., **"Effect of Section Size on Transition Temperature Behavior of Structural Steels,"** *Fracture Mechanics: Fifteenth Symposium, ASTM STP 833*, R. J. Sanford, Ed., American Society for Testing and Materials, Philadelphia, 1984, pp. 378–392.

ABSTRACT: The transition temperature behavior of structural steels was evaluated using fracture mechanics specimens and elastic-plastic evaluation techniques. In the transition temperature region, the ductile tear and J_R-curve behavior of structural steels was relatively independent of test temperature and specimen geometry effects. Consistent J_R-curve development was interrupted by J_c instability events, however, and it is this property that showed high sensitivity to test temperature and constraint variations. The temperature at which the J_c instability toughness rises rapidly was dominated by specimen thickness but within this behavior was a subtle interaction with constraint variations due to ligament length variations.

Specimens of fixed proportionalities can be used with Weibull analysis and extremal statistics methodology to demonstrate why the transition temperature behavior of steels is specimen size dependent. A distribution of J_c cleavage instabilities on small specimens can be used to predict J_c instability distributions on larger specimens.

KEY WORDS: transition temperature, ductile tear, J_R-curve, instability, Weibull analysis

The transition temperature behavior of structural steels has been intensively studied for more than 50 years, and it must seem absurd to suggest that more work and understanding are needed. However, most of the previous work has been empirical in nature, involving impact tests, using a variety of specimen designs and measurement schemes. A general understanding of the correlation between specimen behavior and in-service failure experiences has resulted in reasonably workable general guidelines for fracture control [1,2]. The more recent developments in elastic-plastic fracture mechanics methodology using *J*-integral as a characterizing parameter has provided a new and viable quantitative tool for evaluating toughness relative to in-service perfor-

[1]Manager, Mechanics of Materials Research, and Senior Engineer, respectively, Westinghouse R&D Center, Pittsburgh, Pa. 15235.

mance of materials [3]. Because of this, the emphasis now can be more toward the simulation of in-service loading conditions, relying less on the failure experience factor for fracture control. Despite this fundamental improvement, however, the complexity of specimen dimensions, variable constraint, and temperature-cleavage interaction behavior remains to be appropriately modeled. A working model of material behavior which is based on observations during fracture mechanics type tests made in the transition temperature region is illustrated in Fig. 1 [4]. Much of this has been observed from tests that were designed to determine J_{Ic}, where J_R curves are used to characterize stable-crack growth. The J_{Ic}-values are defined at the onset of crack growth. In the lower part of the transition region there sometimes is plasticity development followed by cleavage instability, but with essentially zero prior crack growth. Within this region, the J-value at instability, if we are permitted to redefine J_{Ic}, can be related to K_{Ic} through

$$K_{Ic}^2 = J_{Ic}E/(1 - \mu^2) \tag{1}$$

For steels tested in the mid to upper transition temperature range, there will be slow stable-crack growth (ductile tear) followed by brittle instability at J_c. The J_{Ic} at crack growth initiation then becomes a conservative design criterion unrelated to instability. Finally, at upper-shelf temperatures, the cleavage toughness becomes too high to cause instability in laboratory tests or in service.

The object of the present paper was to study the special geometry effect

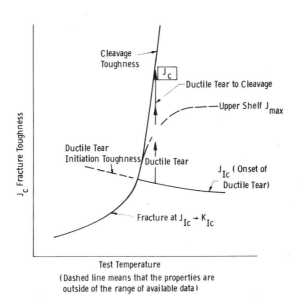

FIG. 1—*Schematic of* J_c *transition temperature curve.*

complexities associated within this region of mixed ductile growth and cleavage fracture interactions where values of J_c are known to be highly sensitive to geometry and other conditions. The committee responsible for the preparation of the British standard for crack tip opening displacement (CTOD) testing has suggested that constraint conditions are completely simulated in test specimens by using the full in-service thickness in the test specimens [5]. On the other hand, other investigations have shown that ligament length is an important factor to evaluate in constraint considerations [6, 7]. In addition, when dimensions are completely fixed, there can be statistical distribution to J_c instabilities in replicate tests [8]. Relatively small (1T size) compact specimens ($W = 2$ in.) develop a scatter band of J_c instabilities that can be fit by two-parameter Weibull distribution functions. At the same temperature, larger specimens of 4T size ($W = 8$ in.) tend to develop a narrower J_c scatter band but with consistently low J_c instability values clustered at the low end of the 1T scatter band. Extremal statistics with the associated weakest link rationalization can then be used to explain specimen size effects and to predict new statistical distributions for different sizes of specimens.

In this paper, the various rationale associated with J_c instabilities will be revaluated, and new data will be presented to compare with several results reported in the past, In addition, information will be given that shows relatively consistent material ductile tear property behavior in the transition temperature range. This behavior has proved to be completely consistent with the Fig. 1 model of material response.

R-Curve Behavior in the Transition Temperature Range

R-curve technology is a fracture mechanics based methodology that utilizes the geometry-independent features of ductile tearing in structural materials [9]. Fracture toughness development is expressed in terms of J_R and is plotted against the observed ductile tearing crack growth. The J_R-curves can be used to predict crack instability under fully ductile conditions and to compare the relative toughness of various materials.

Studies of upper-shelf toughness of structural steels have shown *R*-curve slope and J_{Ic} decrease with increased test temperature, but this is usually at very high temperatures [10]. The present evaluation of J_R-curve behavior is concentrated more at the moderate test temperatures, within the transition range. An example for the behavior of A471 Ni-Cr-Mo-V steel at three temperature levels and for two sizes of compact specimens is given in Fig. 2. Mechanical properties are reported in Table 1. The J_R-curve behavior appears to be independent of geometry effects within normal specimen-to-specimen variability and independent of test temperature. The J_R-curve development is frequently terminated by a sudden onset of cleavage fracture.

Perhaps a more convenient form for representing J_R-curves where the effects of material flow properties are amplified is to plot the crack growth on

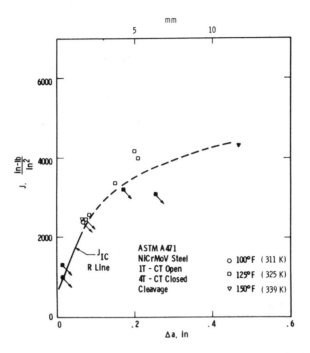

FIG. 2—J versus Δa for crack extension beyond the J_{Ic} bounds (ASTM A471 steel).

the abscissa in terms of effective Δa [11]. Here, crack growth by ductile tear is augmented with a linear dimension corresponding to the plastic zone size which represents its effect on pseudo-elastic displacement. The crack growth variable on the abscissa then effectively incorporates the material flow property effects into the J_R-curve.

Tests were conducted on a series of compact specimens made of A533B steel where specimen plan view size was varied from 0.4T to 4T and thickness was varied from 0.4 to 4 in. as permitted by plan view size. The test specimen matrix is illustrated in Fig. 3. Mechanical properties are reported in Table 1. Test temperatures were 144 K (−200°F), 273 K (32°F), 297 K (75°F), and 366 K (200°F).

TABLE 1—*Mechanical properties.*

Material	Yield Strength, MPa (ksi)	Tensile Strength, MPa (ksi)	Elongation, %	HRC	CVN, FATT
A471 (Ht 388)	765 (112)	900 (130)	21	29 R_c	294 K (70°F)
A533B Class 2	462 (67)	606 (88)	25	88	297 K (75°F)
A508 Class 3	480 (70)	635 (92)	27	. . .	266 K (+20°F)
A533A (Ht 4294)	550 (80)	700 (101)	25	. . .	244 K (−20°F)

FIG. 3—*Specimens covered in plan view study.*

Two sample J_R-curves are shown in Figs. 4 and 5. In the former case, plan view size was fixed at 4T and test temperature at 297 K (75°F). Thickness was varied from 0.4 to 4-in. For this particular material, constraint increase due to increased thickness did not detectably affect the ductile behavior part of the J_R-curve toughness development. Here as before, the ductile J_R was terminated by onset of cleavage fracture at J_c. In the present case, cleavage instability appears to be far more sensitive to constraint variations due to increased thickness. Figure 5 is an example for fixed specimen geometry at 4T, 1-in. thick, but with test temperature varied through the transition range. The ductile tear characteristics are again relatively unaffected by test temperature and the value of brittle cleavage toughness (J_c) controls.

A summary of J_c instability developments for the various specimens in Fig. 3 is presented in a familiar transition temperature curves format in Fig. 6. The scatter band of J_c events is covered by the length of the bars, and the width is

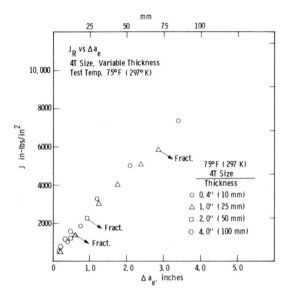

FIG. 4—*Effect of specimen thickness on R-curve behavior at 297 K (75°F).*

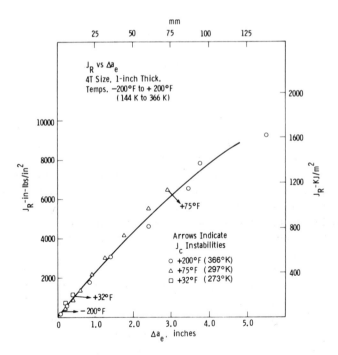

FIG. 5—*Effect of test temperature on R-curve behavior (4T specimens, 1 in. thick).*

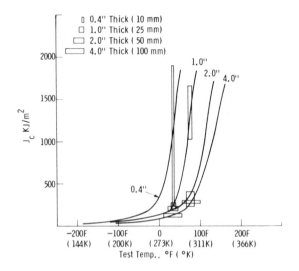

FIG. 6—J_c *instability versus test temperature (separated according to thickness).*

scaled to represent the specimen thicknesses. Plan view size is not identified in this case. A separate transition range is identified for each thickness of material, which explains the unique J_c scatter band characteristics that tend to develop for every given specimen size. Figure 6 suggests, therefore, that specimen thickness exerts the principal control over the transition temperature behavior. Hidden within the scatter band is a subtle ligament size influence, however, the details of which will be covered in a later part of the present report.

Statistical Characterization of J_c Instabilities

Populations of J_c instability data can be modeled using Weibull distribution functions [8]. Since J_c-values of large specimens tend to show less variability than small specimens and tend to concentrate at the low end of the small specimen scatter bands, the logic of extremal statistics seems applicable. The rationale is that a given heat of material is locally variable in fracture toughness. An imperfection or inhomogeneous toughness distribution exists such that small specimens, on the order of 1T specimens, that sample a small volume of material will display a wider range of J_c-values than large specimens of 4T size which sample a larger volume of material and have consistently lower J_c. Figures 6 and 7 are classical data presentation methods that support this material sampling size model. Whether this behavior is largely due to varied constraint or due to inhomogeneity of material is not overly important to the mathematical modeling used, however.

A common two-parameter form of Weibull distribution function was originally used by Landes and Shaffer and represented by

$$F(x) = 1 - e^{-(x/\theta)^b} \qquad (2)$$

where

$F(x)$ = the probability that a given specimen has a property less than level x,
x = the measured variable (J_c in the present case),
θ = a scale parameter, $\theta = x$ when $F(x) = 0.632$, and
b = Weibull slope.

For a 1T specimen, the probability that J_c toughness is greater than J is given by

$$[1 - F_1(J)] = \exp - [J/\theta]^b \qquad (3)$$

If one tests a specimen that has N times 1T thickness, the probability of $J_c > J$ is

$$[1 - F_1(J)]^N = \exp\{-[(J/\theta)^b]N\} \qquad (4)$$

$$= \exp - \left[\frac{(N)^{1/b}J}{\theta_1}\right]^b \qquad (5)$$

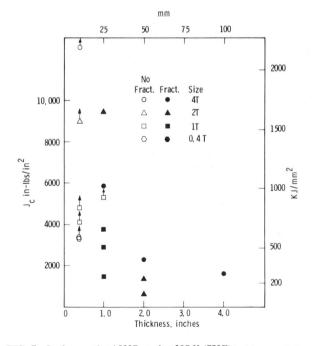

FIG. 7—J_c *cleavage for A533B steel at 297 K (75°F) test temperature.*

This is rewritten for a 4T specimen as

$$[1 - F_4(J)] = \exp - \left[\frac{J}{\theta_1/4^{(1/b)}} \right]^b \qquad (6)$$

It can be seen that the 1T distribution scale parameter θ_1 is reduced by a factor $N^{1/b}$. However, this model has the unfortunate feature of predicting mean J_c toughness distributions that tend toward zero for extremely large specimens. It is far more likely that at a given test temperature, J_c toughness tends toward a lower limiting value as the specimen size increases toward infinity. A lower limiting J_c feature can be easily accommodated in Weibull functions by adding a third parameter, J_0. Here it is assumed that a straightforward plot of J_c versus $F(J_c)$ on Weibull paper is nonlinear such that a straight line fit with slope b of Eq 6 does not optimize the mathematical fit to the data. Instead of Eq 3 for 1T specimens, we substitute

$$[1 - F_1(J)] = \exp - \left[\frac{(J - J_0)}{(\theta - J_0)} \right]^b \qquad (7)$$

With this model, J_0 can be adjusted to optimize the linearity of the Weibull fit. This is illustrated in Fig. 8 for three levels of J_0 using data taken from the work of Iwadate and co-workers [12]. About thirty 1/2T specimens of A508 Class 3 were tested, and many data points were omitted in the plotting for the purpose of clarity of presentation. The best straight line behavior was for J_0 equal to 13.73 KJ/m². This represents a lower bound J_c-value for the data set, and its implication is that an infinite number of tests will never give a J_c-value below this level. Of course, with statistics there is a finite probability that the statement is incorrect, but this aspect was not further pursued in the present work.

In the extremal analogy, it is hypothesized that an infinitely large specimen will also have a lower limiting J_c toughness at J_0 and that the Weibull slope remains the same over all specimen size distributions. In a similar manner to Eq 6, the exceedence probability of the $N \times 1T$ specimens is given by

$$[1 - F_N(J)] = \exp - \left[\frac{(J - J_0)}{(\theta - J_0)/(N)^{1/b}} \right]^b \qquad (8)$$

To illustrate the effect of Eq 8 on distribution functions, the Iwadate data was converted into a 4T specimen distribution ($N = 8$) and plotted as density functions in Fig. 9. The experimental median J_c for 1/2T size specimens was 39.9 KJ/m² and a median J_c for 4T specimens is predicted to be 21 kJ/m². A transition temperature curve in the Iwadate report indicated that an average 4T value at 213 K ($-75°F$) would be about 17 KJ/m².

A computer program was set up to automatically determine best linear fits

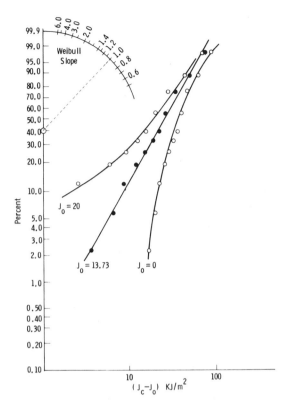

FIG. 8—*Cumulative distribution function* F(x) *versus* (J_c − J_0); *Weibull plot of Iwadate data* [12] *for A508 Class 3 at* −59°C.

of three-parameter Weibull distributions. Seven sets of data were found in various sources. Some of the sets were previously reported in the literature and others are new and unreported. The results are summarized in Table 2. Mechanical properties on these materials are given in Table 1.

Extremal predicted mean J_c-values for 4T specimens were then calculated from the Weibull fits to the small specimen data, and these are compared to experimentally observed J_c data in Table 3. The predictions are generally reasonably good but things become a bit tenuous in cases where specimen dimensions were not controlled to the standard compact specimen proportionalities (sized according to thickness dimension). For compact specimens of $a/W = 0.5$ and $W = 2B$, the cross section of the remaining ligament is square over all sizes and constraint conditions are maximized to near plane strain over an appreciable range of specimen sizes. Aside from the proportionality and constraint issue, it appears that a weakness develops when the toughness of the small specimens becomes quite high. This suggests that one of three possibilities exist which could introduce prediction inaccuracies. One

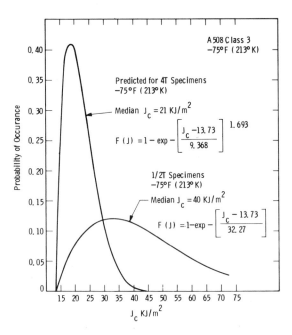

FIG. 9—*Distribution of* J_c *instabilities for 1/2T and 4T compacts; Iwadate data* [12].

is that there could easily have been an insufficient number of small specimens tested to properly establish the lower bound J_0 and the proper small specimen distribution function shape. The number of specimens required is most probably a function of specimen size and the relative position in the transition range. Another possibility is that accurate determination of high J_c toughness was not obtained in small specimens because of size limitations on the validity of J. Extreme plasticity and crack growth can result in violation of conditions for J-controlled growth. The third possibility, of course, is that the extremal statistic model does not accurately represent the size difference phenomenon. It is by its very nature not well suited to handle potential constraint variations between large and small specimens, and, in particular, the plane strain condition associated with square ligament cross sections can break down for small specimens at high toughness levels.

The present data analysis has served to point out that much remains to be tried in establishment of the extreme-value prediction model. In addition to the aforementioned concerns, two basic assumptions that should be proved are (1) that J_0 is in fact a lower bound J_c behavior over all specimen sizes and (2) the Weibull slope is essentially constant over all specimen sizes. Contingent upon these features being established as valid, another major task would be to establish a strong statistical justification for the numbers of small specimens required to develop accurate Weibull representation of the J_c popula-

TABLE 2—*Weibull three-parameter fits to statistical J_c instability distributions.*

Source	Material	Test Temperature	Specimen Size	No. of Data Points	Weibull Equation	Median J_c, KJ/m²
Iwadate [12]	A508 Class 3	213 K (−75°F)	1/2T	30	$F(x) = 1 - \exp - \left[\dfrac{(J - 13.65)}{(46 - 13.65)}\right]^{1.702}$	40
Iwadate [12]	A508 Class 3	253 K (−4°F)	1/2T	26	$F(x) = 1 - \exp - \left[\dfrac{(J - 51.6)}{(130 - 51.6)}\right]^{0.941}$	105
Westinghouse	A533A HT4294	233 K (−40°F)	1T	13	$F(x) = 1 - \exp - \left[\dfrac{J}{487}\right]^{2.463}$	420
Westinghouse	A533A HT4294	233 K (−40°F)	4T	4	$F(x) = 1 - \exp - \left[\dfrac{(J - 40.95)}{(160 - 40.95)}\right]^{3.280}$	147
Landes et al [8]	A471 HT388	294 K (70°F)	1T	9	$F(x) = 1 - \exp - \left[\dfrac{(J - 84.8)}{(163.5 - 84.8)}\right]^{2.406}$	152
Landes et al [8]	A471 HT388	311 K (100°F)	1T	8	$F(x) = 1 - \exp - \left[\dfrac{(J - 175.6)}{(260 - 175.6)}\right]^{0.851}$	230
Westinghouse	A533B Class 2	273 K (32°F)	0.4 in. B	7	$F(x) = 1 - \exp - \left[\dfrac{(J - 152.9)}{(782 - 152.9)}\right]^{1.117}$	605
Westinghouse	A533B Class 2	273 K (32°F)	1.0 in. B	5	$F(x) = 1 - \exp - \left[\dfrac{J}{241}\right]^{3.765}$	219
Westinghouse	A533A HT4294	233 K (−40°F)	4T to 1 in. B	4	$F(x) = 1 - \exp - \left[\dfrac{(J - 442)}{(970 - 442)}\right]^{1.823}$	874

TABLE 3—*Prediction of median J_c-values from small specimen distributions.*

Data	Test Temperature	Size Differential	Median J_c, KJ/m^2			
			Small Specimens	Predicted 4T	Experimental J_c	J_0
Iwadate [12] A508	213 K ($-75°$F)	8	40	21.5	17	13.5
Iwadate [12] A508	253 K ($-4°$F)	8	105	57.5	65	51
Landes [8] A471	294 K (70°F)	4	152	123	126	85
Landes [8] A471	311 K (100°F)	4	230	186	175	175
A533B	273 K (32°F)	10	605^a	210	120	153
A533B	273 K (32°F)	4	219^a	151	120	0
A533A	233 K ($-40°$F)	4	420	239	147	0

^{a}Nonstandard ligament length-to-thickness ratios, $(W - a)/B$, included in distribution.

tion. Iwadate and coworkers [12] have made an empirically based suggestion for the number of specimens required to establish J_0 using

$$(\# \text{ of specimens}) > 3000 J_c/B\sigma_Y \qquad (9)$$

A difficulty with this is that when J_c toughness is reasonably good, the numbers of specimens called for can be large to the point of impracticality.

Ligament Dimensions versus Constraint

Earlier indications (Fig. 6) suggest that specimen thickness tends to control J_c instability behavior. This was quite evident, but within these particular scatter bands existed subtle effects due to remaining ligament lengths. Table 4 lists data that clearly demonstrate this suggestion. The A533B data for the 0.4-in.-thick specimens (see also Figs. 3 and 6) had ligament length-to-

TABLE 4—*Two examples of ligament dimension ratio effects* $[(W - a)/B]$.

Material	Test Temperature	Thickness, mm	Ligament Size, mm	$(W - a)/B$	J_c, KJ/m^2
A533B	273 K (32°F)	10	10	1.0	217
	273 K (32°F)	10	25	2.5	869
	273 K (32°F)	10	50	5.0	599
	273 K (32°F)	10	100	10.0	1912
A533A	233 K ($-40°$F)	25	25	1.0	430
		25	100	4.0	895

thickness proportions, $(W - a)/B$, that varied drastically, and the effect on J_c is evident. The smallest specimens having square ligaments gave the lowest J_c. Statistical models become biased by these constraint variations, thereby weakening the predictive capabilities.

The A533A specimens were also affected by this loss of constraint in the 1-in. versus 4-in. long ligaments. As can be seen in Fig. 10, the specimens clearly generated two distinctly different populations of J_c instabilities. Extremal statistics with an inhomogeneity and a relative material volume rationale involved cannot be used to bridge such a gap between the two distinct experimental J_c populations.

Conclusions

Steels that have fracture mode transition behavior exhibit strong geometry effects in J_c instability behavior. Ductile crack growth resistance development remains relatively unchanged in specimens tested through the transition temperature range. However, instability J_c with onset of a running crack by cleavage cracking is far more geometry and temperature dependent. Specimen thickness tends to dominate the J_c transition temperature, but within a thickness-dominated J_c data scatter band, more subtle effects exist which are influenced by ligament length-to-thickness ratios, $(W - a)/B$. Long-thin ligaments tend to reduce constraint and instability J_c is increased.

Data scatter developed in small specimens in the critical transition range can be characterized by three-parameter Weibull statistical distributions. The Weibull slope, scale parameter, and lower bound J_c-value is identified on small specimens. Extremal statistics can then be used to predict mean J_c-values for larger specimens. In many cases the prediction capability was ex-

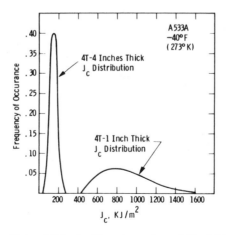

FIG. 10—*Distribution of* J_c *instabilities for 4T specimens (one set 4 in. thick, the other 1 in. thick).*

cellent. Where strict proportional dimension control or constraint control was not used, the predictions were of marginal accuracy.

Data sets that were developed without confining the conditions to dominant plane strain tended to have distinctly different Weibull J_c populations which were not amenable to an extremal statistics rationale.

Acknowledgments

The authors would like to thank Mr. F. X. Gradich of the Mechanics of Materials Laboratory, who prepared specimens, and Messrs. M. G. Peck and L. F. Burtner, who conducted the majority of tests reported herein.

References

[1] Jonassen, F., "A Resumé of the Ship Fracture Problem," *Welding Journal*, Research Supplement, June 1952, pp. 316–318S.

[2] Pellini, W. S., "Design Options for Selection of Fracture Control Procedures in the Modernization of Codes, Rules and Standards," presented at U.S.-Japan Symposium on Application of Pressure Component Codes, Tokyo, 1973.

[3] Paris, P. C. in *Flaw Growth and Fracture, ASTM STP 631*, American Society for Testing and Materials, 1977, pp. 3–27.

[4] Logsdon, W. A. in *Elastic-Plastic Fracture, ASTM STP 668*, J. D. Landes, J. A. Begley, and G. A. Clarke, Eds., 1979, pp. 515–536.

[5] "Methods of Crack Opening Displacement (COD) Testing," British Standard BS 5672, London, 1972.

[6] McCabe, D. E., Landes, J. D., and Ernst, H. A. in *Elastic-Plastic Fracture: Second Symposium, Volume II—Fracture Resistance Curves and Engineering Applications, ASTM STP 803*, C. F. Shih and J. P. Gudas, Eds., American Society for Testing and Materials, 1983, pp. II-562–II-581.

[7] Davis, D. A., Vassilaros, M. G., and Gudas, J. P. in *Elastic-Plastic Fracture: Second Symposium, Volume II—Fracture Resistance Curves and Engineering Applications, ASTM STP 803*, C. F. Shih and J. P. Gudas, Eds., American Society for Testing and Materials, 1983, pp. II-582–II-610.

[8] Landes, J. D. and Shaffer, D. H. in *Fracture Mechanics: Twelfth Conference, ASTM STP 700*, American Society for Testing and Materials, 1980, pp. 368–382.

[9] *Fracture Toughness Evaluation by R-Curve Methods, ASTM STP 527*, American Society for Testing and Materials, 1973.

[10] Logsdon, W. A. and Begley, J. A., *Engineering Fracture Mechanics*, Vol. 9, 1977, pp. 461–470.

[11] McCabe, D. E. and Ernst, H. A. in *Fracture Mechanics: Fourteenth Symposium—Volume I: Theory and Analysis, ASTM STP 791*, J. C. Lewis and G. Sines, Eds., American Society for Testing and Materials, 1983, pp. I-561–I-584.

[12] Iwadate, T., Tanaka, Y., Ono, S.-I., and Watanabe, J. in *Elastic-Plastic Fracture: Second Symposium, Volume II—Fracture Resistance Curves and Engineering Applications, ASTM STP 803*, C. F. Shih and J. P. Gudas, Eds., American Society for Testing and Materials, 1983, pp. II-531–II-561.

K. Ogawa,[1] *X. J. Zhang,*[1] *T. Kobayashi,*[1] *R. W. Armstrong,*[1] *and G. R. Irwin*[1]

Microstructural Aspects of the Fracture Toughness Cleavage-Fibrous Transition for Reactor-Grade Steel

REFERENCE: Ogawa, K., Zhang, X. J., Kobayashi, T., Armstrong, R. W., and Irwin, G. R., **"Microstructural Aspects of the Fracture Toughness Cleavage-Fibrous Transition for Reactor-Grade Steel,"** *Fracture Mechanics: Fifteenth Symposium, ASTM STP 833,* R. J. Sanford, Ed., American Society for Testing and Materials, Philadelphia, 1984, pp. 393–411.

ABSTRACT: Particularly in the transition temperature range, reliable use of toughness data, from small specimens of structural steel can be helped by improved understanding of the initiation of cleavage. For this purpose, fracture details in local regions were studied. Methods of examination were scanning electron microscopic (SEM) fractography, microstructural analysis, hardness tests, and topographic examination of fracture surfaces.

The results indicate that examination of fracture surface morphology through SEM fractography is essential for improved understanding of cleavage fracturing in the transition range. Fracture surface details reflect influences of both ferrite and prior-austenite grain size. In addition, for some specimens, fracture morphology is also influenced by a large size inhomogeneity due to dendritic solidification. In heavy section A533B steel, explanations for large scatter of small specimen K_{Ic} estimates may be closely related to dendritic inhomogeneity.

KEY WORDS: cleavage fracture, fibrous fracture, fracture toughness, A533B steel, fractography, fracture surface topography, microstructure, ferrite grain structure, prior-austenite grain structure, dendritic structure, microhardness test

The tendency of structural steel toughness values to show a large scatter in the transition temperature range is well known. For example, Fig. 1 [*1*] shows estimates (essentially *K* from *J*-values) of the fracture *K*-value obtained using 25-mm-thick compact specimens from a 152-mm (6-in.)-thick section of

[1] Mechanical Engineering Department, University of Maryland, College Park, Md. 20742. Dr. Kobayashi is currently at Stanford Research Institute, Menlo Park, Calif. 94025. Dr. Ogawa is currently at Department of Aeronautical Engineering, Kyoto University, Kyoto, Japan.

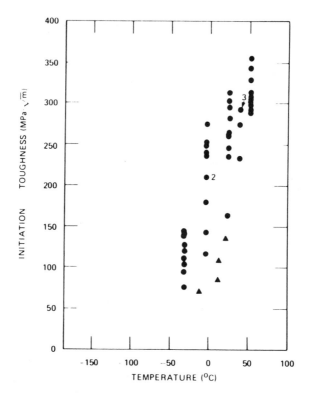

FIG. 1—*Initiation toughness data for A508 steel used by ORNL for TSE-5A. Symbols shown as* ▲ *indicate* $K_{calculations}$ *for initiation of long-crack-front cracks in the test cylinder.*

A508 steel. In the thermal shock fracture tests at ORNL which employed this material, the crack front was quite long and the test results corresponded to cleavage initiation at values of K near but below the lowest K-values obtained with small specimens [1]. Examination of small specimen fracture surfaces reveals that the onset of cleavage fracture is preceded by various degrees of fibrous fracture extension, which suggests that the time during fibrous cracking for initiation of rapid cleavage fracture is variable and that this contributes to the scatter. The continuum fracture mechanics viewpoint alone does not explain the observed scatter of toughness values. It is thus necessary to study fracture details in local regions and to relate variability of these to the variability of cleavage initiation. In order to accomplish this goal fracture surface analyses with a scanning electron microscope, microstructural analyses and hardness tests of fractured A533B steel specimens were performed. The results provided indications of the cause of scatter in the estimated K_{Ic}-values.

TABLE 1—*Mechanical properties of ASTM A533B at room temperature.*

Material/ Code	0.2% Yield Strength, MPa	Ultimate Tensile Strength, MPa	Elongation in 50 mm, %	Reduction in Area, %	CVN Impact Energy, J
ASTM A533B (HSST-03)	489	634	23	60	19

Material and Fracture Test Conditions

The material investigated was A533B steel from Oak Ridge National Laboratory HSST A533-03 plate. Mechanical properties and chemical composition of the material are listed in Tables 1 and 2 respectively.

The fractured specimen was of the compact tension type with dimensions of $W = 50.8$ mm and $B = 25.4$ mm. In order to control the formation of shear-lips, side-grooves were machined to the depth of 10% of the thickness from each side. Specimens were tested using a spring-in-series loading system. This arrangement causes an acceleration of crack extension speed after the start of a "tearing instability." The tendency for the resulting high strain rates to initiate cleavage fracture was of special interest. Loading records and broken specimens for examination were furnished by the Naval Ship Research and Development Center at Annapolis.

Examination of Fractured Specimens

In order to obtain microstructural information we employed four examination methods: fractography, vertical sectioning and etching of cut and polished surfaces, hardness characterization of microstructure, and topographical characterization of fracture surfaces.

Fractography

Fracture surfaces of the specimen were examined with a scanning electron microscope (SEM). A pair of stereographic photographs were taken by rotat-

TABLE 2—*Chemical composition of ASTM A533B.*

Material/ Code	Chemical Constituents, wt%								
	C	Mn	P	Si	Ni	Mo	S	Cr	Al
ASTM A533B (HSST-03)	0.20	1.26	0.011	0.25	0.56	0.45	0.018	0.10	0.34

ing the sample. A stereoscopic examination of fracture surfaces provides a better understanding of fracture features such as surface roughness, undercutting, and orientation of cleavage planes. These photographs were also used to analyze surface topography and to obtain quantitative information of crack opening displacement [2].

Vertical Sectioning of Specimen and Etching of Cut-Surfaces

The specimen was sectioned perpendicular to the fracture surface and along the crack propagation direction for examination of the microstructure underneath the fracture surface. A specific etchant was used to reveal selective microstructures such as ferrite grains, prior-austenite grains, and dendritic structures. Etchant compositions and etching procedures are given in Table 3.

Hardness Characterization of Microstructures

Microhardness tests were performed to reveal possible correlations between hardness reading and microstructures.

Topographical Characterization of Fracture Surfaces

Fracture surface height measurements were made from a stereo pair of fracture surface photographs. A more detailed description of height measurement from stereophotographs is presented elsewhere [2]. These measurements were made for matching points of top and bottom surfaces. Crack opening displacement estimates were made from the top and bottom surface contours. Furthermore, these curves were used to characterize the fracture

TABLE 3—Composition and procedures used for Etchants A, B, and C.

Etchant	Composition	Procedure	Structures
A	2-mL nitric acid + 100-mL ethyl alcohol	• Immerse for 30 to 60 s • Rinse in distilled water	ferrite grain
B	0.5-g picric acid + 2.0-g cupric chloride + 30-mL detergent + 50-mL distilled water	• Heat etchant to 65 to 75°C • Immerse for 30 to 40 s • Swab in $NH_4(OH)$ to remove brown stain • Polish lightly with Al_2O_3	prior-austenite grain
C	Saturated solution of picric acid in ethyl alcohol	• Immerse for 50 to 60 s • Rinse in distilled water	dendritic structure

processes, that is, microcrack initiation ahead of main crack tip and mode of crack propagation.

Results and Discussion

Results obtained from two specimens of A533B steel are discussed in this paper: one tested at room temperature and the other at 54°C. The location of these tests relative to the transition temperature region, as indicated by Charpy V-notch (CVN) impact tests, can be seen by examination of Fig. 2. The specimen tested at room temperature exhibited approximately 0.5-mm crack extension before the occurrence of cleavage fracture. Figure 3 shows a SEM photograph of the fracture surface of the specimen tested at room temperature. The photograph covers the area of initial fatigue crack, short fibrous extension, and dominant cleavage fracture. An examination of the cleavage-dominated region reveals networks of fibrous ridges in that area. These networks appeared to have structures on two levels, coarse and fine, as marked in Fig. 3. Coarse networks, which are circular patches surrounded by fibrous ridges, are approximately 0.3 to 0.5 mm in diameter. Fine structures are approximately 0.01 to 0.03 mm in diameter.

In order to understand the mechanism of formation of these networks, microstructural examination was performed. Figures 4a to 4c exhibit micro- and macro-structures found in A533B steel. Ferrite grains shown in Fig. 4a are approximately 0.01 to 0.03 mm in diameter. This size corresponds well with the fine scale network found on the fracture surface (Fig. 3). Prior-aus-

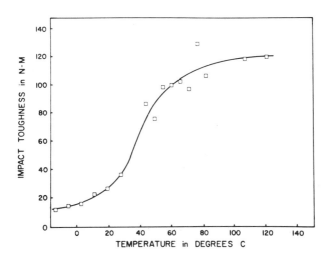

FIG. 2—*Charpy V-notch impact energy as a function of test temperature for the A533B material studied in this paper.*

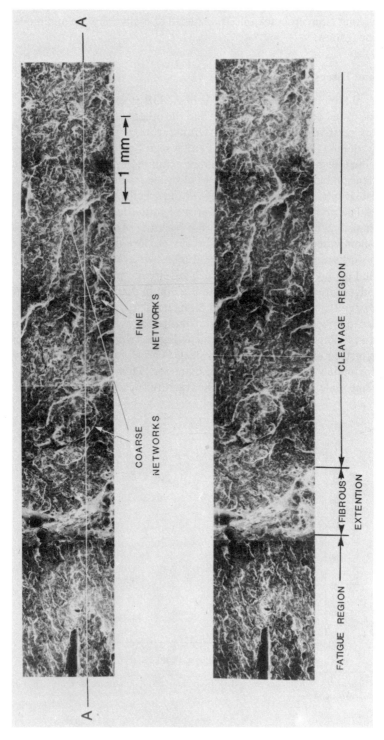

FIG. 3—SEM stereo pair of the fracture surface of the specimen tested at room temperature.

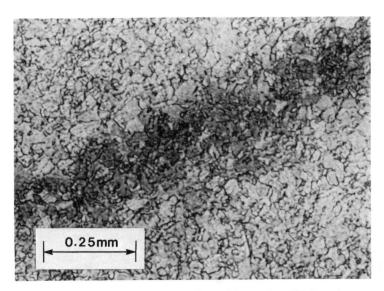

FIG. 4a—*Feature of ferrite grains obtained from using of Etchant A.*

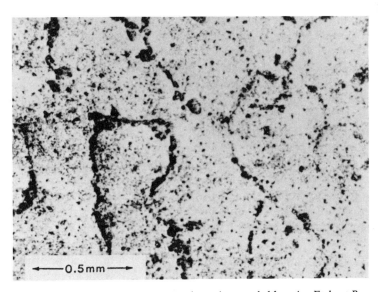

FIG. 4b—*Appearance of prior-austenite grains revealed by using Etchant B.*

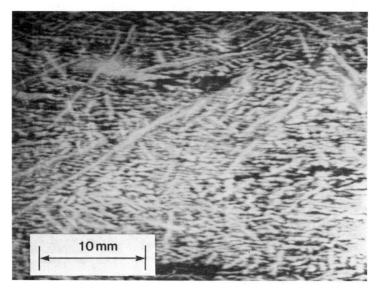

FIG. 4c—*Appearance of dendritic structures revealed by using Etchant C.*

tenite grains shown in Fig. 4*b* are approximately 0.3 to 0.5 mm in diameter. Again the size of prior-austenite grains corresponds well with the coarse network of fibrous ridges found on the fracture surface (Fig. 3). In addition to these microstructures we have found macroscopic dendritic structures as shown in Fig. 4*c*. The dark bands indicate carbide-rich regions and the light bands indicate carbide-poor regions. Figures 5*a* and 5*b* show the carbide density appearances and ferrite grain sizes in dark and light regions. The average ferrite grain size was measured and found to be 0.019 mm in a dark band and 0.024 mm in a light band. The obvious difference in carbide density (possibly assisted by differences of grain size) would be expected to cause differences between light and dark regions in terms of mechanical behavior.

To investigate how these structures at three size levels interact with the fracture process, the fracture surface was sectioned vertically. Figure 6 exhibits a polished vertical section of the sample. A close examination of the section along the fracture surface reveals that numerous needle-like areas are deformed outward from the fracture surface, indicating significant plasticity during fracture. Examination of this sideview of the fracture surface alone may give a false impression that dominantly ductile fracture has occurred, whereas the fracture surface photograph reveals dominant cleavage fracture characteristics.

The polished section shown in Fig. 6 was etched to reveal prior-austenite grain boundaries and the result is shown in Fig. 7. It is apparent that fracture surface contours correlate well with prior-austenite grain boundaries. In or-

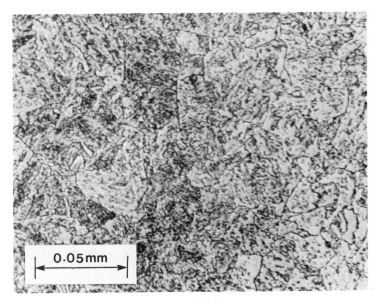

FIG. 5a—*Features of ferrite grains in dark regions in Fig. 4c revealed by using Etchant C.*

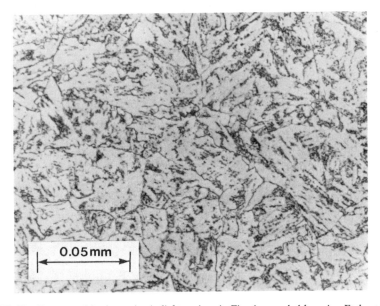

FIG. 5b—*Features of ferrite grains in light regions in Fig. 4c revealed by using Etchant C.*

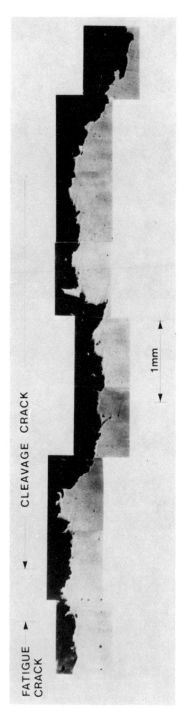

FIG. 6—*Vertically sectioned view of fracture surface of the specimen tested at room temperature.*

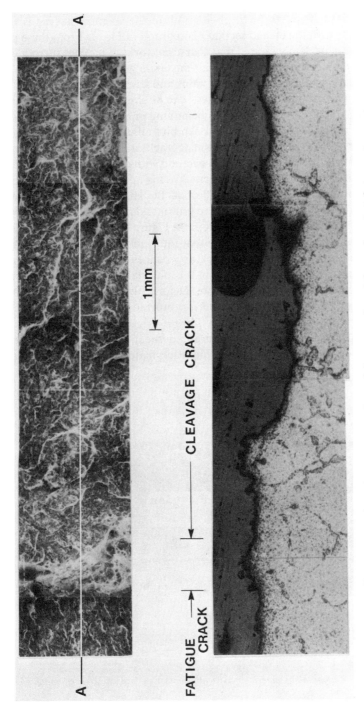

FIG. 7—*Prior-austenite grain structures underneath fracture surface of the specimen tested at room temperature.*

der to show the relation of prior-austenite grain boundaries to the fracture surface morphology, an SEM photograph of the fracture surface is placed side-by-side with the etched section photograph in Fig. 7. From these photographs it can be demonstrated that coarse networks of patches surrounded by fibrous ridges correspond to the prior-austenite grain boundaries.

The relationship between ferrite grains and fracture surface morphology is more difficult to show clearly. Figures 8a to 8c show features observed underneath the fracture surface near the beginning of a cleavage fracture zone at different magnifications. Figure 8a exhibits minute crack segments underneath the fracture surface. Similar small cracks underneath the fracture surface have been often observed in association with cleavage fracturing. Figure 8b is an enlargement of a minute crack. It is apparent that this particular section of the crack did not grow beyond the length shown here; however, other sections of the crack might have grown and might be joined to the main crack tip resulting in a zig-zag crack path. Figure 8c is a further enlargement of the crack. It is of interest to note that the crack becomes more nearly planar and tight as the exhibited crack approaches the size of a ferrite grain. Furthermore, bending and tearing of the material between planar crack segments are demonstrated in this figure. Deformed connections of this kind, in addition to the ferrite grain boundaries, provide a plausible mechanism for the observed small-scale network of fibrous ridges.

Figure 9 reveals dendritic structure underneath the fracture surface. The dark regions correspond to the carbide-rich material, and their mechanical

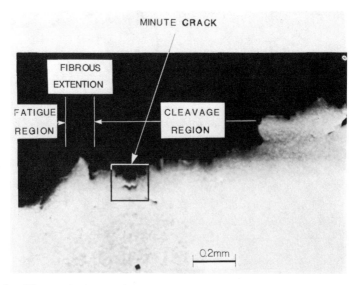

FIG. 8a—*Micrograph of material underneath fracture surface near the beginning of cleavage crack extension.*

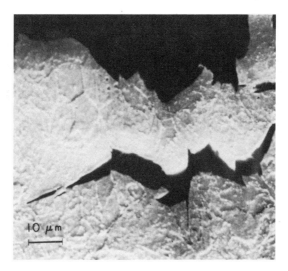

FIG. 8b—*Enlargement of section of Fig. 8a.*

properties are presumed to be different from those light regions where carbide density is lower. Comparison of dendritic features shown in Fig. 9 with the prior-austenite grain features shown in Fig. 7 reveals that prior-austenite grains formed in dark, carbide-rich dendritic structures are smaller than those formed in light, dendritic structures. Furthermore, the appearance as-

FIG. 8c—*Further enlargement of section of Fig. 8a.*

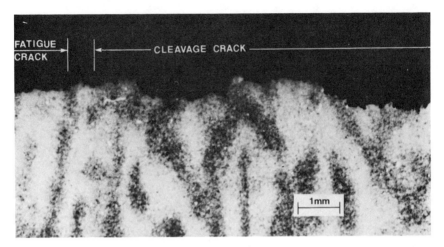

FIG. 9—*Dendritic structures underneath fracture surface revealed by using Etchant C.*

pects of prior-austenite grain boundaries in carbide-rich dendritic structures exhibit broader width and much more complicated structures than those in lower carbide dendritic structures. We have not yet established exactly how dendritic structures influence fracture morphology; however, because of its large size scale, this type of inhomogeneity seems likely to be of special interest in relation to scatter in K_{Ic}-values.

To study how these micro- and macrostructures influence fracture processes, it is necessary to know their mechanical properties. Microhardness measurements provide a mechanical property of one kind. Figure 10 exhibits the results of hardness measurements across several prior-austenite grains and at grain boundaries. The indented surface had the orientation shown in Fig. 9. It is interesting to see that the hardness of individual prior-austenite grains differs significantly. These differences in hardness can be explained by the carbide density in dendritic structures. Figure 11 shows hardness measurements across dendritic structures. Dark, carbide-rich regions show higher hardness readings than light regions. This suggests that the deformation behavior of dark regions may be quite different from that of light regions. As discussed previously, further study is needed to establish the influence of dendritic structures on fracture morphology and fracture processes.

The A533B specimen discussed next was fractured at 54°C. About 1 mm of fibrous crack extension corresponded to a condition of maximum load. A spring-in-series loading arrangement was used in order to drive the fibrous separation rapidly after development of "tearing instability." The position of the crack front at the start of this instability is uncertain, but was probably in the range of 1.5 to 2.5 mm beyond the fatigue precrack. The load versus opening displacement record showed clear evidence of the initial rapid in-

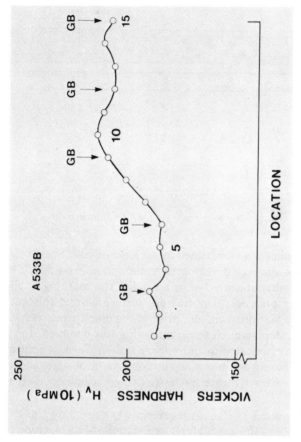

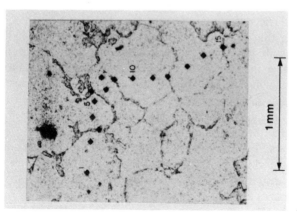

FIG. 10—*Hardness readings across prior-austenite grains and at grain boundaries.*

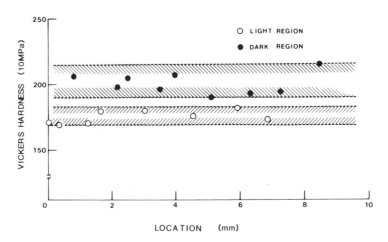

FIG. 11—*Hardness readings across dendritic structures.*

crease of displacement accompanied by decrease of load. Specimen behaviors after the start of cleavage development could not be recorded. The fracture surface showed what appeared to be a crack arrest marking at about 8 mm from the fatigue precrack. The fracture surface beyond this marking was dominant cleavage. Bordering and prior to the crack arrest marking, about 1 mm or less of dominant cleavage fracturing was observed. In a region of several millimetres adjacent to and prior to this cleavage, islands of cleavage surrounded by bands of fibrous separations could be seen and one might question which of the fracture modes (cleavage or hole-joining) was dominant.

The spring-in-series testing arrangement was provided by a 38-mm-thick titanium alloy plate (159 mm wide) fastened to the MTS machine columns at the top of the machine and, at its center, to the loading device by means of a 1.2-m-long rod. The spring plate flexure span was 1 m.

It is plausible that the increase of fibrous separation speed, due to the spring driven instability, would produce local crack front conditions suitable for cleavage. Due to inertia effects in the rod and spring plate and to the usual rapidity of cleavage fracturing, it is also plausible that the separation speed became large enough to nearly isolate the specimen load from the spring plate, resulting in enough specimen load drop for crack arrest followed by rapid re-loading and re-initiation of crack extension.

Figure 12 shows SEM stereo pair photographs of the fracture surface near the crack arrest marking. In the upper photo, the right hand "cleavage" arrow indicates a region where the cleavage prior to crack arrest was about 1 mm. Other arrows indicate islands of cleavage surrounded by bands of fibrous separation. The approximate regions where the fracture appearances are fibrous, mixed, and cleavage are marked in the figure. In order to verify

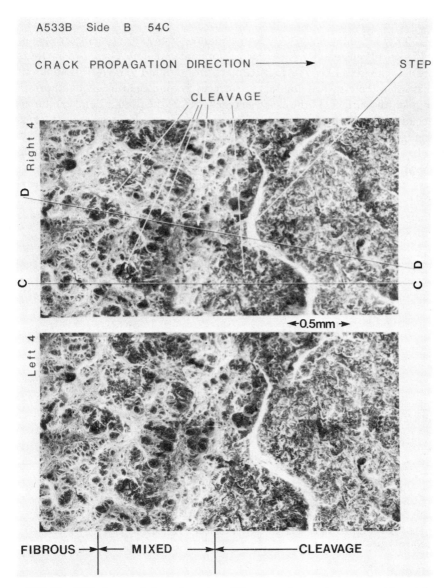

FIG. 12—*SEM photograph from the region of fibrous to cleavage fracture transition of the specimen tested at 54°C.*

that the curved white band in the cleavage region pertains to crack arrest and to illustrate other features of this fracture region, topological measurements along the line *C-C* were made using both top and bottom sides of the fracture. This was done using stereo pair SEM photographs and a parallax bar. The method, similar to areal contour determinations, is described in Ref 2.

Figures 13*a* to 13*c* show the top and bottom fracture surface contours measured along the line *C-C*. These contours exhibit a steep step corresponding to the fibrous band in the cleavage-dominated region. These two contours are now closed until two surfaces contact as shown in Fig. 13*b*. It is apparent that when the cleavage-dominated surfaces beyond the fibrous band are in contact, the surfaces of the cleavage region before the fibrous band exhibit an opening. This finding corresponds to the view that the steep step (fibrous band) was the result of temporary crack arrest and re-initiation.

In Fig. 13*c*, the region marked (1) evidently broke later than did the region shown in contact. The opening movement during re-initiation after crack arrest caused the separation of region (1). Region (2) has a cleavage appearance, although there are many small ridges not detailed in the topographical analysis. Clearly region (2) must have separated well in advance of the average crack front position.

Microstructural aspects of fracture processes which may contribute to the scatter of K_{Ic} were discussed previously. Further study is needed. However, observations of micro- and macrostructures and comparisons of these to fracture morphology show that close relationships exist.

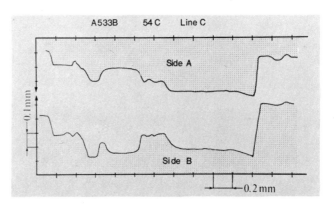

FIG. 13a—*Top and bottom fracture surface contour measurements along the line* C-C *of the specimen tested at 54°C.*

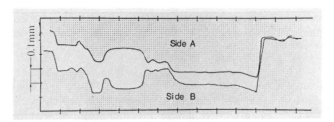

FIG. 13b—*Closing of fracture surface contours along the line* C-C *shown in Fig. 13a until cleavage surfaces beyond crack arrest are in contact.*

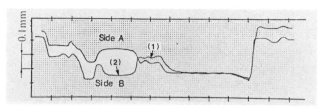

FIG.13c—*Further closing of fracture surface contours along the line C-C until cleavage surfaces just prior to crack arrest are in contact.*

Conclusions

1. Direct observation of fracture morphology is essential for improved understanding of cleavage fracturing in the transition range.

2. Fracture surface details reflect influences of both ferrite and prior-austenite grain size. In specimens from a 305-mm-thick plate of A533B steel, fracture morphology is also influenced variously by a larger size inhomogeneity due to dendritic solidification.

3. The sequence of separational events can be revealed by topographic examination using stereo SEM photography.

4. Using specimens from a thick plate of A533B steel and at a temperature in the upper portion of the transition temperature range, fast cleavage can be introduced by increasing the speed of a fibrous separation.

Acknowledgments

This work was supported by a contract from the Oak Ridge National Laboratory Heavy-Section Steel Technology Program. The authors wish to thank J. P. Gudas and Dan Davis, Naval Ship Research and Development Command, Annapolis, Maryland, for supplying fracture specimens and information on test conditions.

References

[1] Whitman, G. D. and Bryan, R. H., "HSST Program Quarterly Progress Report for October–December 1980," NUREG/CR-1941, Nuclear Regulatory Commission, pp. 37–49.
[2] Kobayashi, T., Irwin, G. R., and Zhang, X. J. in *Fractography of Ceramic and Metal Failures, ASTM STP 827*, J. J. Mecholsky, Jr., and S. R. Powell, Jr., Eds., American Society for Testing and Materials, 1984, pp. 234–251.

Alexander D. Wilson[1]

Influence of Inclusions on the Fracture Properties of A588A Steel

REFERENCE: Wilson, A. D., **"Influence of Inclusions on the Fracture Properties of A588A Steel,"** *Fracture Mechanics: Fifteenth Symposium, ASTM STP 833,* R. J. Sanford, Ed., American Society for Testing and Materials, Philadelphia, 1984, pp. 412–435.

ABSTRACT: The influence of nonmetallic inclusions on the fracture properties of A588 Grade A steel plate is investigated by evaluating one plate produced by conventional steelmaking practices and another using calcium treatment. The two plates had differing inclusion structures and their characterization included: tensile, Charpy V-notch, dynamic tear, and J-integral testing; J_{Ic}, tearing modulus, and J–Δa and J–T curve determinations on the upper shelf; and fatigue crack propagation and near-threshold fatigue evaluations. Testing of up to six specimen orientations was used, depending on the testing method. Of all the test methods used in this investigation the J_{Ic} and J–T curve determinations were found to be the most sensitive to changes in inclusion structure in all testing orientations. The fatigue crack growth rate of the calcium-treated steel was primarily improved in the through-thickness (SL) orientation at higher ΔK levels. Testing at higher load ratios accentuated these differences. Oxide and roughness-induced crack closure had a significant effect on the near-threshold fatigue behavior of the conventional steel as a result of the inclusion-induced roughness on the fracture surface. Fractography and crack closure measurements were used to explain this behavior.

KEY WORDS: inclusion, *J*-integral, tearing modulus, fatigue crack propagation, fatigue threshold, dynamic tear, calcium treatment, steel properties, isotropy

A588 Grade A is a high-strength, low-alloy plate steel used in a number of structural applications. This 345 MPa (50 ksi) minimum yield point, 483 MPa (70 ksi) minimum tensile strength steel is extensively used in bridge and building construction and also is applied for nuclear component supports.[2] In these applications there are often mechanical design details where lamellar tearing during welding is a concern or improved fatigue and toughness properties are required. One way to meet these needs is through control of the nonmetallic inclusion structure during steelmaking by calcium treatment [1].

The development of fracture mechanics test methods and their use in mate-

---

[1]Senior Research Engineer, Lukens Steel Company, Coatesville, Pa. 19320.
[2]These levels apply for up to 102-mm (4-in.)-thick plate; somewhat lower levels are specified for thicknesses up to 203 mm (8 in.) maximum.

rial characterization has played an important part in the evaluation and implementation of fracture mechanics technology. For this reason previous studies were performed to demonstrate the benefits of inclusion control by calcium treatment on the J_{Ic} fracture toughness [2] and the fatigue crack propagation properties [3] of steels. Additional testing techniques are now available which can also have an impact on the design of steel structures. These include the elastic-plastic fracture toughness determinations using the J-integral resistance curve and the tearing modulus [4] and the near-threshold fatigue crack propagation behavior [5] of steels.

For the aforementioned reasons the fracture properties of two 76-mm-(3-in.)-thick A588A plates were investigated. One plate was conventionally produced and had a 0.020% sulfur level, while the other was produced using calcium treatment with a 0.003% sulfur content. The lower sulfur level in the calcium-treated plate resulted in fewer and different nonmetallic inclusions. Both plates were evaluated using tension, Charpy V-notch, dynamic tear, J_{Ic}, J–R curve, tearing modulus, fatigue crack growth rate, and near-threshold fatigue testing. Depending on the test method a number of testing orientations were used. This was done to assist in showing the influence of the inclusion structure on the micromechanisms causing fracture in the test method. One important question to be addressed by this study was: Does changing the inclusion structure have similar effects on these fracture mechanics properties?

Experimental Procedures

Materials

The chemical composition of the two test plates is given in Table 1. The sulfur level is the principal difference noted. A588A steel can be obtained in the as-rolled, normalized, or quenched-and-tempered heat treated conditions. Both of the 76-mm (3-in.)-thick test plates in this study were normalized at 899°C (1650°F) for 3 h and air cooled; the microstructures were both ferrite-pearlitic. The major microstructural difference between the two plates was in their nonmetallic inclusion structure.

The conventionally produced plate (CON) with the higher sulfur level had Type II manganese sulfide inclusion "stringers," which are elongated and

TABLE 1—*Chemical composition (weight percent).*

	C	S	Mn	P	Cu	Si	Ni	Cr	Mo	V	Al	Ca
CON	0.15	0.020	1.07	0.006	0.27	0.24	0.16	0.58	0.065	0.051	0.026	ND[a]
CaT	0.15	0.003	1.11	0.011	0.29	0.46	0.19	0.59	0.048	0.081	0.030	0.0041

[a]ND = none detected.

pancaked as a result of hot rolling. They are shown in Fig. 1a. Clusters of aluminum oxide inclusions were also present in the CON steel. Both of these groups of inclusions are minimized by calcium treatment through desulfurization and inclusion shape control [1]. The effect on the inclusions in the calcium-treated plate (CaT) is indicated in Fig. 1b. Calcium treatment was accomplished by the injection of calcium compounds (using an argon carrier gas) into the molten steel after tapping from the electric arc furnace. This process results in sulfur level reduction of greater than 50%, depending on the sulfur level on tapping and other processing parameters. The duplex, calcium-modified inclusions in the CaT plate are also more resistant to deformation during hot rolling. The lower overall content of inclusions in the CaT steel as a result of desulfurization and alumina cluster minimization and the inclusion shape control exhibited by most of the remaining inclusions leads to the improved level and isotropy of ductility, toughness, and fatigue properties for CaT steels [1-3]. Inclusion control is particularly effective for A588A plates because they often are very heavy and long, thus requiring a significant amount of rolling reduction and elongation.

Mechanical Testing

The test specimens were removed from each plate at the center-gage location. The testing orientations used in this investigation are depicted in Fig. 2. Up to six different testing orientations were used, depending on the test method. Two tension specimens were taken in each of the longitudinal (L), transverse (T), and through-thickness (S) testing orientations. The strain-hardening exponent, as well as the conventional strength and ductility parameters, were determined for these tests.

Toughness—The basic toughness properties of the two plates were established using Charpy V-notch (CVN), dynamic tear (DT), and nil-ductility temperature (NDT) determinations. All these tests were performed to the applicable ASTM specifications: Notched Bar Impact Testing of Metallic Materials (E 23), Dynamic Tear Testing of Metallic Materials (E 604), and Drop-Weight Test to Determine Nil-Ductility Transition Temperature of Ferritic Steels (E 208). The CVN tests were performed over the full transition range in each of the six major testing orientations using 15 specimens per curve. The DT transition curves were obtained using 10 specimens per curve in four testing orientations (LS, LT, TS, and TL). The NDT of each plate was determined in the TL orientation.

Two J-integral fracture tests per test orientation were conducted on each plate at room temperature in accordance with ASTM Test for J_{Ic}, a Measure of Fracture Toughness (E 813) using the compliance-unloading, single-specimen technique [6]. 1T-CT specimens, 25 mm (1 in.) in thickness with 25% side-grooves, were evaluated in the three major testing orientations (LT, TL, and SL). All cracking during the testing was ductile (noncleavage) and was

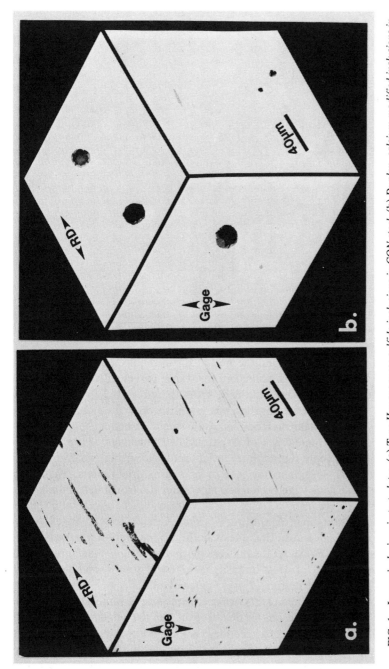

FIG. 1—*Larger inclusions in test plates.* (a) *Type II manganese sulfide inclusions in CON steel.* (b) *Duplex, calcium modified inclusions in CaT steel.*

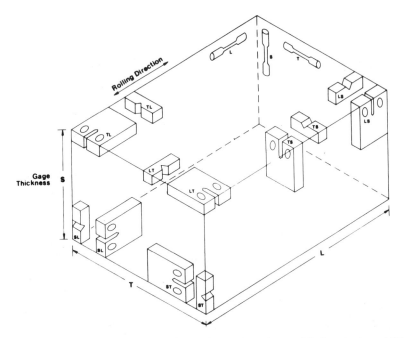

FIG. 2—*Schematic drawing showing the specimen orientations and designations per ASTM E 399. Actual test location was at center-gage for this investigation.*

continued significantly beyond the ASTM E 813 requirements for determining J_{Ic}, so that J–Δa (crack extension) resistance curves with over 5 mm (0.2 in.) of crack extension were obtained. From these results J_{Ic} values were established. A best-fit power law line was established for the J–Δa curves, and plots of J versus the tearing modulus (J–T) were obtained.

Fatigue—The fatigue crack propagation (FCP) behavior of the test plates was identified at room temperature in laboratory air using two 13-mm (0.5-in.)-thick 1T-CT specimens for each of six testing orientations. These constant-amplitude tests were performed at a load ratio (R) of 0.1 and a testing frequency of 10 Hz using a triangular waveform in accordance with ASTM Test for Constant-Load-Amplitude Fatigue Crack Growth Rates Above 10^{-8}m/Cycle (E 647). Additional tests in the SL orientation were performed at a load ratio of 0.7. All FCP data were plotted on fatigue crack growth rate (da/dN) versus range of stress intensity factor (ΔK) graphs, and best-fit lines using the Paris FCP equation [7] were determined.

The FCP near-threshold testing was performed on 6.4-mm (0.25-in.)-thick 1T-WOL specimens at room temperature in air; the relative humidity was between 33 and 54%. One test specimen for each of three testing orientations (LT, TL, and SL) was conducted at $R = 0.1$ and testing frequencies of 120 to 170 Hz using a computer-controlled, load-shedding technique described in

Ref. *8*. Additional tests at $R = 0.7$ were performed in the SL orientation. The FCP threshold was established at a fatigue crack growth rate of 1×10^{-7}mm/cycle (4×10^{-9} in./cycle) following the guidelines of Hudak et al [9]. The occurrence and level of crack tip closure [10] during the tests was detected and recorded on an oscilloscope using electronic signal cancellation [6]. This allowed detecting the first occurrence of closure within about 10% for effective ΔK and more importantly made closure load determinations possible *during* the test at the testing frequency.

Results

Basic Testing

The tension test results are exhibited in Table 2. Although the CaT plate has a slightly higher strength, the primary difference between the two plates is in the tensile ductility, particularly in the T and S orientation. Also note that there is no significant difference in the strain-hardening exponent between plates or with testing orientation.

The CVN testing results are presented in Fig. 3. The extensive anisotropy in the CON plate, especially on the upper shelf, is evident. The DT transition curves are given in Fig. 4. Once more the differences between the two plates are evident. Since the largest effect of inclusions is noted on the upper shelf where fully ductile behavior is taking place (in contrast to cleavage fracture), the upper shelf energies (USE) for both the CVN and DT testing are summarized in Table 3. Aside from the LS orientation, the CaT steel has a significantly higher USE than the CON in all testing orientations for both CVN and DT testing. The NDT for the CON steel was determined to be $-40°C$ ($-40°F$) and for the CaT steel $-51°C$ ($-60°F$).

TABLE 2—*Tension test results.*[a]

	Orientation	σ_y	σ_u	% El	% RA	m
CON	L	348 (50.5)	525 (76.1)	31.8	74.9	0.242
	T	335 (48.6)	512 (74.2)	29.4	60.0	0.234
	S	336 (48.8)	468 (67.9)	10.1	8.2	0.247
CaT	L	367 (53.2)	552 (80.0)	33.3	76.7	0.241
	T	367 (53.2)	543 (78.7)	32.3	74.4	0.241
	S	362 (52.5)	538 (78.0)	31.4	66.1	0.236

[a]$\sigma_y = 0.2\%$ offset yield strength, MPa (ksi),
σ_u = ultimate tensile strength, MPa (ksi),
% El = percent elongation,
% RA = percent reduction of area, and
m = strain-hardening exponent where $\sigma = K\epsilon^m$ (σ is true stress, ϵ is true strain).

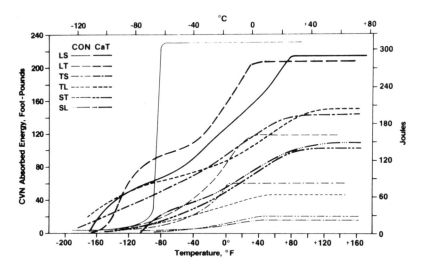

FIG. 3—*Charpy V-notch impact transition curves for tested plates.*

J-*Integral Testing*

The results of the J_{Ic} determinations are summarized in Table 4. Also given in Table 4 are the J values at the point of maximum load (first load drop) J_{FLD}. Note that the higher toughness levels of the CaT steel in the LT and TL orientations resulted in not meeting the minimum specimen size require-

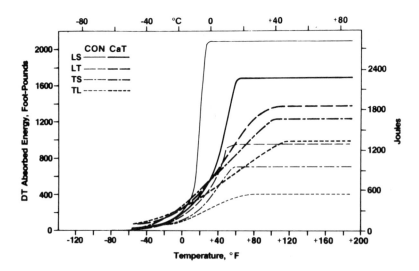

FIG. 4—*Dynamic tear transition curves for tested plates.*

TABLE 3—*Upper shelf impact results.*

Orientation	CVN USE[a]		DT USE[b]	
	CON	CaT	CON	CaT
LS	313 (231)[c]	290 (214)	2841 (2095)[c]	2275 (1678)
LT	160 (118)	279 (206)	1296 (956)	1862 (1373)
TS	81 (60)	195 (144)	960 (708)	1658 (1223)
TL	66 (49)	206 (152)	556 (410)	1372 (1012)
ST	30 (22)	140 (103)	. . .	. . .
SL	23 (17)	149 (110)	. . .	. . .

[a]CVN USE = Charpy V-notch upper shelf energy per ASTM E 23; average of 4 to 6 tests; J (ft-lb).

[b]DT USE = dynamic tear upper shelf energy (16-mm test) per ASTM E 604; average of 3 to 4 tests; J (ft-lb).

[c]This average contains test results that exceeded the capacity of the testing machines.

ment. The K_{Ic} values determined from J_{Ic}, K_J are also given in this table where

$$K_J = \sqrt{J_{Ic}E} \tag{1}$$

The K_{Ic} values calculated from the CVN data using the Rolfe-Novak-Barsom correlation (K_{RNB}) are given for comparison [11,12]:

$$K_{RNB} = \sigma_y \left[5 \left(\frac{CVN\ USE}{\sigma_y} - 0.05 \right) \right]^{1/2} \tag{2}$$

TABLE 4—*J-integral testing results.*[a,b]

	Orientation	J_{FLD}	J_{Ic}	K_J	K_{RNB}	T_i
CON	LT	400 (2282)	227 (1296)	217 (197)	188 (171)	182
	TL	113 (644)	121 (693)	158 (144)	116 (106)	72
	SL	37 (212)	40 (227)	91 (83)	66 (60)	44
CaT	LT	771 (4402)[c]	1450 (8280)[c]	547 (498)[c]	256 (233)	344[c]
	TL	509 (2906)[c]	1028 (5870)[c]	462 (420)[c]	219 (199)	221[c]
	SL	396 (2264)	306 (1745)	252 (229)	185 (168)	185

[a]Average results per ASTM E 813 of two 25-mm (1-in.)-thick CT specimens with 25% side-grooves; single-specimen, compliance-unloading technique.

[b] J_{FLD} = Value of J at point of maximum load (first load drop), KJ/m² (in.-lb/in.²),

J_{Ic} = critical value for crack initiation per ASTM E 813 graphical R-curve determination, KJ/m² (in.-lb/in.²),

K_J = value of K_{Ic} calculated using J_{Ic}, MPa√m (ksi√in.),

T_i = tearing modulus determined from best-fit straight line used to determine J_{Ic}, and

K_{RNB} = K_{Ic} calculated using Rolfe-Novak-Barsom CVN correlation [11,12].

[c]For both test results, specimen sizes did not meet ASTM E 813 requirements.

where σ_y is 0.2% offset yield strength in ksi, CVN USE is in ft-lb, and K_{RNB} is in ksi$\sqrt{\text{in}}$.

The value of the tearing modulus at initiation (T_i) is given in Table 4 and is calculated from the equation established by Paris et al [4]:

$$T_i = \left(\frac{dJ}{da}\right)_i \frac{E}{\sigma_Y^2} \qquad (3)$$

where $(dJ/da)_i$ is the slope of the straight line used in the J_{Ic} determination and $\sigma_Y = \frac{1}{2}(\sigma_y + \sigma_u)$. For all five parameters given in Table 4 the higher toughness of the CaT steel is evident in all three testing orientations.

The J-Δa resistance curves for these plates are given in Fig. 5. Only data greater than 0.15 mm from the ASTM E 813 "blunting line":

$$J = 2\sigma_Y \Delta a \qquad (4)$$

are plotted in Fig. 5 and used in the best-fit curve determinations. A family of equations was used to fit the J-Δa curves, but the best overall fit was given by the power law

$$J = A(\Delta a)^B \qquad (5)$$

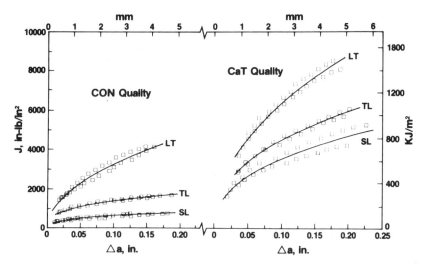

FIG. 5—J-Δa resistance data 0.15 mm (0.006 in.) beyond the ASTM E 813 "blunting-line" for each specimen (two specimens per orientation). Overall best fit power law lines are shown; parameters are given in Table 5.

where A and B are the constants determined for each material. Others have also found this form to be most applicable [13]. These best-fit lines are given in Fig. 5, and the constants A and B are summarized in Table 5. The good statistical basis of this equation is revealed by the high multiple correlation coefficients (R^2). In comparing the two steels, once more the CaT steel displays higher J values at any Δa.

From the J-Δa data for each test specimen, a J-T graph was established using

$$T = \frac{dJ}{da} \frac{E}{\sigma_Y^2}$$ (6)

In this comparison the dJ/da value at any particular J data point was taken as the slope between the immediately preceding and succeeding adjacent data points. The resulting J-T graphs are given in Fig. 6 with best-fit power law curves shown of the form

$$J = \beta \, T^\alpha$$ (7)

The values of β and α are summarized in Table 6. Figure 6 shows that the CaT steel tends to have higher tearing moduli for each testing orientation.

Fatigue Testing

The FCP testing results at $R = 0.1$ are given in Fig. 7 in the form of the best-fit straight lines from the Paris equation [7]

$$da/dN = C_0 \, \Delta K^n$$ (8)

TABLE 5—J–R *curve results.*[a]

	Orientation	B	KJ/m²/mm	(in.-lb/in.²/in.)	R^2
			\multicolumn{2}{c}{A}	
CON	LT	0.497	359	(10 235)	0.968
	TL	0.322	178	(2 883)	0.986
	SL	0.364	74.9	(1 388)	0.943
CaT	LT	0.470	724	(18 899)	0.974
	TL	0.406	538	(11 414)	0.988
	SL	0.411	422	(9 113)	0.948

[a]B and A are parameters for power law fit to J-Δa data greater than 0.15 mm from blunting line where $J = A(\Delta a)^B$. A value determined where J and Δa had the units indicated.

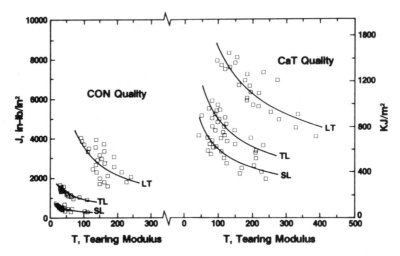

FIG. 6—*J-T data determined for each J-Δa test record (two per orientation). J-T best fit lines using power law are shown; parameters are given in Table 6.*

The values of C_0 and n are summarized in Table 7. The results of the $R = 0.7$ testing are given in Fig. 8 and Table 7. The higher fatigue crack growth rate of the CON material at higher ΔK levels is particularly evident in the SL orientation. The difference with the CaT steel appears to be accentuated by testing at the 0.7 load ratio (Fig. 8).

The results of near-threshold FCP testing are displayed in Fig. 9 for both steels. The FCP threshold (ΔK_{th}) values for each test are summarized in Table 8. The considerable effect of R ratio is evident in both orientations. Also given in Table 8 are the ΔK_{th} values established using data where a correction for the closure load was used to determine the effective ΔK. The effective load was determined by subtracting the closure load from the applied load. The closure load was established for each data point by assuming a straight line

TABLE 6—*J-T curve results.*[a]

	Orientation	α	β KJ/m^2	β (in.-lb/in.^{2})	R^2
CON	LT	−0.721	1 736	(99 132)	0.368
	TL	−0.396	996	(5 688)	0.843
	SL	−0.441	392	(2 237)	0.687
CaT	LT	−0.450	11 969	(68 342)	0.707
	TL	−0.474	7 783	(44 443)	0.773
	SL	−0.459	5 047	(28 821)	0.612

[a] α and β are parameters for power law fit to J-T data where $J = \beta T^\alpha$; β values determined where J has units indicated.

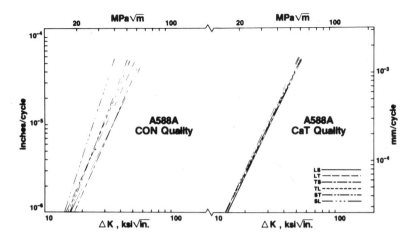

FIG. 7—*Fatigue crack propagation best fit lines using Paris power law equation [7] for R =
0.1 testing. Improved isotropy of CaT steel is evident.*

TABLE 7—*Fatigue crack propagation results.*[a,b]

		n	C_0 MPa$\sqrt{m}$, mm/cycle	(ksi$\sqrt{in.}$/cycle)	R^2
CON	LS	3.00	4.71×10^{-9}	(2.46×10^{-10})	0.982
	LT	3.17	3.67×10^{-9}	(1.95×10^{-10})	0.989
	TS	3.28	1.68×10^{-9}	(9.00×10^{-11})	0.987
	TL	3.50	1.37×10^{-9}	(7.52×10^{-11})	0.996
	ST	3.81	2.28×10^{-9}	(2.59×10^{-11})	0.961
	SL	4.39	1.40×10^{-10}	(8.34×10^{-12})	0.977
	SL[c]	4.68	6.78×10^{-10}	(4.15×10^{-11})	0.925
CaT	LS	3.00	6.50×10^{-9}	(3.40×10^{-10})	0.987
	LT	2.97	7.38×10^{-9}	(3.85×10^{-10})	0.986
	TS	3.08	4.84×10^{-9}	(2.55×10^{-10})	0.980
	TL	2.99	6.87×10^{-9}	(3.59×10^{-10})	0.987
	ST	3.18	3.72×10^{-9}	(1.98×10^{-10})	0.982
	SL	3.10	5.24×10^{-9}	(2.77×10^{-10})	0.988
	SL[c]	3.84	1.80×10^{-10}	(1.02×10^{-11})	0.994

[a]12.7-mm (0.5-in.)-thick 1T-CT specimens tested per ASTM E 647 in air at room temperature
at 10 Hz and $R = 0.1$ except as noted.

[b] n = Value of exponent in Paris equation $da/dN = C_0 \Delta K^n$,
   C_0 = coefficient in Paris equation where ΔK and da/dN have the units indicated, and
   R^2 = statistical parameter indicating accuracy of best-fit straight line to data; for a perfect
   fit $R^2 = 1.000$.

[c]$R = 0.7$.

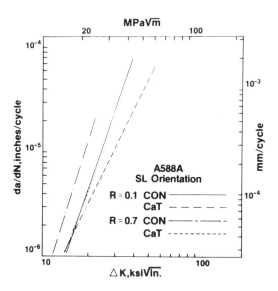

FIG. 8—*Fatigue crack propagation best fit lines for SL orientation for load ratios of 0.1 and 0.7.*

relationship from the several results obtained from the oscilloscope traces. The uncorrected, as-tested ΔK_{th} values indicated that the CON steel has higher values than the CaT; however, when a correction is made for closure, the reverse appears to be the case.

To show the overall variation in FCP behavior of the two steels, the FCP and near-threshold results for the SL orientation for both load ratios are plotted in Fig. 10. The cross-over, CON versus CaT, in behavior between the two steels at $R = 0.1$ is of most interest.

Discussion

These two A588A plates with dissimilar inclusion structures show varying degrees of property differences, depending on the test method being used. These comparisons are summarized in the bar graphs in Fig. 11. Whereas the tensile reduction of area shows the greatest difference in the S orientation, the CVN and DT show differences in all testing orientations (Figs. 11a–c). It has been previously reported that CVN testing is generally more sensitive to inclusion structure changes than tensile ductility [14]. In all orientations the CaT displays higher CVN and DT toughness, except in the LS orientation. In the LS orientation the elongated, planar inclusion clusters deflect the propagating crack and absorb a significant amount of energy. In the other orientations the inclusions act as preferred sites for void nucleation, and because of their close spacing, void growth is minimized and lower toughness levels result. It

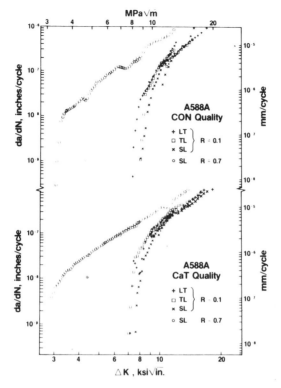

FIG. 9—*Near-threshold fatigue crack propagation results in noted orientations and load ratios (R).*

TABLE 8—*Fatigue crack propagation threshold results.*[a]

		R[b]	ΔK_{th} (As-Tested)[c]		ΔK_{th} (Closure Corrected)[d]	
			MPa√m	(ksi√in.)	MPa√m	(ksi√in.)
CON	LT	0.1	8.29	(7.54)	5.56	(5.06)
	TL	0.1	8.99	(8.18)	4.48	(4.08)
	SL	0.1	9.22	(8.39)	3.32	(3.02)
	SL	0.7	3.44	(3.13)	...	...
CaT	LT	0.1	7.69	(7.01)	5.53	(5.03)
	TL	0.1	7.92	(7.21)	6.41	(5.83)
	SL	0.1	8.48	(7.72)	5.80	(5.28)
	SL	0.7	3.12	(2.84)	...	...

[a]Determined using 6.4-mm (0.25-in.)-thick 1T WOL specimens at room temperature in air; relative humidity 33 to 54%; 120 to 170 Hz.
[b]Ratio of minimum to maximum load.
[c]Determined per Ref 9. ΔK at da/dN of 10^{-7} mm/cycle.
[d]Closure correction made by measuring closure load on oscilloscope using electronic signal cancellation during testing to determine effective ΔK.

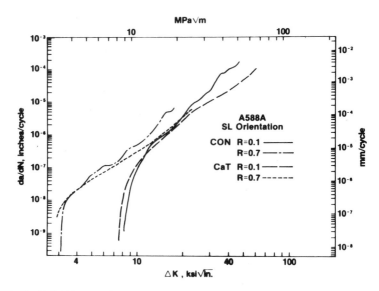

FIG. 10—*Full fatigue crack propagation plots in SL orientation for both load ratios* (R). *Note cross-over in both* R = 0.1 *and* 0.7 *comparisons.*

appears that the CVN may be slightly more sensitive than the DT to inclusion structure changes. This may be due to the inclusion influence on initiation being more important in the blunter notched CVN specimens.

J-*Integral Toughness*

The J_{Ic} results for these two steels are dramatically different (Fig. 11*d*). The J_{FLD} results are also significantly different, although not as much as the J_{Ic}. As noted in this figure and in Table 4, the J_{Ic} and J_{FLD} values for *valid* tests are similar. When the specimen size requirements are not met they are vastly different. This has been reported previously for other plate steels [2]. Also, J-integral results had previously been found to be more sensitive to inclusion structure changes than CVN and DT tests [2]. This is also shown in this study for the initiation results. The additional constraint provided by the thicker J-integral specimens contributes to void nucleation [15] and growth [16] at inclusions and thus accentuates material differences. It is also worth noting, however, that the K_{Ic} results predicted from CVN data are still conservative when compared to those obtained from J-integral testing as exhibited in Table 4. This is in conflict to previous work where the reverse was found for valid J_{Ic} results [2].

The elastic-plastic behavior of the CaT steel continues to be tougher after initiation, during slow stable crack growth. This was shown in the J–T curves in Fig. 6 and is summarized in Fig. 11*f* where the J-integral values at a tearing

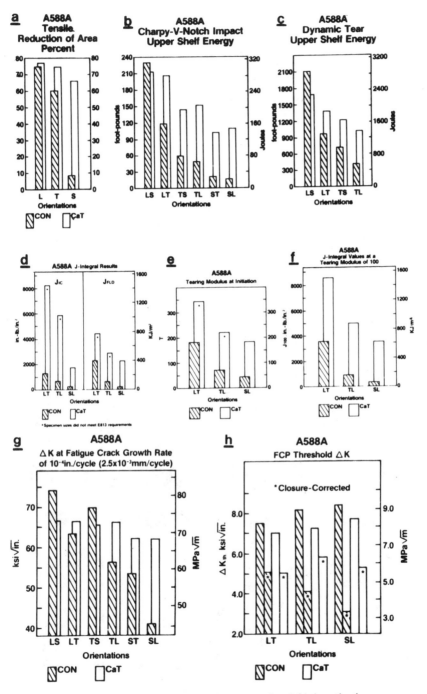

FIG. 11—*Summary bar graphs of key test results of this investigation.*

modulus of 100 are given for each steel and orientation. This demonstrates that inclusion control has a dramatic influence on both the ductile crack initiation and growth behavior of A588A steel.

The side-grooving of J-integral test specimens has often been used to help maintain a straight (nonbowed) crack front during the test. However, in the CON steel a completely straight crack front was not possible in any of the three orientations because of the inclusion structure interaction. Figure 12a shows that all three testing orientations of the CON steel have an irregular crack front, particularly the SL orientation. This behavior resulted in greater differences between the crack length measurement made on the specimen using the heat-tinted marking and the crack length determined from compliance measurements. For the CON steel the heat-tinted measurement was always greater, by an average of 13, 15, and 40% for the LT, TL, and SL orientations respectively, even when using the nine-point measuring method. The same differences for the CaT steel were 7, -2, and 7% respectively. This indicates that there may be larger differences between data determined by multiple-specimen testing when compared with that determined by single-

FIG. 12—*Typical* J-*integral fracture surfaces for each plate.* (a) *CON tests.* (b) *CaT tests. Note extremely uneven crack fronts in CON specimens.*

specimen compliance techniques for steels with higher inclusion contents. In this program, however, the increase of crack extension levels would have had little if any effect on J_{Ic} and would have led to even lower tearing modulus values for the CON steel in Fig. 6.

Fatigue Crack Propagation

The FCP results of Fig. 7 are summarized in Fig. 11g by giving the ΔK level when a fatigue crack growth rate of 2.5×10^{-3} mm/cycle (1×10^{-4} in./cycle) is attained for each material and orientation.[3] In all but the LS and TS orientations the CON material reaches a specified da/dN at a lower ΔK level. The LS and TS orientations show an opposite effect due to the inclusion clusters locally deflecting the fatigue crack and slowing growth [3]. In the other testing orientations at higher fatigue crack growth rates the inclusion clusters increase the growth rate as a result of the ductile fracture of material between inclusions of a particular cluster [3]. These local accelerations of the crack front lead to an overall faster growth rate, particularly in the SL orientation. The effect of load ratio accentuates this ductile fracture process and resulted in the larger material differences shown in Fig. 8 between the two steels. However, the effects of inclusion structure on the FCP behavior are not nearly as great as that shown previously in any of the fracture toughness test methods.

The FCP testing results for the CaT steel in all testing orientations and for the CON steel in the LT orientation were similar to those reported for Barsom [17] for a number of ferritic-pearlitic steels and for A516-70 [18] and A633C [19] carbon steels. The CON results in the TL, ST, and SL orientations, however, indicate faster fatigue crack growth rates at any particular ΔK. This suggests that the use of a generalized FCP behavior is neither conservative nor appropriate when the orientation of concern can be significantly influenced by inclusion structure.

The results of the near-threshold FCP testing reported in Table 8 and Fig. 9 are summarized in Fig. 11h. The as-tested results indicated that the CON steel has higher ΔK_{th} values in all three testing orientations. However, there was evidence of significantly greater amounts of crack tip closure in all the CON tests. When this was accounted for, the CaT steel gave higher ΔK_{th} levels. The higher closure measurements for the CON specimens were accompanied by heavy, very adherent oxide deposits at the near-threshold region of these rougher fracture surfaces, particularly for the TL and SL orientations. This is exhibited in Fig. 13. This oxide also could not be stripped-off with methyl acetate replicating tape.

An examination of the fractures using the scanning electron microscope (SEM) showed the adherent oxide more dramatically (Fig. 14). In all the $R =$

---

[3]Although this level of growth rate was not attained in all tests, this value was used for comparison to accentuate differences between materials and orientations. The values in Table 7 were used to make these extrapolations.

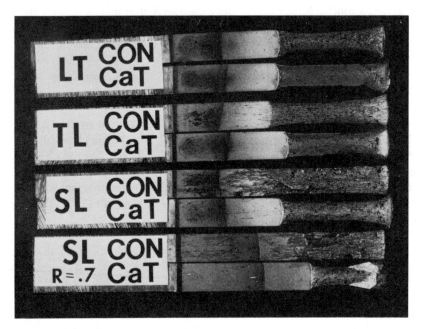

FIG. 13—*Fracture surfaces of all near-threshold FCP tests. Note oxide on* R = 0.1 *tests, particularly in TL and SL orientations of CON steel. Very little oxide is shown in* R = 0.7 *tests.*

0.1 threshold tests there was some oxide product at the near-threshold region. This has been noted by others [5,19-22]. In the CON tests, however, particularly the TL and SL orientations, there appeared to be an additional heavier, more adherent oxide which shows up in the SEM photo in Fig. 14a. Further examination disclosed that this heavy oxide was associated with rougher surface details on the fracture. The surface roughness was a result of the interaction of the crack front with inclusion formations in the CON steel. The heavy oxide forming at the ledges is particularly evident (Fig. 15). It was also noted that there was very little if any oxide present on either of the $R = 0.7$ fractures, even though the CON fracture surface was still rougher than the CaT.

The foregoing evidence indicates that the near-threshold behavior of these steels is being influenced by a combination of the roughness-induced crack closure and oxide-induced crack closure phenomena promoted by Ritchie and others [20-23]. The additional surface roughness of the CON specimens, as a result of the inclusion interaction, promotes a fretting oxidation starting at the ledges created by the inclusion clusters. Continual production of this oxide leads to the spread of the fretting reaction to the surrounding fracture surface. The buildup of corrosion product then wedges open the crack and prevents the crack tip from seeing the full range of loading. This leads to the higher apparent ΔK_{th} for the CON steel and to the apparent inconsistency

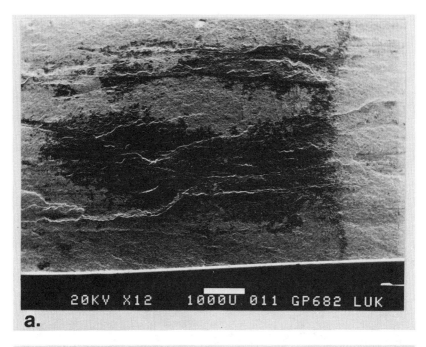

FIG. 14—*Scanning electron microscope photographs of near-threshold, SL specimens after cleaning with several applications of methyl acetate replicating tape. (a) CON test. (b) CaT test. Note the rougher surface and heavy oxide present in the CON test.*

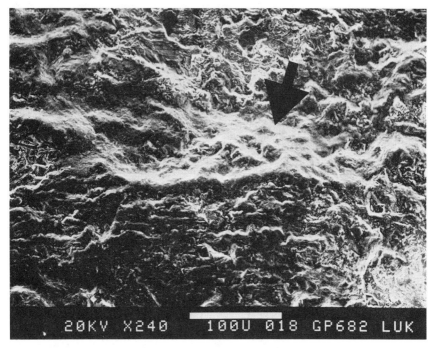

FIG. 15—*Scanning electron microscope view of CON near-threshold, SL oriented, CON steel specimen. Heavy oxide on ledge noted by arrow.*

with the recognized adverse effect of inclusions on fatigue under low load, high cycle conditions.

The benefits of inclusion control towards the fatigue endurance limit are well recognized [18,24], particularly in fatigue crack initiation. The results of this program do not suggest an opposite effect; rather, they indicate that the micromechanisms of the fatigue process can be disturbed by other mechanisms, namely oxide and roughness-induced crack closure. When crack-closure is accounted for, the effective ΔK_{th} results appear to be more in line with expected behavior. Which values should be used in design depends on the environment, the stress intensity, and the physical crack length of concern. The load-shedding experimental technique and crack lengths used in this study may not represent real-world situations, and the oxide and roughness-induced crack closure may not respond in a similar fashion.

The ΔK_{th} results of this A588A steel appear slightly different from other ferritic-pearlitic steels. A516-70 and A633C normalized, carbon, ferritic-pearlitic steels have been found to have ΔK_{th} levels of 8.88 MPa√in. (8.08 ksi√in.) and 8.73 MPa√in. (7.94 ksi√in.) respectively in the calcium-treated TL orientation and in the as-tested conditions [19]. This compares with the 7.92 MPa√in. (7.21 ksi√in.) values found for the A588A CaT steel in the TL orientation. All three steels showed oxide-induced crack closure of about the same level. The foregoing differences, therefore, must be based on metallurgical variables. A588A steel is also known as a "weathering steel" because it forms a very adherent oxide in general atmospheric corrosion applications. This does not appear to have had any influence on the general ΔK_{th} properties. However, the very heavy oxide formed in the CON steel ΔK_{th} tests may be related to this behavior of A588A.

Discussion Summary

The previous discussions on toughness and fatigue behavior have shown that nonmetallic inclusion control can have a range of influences on fracture properties. The most significant improvement was noted in the J_{Ic} and J-T determinations, while the behavior in the near-threshold region appeared to be reversed. This characterization further points out that caution must be exercised in making micromechanistic conclusions from macroscopic measurements. Furthermore, care should be taken in predicting the effect of a certain metallurgical variable on a new fracture property. The effect of inclusions in steel on the corrosion fatigue crack propagation, fatigue crack initiation, and dynamic crack toughness behavior may be different in extent than what has been shown in this characterization.

Conclusions

The fracture mechanics testing in a number of orientations of a conventionally produced 0.020% sulfur A588A steel (CON) and a 0.003% sulfur, calcium-treated A588A steel (CaT) with improved inclusion structure has led to the following conclusions:

1. The presence of improved inclusion structure in the CaT steel led to varying property differences between the two steels, depending on the test method.

2. The CaT steel had higher Charpy V-notch and dynamic tear upper shelf energies except in the LS testing orientation. The elastic-plastic fracture toughness was also significantly improved as measured by J_{Ic}, J-Δa, and J-T curves in the LT, TL, and SL testing orientations.

3. The J_{Ic} and J-T behavior were found to be more sensitive to inclusion structure changes than any of the other methods used in this investigation.

4. The fatigue crack propagation (FCP) behavior of the CaT steel was improved at higher ΔK levels. The largest difference was found in the SL orientation, where differences were also accentuated by testing at higher load ratios. These results demonstrate that generalized FCP behavior for ferritic-pearlitic steels is not conservative when the orientation of concern may be influenced by inclusion structure.

5. Oxide and roughness-induced crack closure had a significant effect on the near-threshold FCP behavior of the CON steel. This resulted in higher ΔK threshold levels for the CON steel in the as-tested condition. When crack closure was accounted for, the results were reversed.

Acknowledgments

The contributions of L. P. Kerr to the testing and analysis efforts of this program are gratefully acknowledged. The assistance of J. R. Lohr and R. L. Urbine is also noted. The critical review of the manuscript by J. A. Gulya is appreciated.

The author also wishes to appreciatively acknowledge the assistance of the Professional Services Group, Hellertown, Pennsylvania, in performing the single-specimen J-integral tests and the computer-controlled near-threshold tests.

References

[1] Wilson, A. D., *Metal Progress*, April 1982, Vol. 121, No. 5, pp. 41-46.
[2] Wilson, A. D. in *Elastic-Plastic Fracture, ASTM STP 668*, American Society for Testing and Materials, 1979, pp. 469-492.
[3] Wilson, A. D. in *Fractography and Materials Science, ASTM STP 733*, American Society for Testing and Materials, 1981, pp. 166-186.
[4] Paris, P. C., Tada, H., Zahoor, A., and Ernst, H. in *Elastic-Plastic Fracture, ASTM STP 668*, American Society for Testing and Materials, 1979, pp. 5-36.
[5] Ritchie, R. O., *International Metals Review*, No. 5 and 6, 1979, pp. 205-230.
[6] Clarke, G. A., Andrews, W. R., Paris, P. C., and Schmidt, D. W. in *Mechanics of Crack Growth, ASTM STP 590*, American Society for Testing and Materials, 1976, pp. 27-42.
[7] Paris, P. C., "The Fracture Mechanics Approach to Fatigue," in *Fatigue: An Interdisciplinary Approach*, Syracuse University Press, N.Y., 1964, pp. 107-132.
[8] Saxena, A., Hudak, Jr., S. J., Donald, J. K., and Schmidt, D. W., *Journal of Testing and Evaluation*, Vol. 6, No. 3, May 1978, pp. 167-174.
[9] Hudak, Jr., S. J., Saxena, A., Bucci, R. J., and Malcolm, R. C., "Development of Standard Methods of Testing and Analyzing Fatigue Crack Growth Rate Data," Air Force Materials Laboratory Report AFML-TR-78-40, Wright-Patterson AFB, Ohio, May 1978.
[10] Elber, W. in *Damage Tolerance in Aircraft Structures, ASTM STP 486*, American Society for Testing and Materials, 1971, pp. 230-242.
[11] Rolfe, S. T. and Novak, S. R. in *Review of Developments in Plain Strain Fracture-Toughness Testing, ASTM STP 463*, American Society for Testing and Materials, 1970, pp. 124-159.
[12] Barsom, J. M. and Rolfe, S. T. in *Impact Testing of Metals, ASTM STP 466*, American Society for Testing and Materials, 1970, pp. 281-302.
[13] Loss, F. J., Menke, B. H., Hiser, A. L., and Watson, H. E. in *Elastic-Plastic Fracture: Second Symposium, Volume II—Fracture Resistance Curves and Engineering Applications, ASTM STP 803*, American Society for Testing and Materials, 1983, pp. II-777-II-795.
[14] Wilson, A. D. in *Through-Thickness Tension Testing of Steel, ASTM STP 794*, American Society for Testing and Materials, 1983, pp. 130-146.
[15] Fisher, J. R. and Gurland, J., *Metal Science*, May 1981, pp. 193-202.
[16] Rice, J. R. and Tracey, D. M., *Journal of the Mechanics and Physics of Solids*, Vol. 17, 1969, pp. 201-217.

[*17*] Barsom, J. M., *Journal of Engineering for Industry, Transactions of ASME*, Vol. 93, Series B, No. 4, Nov. 1971.

[*18*] Wilson, A. D., *Journal of Engineering Materials and Technology, Transactions of ASME*, Vol. 101, July 1979, pp. 265-274.

[*19*] Wilson A. D., *Journal of Engineering Materials and Technology, Transactions of ASME*, Vol. 102, July 1980, pp. 269-279.

[*20*] Suresh, S., Zamiski, G. F., and Ritchie, R. O., *Metallurgical Transactions A*, Vol. 12A, Aug. 1981, pp. 1435-1443.

[*21*] Ritchie, R. O. and Suresh, S., *Metallurgical Transactions A*, Vol. 13A, May 1982, pp. 937-940.

[*22*] Stewart, A. T., *Engineering Fracture Mechanics*, Vol. 13, 1980, pp. 463-478.

[*23*] Mayes, I. C. and Baker, T. J., *Metal Science*, July 1981, Vol. 15, pp. 320-322.

[*24*] *Metals Handbook*, 9th ed., Vol. 1, American Society for Metals, Metals Park, Ohio, 1978, pp. 672-677.

J. D. Landes[1] and T. R. Leax[1]

Load History Effects on the Fracture Toughness of a Modified 4340 Steel

REFERENCE: Landes, J. D. and Leax, T. R., **"Load History Effects on the Fracture Toughness of a Modified 4340 Steel,"** *Fracture Mechanics: Fifteenth Symposium, ASTM STP 833*, R. J. Sanford, Ed., American Society for Testing and Materials, Philadelphia, 1984, pp. 436–448.

ABSTRACT: The effect of load history on fracture toughness was examined for a modified 4340 steel where toughness was evaluated in the elastic-plastic regime by the J_R-curve test method. Four types of load histories were applied: (1) cyclic loading during R-curve development; (2) fatigue precracking at high load levels; (3) bulk prestraining, both monotonic and cyclic; and (4) slow loading during the R-curve test. These load histories were applied in four separate sets of tests. The resulting J_R-curves were compared with R-curves generated for specimens tested in a normal monotonic toughness test. Raising the R-curve was judged as an increase in toughness and lowering it as a decrease. The results showed that (1) the R-curve developed during a cyclic load history has both a monotonic component and a cyclic component, (2) fatigue precracking at high load levels does not affect the R-curve, (3) monotonic bulk prestraining lowers the R-curve whereas cyclic bulk prestraining raises it, and (4) the R-curve may be lowered by decreasing the monotonic loading rate.

KEY WORDS: fracture toughness, J_R-curves, load history, prestrain, cycles, loading, 4340 steel, stress intensity

The loading history that a structural component experiences in service can often be complex. However, the load history assumed in making a fracture mechanics evaluation of a cracked component is usually very simple. Very often the crack is assumed to grow subcritically under the influence of cyclic loading until the fracture toughness value is reached, where fracture toughness is measured under an applied monotonically increasing load. This approach is used because the data needed to handle a more complex load history do not exist.

Some studies have been made to determine the effects of a cyclic load history on the fracture toughness of metals in the linear elastic regime [1,2] and the elastic-plastic regime [3]. These studies showed that fracture toughness

---

[1]Metallurgy Department, Westinghouse R&D Center, Pittsburgh, Pa. 15235.

measured by cycling to failure may be different from that measured by mono-tonically loading to failure. Therefore fracture toughness may be different for different loading histories.

Cyclic loading to failure is one load history that may influence toughness; other types of load histories may also have an effect. In this paper the load his-tories considered are:

1. Cyclic loading during R-curve development.
2. Fatigue precracking at high load levels.
3. Bulk prestraining of a material under monotonic and cyclic loading.
4. Slow rates of monotonic loading during the R-curve test.

These load histories were imposed on a modified 4340 steel where fracture toughness was measured in the elastic-plastic regime using the J_R-curve approach.

The results presented in this paper are of a survey nature. The load histories listed were each applied in separate series of tests. The effect of each load his-tory was noted empirically by comparing the resulting R-curve to a reference R-curve for the material. In each case a simple conclusion was drawn regard-ing the effect of that load history on the R-curve behavior of the test material. No attempt was made to explain the noted effect through related models or mechanistic considerations and no attempt was made to extrapolate the be-havior noted here to other materials. This simple approach was taken because the amount of data generated in these tests was too much to consider fully in one paper. More detailed examinations of the effect of each load history will be subjects for future papers.

Test Technique

The material used for this study was a modified 4340 steel given the desig-nation 4335 by the steel industry. The chemical composition and mechanical properties are listed in Tables 1 and 2 respectively. It differs slightly from modern 4340 specifications in having higher nickel and lower carbon content and is used in heavy section forgings.

The specimen was of the compact type (1T-CT). It had a width of 50 mm (2 in.) and a thickness of 25 mm (1 in.) and was not side grooved.

The J_R-curve tests were conducted at room temperature using the unload-ing compliance method [4] where crack length changes are determined from elastic unloading slopes. The J used as the characterizing parameter in the

TABLE 1—*Chemical composition (weight percent) of modified 4340 steel.*

C	Mn	P	S	Si	Cr	Ni	Mo	V
0.32	0.69	0.007	0.005	0.26	0.91	2.72	0.42	0.09

TABLE 2—*Mechanical properties of modified 4340 steel.*

Yield Strength, MPa	(ksi)	Tensile Strength, MPa	(ksi)	Elongation, %	Reduction of Area, %
1041	(151)	1124	(163)	16	53

R-curves is the deformation *J* as given by Ernst et al [5]. This method of testing and analysis is similar to the one recommended by the ASTM working group on the J_R-curve test method [6].

Two tests were conducted on virgin material specimens. These are specimens tested monotonically in the recommended manner [6] with no unusual load history applied. The J_R-curve results from these tests are given in Fig. 1. The scatter in results is attributable to material property scatter and test technique and is typical of the scatter reported in an ASTM round robin [7]. A scatterband was drawn around these data and labeled "virgin specimen bounds." These bounds were then used to judge the effect of the various load histories. A load history which resulted in the J_R-curves falling within these bounds was judged to have no effect. For those that fell outside the bounds the given load history was judged to have an effect.

Each type of load history mentioned previously is treated in a separate section. Details of the test technique that apply to a given load history are discussed in that section.

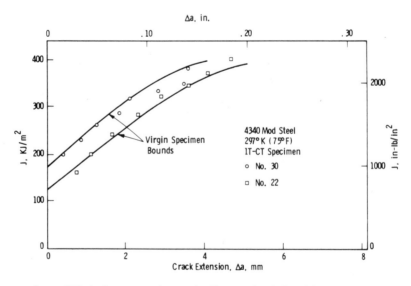

FIG. 1—*J versus crack extension R-curves for virgin specimens.*

Cyclic R-Curve

The effect of cycling the load in the elastic-plastic regime during R curve development has been previously discussed for two steels [3]. The same types of tests were conducted for the present material. The cyclic loading applied is shown in Fig. 2. The ratcheting crack and the elastic dominance cases were used for these tests. A fuller explanation of these types of loading is given in Ref 3. To make the R-curve evaluation an envelope is drawn over the load versus displacement record and the analysis is then made in a manner similar to the usual unloading compliance test evaluation [4].

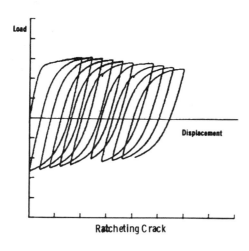

Ratcheting Crack

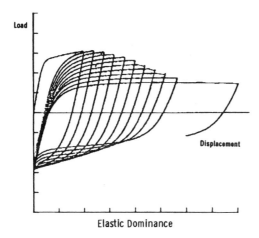

Elastic Dominance

FIG. 2—*Examples of load histories used for developing cyclic R-curves.*

Each cycle contains an unloading portion from which crack extension can be determined. The value of J is determined from the area under the envelope curve. The crack extension resulting from this type of test is thought to have two components, one from the monotonic loading and one from the cyclic loading. A suggestion has been made [3] that these two components of crack growth may be evaluated by a linear summation (Fig. 3). The monotonic component would be evaluated from a monotonic R-curve test and the cyclic component from a fatigue test that had very high cyclic loads.

The J_R-curve results are given in Fig. 4. Although many tests were conducted, only four were selected for presentation here. They represent both types of cyclic histories, the ratcheting crack (RC) and elastic dominance (ED), and two values of applied cycles (10 and 45). In all cases the results fall below the virgin specimen bounds, which illustrates that the cycling has the effect of giving an additional component of crack growth to the monotonic R-curve growth.

In keeping with the guidelines of simply conducting the test and observing behavior no additional attempt is made here to rationalize these results. It is simply concluded that the cyclic loading imposed in these tests appears to result in two components of crack growth, one due to monotonic loading and one due to cyclic loading. Further attempts to evaluate the results in Fig. 4 and the additional tests in terms of a linear summation model will be the subject of a future paper.

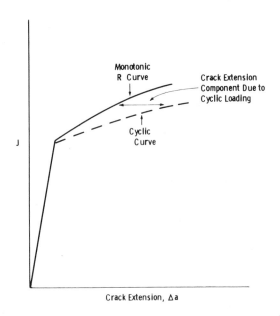

FIG. 3—*Schematic showing linear summation principle for developing a cyclic R-curve.*

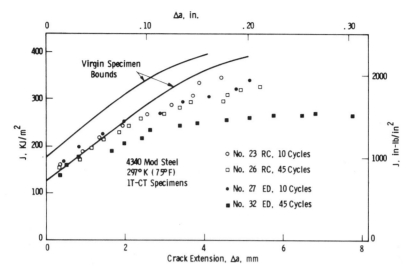

FIG. 4—J *versus crack extension R-curves for cyclic loading during R-curve development.*

High Precracking Loads

Fracture toughness test specimens are always precracked in fatigue at prescribed load levels before testing. Standard test methods prescribe low precracking loads to ensure that the precracking procedure will not interfere with the subsequent fracture toughness test [6,8]. It is likely, however, that a cracked structural component may encounter high cyclic loads in service which extend the crack prior to the final load which in turn causes the component to reach the fracture toughness limit. These high cyclic loads could influence the toughness level and cause the standard toughness to be incorrect for evaluation of the component.

Tests were therefore conducted where the precrack was introduced at very high load levels. These tests differ from the previous ones in that the crack was first introduced at high cyclic load levels and then a monotonic test was conducted. The precracking load levels are given in Table 3, where P_c represents

TABLE 3—*High precracking load levels.*

Specimen	Precrack Load (P_c), MN	(kips)	Nominal K_{max}, MPa $\sqrt{mm}$	(ksi $\sqrt{in.}$)	P_c/P_{max}
21	67	(15)	178	(162)	1.03
36	44	(10)	101	(92)	0.63
37	53	(12)	132	(120)	0.77
38	62	(14)	157	(143)	0.88

the maximum load and $R = 0.1$. Also given in this table are a nominal value of K for precracking calculated by elastic formulas from the maximum precracking load and a comparison of the maximum precracking load (P_c) to the maximum load attained in the monotonic toughness test (P_{max}).

The J_R-curve results for these tests are given in Fig. 5. For all the precracking load levels the J_R-curve results fall predominantly within the virgin specimen bounds.

An additional evaluation of precrack load level was made by testing some of the specimens from the previous section on cyclic R-curve evaluation. These specimens were tested as previously discussed by developing crack extension with combined monotonic and cyclic components. Subsequent to this test a normal monotonic R-curve test was conducted. The effect is the same as precracking at the maximum attainable load for these specimens, namely the limit load. The J_R-curve results for these tests are given in Fig. 6. These results also fall predominately within the virgin specimen bounds.

The conclusion from this series of tests is that the precracking load level has no effect on subsequent J_R-curve toughness for this steel.

Bulk Prestraining

The previous series of tests were conducted by imposing a given load history on a specimen already containing a crack. This set of tests was conducted by imposing a loading on the material before the crack was introduced. Tension specimens were made with large cross sections so that compact specimens could

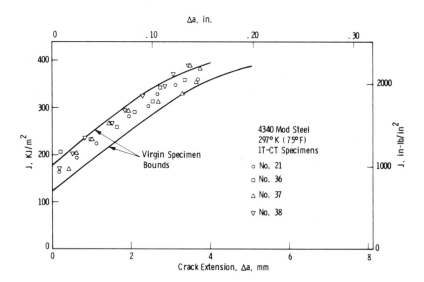

FIG. 5—*J versus crack extension R-curves for high load precracking.*

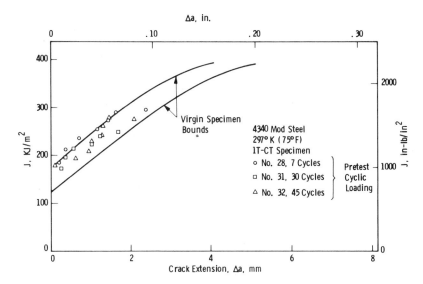

FIG. 6—J *versus crack extension* R-*curves for cyclic* R-*curve specimens (monotonic loading after cyclic test).*

be taken from them after the bulk prestraining was applied (Fig. 7). The prescribed prestrain history was given to the tension specimen, which was then remachined into two of the IT-CT specimens (Fig. 7) so that duplicate R-curve tests could be conducted.

The bulk prestrain histories were both monotonic and cyclic in nature. Five tension specimens were prestrained. The first was subjected to a 2% monotonic plastic prestrain. The other four were given a cyclic prestrain (Fig. 8). Blocks of decreasing and increasing cycles were imposed on the specimens. Three different points were taken for stopping the cyclic loading (Fig. 8). The first (1) is at the zero point on the cyclic envelope, labeled *zero*. The second (2) is at the maximum load point, labeled *up*. The third (3) is at the minimum load point, labeled *down*. These different stopping points give the tension specimen different stress strain properties and may influence the toughness. A summary of bulk prestrain load histories is given in Table 4.

After the bulk prestrain was given to the tension specimens, they were machined into two compact specimens and precracked at low ΔK levels. Duplicate J_R-curve tests were conducted using the normal unloading compliance method. Results from tension specimens 1, 2, and 3 are given in Fig. 9. These are the 2% plastic monotonic prestrain and the 1½ and 2% cyclic prestrain with the stopping point up (No. 2). Because there is little scatter between duplicate tests, only one of the duplicate results is given here for clarity of presentation. These results show that the monotonic prestrain lowers the R-curve toughness, whereas the cyclic prestrain tends to increase it. Also the toughness increases with increasing cyclic prestrain level.

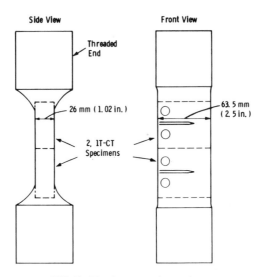

FIG. 7—*Tension prestrain specimens.*

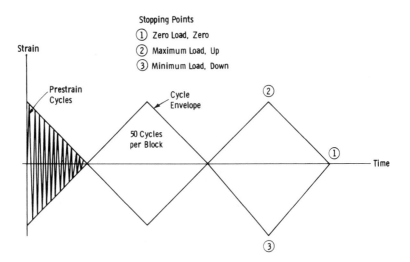

FIG. 8—*Load history for cyclic prestraining of tension specimens.*

Results illustrating the effects of stopping point for 1½% cyclic prestrain are given in Fig. 10 (tension specimens 2, 4, and 5). These show that the R-curve is raised for all the imposed cyclic prestrains. There is a definite order to the increase, however, with the progression going up-zero-down in terms of stopping point. Again, in Fig. 10 only one of the duplicate specimen results was plotted for the sake of clarity; however, both specimens showed the same trend.

TABLE 4—*Bulk prestrain load histories.*

Tension Specimen	Load Type	Strain Amplitude, %	Stopping Point	No. of Blocks
1	monotonic	2	...	...
2	cyclic	$\pm 1\frac{1}{2}$	up	2
3	cyclic	± 2	up	2
4	cyclic	$\pm 1\frac{1}{2}$	zero	$2\frac{1}{2}$
5	cyclic	$\pm 1\frac{1}{2}$	down	2

The conclusion from the bulk prestrain tests is that monotonic prestrain lowers the R-curve whereas cyclic prestrain raises it. The type of cyclic prestrain also appears to influence the amount that the toughness is increased.

Slow Loading

The last type of load history considered was that of slow loading. Structures which are loaded slowly or spend a long period of time at a constant load may have a toughness value different from that measured in a normal short time fracture toughness test. The slow loading rates imposed on the specimens are given in Table 5. All specimens were tested in the same manner as the virgin specimens; the only difference was the rate of loading or total test time. The

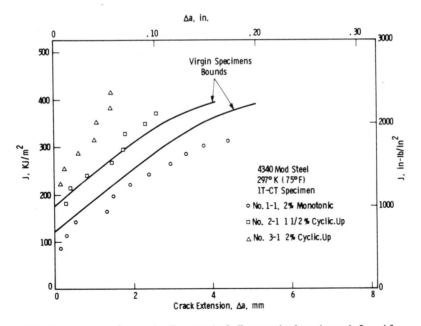

FIG. 9—*J versus crack extension R-curves for bulk prestrained specimens 1, 2, and 3.*

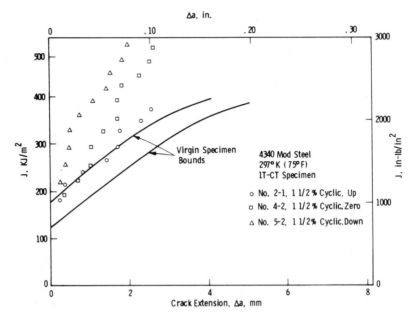

FIG. 10—*J versus crack extension R-curves for bulk prestrained specimens 2, 4, and 5.*

approximate loading rate for a normal J_R-curve test is also given in Table 5. The slow loading rates were up to three orders of magnitude slower than the normal rate; the longest test lasted about ten days.

As shown in Table 5, tests approximately one, two, and three orders of magnitude slower than normal were conducted. The J_R-curve results are shown in Fig. 11. In all cases the slow loading rates resulted in lower R-curve toughness. There is an apparent trend to decreasing toughness with decreasing loading rate, though this trend is not very strong. The conclusion for this set of tests is that the slower loading rates decrease the R-curve toughness.

TABLE 5—*Slow loading rates.*

Specimen	Displacement Rate, mm/h	(in./h)	Equivalent $\dot{K}$, MPa/$\sqrt{m}$/h	(ksi$\sqrt{in.}$/h)	Test Duration, h	Approximate Equivalent Speed
20	0.5	(0.02)	110	(100)	2.5	10^{-1}
33	0.05	(0.002)	11	(10)	23	10^{-2}
34	0.005	(0.0002)	1.1	(1)	230	10^{-3}
35	0.013	(0.0005)	2.7	(2.5)	90	10^{-3}
Normal rate [7]	10	(0.4)	2200	(2000)	0.25	1

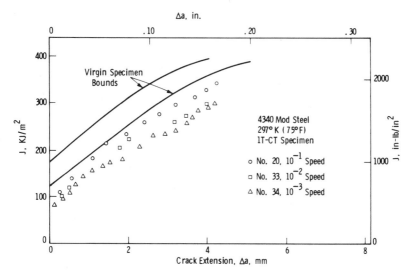

FIG. 11—*J versus crack extension R-curves for slow loading rate specimens.*

Summary

The effect of load histories on J_R-curve toughness was examined for a modified 4340 steel. Four types of load histories were considered:

1. Cyclic loading during R-curve development.
2. Fatigue precracking at high load levels.
3. Bulk prestraining of the material under both monotonic and cyclic loading.
4. Slow loading rates during the J_R-curve test.

Tests were conducted for each category of loading history and the resulting effect on the R-curve toughness was observed. No attempt was made to explain the results or extrapolate them to other materials.

For the four load histories imposed the following effects were observed for the modified 4340 steel used in this study:

1. Crack growth caused by cycling a compact specimen to failure appears to have both a monotonic component and a cyclic component. These components can be observed by plotting them in an R-curve format.
2. Precracking at high load levels does not appear to affect the J_R-curve toughness.
3. Bulk prestraining of the steel prior to introducing a crack affects the toughness in the following ways:

    (*a*) Monotonic bulk prestraining lowers the R-curve.

(*b*) Cyclic bulk prestraining raises the R-curve. The amount of increase in toughness appears to depend on the percent of cyclic strain imposed and the manner in which the cyclic prestrain is given (that is, the stopping point for the prestrain cycle).

4. Conducting the J_R-curve test at very slow loading rates (tests of up to ten days) lowers the R-curve.

Acknowledgments

Dr. N. E. Dowling provided helpful discussions on the cyclic prestraining. M. G. Peck of the Mechanics of Materials Laboratory assisted with the experimental aspects of this work. Donna Shearer typed the manuscript. The operators of the Word Processing Department helped with the preparation and handling of the manuscript.

References

[1] Dowling, N. E. in *Flaw Growth and Fracture, ASTM STP 631*, American Society for Testing and Materials, 1977, pp. 139–158.
[2] Troshchenko, V. T., Pokrovsky, V. V., and Prokopenko, A. V., "Investigation of the Fracture Toughness of Constructional Steels in Cyclic Loading," in *Fracture 1977*, Vol. 3, ICF4, Waterloo, Canada, June 1977.
[3] Landes, J. D. and McCabe, D. E. in *Elastic-Plastic Fracture: Second Symposium, Volume II— Fracture Resistance Curves and Engineering Applications, ASTM STP 803*, C. F. Shih and J. P. Gudas, Eds., American Society for Testing and Materials, 1983, pp. II-723–II-738.
[4] Clarke, G. A., Andrews, W. R., Paris, P. C., and Schmidt, D. W. in *Mechanics of Crack Growth, ASTM STP 590*, American Society for Testing and Materials, 1976, pp. 27–42.
[5] Ernst, H. A., Paris, P. C., and Landes, J. D. in *Fracture Mechanics: Thirteenth Conference, ASTM STP 743*, R. Roberts, Ed., American Society for Testing and Materials, 1981, pp. 476–502.
[6] Albrecht, P. et al, *Journal of Testing and Evaluation*, Vol. 10, No. 6, Nov. 1982, pp. 245–251.
[7] Clarke, G. A., *Journal of Testing and Evaluation*, Vol. 8, No. 5, Sept. 1980, pp. 213–220.
[8] Clarke, G. A. et al, *Journal of Testing and Evaluation*, Vol. 7, No. 1, Jan. 1979, pp. 49–56.

Brian N. Leis[1]

Microcrack Initiation and Growth in a Pearlitic Steel—Experiments and Analysis

REFERENCE: Leis, B. N., **"Microcrack Initiation and Growth in a Pearlitic Steel—Experiments and Analysis,"** *Fracture Mechanics: Fifteenth Symposium, ASTM STP 833*, R. J. Sanford, Ed., American Society for Testing and Materials, Philadelphia, 1984, pp. 449–480.

ABSTRACT: Results of experiments indicate that crack growth from a notch root in a 1080 pearlitic (rail) steel exhibits anomalous behavior when analyzed in terms of linear elastic fracture mechanics (LEFM). (Cracks as long as 0.4 cm exhibit anomalous behavior.) It is shown that the length of crack exhibiting anomalous behavior corresponds with the size of the plastic zone at the notch root. A model is presented which characterizes the crack growth driving force in terms of the displacement (strain) control condition that this inelastic notch field is postulated to impose on cracks contained within it. It is shown that the model reasonably predicts the growth rate behavior within the inelastic notch field. Notched component life predictions, based on this model, coupled with a nonlinear scheme to predict the formation of a 125 μm crack and LEFM analyses to continue growth beyond the inelastic field, are presented. It is shown that, for the pearlitic steel examined, the life of the components is equally divided between initiation and growth within the inelastic notch field at higher stresses. As the applied cyclic stress applied decreases, initiation becomes increasingly dominant. The portion of life controlled by LEFM behavior is negligible except within a narrow window of cyclic stress. The implications of this observation are discussed with respect to other materials and notch severities.

KEY WORDS: 1080 pearlitic steel, linear elastic fracture mechanics (LEFM), short (small) cracks, inelastic notch field, model, prediction, fatigue life, crack initiation, crack propagation, plasticity

Results of numerous experiments have shown that problems may be encountered in characterizing the growth of cracks within the stress field of a notch [*1–8*]. Past papers that explored these problems suggest that two aspects are of consequence. On the one hand, by analogy to the behavior of smooth specimens [*9,10*], it has been argued that the growth of cracks in notch fields is controlled by microstructural features. This view ignores the influence of possible mechanics-related plasticity effects that can develop in

[1] Senior Scientist, Battelle Columbus Laboratories, Columbus, Ohio 43201.

the notch field. It also has been argued in various ways that the mechanics of the crack-driving forces at crack tips within inelastically strained notch fields and beyond notch fields is not adequately characterized by linear elastic fracture mechanics (LEFM) [3,5,7,11]. This argument remains even when inelastic analysis based on estimates of ΔJ [12] are used in lieu of LEFM [13].

A detailed review examining the growth of cracks in notch fields [14] suggests that mechanics *and* microstructural arguments may be invoked to rationalize observed data, depending on the circumstances. Microstructural features appear to dominate short crack growth in situations where very high strength materials are involved and the stresses are low and notch plasticity is small and confined within a few grains. Crack growth within that region is associated with the transition from the transient state of initiation to the steady-state condition associated with a continuous well-defined crack front. Microstructural features including anisotropy and grain boundaries control this transient behavior, along with the microcrack closure due to the plane stress surface flow constrained by the plane strain flow of interior grains. Lack of similitude between the transient microcrack and its longer steady-state counterpart thus explains the differences in their respective growth rates. Mechanics explanations are most appropriate where cracks are initiating and growing in an inelastic field controlled by the notch root. In such cases, however, it is questionable whether LEFM provides an appropriate analysis framework. Cracks in these cases are often many grains long and may have well-defined continuous fronts, still show anomalous growth in terms of LEFM analyses. Furthermore, although reduced, the anomalous growth remains in some cases even after the inelastic action is accounted for by estimates of ΔJ (e.g. [13]). In this respect, there remains a problem in the analysis of cracks at notches that require mechanics analyses beyond the usual explanations related to crack tip macroplasticity or microstructural effects.

The purpose of this paper is to present and discuss the results of a study recently completed for the Department of Transportation (DOT) [15] whose purpose was to characterize the initiation and growth of short cracks in AISI 1080, fully pearlitic, railroad rail steel. Data developed for this steel addressed microcrack initiation and growth up to and through the macrocrack domain, for both constant amplitude and hi-lo block cycle histories.[2] Microcrack initiation predictions are made following well-documented procedures (e.g. [16]). Microcrack growth predictions within the inelastic notch field are made using long crack data, coupled with the hypothesis that the usual stress intensity factor solution does not account for the inelastic action. It is postulated that the crack tip, when in the plastic zone of the notch, grows in a field which is displacement controlled by the surrounding elastic field [7,17].

---

[2] Although block cycle data were included in the original study [15], they are not addressed herein to keep the paper focussed on the problems of characterizing the growth of cracks from notches.

Experimental Aspects

Phenomenological evaluation of the crack growth behavior at a notch root requires systematic observation of the cracking process. This has been accomplished by periodic photographic tracking at cracking sites. The study has considered a range of nominal stresses from very confined yield to more extensive but still confined flow at the notch root. To characterize the notch field, notch root and transverse section strain have been measured.

Material and Specimens

The material used in this study was AISI 1080 (eutectoid) steel provided by the DOT. The structure was fully pearlitic with a grain size of about 30 to 50 μm and was somewhat pancaked through the thickness. The hardness was R_c19. Specimens were cut from the web of a single section of 132 lb/yd rail, with the axis of testing parallel to the longitudinal axis of the rail. Mechanical properties developed for this material are presented in Table 1. All samples were machined symmetrically about the midthickness of the web, using comparable procedures, to produce a uniform high quality surface finish (0.41 μm, RMS) and low residual stress. All samples received a light polish perpendicular to the direction of maximum principal stress prior to testing. Notched plates were fabricated as shown in Fig. 1a, whereas smooth specimens were made with either a round cross section as per Fig. 1b or a rectangular cross section as per Fig. 1c.

All notched plates were fitted with foil crack detectors located 2.5 mm from the notch roots, on both sides. These detectors were used to stop the cycling when the crack grew beyond the field of the camera. At this point several plates were fitted with KRAK gages to develop long crack data. Further, one half of the notched specimens were fitted with strain gages at each notch root. Two of these were also fitted with five additional gages located across the transverse net section. Three gages were laid close to the hole, while of the remaining two, one was laid near the plate's edge and the second was placed equidistant between the hole and the edge. Five smooth specimens were also

TABLE 1—*Typical mechanical properties.*

	Monotonic	Cyclic
Proportional limit, MPa	310	207
0.2% offset yield, MPa	359	400
Ultimate tensile strength, MPa	820	...
True fracture strength, MPa	965	...
True fracture strain	0.251	...
Modulus of elasticity, MPa	2.07×10^5	1.89×10^5

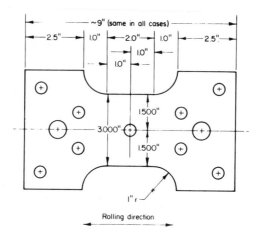

FIG. 1a—*Notched specimen geometry; $K_t = 2.5$.*

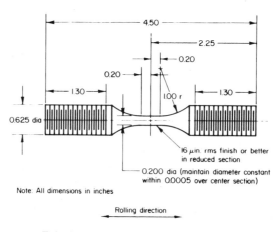

FIG 1b—*Round smooth specimen geometry.*

gaged, three rectangular and two round. In all cases, the gages had a 0.4-mm-square uniaxial grid made of a stabilized constantan foil.

Experimental Procedure and Techniques

All experiments were conducted in closed-loop servocontrolled facilities under either load or strain control. The procedures employed to calibrate these facilities were in accordance with the appropriate ASTM standards.

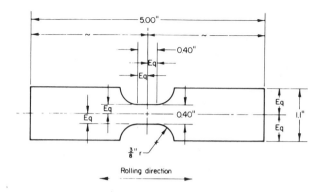

FIG. 1c—*Rectangular smooth specimen geometry.*

Likewise, applicable standards were followed in data development.[3] All tests were started in tension and consisted of fully reversed cycling at convenient frequencies ($0.1 \leq f \leq 20$). Further details of the apparatus and procedure may be found in Ref *18* for smooth specimens.

Test procedure for the notched specimen involved the usual installation of specimens and implementation of test conditions. As mentioned earlier, crack detector foils at notch roots were wired in series and connected so as to interrupt cycling when a crack 2.5 mm long developed. Gaged specimens were wired into a commercial signal conditioner, the output of which was recorded on a time base chart, and on an *X-Y* plotter, as a function of load. Finally, cameras with dioptric lenses were focussed on both notch roots on both sides of the plate. Thereafter their automatic trigger and flash and their advance motors were programmed. The cameras provided $5\times$ magnification at the film plane, typical results for which are shown in Fig. 2. The film was interpreted and data reduced in an enlarger, back tracking crack size in time until no crack was visible. A magnification of $8\times$ was found satisfactory for this purpose, giving a total magnification of $40\times$. Cycling commenced and continued until a crack 2.5 mm long developed. Samples were then removed and KRAK gages installed on selected specimens. Samples were again installed and the KRAK gages connected to their signal conditioning, and then cycling continued to specimen separation. Other than as just described, the apparatus and procedure were similar to smooth specimen tests.

Since the purpose here was to detect the initiation and track the growth of small cracks, it is appropriate to discuss the accuracy of the measured crack length in the short crack domain. Crack length has been determined from a permanent record of the growth process at $40\times$ magnification. It has been found that the crack tip is rather easily located and measured down to a

---

[3] Crack growth rate standards were followed *in spirit*, since strict adherance to editing and growth rate calculation procedures radically reduces data available for the short crack regime. Consideration is given to possible errors introduced by this exception later in the paper.

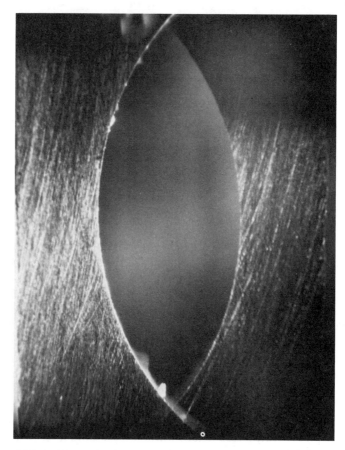

FIG. 2—*View of notch root at the film plane through the dioptric lense.*

length of about 25 μm. At 40 × this crack appears as a crack 1000 μm long (1 cm). Measurement precision is about 0.1 mm,[4] so that the absolute error is on the order of 1%. Because the record is permanent, the film can be back-tracked until microcrack formation is detected and then advanced until 25 μm (1 cm at 40×) of growth is indicated and the associated cycle number recorded. Repeating this incremental advance process thus leads to very precise estimates of crack growth increment at a given cycle number.

Smooth Specimen Results

Results of the smooth specimen tests are reported graphically in Figs. 3a to 3c and in a tabular format in Ref 15.

---

[4] Based on the value of the standard deviation of six replicate independent measurements of the same crack.

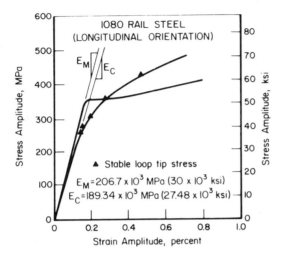

FIG. 3a—*Monotonic and cyclic deformation behavior for smooth specimens (*$E_M \equiv$ *modulus of elasticity;* $Ec \equiv$ *stable cyclic value of* E_M*).*

Figure 3a presents the monotonic and cyclic deformation behavior of the rail steel examined. Note that as with many other pearlitic steels this material cyclically softens at lower strains and cyclically hardens at higher strains. Further, the stable loop tip stresses developed under constant amplitude cycling obtained from the data shown in Fig. 3b lie on the cyclic curve developed from incremental step cycling, as evident in Fig. 3a. This indicates that the same cyclic stress-strain curve develops under significantly different cyclic histo-

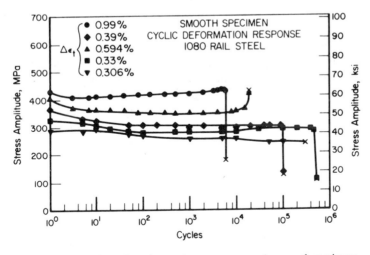

FIG. 3b—*Cyclic dependent changes in stress response for smooth specimens.*

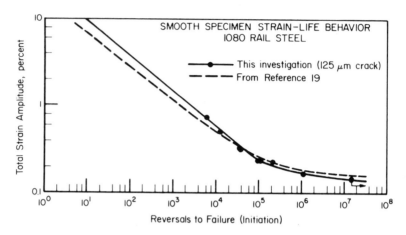

FIG. 3c—*Strain-based comparison of fatigue resistances for smooth specimens.*

ries. However, the cyclic curve for this material falls significantly below that reported in the literature for other similar rail steels (e.g. [*18,19*]), as does the monotonic behavior, a trend consistent with the lower hardness of this rail compared to others.[5]

The fatigue resistance of the rail steel is presented on strain life coordinates in Fig. 3c. Note here that life is presented in terms of cycles to form a small crack (about 125 μm deep). At long lives there is little difference between formation of this crack size and failure, whereas at shorter lives this difference is about 15% in life. As evident in Fig. 3c, which also plots other published data for rail steels, the results of this investigation are comparable to those previously published. Finally, the stress at 10^6 reversals is similar to that predicted based on the hardness of the current material in terms of trends presented in the literature for fully pearlitic steels [*18*]. Compared with data developed for a similar steel reported in Ref *19*, the present rail steel exhibits a reduced fatigue resistance at longer life and an increased resistance at very short lives, as expected based on their relative strengths and ductilities.

For strains measured via strain gages under controlled strain tests, the present results indicate that strain gage ranges correspond to controlled values within 5% over the range of strains considered in this investigation. Therefore measured strains provide a reasonably accurate description of the strain field at the sites of concern.

---

[5]Note that while it is often believed that the same cyclic curve is obtained for a given class of materials regardless of the monotonic behavior, this situation only develops for certain classes of material. For rail steels differences in strength relate to subtle differences in the same basic microstructure, and as such different cyclic stress strain curves can be expected.

Notched Specimen Results

Figures 4a to 4c detail the results of the deformation studies on the plates. Figure 4d presents "initiation" results, and Figs. 4e and 4f present crack growth data. Tabular results are included in Ref 15.

The deformation behavior at the notch root is plotted as a function of applied stress on coordinates of maximum principal strain and nominal stress in Fig. 4a. For the plate geometry used there exists only a slightly biaxial local stress state. Analysis indicates that the ratio of the through thickness stress to the hoop stress is about 0.04 [20]. Given that the multiaxial octahedral equivalence criterion adopted in studies of pearlitic steels [21] is valid, there is only a 1% difference between the value of K_t based on the maximum principal stress and the corresponding value in fatigue governed by an octahedral criterion K_t^f [20]; that is, $K_t^f = 0.99\,K_t$. It can be shown for this plate geometry that the corresponding ratios of the principal notch root strains are essentially equal to Poisson's ratio, the same as for the uniaxial test specimen [22]. Thus there is a one-to-one correspondence between the maximum principal values of stress and strain in both smooth and notched specimens. Consequently, comparisons will be made directly between these maximum principal quantities without further reference to multiaxial equivalence conditions.

With regard to Fig. 4a, note that there are trend curves marked as monotonic and cyclic. These curves represent measured local strains developed under monotonic loading and under incremental step loading in force-local strain space [23]. In both cases the initial slope corresponds to an experimentally determined value of $K_{exp}^t = 2.5$. In the case of the stable cyclic curve, note that values of loop tip strains developed under constant amplitude condi-

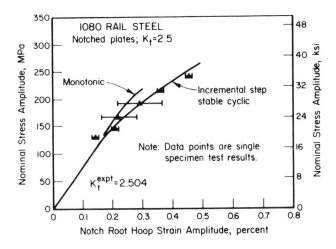

FIG. 4a—*Local deformation as a function of applied stress for notched specimens.*

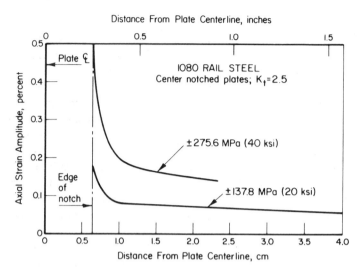

FIG. 4b—*Typical strain distributions across the transverse net section for notched specimens.*

tions correspond closely with the incremental step result. This suggests that the same stable local strain develops under the action of a given stress under variable amplitude loadings.

Figure 4b presents results of surface strains measured across the transverse net section at applied stress amplitudes of 138 and 276 MPa. Given that yield occurs at 0.15% strain based on the monotonic proportional limit (0.01% offset),[6] these and other measured distributions can be used to estimate the size of the notch root plastic zone. The plastic zone so determined is shown as a function of nominal stress in Fig. 4c. The fatigue resistance of the notched plates is compared with that for smooth specimens in Fig. 4d. The ordinate of this figure contains the product of stress and strain. This product has a functional form identical to the currently popular damage parameter postulated by Smith et al [24] and subsequently derived from energy considerations [25]. It has been shown valid for pearlitic steels for a range of conditions [26]. For convenience this product has been multiplied by the modulus and the square root extracted to retain units of stress.

Consider the fatigue crack growth behavior of the notch plates, results for which are presented in Fig. 4e on coordinates of crack length and cycles. In many cases multiple cracks developed, requiring many cracks to be tracked until a dominant crack developed. When corner cracking did occur, its growth was initially through thickness, and link-up with midthickness cracks resulted in an essentially plane-fronted crack reasonably characterized by its

---

[6] The monotonic behavior is used since stresses within the plastic zone cyclically decrease for softening materials to values less than the surrounding material, which concentrates strain in that zone. Therefore the monotonic criterion is appropriate.

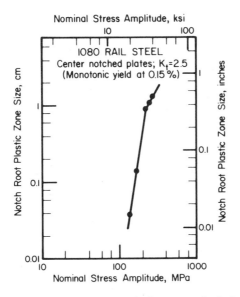

FIG. 4c—*Notch plastic zone as a function of nominal stress amplitude for notched specimens.*

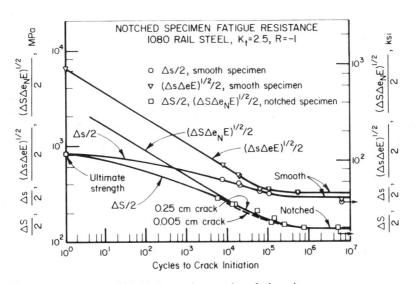

FIG. 4d—*Fatigue resistance of notched specimens.*

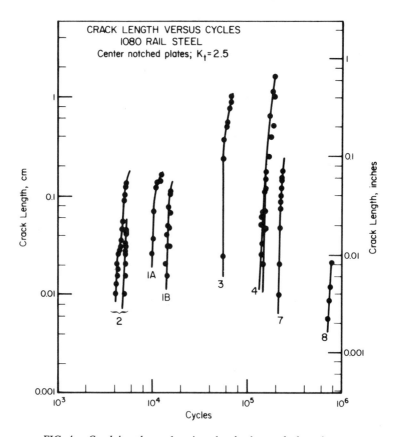

FIG. 4e—*Crack length as a function of cycles for notched specimens.*

surface length. Short crack growth rates shown in Fig. 4e are essentially constant with increasing crack length; that is, crack length is a power-law function of cycles. These crack length versus cycle curves are generally quite steep. Such data can be extrapolated to shorter lengths without much fear of bias to lower than actual rates for shorter cracks, particularly since these trends develop for actual data down to lengths as short as 50 μm.[7] (Thus the errors in measured crack length discussed earlier do not significantly affect the observed growth rates in this investigation.)

Consider growth rate as a function of stress intensity to compute growth rates from discrete data. Values of growth rates can be compared on a speci-

[7] There is a limit below which extrapolation may be questionable. In general, factors such as changes in crack aspect ratio, microstructural features, local notch gradients, etc., may contribute to the growth process in a manner different than represented by the trends. For the present data, this extrapolation is sensible down to crack sizes on the order of 50 μm; below this size microstructural features tend to control growth and the local aspect ratio may be expected to vary from that represented by the data trends.

men-to-specimen basis with a universal measure of crack-driving force. For long cracks, the LEFM stress intensity factor (K) [27,28] serves as such a parameter.

In a simple functional form, K is defined for Mode I cracking as

$$K = \beta\left(\frac{\ell}{W}\right) S\sqrt{\pi\ell} \tag{1}$$

where S is the far field stress, ℓ is the semicrack length measured from the notch root, and $\beta(\ell/W)$ is a function of the geometry and crack length. It has been postulated [29,30] that the fatigue crack growth rate is a unique function of the change in K and other constant parameters that pertain to the loading, specifically the ratio of the minimum to maximum stress intensity factor (R); that is,

$$\frac{d\ell}{dN} = g(\Delta K, R) \tag{2}$$

In this equation, ΔK is defined from the cyclic range of stress (ΔS) inserted into Eq 1. It should be emphasized that in LEFM the confined crack tip plastic zone size (r_p) is a unique function of K; that is,

$$r_p \, \alpha K^2$$

Consequently, care must be exercised in interpreting data from stress histories where r_p is history dependent and therefore not uniquely related to the current value of ΔK (or K_{max}). Likewise, a significant limitation to the utility of LEFM in the present context is that it is valid only so long as the plastic zone is small compared to the crack length. Although it is recognized that this limitation can cause difficulties in using Eq 1 [3], the LEFM ΔK will nevertheless be used for the time being.

Stress intensity solutions for the present case have been derived to establish $\beta(\ell/W)$ for the geometry of interest. Specifically, the solution for a center notched finite plate from Ref 31 has been combined with the secant finite width correction to calculate K. For the case when cracks initiated at each of the notch roots, K was calculated assuming the cracking behavior developed symmetrically, an assumption which has little effect on K for small cracks. In cases where multiple cracks initiate from the same notch root, analysis for the second or third to initiate cracks ignored the presence of cracks which initiated earlier. Since such cracks relieve local stresses to values below those based on elastic analysis, the value of K so calculated is an upper bound value.

Crack growth rate is plotted as a function of the maximum value of K in a given cycle (K_{mx}) in Fig. 4f for several specimens. These results are typical for

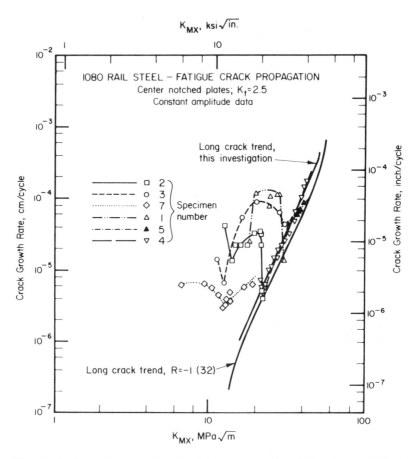

FIG. 4f—*Crack growth rate as a function of the maximum value of* K *in a given cycle* (K_{mx}) *for notched specimens.*

all tests in which crack growth for short cracks has been captured photographically (Specimens 1A, 1B, 2, 3, 4, 7, 8, and 9). First, with regard to the long crack trend developed from Specimen 4 for this investigation, note that this trend lies within a factor of about two of the trend for $R = -1$ developed for a similar pearlitic steel by Broek and Rice [32]. Furthermore, with respect to the data available, this trend lies towards the center of the scatter band for the data for pearlitic steels summarized by Broek and Rice [32]. One can conclude that the long crack growth rate behavior is typical of these steels. As indicated in Fig. 4f, however, physically short cracks grow at rates up to two orders of magnitude greater than the long crack trend. For the present study, this is true for cracks more than 4 mm long. Significantly, too, the results show short cracks initiate at stress intensity levels well below the apparent threshold for this material.[8] In view of these trends, LEFM fails to portray the driving force for short cracks in pearlitic (rail) steels in a mathematically unique way. Further, LEFM provides a significantly nonconservative estimate of both the threshold (if indeed a threshold exists) and the growth rate.

Life and Growth Rate Predictions

This section considers life prediction of the notched plate in terms of mechanics analysis and materials data such as typically are available in handbooks. The approach used considers the life of the plate specimen to be composed of (1) cycles to form or "initiate" a 125-μm-long crack, (2) cycles to grow that crack beyond the inelastic notch field, and (3) cycles to traverse the remainder of the specimen [17].

The transition from the transient process of "initiation" to the process of microcrack growth at a crack length of 125 μm is adopted as follows. First, as noted in the introduction, the growth of microcracks tends to be dominated by microstructural features within the first few grains. The length of 125 μm, adopted as the "initiation" transition, embraces several grains and so should avoid the early growth transients controlled by the microstructure. Furthermore, smooth specimen results are used to predict the formation of that size crack. It is reasonable, therefore, to assume that transients in forming small cracks would be similar in material elements subjected to the same strain (damage history), whether at the surface of a smooth specimen or a notch root. It has long been observed that smooth and notched specimens do indeed form cracks at length equal to 125 μm at statistically equal lives when subjected to the same strain (damage history) [16,33]. Thus, based on this extensive data base, the 125 μm length appears to implicitly account for the microstructurally dominated short crack growth behavior for both smooth and notched specimen geometries. In turn, this means that if the number of cycles to "initiate" a 125-μm-long crack can be predicted using smooth specimen data, the microstructurally dominated growth at the notch is inherent in that prediction. There is therefore no need to directly predict this complex behavior at the notch root. Furthermore, since the aforementioned observations of Ref 16 and 33 involved a range of metals over lives bounded from 10^2 to 10^7 cycles and K_t bounded from 1 to 15, it is reasonable to anticipate that the 125 μm length would provide a viable "initiation" criterion for the present study.[9]

[8] Cracks grow in smooth specimens from initiated microcracks, since in the limit these cracks have zero length, they grow from a lower bound stress intensity of zero. The concept of a "threshold" is therefore one of dubious physical significance, since in the limit its value is zero.

[9] A precise length of 125 μm is adopted herein. Data analyses indicates that lengths smaller than this could be used; so also could lengths as great as several hundred microns. The length of 125 μm, however, is used because the literature [16,33] indicates that the equal damage/equal life assumption is valid for a wide range of data when this specific length is used.

Note that LEFM analysis (e.g. [36,37]) suggests that the notch field controls the growth over a small fraction of the root radius of the hole (~ radius/10). In the limit, as the radius goes to zero

Adopting this criterion circumvents the need to directly predict microstructurally controlled growth at the notch root.

Initiation

The term *initiation* herein denotes the formation of a crack about 125 μm deep. It involves the assumption that if equal amounts of damage (strain) are imposed at crack initiation sites, the formation of a 125-μm-deep crack will occur in a statistically equal number of cycles.[10] Figure 5 presents results developed in this investigation which suggest this assumption is valid for the notched pearlitic steel plates. As shown in the figure, 125-μm-deep cracks formed at the notch roots in a statistically equal number of cycles for smooth specimens. Thus the microcrack initiation and early transient growth behavior are similar in material elements subjected to the same damage history during the formation of a 125-μm-long crack for the pearlitic steel considered herein.

The results of Fig. 5 suggest that predictions of crack initiation at notches follow directly from the results of mechanics analysis to estimate local strains and stresses from nominal stresses using smooth specimen procedures. Since this approach is well established for notched coupons [*16,33*] as well as for actual hardware [*16,34*], the approach will not be detailed. The interested reader should pursue the just-noted references.

The results of Fig. 4a have been used to estimate local strain from nominal stress. Used in conjunction with the smooth specimen data shown in Fig. 3c, the life to form a 125-μm-long crack is shown as a function of nominal stress as the broken line in Fig. 6.[11] As expected in view of Fig. 5, this line corresponds closely with the observed formation of cracks shown by the open squares.

(that is, as K_t increases), this analysis suggests the crack length at initiation should decrease as the root radius decreases. However, there is a lower limit to this trend associated with the use of continuum fracture mechanics. This limit is related to the volume of material needed to develop bulk, polycrystalline response. Consideration of materials flow behavior suggests a lower bound crack size from 3 to 10 grain diameters deep. Even in very fine grained materials (say, ASTM 10 ~ 10 μm intercept), this "continuum" limit restricts the practical applicability of LEFM results to crack lengths or depths greater than about 100 μm. Consequently, so long as one is dealing with continuum fracture mechanics analysis, definitions of initiation are bounded below by a length of about 100 μm even for very sharp notches. For this reason the definition of initiation adopted as 125 μm provides a criterion consistent with both the equal deformation/equal life assumption and the subsequent prediction of growth using continuum fracture mechanics.

[10] This is limited to situations where the root radius is large compared with metallurgical features. Metallurgical size effects become a factor thereafter and micromechanics analyses may be required.

[11] Nominal stress here is net section. Net section stress is used to report all results even though calculations for growth have been made using the corresponding gross section stress.

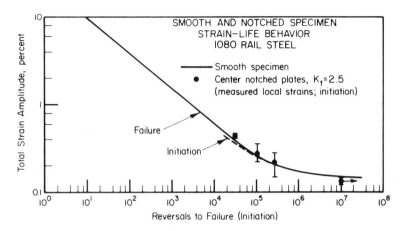

FIG. 5—*Comparison of smooth and notched specimen strain-life behavior.*

Growth in the Inelastic Field of the Notch

A brief review of the current LEFM scheme for predicting notch root crack growth serves to put the problem of predicting crack growth rate in an inelastic notch field into perspective. Long cracks have been typically studied under load-controlled conditions. Similitude [35] is invoked in the growth rates through the LEFM K, which for equal values means equal plastic zones at crack tips. Clearly, similitude can be expected only so long as the crack tips exist in a load-controlled domain, such as develops in the reference "long crack" geometries. But when examining short cracks which grow out of notches, the notch stress field must be accounted for. In the context of LEFM this is tantamount to including the notch gradient in K through $\beta(\ell/W)$ in Eq

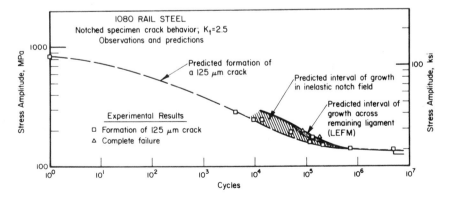

FIG. 6—*Cracking behavior of notched plate specimens—observations and predictions.*

1, an approach which has been used for some time. What is missed in such a formulation is the fact that the notch may have its own yield zone that completely contains the crack. Thus the gradient derived from elasticity theory is inappropriate. Further, the plastic zone of the notch is largely displacement controlled by the surrounding elastic field,[12] as suggested in Fig. 7a [17]. This displacement control leads to a constant or possibly decreasing driving force for growth until the crack nears the elastic-plastic boundary of the notch field, at which point there is a gradual shift from predominately displacement control to load control. Thereafter the crack behaves as the so-called long crack in that it grows in a load-controlled field.

The significance of the local control condition can be appreciated in terms of differences in the magnitude of the local stress field as compared to the LEFM value and the effects of crack closure. For short cracks the displacement field (the plastic deformation in the wake of the crack) does not develop closure; thus the value of R differs from the far field value. As the crack grows, however, the value of R approaches a stable value, and once the crack tip enters the load-controlled domain, the local value of R duplicates that of the far field condition inherent in reference long crack data. Only then is there true similitude between the K fields of the notched and center cracked plates, and thus only then can K be expected to correlate cracking.

Predictions of the rate of growth of short cracks based on the foregoing scheme follow from the limiting considerations derived from it. These include: (1) the value of R is -1 at initiation in that there are no crack faces to close, (2) the value of R at the transition to long crack behavior corresponds to that of the far field, and (3) the transition to long crack behavior occurs when the length of the crack plus its plastic zone equals the length of the notch root plastic zone. (For the present, the length of the crack plus the length of its plastic zone are approximated by the length of the crack.)

Figure 4c presents results from which the transition crack depth from the displacement-controlled notch field to the load-controlled LEFM domain can be determined as a function of nominal stress. Predictions of the transition depth based on the measured depth of the inelastic field of notch are compared with the corresponding crack lengths associated with the transition to long crack behavior in Fig. 7b. A one-to-one correspondence in this figure means the prediction is exact. Observe from this figure that the predicted transitions correspond reasonably with the observed behavior. This one-to-one correspondence can be expected only so long as a transition to LEFM growth occurs. Such a transition will not occur if the critical crack size is smaller than the plastic zone at the notch root. Likewise, a transition will not occur when net section yield occurs. With reference to Fig. 7b, data for the larger plastic zone sizes show that critical cracks were developed before the

---

[12] Note that displacement control as discussed develops only for the case where the elastic field is large compared to the inelastic notch field. Thus application of this displacement control concept tacitly assumes confined notch plasticity.

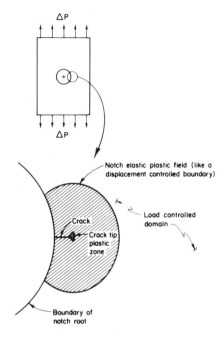

FIG. 7a—*Postulate to rationalize crack growth in the inelastic field of the notch.*

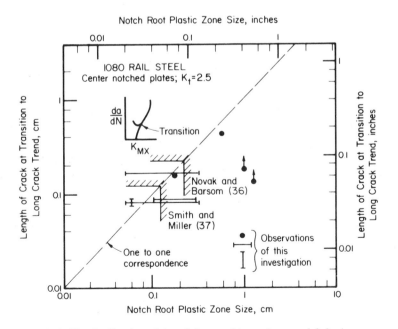

FIG. 7b—*Predicted crack length for transition to long crack behavior.*

plastic zone was traversed. Thus these points, which are denoted by arrows, tend towards the one-to-one correspondence but never reach it.

Note that the crack lengths associated with the transition to growth correlated by LEFM include cracks which lie within as well as much beyond the various LEFM definitions of the notch field. Using the notation of D equal to the semimajor axis of the notch and ρ equal to the root radius, the transition length based on the 1976 definition of Novak and Barsom $(0.25\,(D\rho)^{1/2}$ [36]) is 6.35 mm. The corresponding result based on the 1977 definition of Smith and Miller $(0.13\,(D\rho)^{1/2}$ [37]) is 3.3 mm. In view of Fig. 4c, the prediction of Smith and Miller becomes nonconservative beyond a net section stress of about 165 MPa and that of Novak and Barsom beyond about 179 MPa. As ratios of gross section stress to yield stress, these values are 0.38 and 0.41, respectively. In this respect, the LEFM analysis portrayed in these measures of the notch behavior becomes nonconservative at ratios of gross stress to yield stress beyond about 0.40. Figure 7b thus suggests that constant LEFM predictions of the notch field are misleading and may develop nonconservative estimates of the notch field.

Analyses of the crack growth behavior in the inelastic domain of the notch requires some form of pseudoplastic or inelastic measure of the crack-driving force. At the transition from the displacement-controlled inelastic field to the load-controlled far field, this form must reduce to the usual LEFM description of the crack tip field (cf. Eq 1). For the present, the pseudoplastic form

$$K_{mx}^{p} = 1.12\ \epsilon_{mx}\,E_{m}\sqrt{\pi \ell} \qquad (3)$$

where 1.12 is the usual free surface correction, ϵ_{mx} is the stable maximum strain in the material element at the depth of interest, E_{m} is the monotonic modulus, and ℓ is the crack length from the notch root, is adopted. Note that the product $E_{m}\epsilon_{mx}$ is used to estimate stress because this equation must assume the form of Eq 1 at the boundary of the inelastic field.[13] Other pseudoplastic forms could be used (e.g. [3]); however, the results of a recent review of such forms failed to indicate one is more appropriate than another [13]. Note that ϵ_{mx} is measured in the absence of the crack. It reflects the influence of the inelastic action in the notch field. By analogy to Eq 1, the value of $\epsilon_{mx}E_{m}$ is a pseudoplastic measure of the remote stress S_{mx} mapped through the inelastic action at the notch root.

[13] For small values of crack length/notch root radius this equation provides an almost exact match at the elastic plastic boundary. For larger values, the equation first slightly underestimates and then slightly overestimates crack-driving force as the notch plastic field expands. Because, as shown in Fig. 7b, this paper deals with a range of plastic zone sizes, no single equation is exact. Equation 3, however, is exact for small plastic zones. Moreover, it reasonably matches the LEFM result at the elastic plastic interface as the plastic zone size increases over the range of notch fields covered in this study. Finally, the absolute error in Eq 3 over the range of crack lengths studied is on the same order as the error in measured strain used in the equation. For these reasons, sophistication beyond a single equation is not warranted. Therefore this single equation has been adopted.

Equation 3 has been used to estimate values of K_{mx} at the transition from initiation to growth in the notch field and the growth rate through the inelastic field of the notch. Beyond the inelastic field, LEFM is valid so that Eq 1 is used to calculate K_{mx}. Values of K_{mx} taken in conjunction with long crack results for this rail steel shown in Fig. 4f generate predicted growth rates as a function of crack length for each specimen. Plotting these growth rates as a function of the corresponding LEFM value of K_{mx} results constitutes a prediction of the behavior shown in Fig. 4f. Such a plot is shown in Fig. 7c.

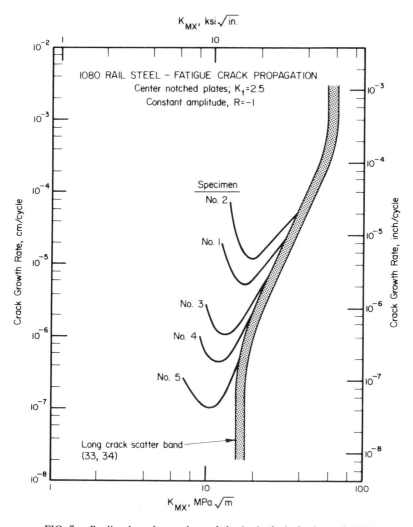

FIG. 7c—*Predicted crack growth rate behavior in the inelastic notch field.*

The trend developed using Eq 3 indicates that growth rates are much higher within the inelastic notch field than would be predicted using LEFM in terms of Eq 1. Comparing the trends in Fig. 7c with the corresponding experimental growth beyond the initiation transition shown in Fig. 4f shows that the experimental trends are reasonably predicted. Likewise, the decrease in growth rates approaching the elastic-plastic transition is also predicted if the offsetting decreasing value of ϵ_{mx} and the increases in $\sqrt{\ell}$ develops this condition.

While the behavior predicted at the two transitions corresponds well with observations, the predicted growth between these trends tend to underestimate that observed. It is probable that this situation arises because of local closure effects which are reasonably characterized by the assumed limiting values of R at the transitions but poorly characterized in between. Because such local closure appears to be a major factor in producing anomalous growth of short cracks [14], steps are being taken to expand this approach to account for it.

Integrating the crack growth trends in Fig. 7c between the initiation and the plastic-elastic notch field transitions results in the number of cycles predicted to traverse the inelastic notch field. The results of this integration are shown in Fig. 6 as the cross-hatched domain beyond "initiation". At about 190 MPa, growth through the inelastic domain takes about as many cycles as does the formation of the 125-μm-long crack. Above 190 MPa, the number of cycles to develop a 125-μm-long crack remains approximately equal to the number of cycles to grow through the inelastic field. Below 190 MPa, the number of cycles to initiate the 125-μm-long crack becomes increasingly dominate. Near the fatigue limit the number of growth cycles becomes negligible. This is to be expected since the notch plastic zone decreases in size as nominal stress decreases and ultimately is contained within the 125 μm length defined as initiation.

Macrocrack (LEFM) Growth

Crack growth from the plastic-elastic transition until specimen separation occurs is correctly characterized by LEFM (Eq 1). For the present, separation is assumed to occur when the fracture toughness is exceeded. The predicted number of cycles required to cause separation is shown in Fig. 6 as the shaded region beyond the results for growth through the inelastic field. As evident in the figure, this type of growth is of consequence only below about 190 MPa and above about 175 MPa. In this narrow window of stresses, LEFM growth accounts at most for only 20% of the total life for the specimen geometry used. At stresses above 190 MPa, the total life is split almost equally between "initiation" and growth in the inelastic notch field. At stresses below 175 MPa, the life is increasingly dominated by "initiation".

Discussion

This investigation has shown that the fatigue behavior of notched components can be reasonably predicted. Care has been taken to accommodate much of the uncertainty regarding predictions of the growth rate of physically small cracks by predicting life in terms of "blocks of behavior".[14]

The first block is associated with the formation of a small crack 125 μm long and has been termed *initiation*. The literature shows that cracks grow to this length in a statistically equal number of cycles, provided the damage histories imposed are equal [33]. Figure 5 showed that the pearlitic steel examined also followed this definition of initiation. Smooth specimen strain-life data which embody the microcrack growth process associated with the formation of a 125-μm-long crack have therefore been used to estimate the number of cycles spent in the first block. Therefore the complicated effects of microstructure and free surface on the microcrack initiation and growth process at the notch are implicit in this life interval. Established predictive procedures (e.g., see [16]) have been used in this block. The second block involved the growth of the crack through the "inelastic notch field". A simple model [7,17] was used to estimate the growth rate behavior in this block. As evident in Fig. 7b, close correspondence of the transition crack length from this field to the elastic field with the observed growth rate trends shown in Fig. 4a lends creedence to this approach. Further, trends in predicted growth rates presented in Fig. 7c correspond reasonably with the observed behavior shown in Fig. 4f. This too supports the use of such an approach to predict the growth of cracks in the inelastic notch field wherein LEFM is of questionable validity.

The final block involves growth through the "macrocrack (LEFM) field", the domain in which LEFM is by definition valid. Usual procedures are employed therein, as set forth in Eqs 1 and 2.

---

[14] It could be argued that the results of these predictions depend on the crack sizes used to define the "blocks". While in general this is true, it should be remembered that these block sizes are defined with a view to satisfying conditions for the validity of the analyses procedures employed. That is, the assumptions implicit in the analyses have been identified and the "blocks" defined on that basis. Therefore the definition of each block is not arbitrary, neither can the definition of a block be chosen to yield accurate predictions. For these reasons the same definitions of blocks can be applied to the analyses of other geometries and materials with the anticipation of equally accurate predictions. Recall that the first block is bounded above by a crack size that satisfies a similitude criterion to predict initiation using smooth specimen data [16]. Phenomenology indicated that the equal damage/equal life assumption is valid for many materials and notch configurations for "initiation" defined by 125 μm. Moreover, a length of 125 μm is a reasonable transition to continuum fracture mechanics. (For the present material it corresponds to about 3 to 5 grain diameters.) Thus the choice of the first block as 0 to 125 μm appears to satisfy mechanics and materials considerations related to microcrack initiation. The second block relates to the interval of growth through the inelastic field. It is therefore rationally defined as the crack size at the elastic plastic field of the notch. Again the block size selected is consistent with the mechanics used to characterize the growth rate process. The final block is simply the remaining ligament consistent with the physical constraints of the problem. In this respect, while the results depend on the specific definition of the blocks, the analyst is reasonably constrained in regard to the crack interval he assigns to each block.

Predicted results are summarized in Fig. 6. For the bluntly notched samples studied, "initiation" dominates the total life over the range of stresses for which LEFM is valid (<190 MPa). Above that stress, the life tends to be evenly split between "initiation" and growth through the "inelastic notch field". Thus LEFM has a very limited range of utility for the bluntly notched pearlitic steel samples. For similar bluntly notched components, harder steels (or materials with more confined plasticity) would show a decreased significance of "initiation" and "inelastic notch fields", and LEFM would tend to dominate. In contrast, in softer steels (or materials with less confined plasticity), LEFM would be less significant than suggested by Fig. 6. Such materials would tend to show "initiation" and the "inelastic notch field" blocks equally significant. By analogy, notch profiles which confine notch deformations would show trends comparable to hard steels. Specimens or components which tend to spread plasticity by virtue of their geometric details would also tend to have pronounced "initiation" and "inelastic notch field" blocks.

The point to be made is that LEFM should not be casually adopted as the basis for fatigue life prediction or crack growth analyses. Analyses that predict component life via LEFM when inelastic notch action occurs may be in error by more than an order of magnitude [13]. Until rigorous nonlinear methods of crack growth analyses evolve, care must be exercised in developing and applying published analyses procedures. Clearly, the traditionally independent nonlinear approach for predicting initiation and LEFM approach for predicting macrocrack growth cannot be simply married for the purposes of predicting total life, except for a very few restricted cases.

Summary and Conclusions

It has been shown that LEFM does not consolidate microcrack behavior in the inelastic field of a notch. An approach has been suggested to rationalize microcrack growth within the inelastic notch field. Coupled with predicted initiation and macrocrack growth behavior this approach results in accurate predictions for notched components. Finally, it was shown that present technologies used to predict initiation and macrocrack growth cannot be simply married to predict the life of notched components—microcrack growth in the inelastic notch field must be addressed.

A number of conclusions may be drawn on the basis of the data developed and the analyses presented. Among the more significant are:

1. Rail steel exhibits the so-called short crack effect. The observed growth rate behavior can be explained in terms of inelastic action at the notch root when crack growth is contained within the inelastic notch field.

2. Crack initiation and LEFM macrocrack growth predictions cannot be simply added to predict life of notched components except at very long lives. The displacement-controlled growth within the inelastic notch field must be accounted for.

3. Rational predictions of notched component cracking behavior are possible using continuum mechanics coupled with appropriate materials data.

Acknowledgments

This work has been performed in part under the sponsorship of the Transportation Systems Center, Department of Transportation; Dr. O. Orringer served as the contract monitor.

References

[1] Frost, N. E., Pook, L. P., and Denton, K., *Engineering Fracture Mechanics*, Vol. 3, 1971, pp. 109-126.

[2] Broek, D., "The Propagation of Fatigue Cracks Emanating from Holes," NLR TR 72134 U, National Aerospace Laboratory, The Netherlands, Nov. 1972.

[3] Gowda, C. V. B., Leis, B. N., and Topper, T. H., "Crack Initiation and Propagation in Notched Plates Subject to Cyclic Inelastic Strains," in *Proceedings*, International Conference on Mechanical Behavior of Materials, Kyoto, Japan, Vol. II, 1972, pp. 187-198.

[4] Pearson, S., *Engineering Fracture Mechanics*, Vol. 7, 1975, pp. 235-247.

[5] Ohuchida, H., Usami, S., and Nishioka, A., *Bulletin of the JSME*, Vol. 18, No. 125, Nov. 1975, pp. 1185-1193.

[6] El Haddad, M. H., Smith, K. N., and Topper, T. H. in *Fracture Mechanics, ASTM STP 677*, American Society for Testing and Materials, 1979, pp. 274-289.

[7] Leis, B. N. and Forte, T. P. in *Fracture Mechanics: Thirteenth Conference, ASTM STP 743*, American Society for Testing and Materials, 1981, pp. 100-124.

[8] Leis, B. N. and Galliher, R. D. in *Low-Cycle Fatigue and Life Prediction, ASTM STP 770*, American Society for Testing and Materials, 1982, pp. 399-421.

[9] Lankford, J., *Engineering Fracture Mechanics*, Vol. 9, 1977, pp. 617-624.

[10] Sheldon, G. P., Cook, T. S., Jones, J. W., and Lankford, J., *Fatigue of Engineering Materials and Structures*, Vol. 3, 1981, pp. 219-228.

[11] Hammouda, M. M. and Miller, K. J. in *Elastic Plastic Fracture, ASTM STP 668*, American Society for Testing and Materials, 1979, pp. 703-719.

[12] Dowling, N. E. in *Cyclic Stress-Strain and Plastic Deformation Aspects of Fatigue Crack Growth, ASTM STP 637*, American Society for Testing and Materials, 1977, pp. 97-121.

[13] Leis, B. N., *International Journal of Pressure Vessels and Piping*, Vol. 10, 1982, pp. 141-158.

[14] Leis, B. N., Ahmad, J., Broek, D., Hopper, A. T., and Kanninen, M. F., "A Critical Review of the Fracture Mechanics of Small Cracks", AFWAL-TR-83-4019, Wright-Patterson AFB, Ohio, Jan. 1983.

[15] Leis, B. N., "Microcrack Growth Behavior of a Fail Steel," Item 13, in *Analysis of Service Stresses in Rails*, Final Report on Contract DOT/TSC 1663 from Battelle Columbus Laboratories (BCL) to the Transportation Systems Center (TSC), Jan. 1982.

[16] Leis, B. N., "Predicting Crack Initiation Fatigue Life in Structural Components," in *Methods of Predicting Fatigue Life*, ASME, 1979, pp. 57-76.

[17] Leis, B. N., "Displacement Controlled Fatigue Crack Growth in Elastic-Plastic North Fields and the Short Crack Effect," submitted for publication in *Engineering Fracture Mechanics*.

[18] Leis, B. N. and Laflen, J. H., "Cyclic Inelastic Deformation and Fatigue Resistance of a Rail Steel—Experimental Results and Mathematical Models," Interim Report on Item 11/Contract DOT-TSC-1076, June 1977.

[19] Leis, B. N. in *Rail Steels—Developments, Processing and Use, ASTM STP 644*, American Society for Testing and Materials, 1978, pp. 449-468.

[20] Leis, B. N., and Topper, T. H., *Journal of Engineering Materials and Technology, Transactions of ASME*, Vol. 99, No. 3, July 1977, pp. 215-221.

[21] Leis, B. N. and Rice, R. C., "Selection of Failure Criteria," in *Analysis of Service Stresses*

in Rail, Final Report on Contract DOT/TSC 1663, from Battelle Columbus Laboratories (BCL) to the Transportation Systems Center (TSC), Jan. 1982.

[22] Leis, B. N., "Fatigue Analysis to Assess Crack Initiation Life for Notched Coupons and Complex Components," Ph.D. thesis, University of Waterloo, Ont., Canada, Sept. 1976.

[23] Williams, D. P., Lind, N. C., Conle, F. A., Topper, T. H., and Leis, B. N., "Structural Cyclic Deformation Response Modeling," in *Proceedings*, ASCE Speciality Conference on Engineering Mechanics, May 1976, in *Mechanics in Engineering*, University of Waterloo Press, Ontario, Canada, 1977, pp. 291-311.

[24] Smith, K. N., Watson, P., and Topper, T. H., *Journal of Materials*, JMLSA, Vol. 5, No. 4, Dec. 1970, pp. 767-778.

[25] Leis, B. N., *Journal of Pressure Vessel Technology, Transactions of ASME*, Vol. 99, No. 4, Nov. 1977, pp. 524-533.

[26] Leis, B. N. and Rice, R. C., "Rail Fatigue Resistance-Increased Tonnage and Other Factors of Consequence," in *Proceedings*, 2nd International Heavy Haul Conference, ASME, Sept. 1982, pp. 82-HH-12.

[27] Irwin, G. R., "Analysis of Stresses and Strains Near the End of a Crack Traversing a Plate," *Journal of Applied Mechanics, Transactions of ASME*, 1957.

[28] Tada, H., Paris, P. C., and Irwin, G. R., *The Stress Analysis of Cracks Handbook*, Del Research Corp., revised ed., 1977.

[29] Paris, P. C., Gomez, M. P., and Anderson, W. E., *The Trend in Engineering*, Vol. 13, 1961, pp. 9-14.

[30] Broek, D. and Schijve, J., "The Influence of the Mean Stress on the Propagation of Fatigue Cracks in Aluminum Alloy Sheets," TR-M-2111, National Aerospace Institute, Amsterdam, 1963.

[31] Rooke, D. P. and Cartwright, D. J., *Stress Intensity Factors*, HMSO, London, 1976.

[32] Broek, D. and Rice, R. C., "Fatigue Crack Growth Properties of Rail Steels," Final Report (Part I), and "Prediction of Fatigue Crack Growth in Rail Steels," Final Report (Part II), Battelle Columbus Laboratories for the U.S. Department of Transportation, Contract DOT-TSC-1076, Sept. 1977.

[33] Leis, B. N. and Topper, T. H., *Nuclear Engineering and Design*, Vol. 29, 1974, pp. 370-383.

[34] Leis, B. N., "An Approach for Fatigue Crack Initiation Life Prediction with Applications to Complex Components," in *Fatigue Life of Structures under Operational Loads*, Proceedings of the 9th International Committee on Aeronautical Fatigue Meeting, ICAF Doc. 960, Laboratorium fur Betriebsfestigeit, May 1977, pp. 3.4/1-47.

[35] Broek, D. and Leis, B. N., "Similitude and Anamolies in Crack Growth Rates," in *Materials, Experimentation, and Design in Fatigue*, Westbury House, IPC Science Press, UK, March 1981, pp. 129-146.

[36] Novak, S. R. and Barsom, J. M. in *Cracks and Fracture, ASTM STP 601*, American Society for Testing and Materials, 1976, pp. 409-447.

[37] Smith, R. A. and Miller, K. J., *International Journal of Mechanical Sciences*, Vol. 19, 1977, pp. 11-22.

DISCUSSION

N. E. Dowling[1] (written discussion)—Near the beginning of the paper, the author denounces the ΔJ approach to fatigue crack growth, without stating specific reasons or further discussing the matter. Some comparisons between ΔJ and the strain intensity approach actually adopted (Eq 3) are in order.

[1] Associate Professor, Engineering Science and Mechanics, Virginia Polytechnic Institute and State University, Blacksburg, Va. 24061.

Theoretical Considerations

For monotonic loading of materials with nonlinear stress-strain curves, the J-integral provides a measure of the intensity of the crack tip stress-strain field. It can be used to determine modified values of the stress intensity (K) of linear elastic fracture mechanics (LEFM), so that the nonlinearity is accounted for. In the case of cyclic loading of elastic-plastic materials, such as engineering metals, instantaneous values of J have no meaning. However, there is no fundamental theoretical objection to employing a ΔJ which is based on changes in the stress-strain field during a cycle.[2]

Two problems encountered in the use of ΔJ are that it does not directly account for either crack closure or the effects of cycle-dependent creep (ratchetting) deformation; however, neither do linear elastic fracture mechanics or the strain intensity parameter. Strain intensity is simply an assumption which attempts to extrapolate linear elastic fracture mechanics, so that it lacks the theoretical mathematical basis such as K has for the linear elastic case and J has for the nonlinear case. Use of the strain intensity was certainly reasonable prior to the introduction of J.[3] But its use now seems rather quaint.

Numerical Comparison of Strain Intensity and J

Despite the theoretical difference, values of linear elastic K modified according to strain intensity may not differ significantly from those modified based on J. Consider a cracked body subjected to uniform strain. The strain intensity, as in Eq 3 of the paper, is related to the linear elastic K by

$$K_\epsilon = K \frac{\epsilon}{\epsilon_e} \tag{4}$$

where $\epsilon = \epsilon_e + \epsilon_p$ is total strain, which is the sum of elastic strain (ϵ_e) and plastic strain (ϵ_p).

For uniform strain applied to a center or edge cracked plate, provided the crack length is small compared with the plate width, modified values of K based on J are given approximately by[4]

$$K_J = K \sqrt{1 + \sqrt{n} \frac{\epsilon_p}{\epsilon_e}} \tag{5}$$

---

[2] Additional discussion and references on this point are given in Dowling, N. E., "Growth of Short Fatigue Cracks in an Alloy Steel," ASME Paper 83-PVP-94, ASME 4th National Congress on Pressure Vessels and Piping, Portland, Ore., June 1983.

[3] See Boettner, R. C., Laird, C., and McEvily, A. J. Jr., "Crack Nucleation and Growth in High Strain Low Cycle Fatigue," *Transactions of the Metallurgical Society of AIME*, Vol. 233, Feb. 1965, pp. 379–387.

[4] See the aforementioned ASME Paper 83-PVP-94, where this approximation is derived based on the work of M. Y. He, J. W. Hutchinson, and others.

where n is the exponent relating plastic strain and stress:

$$\epsilon_p \sim \sigma^n$$

The cyclic stress-strain curve for the rail steel from Fig. 3a of the paper was used with Eqs 4 and 5 above to compute K_ϵ and K_J for various values of strain. The results are shown in Fig. 8, where the quantities K_ϵ/K and K_J/K are plotted versus strain. Note that these ratios are correction factors to be applied to K. The factor exceeds two but does not differ significantly between the two calculations, except perhaps at the highest strains of interest. The difference between K_ϵ and K_J would increase and become quite large if larger strains were considered.

Hence K_J could be used in place of K_ϵ and the computations would not change in any important way. This is of course provided that crack closure is handled in the same manner and that the estimate is based on the same strain, neither of which is inconsistent with the use of J.

Effects of Strain Gradient

In the paper, the stress and strain gradients associated with the hole-in-plate geometry are considered by calculating K_ϵ based on the strain at the crack tip location, specifically the strain present before the introduction of a crack. For the elastic case for this particular geometry, such a procedure is probably reasonably accurate. An accurate numerical solution is available for the related case of a pair of cracks growing from a round hole in a wide plate.[5] Here, the approximation used in the paper is within 2% for crack lengths less than $0.1r$, where r is the hole radius. At $\ell = 0.8r$, the approximation is 8% low. And for crack lengths large compared to r, the approximation is 12% high, failing to accurately approach the correct limit due to the 1.12 factor in Eq 3 of the paper.

However, caution is needed in using this approximate method. Consider a notched bending member with a crack having its tip at the location which was the neutral axis prior to the introduction of the crack. The strain for Eq 3 would be zero, giving $K_\epsilon = 0$, which is grossly in error. For this case, more detailed analysis is needed, such as that of Newman in NASA TN D-6376.

Also, there is no guarantee that the approximate method is meaningful for plastic deformation of geometries for which it is reasonably accurate in the elastic case. As strain intensity is an assumption, its appropriateness can only be evaluated empirically, as by study of fatigue crack growth rates, or theoretically, by comparison with J.

Hence it would be useful to perform specific detailed analysis of J for vari-

[5] See Newman, J. C., Jr., "An Improved Method of Collocation for the Stress Analysis of Cracked Plates with Various Shaped Boundaries," NASA TN D-6376, National Aeronautics and Space Administration, Washington, D.C., 1971.

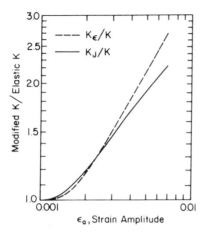

FIG. 8—*Correction factors for plasticity effect on the linear elastic stress intensity* (K) *based on strain intensity and* J-*integral.*

ous cases of cracks growing from stress raisers in plastically deformed members.[6] These results could then be compared to K_ϵ and also to values of K_J estimated from before-crack strain. Successful comparisons would help identify situations where K_ϵ or the approximate K_J could be used without it being necessary to perform specific detailed analysis for J.

Note that values of strain intensity will always reflect the arbitrary choice made in defining the strain to be used. Except for small cracks under uniform strain, there is no clear choice. On the other hand, accurate values of J automatically account for the geometry and loading situation, including such factors as stress raisers, finite width, and loading in bending.

Conclusions

The strain intensity parameter used in the paper has no advantages over values of linear elastic stress intensity (K) modified to include plasticity effects based on the J-integral. In contrast to J, strain intensity lacks a theoretical basis and the generality needed to deal with a variety of geometries and loading situations. Further work in the area of crack growth from stress raisers should include detailed analysis of J for typical cases.

B. N. Leis (author's closure)—A number of interesting issues have been raised by Dr. Dowling in his discussion of the paper, apparently based on his concern that ΔJ has not been adopted as the framework for the analysis pre-

[6] An initial effort in this direction is described in Sumpter, J. D. G. and Turner, C. E., "Fracture Analysis in Areas of High Nominal Strain," in *Proceedings*, Second International Conference on Pressure Vessel Technology, ASME, San Antonio, Tex., 1973, pp. 1095–1103.

sented. Dowling begins by noting that ΔJ has been denounced. This is a rather strong statement in view of the written text in which only one sentence notes that the mechanics of crack growth in inelastically strained notch roots are not adequately characterized by linear elastic fracture mechanics, and that similar difficulties exist "when *estimates* of ΔJ are used". An extensive development is then offered by Dowling showing that the inelastic strain intensity of the paper and ΔJ are comparable.

Although the utility and validity of ΔJ is not an issue in the paper, the fact that Dr. Dowling's discussion focuses on it almost exclusively means this closure of necessity must address ΔJ and Dowling's concerns.[7] Consequently, there is only limited consideration of the intent or detail of the paper itself in this closure.

In conclusion to the second segment of his discussion, Dowling closed by noting "K_J could be used in place of K_ϵ and the computations would not change . . . provided that closure is handled in the same manner". In drawing this conclusion, he has tacitly passed over the key point—one somehow has to determine ΔJ. Evaluation of the maximum value of J, J_{mx}, for a cycle using other than measured strains is uncertain at best given the strain field of the crack sets in the inelastic field of the notch. Even more difficulties are involved in estimating the range of J, ΔJ, given the contribution of the notch plastic field to crack closure, *without* with use of measured strains. Granted detailed inelastic analyses could be done to estimate ΔJ, but from that one could obtain more fundamental measures such as CTOD near the crack tip or more generally valid field parameters such as ΔT. Moreover, given such detailed results one could equally account for closure in these more fundamental and general parameters. In the absence of such detailed analysis, and to avoid the uncertainty in implementing J estimation schemes, one is obligated to infer J at some point in the cycle from measured strains. Because measured strain forms the basis for my calculations, I have chosen to label my crack driving force as *strain intensity factor*. Indeed, one gains absolutely nothing by transforming measured strains into J for present purposes.

Given his comments, Dr. Dowling would rather see measured strains camouflaged as J. Apparently measured strain, which when cast as a strain intensity is quaint by Dowling's standards, takes on new significance (from Dowling's perspective) when the operation of his Eq 5 transforming strain to J is performed. It is noteworthy that J obtained from measured strains has the same attributes as strain intensity. Because measured strains reflect any sequence dependence due for example to an initial overload, J and K_{mx}^P derived

---

[7] It is not the intent of this closure to provide an exhaustive discussion on the suitability of ΔJ in general. Comments on the limitations of ΔJ have been offered elsewhere by many authors. Suffice it to state here that these limitations have provided the impetus for the development of generally valid incremental formulations which under restricted conditions reduce to a J-type formulation. Details of one such general formulation can be found in Atluri, S. N., Nishoika, T., and Nakagaki, M., "Incremental Path Independent Integrals in Inelastic and Dynamic Fracture Mechanics," *Engineering Fracture Mechanics*, to appear.

from this measurement would also show effects of sequence dependence. This is in strong contrast to values of J obtained by other approximate schemes such as J estimation procedures. It is in this context that the opening paragraph of the paper states that crack tip fields are "not adequately characterized ... even when inelastic analysis based on *estimates* of ΔJ are used".

The point to be made is that for cracks in notch plastic fields one can only estimate J in the absence of notch-induced closure effects. The value of ΔJ so obtained has uncertain significance. As noted above, relevant values of ΔJ could be obtained from detailed numerical calculations—or inferred from measure measured strains. It is emphasized again that in the case of detailed analysis more fundamental measurements such as CTOD or more general field parameters such as ΔT are also available and would be used in lieu of J.

In the case of measured strains (the present situation), measured strains are no more useful for the purposes of the paper when manipulated into ΔJ as Dowling suggests; they are still just measured strains. For this reason, the driving force for growth has been presented in the most directly related format—strain intensity. Whether or not it is quaint is not an issue, and concern for its being quaint seems rather trite.

In the last part of his discussion, Dr. Dowling considers the effects of strain gradients. While, in general, the caution he advocates is correct, it is in some cases less relevant to the present paper than his concern for the single reference to estimated values of ΔJ. One example is the subject of K_ϵ based on initial strain gradient in the case of bending. While true for bending, this issue is not of much consequence for the present paper, which uses measured gradients for relatively small crack lengths. Likewise, his comments on the correct limit of Eq 3 of the paper are relevant for an infinite plate subject to large-scale plasticity. But they are not appropriate for the specific finite plate considered, particularly given the confined flow limitation implicit in Eq 3. Indeed Eq 3 has been determined such that it nearly matches the LEFM solution at the elastic plastic boundary of the notch field, and measured strains have only been used within that domain. That is, as noted twice in the paper, once the crack leaves the inelastic notch field, the crack driving force is calculated using Eq 1. Dr. Dowling goes on to consider errors inherent in the use of Eq 3. He notes that for large cracks, Eq 3 is in error by 12%. However, Eq 3 is never, nor can it ever, be applied at or near the limit discussed because the confined flow assumption implicit in Eq 3 is violated in such applications. For the sake of direct evaluation, Dowling also compares other values of K_{mx}^p with "accurate numerical" results for selected values of crack length divided by notch root radius. It is emphasized in this regard that K_{mx}^p is used under conditions of inelastic notch action which preclude sensible comparison of Eqs 1 and 3. Indeed, it is because Eqs 1 and 3 are expected to produce different answers over the inelastic notch field that Eq 3 is introduced. Thus it is not clear why Dowling chooses to compare values of K_{mx}^p to values of LEFM K, since K_{mx}^p expressly applies only to the inelastic field of the notch.

One final point regarding Dowling's comments on strain gradients is appropriate. Dowling notes situations where K_ϵ is appropriate and could be "evaluated . . . theoretically by comparison with J". As this appears in his section on caution, I believe it appropriate to note that many assumptions exist with respect to J and ΔJ. Associated with these are limitations in its practical applicability. Given that assumptions regarding the dominance of the crack strain field may be tenuous in highly strained notch field, comparison with J seems somewhat questionable. Moreover, uncertainties regarding unloading in the notch field raise questions as to the validity of solutions that rest on the deformation theory of plasticity. In this respect I agree with Dowling that detailed numerical evaluation of J will in general be necessary to find accurate values of J. Granted in such cases, J may reasonably characterize the crack driving force. Unfortunately, the need for such detailed calculations plays down a major attraction of J—ease of evaluation. After such calculations, however, a number of physically significant measures such as ΔCTOD[8] are also available, as are field parameters such as ΔT.[7] One wonders, with measures such as ΔCTOD or ΔT available, why resorting to J would even be considered.

In conclusion, it is agreed that the LEFM K solution modified to include plasticity effects as presented in the paper has no advantages over a form based on J integral. It is emphasized, however, that for present purposes nothing is gained by transforming strain to a J format. Moreover, there is sufficient uncertainty with respect to J dominance and closure for cracks contained in notch plastic fields to question the utility and accuracy of J in such applications, unless J is found via detailed geometry and load specific numerical analysis. In such cases, ΔCTOD and field parameters such as ΔT are also available and present viable physically and theoretically valid alternatives to the apparently path-dependent J integral. For this reason, further work should concentrate on detailed numerical analysis that embraces measures more general than J.

[8] For comments on measures such as ΔCTOD and their implementation in a simple framework see, for example, Kanninen, M. F., Ahmad, J., and Leis, B. N., "A CTOD-Based Fracture Mechanics Approach to the Short-Crack Problem in Fatigue," *Mechanics of Fatigue*, ASME AMD, Vol. 47, 1981, pp. 81–90.

J. H. Underwood[1] *and G. S. Leger*[1]

Fracture Toughness of High Strength Steel Predicted from Charpy Energy or Reduction in Area

REFERENCE: Underwood, J. H. and Leger, G. S., **"Fracture Toughness of High Strength Steel Predicted from Charpy Energy or Reduction in Area,"** *Fracture Mechanics: Fifteenth Symposium, ASTM STP 833*, R. J. Sanford, Ed., American Society for Testing and Materials, Philadelphia, 1984, pp. 481–498.

ABSTRACT: Analysis is presented of the results of an extensive mechanical and fracture test program. Ninety-six hollow cylinder forgings of A723 steel with five different manufacturing processes were tested at two locations with different outer diameters. Yield strength, plane-strain fracture toughness, Charpy impact energy, and tensile reduction in area were measured and analyzed for statistical variation and difference of mean value. Linear regression was used to fit lines to the data and to determine correlation coefficients. Conclusions were drawn as to the suitability of Charpy energy and reduction in area as predictors of plane-strain fracture toughness.

KEY WORDS: fracture toughness, alloy steel, statistical analysis, mechanical tests, forged cylinders

The energy to failure in a Charpy V-notch impact test and the percentage reduction in area in a tension test are often used as indicators of fracture toughness for steels. Seldom is enough testing performed to rate with statistical significance how well these two tests predict fracture toughness. This report describes an extensive series of tests, the results of which were analyzed to obtain such a rating. It is not the purpose here to establish a specific correlation between Charpy energy or reduction in area and fracture toughness, but rather to determine, using statistical methods, just how well each of these tests can predict fracture toughness.

Charpy energy and reduction in area, along with yield strength, have been the primary material acceptance tests for Army cannon forgings for many years. Plane-strain fracture toughness is now generally recognized as a basic material property for quantitative description of crack-related failure. It is

[1]U.S. Army Armament Research & Development Center, Watervliet, N.Y. 12189

now used in design and prototype development of cannons, whereas the simpler Charpy energy and reduction-in-area tests are used for production material acceptance. Only about one tenth the time is required for either of these tests compared with that for a plane-strain fracture toughness test. Therefore the prediction of plane-strain fracture toughness by using the simpler tests is important.

Appreciation of the importance of fracture toughness to the function of cannons is useful background for this report. Some appreciation can be gained from a review of failure processes often observed in cannons. Figure 1 shows a section of a cannon tube [1] which failed following actual and then simulated firing. The *erosion and heat checking* processes that damage the inner surface are not known to be related to fracture toughness, but these are the processes which often initate crack-related failure. A particularly severe case of erosion of the inner surface is shown in Fig. 1. For the fracture surface in the photo, erosion serves to initiate the *fatigue crack growth*, which then occurs in the characteristic semielliptical crack shape. Fracture toughness has an important effect on fatigue cracking, particularly in high stress, low cycle fatigue which usually is the loading condition for a cannon. As the maximum applied stress intensity factor (K_I) during the fatigue cycle becomes nearly equal in value to the plane-strain fracture toughness (K_{Ic}) of the material, the rate of fatigue crack growth becomes higher than the usual relation between applied K_I and growth rate [2]. Further, if the ratio of K_{Ic} to yield strength (σ_{ys}) is small enough, then the area of plastic deformation relative to specimen size is small and a *fast fracture* will occur as the final fracture event. The final fracture will be a classic brittle fracture only in the extreme condition of a very small K_{Ic}/σ_{ys} ratio. Fortunately, it is more typical that the final breakthrough to the tube outer surface occurs in a relatively confined area of the tube (Fig. 1). This results in a leak rather than a breakup of the tube, which is the desired leak-before-break condition. The nature of the final fast fracture is highly dependent on the material fracture toughness.

Fracture toughness, Charpy energy, and tension tests were performed using specimens from both ends of cannon forgings. The test results were analyzed to determine quantitatively how good a prediction of fracture toughness can be obtained from the Charpy energy and reduction-in-area tests. The predictive ability of the two tests was determined by comparing results from opposite ends of the forging which have clearly defined differences in fracture toughness due to manufacturing process variables. A statistical test for different mean values was used to measure the ability of the two mechanical tests to differentiate between high and low fracture toughness.

Test Procedures

Ninety-six cannon forgings were tested as engineering support to cannon tube production. Reference *3* describes some of the results and analyses of the

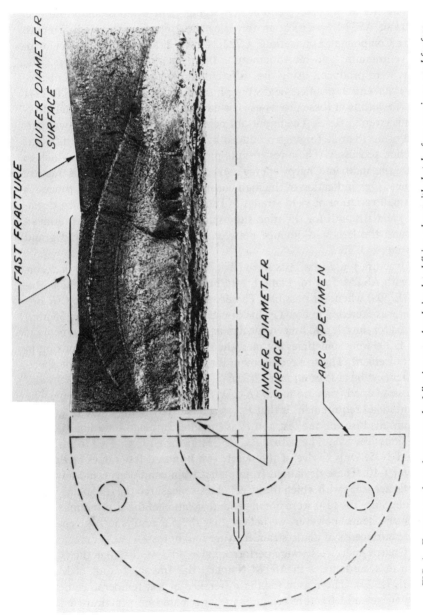

FIG. 1—*Fracture surface of cannon tube following actual and simulated firing, along with sketch of arc specimen used for fracture toughness testing.*

tests. Only certain details will be described here. The forgings were hollow cylinders, about 220 in. (5.6 m) long, 3.5 in. (89 mm) inner diameter, 6.7 in. (170 mm) outer diameter at the muzzle end of the forging, and 9.4 in. (239 mm) outer diameter at the breech end of the forging.[2] The composition of the steel was that of ASTM Specification for Alloy Steel Forgings for High-Strength Pressure Component Application (A 723), Grade 2, a high strength steel with nickel, chromium, and molybdenum as the primary alloying elements. The forgings were produced using five different forging and heat treatment processes, each with a specified yield strength range of 160 to 180 ksi (1103 to 1241 MPa). An outline of these processes is listed in Table 1. Some additional information is given in Ref 3. The important point is that extensive testing was performed using 18 or 20 forgings produced by each of five significantly different production processes. These test results provide excellent data with which to compare the ability of Charpy energy and reduction in area to predict fracture toughness. One indication of the high consistency of the production processes is the small variation of yield strength (Table 1). The largest standard deviation of yield strength for the nine subgroups of results is 1.9% of the mean value, and the largest difference between subgroup mean value and grand mean value is 3.2%.

The fracture toughness tests were performed at $+70°F$ ($+21°C$) in accordance with ASTM Test for Plane-Strain Fracture Toughness of Metallic Materials (E 399) whenever possible. The arc specimen was used in the C-R orientation,[3] as sketched in Fig. 1. Thicknesses of 1.0 and 1.5 in. (25 and 38 mm) were used for muzzle and breech specimens, respectively. Two requirements of ASTM E 399 were sometimes not met, the size requirement and the maximum load requirement. The size requirement is that the crack length and the specimen thickness be at least equal to $2.5 (K_{Ic}/\sigma_{ys})^2$. The actual crack lengths and thicknesses varied from 1.6 to 3.0 $(K_{Ic}/\sigma_{ys})^2$ and averaged 2.5 $(K_{Ic}/\sigma_{ys})^2$. The maximum load requirement is that P_{max}/P_5 be at most 1.10, where P_{max} is the maximum load during the test and P_5 is the intercept load for a line with the elastic slope less 5%. The actual P_{max}/P_5 ratio varied from 1.00 to 1.13 and averaged 1.05. Only twelve of the nearly two hundred test values of P_{max}/P_5 were over 1.10. These deviations from plane-strain conditions could adversely affect the accuracy with which the affected tests measured the true plane-strain K_{Ic} value. Since the tests were not in general violation of the size and maximum load requirements, however, we believe that the K_{Ic} results in this report are good measurements of plane-strain fracture toughness.

The Charpy energy tests were performed at $-40°F$ ($-40°C$) in the C-R orientation in accordance with ASTM Notched Bar Impact Testing of Metallic Materials (E 23). The reason for the difference in test temperature between Charpy energy and fracture toughness tests is based on practical considera-

---

[2]The tests and analyses were performed with English (inch-pound) units, so they are the primary units used.

[3]The C-R orientation is that with the crack plane normal to the circumferential direction and crack growth in the radial direction.

TABLE 1—*Test conditions.*

Process	Number of Forgings	Location	Yield Strength, 0.1% Offset, ksi (MPa) Mean	Standard Deviation
(1) Electroslag refined, Supplier A, rotary forged	20	muzzle	165.8 (1143)	3.1 (25)
		breech	172.8 (1191)	3.0 (24)
(2) Vacuum degassed, Supplier A, rotary forged	20	muzzle	170.4 (1175)	2.2 (17)
		breech	173.5 (1196)	2.3 (18)
(3) Vacuum degassed, Supplier B, rotary forged	20	muzzle	164.9 (1137)	2.2 (17)
		breech	166.0 (1145)	1.6 (13)
(4) Vacuum degassed, Supplier A, press forged	18	breech	168.3 (1160)	2.3 (18)
(5) Vacuum degassed, Supplier B, press forged	18	muzzle	165.0 (1138)	1.3 (10)
		breech	165.8 (1143)	1.7 (13)
		Grand mean	168.1 (1159)	

tions. Since the Charpy energy test is nearly as simple at low temperature as at room temperature, it was performed at $-40°F$ ($-40°C$), a typical low service temperature. The more complex fracture toughness tests were conducted at room temperature to decrease cost and to simplify the testing procedure. The difference in test temperature between that of the Charpy tests and the fracture toughness tests is not considered to be a crucial difference, based on prior test results. Kendall [4] found that the plane-strain fracture toughness of a steel similar to that considered here is essentially independent of both loading rate and temperature in the range of interest of the current work. This agrees with the results of Rolfe and Barsom [5], who showed that steels with strength as high as the 168 ksi (1158 MPa) yield strength in the present program are strain-rate insensitive. For lower strength steels which are strain-rate sensitive, the test temperatures used for comparisons of static and dynamic fracture tests should be carefully considered by using the temperature-shift concept [5].

The tension tests were performed at $+70°F$ ($+21°C$) in the C orientation using a 0.357 in. (9.1 mm) diameter specimen in accordance with ASTM Tension Testing of Metallic Materials (E 81). Only the yield strength and reduction-in-area results from the tension tests are considered in this paper.

Results and Analysis

Frequency Distributions

Frequency distribution plots of all fracture toughness (K_{Ic}), Charpy energy (CVN), and reduction-in-area (RA) data are shown in Figs. 2 to 4. Each data

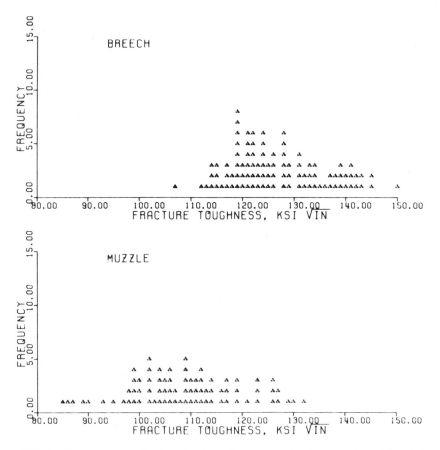

FIG. 2—*Frequency versus fracture toughness from cannon forgings; muzzle and breech locations; ksi · in.$^{1/2}$ × 1.10 = MPa · m$^{1/2}$.*

point is the average of two measured test values. This averaging procedure was used throughout the analyses and reporting of the data, because any given single measured value could be associated only with a 1.0 or 1.5 in. (25 or 38 mm) thick section from a certain end of a forging, not a particular location within that section. Since location within a section was not known, the two measured values were averaged. At most there are 96 data points in the frequency distributions, usually less because of occasional missing data. A separate frequency distribution was plotted for the muzzle and breech ends of the forgings, because the outer diameters are significantly different in size, which is an indication of basic differences in both forging and heat treatment processes.

Two observations can be made concerning the frequency distributions. First, the largest difference in mean values between muzzle and breech results seems

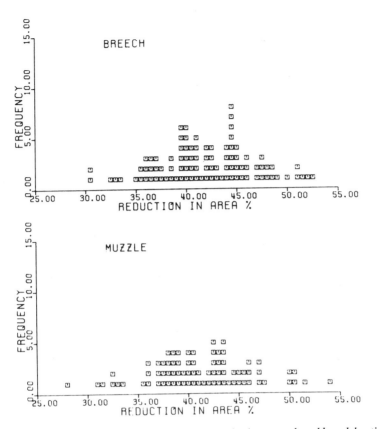

FIG. 3—*Frequency versus reduction in area from cannon forgings; muzzle and breech locations.*

to be with K_{Ic}. Second, the least normal distributions seem to be the CVN distributions, for both muzzle and breech. Upcoming analysis will show that these observations are significant and can be well described quantitatively using statistical tests.

Statistical analysis is simpler if the data are normally distributed. The Kolmogorov-Smirnov test, an accepted method for checking for a normal distribution [6], was applied to the six sets of data of Figs. 2 to 4. The test is a comparison of the actual probability distribution from the measured data with the ideal normal probability distribution having the same mean and standard deviation. If the maximum deviation of the measured distribution from the ideal normal distribution is less than a certain amount for the given sample size, the measured data can be considered to be normally distributed. Figures 5 and 6 compare the measured with the ideal normal probability distributions for the muzzle K_{Ic} and CVN data, respectively. The maximum deviations are

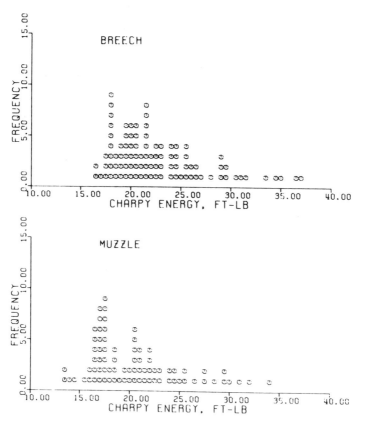

FIG. 4—*Frequency versus Charpy energy from cannon forgings; muzzle and breech locations; ft-lbf × 1.36 = N/m.*

0.039 and 0.138, respectively, in dimensionless probability units. These are to be compared with 0.182, the maximum allowed value [6] for the sample size of 78 in Figs. 5 and 6 and for a 99% confidence level. This comparison shows that both sets of data are normally distributed. In addition, the observation discussed previously, that the CVN distributions appear to be the least normal, is verified by the larger deviation from a normal probability distribution. One reason for this deviation from normal distribution is the fact that CVN distribution is truncated because of rejection of forgings with CVN values below a minimum requirement. Such rejections cause an abrupt drop in probability as the required minimum, 15 ft-lbf (20.4 N/m), is approached. This abrupt drop in CVN frequency and probability can be seen in Figs. 4 and 6, respectively.

The overall results of the tests for normalcy are that the K_{Ic} data are closest to an ideal normal distribution, the RA data are intermediate, and the CVN

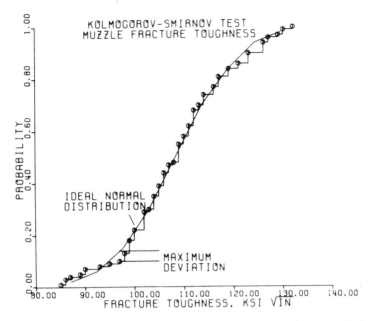

FIG. 5—*Probability versus fracture toughness from cannon forgings; muzzle location;* $ksi \cdot in.^{1/2} \times 1.10 = MPa \cdot m^{1/2}$.

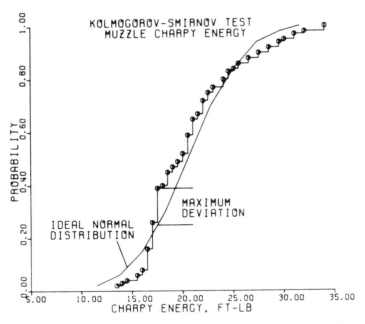

FIG. 6—*Probability versus Charpy energy from cannon forgings; muzzle location; ft-lbf* $\times$ *1.36 = N/m.*

data are the least normal. All six groups of data pass the Kolmogorov-Smirnov test for normalcy.

Probability of Different Means

A statistical summary of all test results is presented in Table 2. Each mean and standard deviation was calculated using nominally 20 average values from 40 tests, as described in the section on Test Procedures. The standard deviation relative to mean averages about 6% for K_{Ic}, 9% for RA, and 12% for CVN, compared with 1 to 2% for yield strength (Table 1). These differences in variation of the test results have an effect on how well CVN or RA can predict K_{Ic}.

There appear to be significant differences in K_{Ic} between muzzle and breech location for each of the four processes for which comparison data was available. The probability of different means statistical test was used to quantify this difference in K_{Ic} and to determine if CVN or RA can detect a difference depending on location. In basic concept the probability of different means depends on (1) the difference between the two mean values, and (2) the amount of variation about both mean values. A dimensionless test statistic d was calculated as [7]

$$d = \frac{|\mu_M - \mu_B|}{(\sigma_M^2 + \sigma_B^2)^{1/2}} \tag{1}$$

where μ_M, μ_B and σ_M, σ_B are the mean and standard deviation for muzzle and breech tests, respectively. Relatively large differences in mean and relatively small standard deviations will produce a large value of d and a high probability that the means are statistically different. The value of d was applied to operating characteristic curves [7] for the appropriate sample size, nominally twenty, and confidence level of 99%, in order to determine the probability of different means (P) shown in Table 2. The P values for the K_{Ic} data confirm the subjective observation that the results differ with location. On average it is 98% probable that the K_{Ic} test distinguishes between a high toughness material at the breech end and a lower toughness material at the muzzle end. This compares with the average 46% probability that the CVN test can distinguish breech from muzzle and the 18% probability that RA can distinguish breech from muzzle.

Results in Table 2 for two of the five processes should be considered further. Process 1 results show an apparent inconsistency. The average K_{Ic} is lower at the muzzle location than at the breech, whereas the CVN and RA results are higher at the muzzle. Since the probability of different means is relatively low for these CVN and RA results, however, this is not a serious inconsistency. Process 3 results show the largest difference in K_{Ic} between muzzle and breech and 99% probability of different means. If CVN or RA are to be useful for predicting K_{Ic}, they should be able to distinguish breech from muzzle in this case. The CVN test does well, with a 96% probability of different means. The RA

TABLE 2—*Summary of fracture toughness, Charpy energy, and reduction-in-area test results.* [a]

| | Fracture Toughness (K_{Ic}) | | | | Charpy Energy (CVN) | | | | | Reduction in Area (RA) | | |
| | μ | | | | μ | | | | | | | |
Process and Location	ksi·m$^{1/2}$	(MPa·m$^{1/2}$)	σ/μ	P	ft·lbf	(N/m)	σ/μ	P		μ, %	σ/μ	P
(1) Muzzle	116.8	(128.0)	0.074		26.7	(36.3)	0.135			47.2	0.064	
Breech	130.2	(142.7)	0.078	0.96	24.5	(33.3)	0.118	0.33		44.6	0.072	0.50
(2) Muzzle	112.7	(123.5)	0.077		19.5	(26.5)	0.108			37.9	0.111	
Breech	124.4	(136.3)	0.047	0.97	21.0	(28.6)	0.105	0.34		40.1	0.097	0.19
(3) Muzzle	96.1	(105.3)	0.055		16.4	(22.3)	0.097			38.6	0.093	
Breech	126.3	(138.4)	0.059	0.99	19.1	(26.0)	0.110	0.96		38.0	0.084	0.02
(4) Breech	134.1	(147.0)	0.064	...	29.6	(40.3)	0.179	...		46.9	0.085	...
(5) Muzzle	107.6	(117.9)	0.049		19.3	(26.2)	0.119			40.8	0.061	
Breech	118.0	(129.3)	0.039	0.99	20.5	(27.9)	0.083	0.22		41.2	0.107	0.02
Average	118.5	(129.9)	0.060	0.98	21.8	(29.6)	0.117	0.46		41.7	0.086	0.18

[a] μ = mean; σ = standard deviation; P = probability of different means.

test does poorly, with a 2% probability of different means and a higher value at the muzzle rather than a lower value as would be expected from the K_{Ic} results.

Correlation

Plots of K_{Ic} versus both CVN and RA for Process 1 are shown in Figs. 7 and 8 for breech and muzzle data, respectively. Standard linear regression of K_{Ic} on CVN and K_{Ic} on RA was performed on a calculator. The regression lines are shown as the solid lines, with associated correlation coefficients. The dashed lines were calculated with the axes reversed, that is, by linear regression of CVN on K_{Ic} and RA on K_{Ic}. The correlation coefficients are unchanged by the reversal of axes but, as can be seen, the regression lines are quite different. The reason for the difference is that the solid line is determined by minimizing the

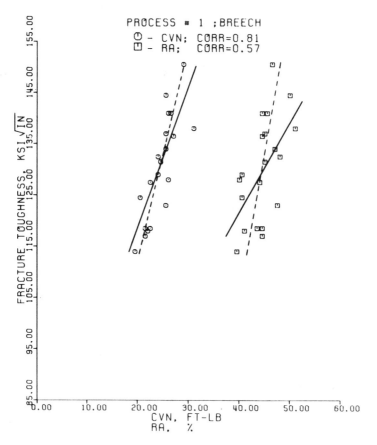

FIG. 7—*Fracture toughness versus Charpy energy (CVN) and reduction in area (RA); breech location; Process 1; ksi · in.*$^{1/2} \times 1.10 = MPa · m^{1/2}$, *ft-lbf* $\times 1.36 = N/m$.

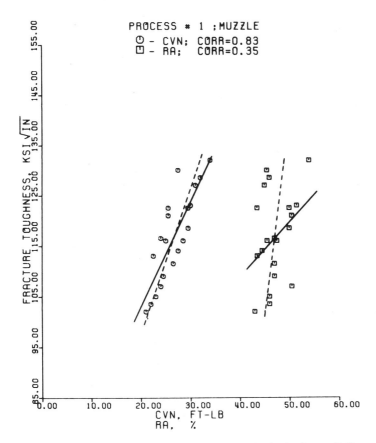

FIG. 8—*Fracture toughness versus Charpy energy (CVN) and reduction in area (RA); muzzle location; Process 1; ksi · in.$^{1/2}$ × 1.10 = MPa · m$^{1/2}$, ft-lbf × 1.36 = N/m.*

square of the difference in K_{Ic} between the data and the line, whereas the dashed line is determined by minimizing the square of the difference in CVN or RA between the data and the line. The solid line, with K_{Ic} as the dependent variable, is the appropriate procedure for making predictions of K_{Ic} by another test result, CVN or RA in this case. With K_{Ic} as the dependent variable the differences in K_{Ic} are minimized; this is appropriate since K_{Ic} is the known quantity which is being predicted by other tests. It should be noted, however, that the appropriate linear regression of K_{Ic} on CVN or RA can result in a straight line which appears to be quite different from an eyeball fit. The solid line in the K_{Ic} versus RA plot in Fig. 8 is an example. One reason for this is that an eyeball fit is determined by minimizing the difference between the data point and the line in a direction perpendicular to the line, whereas the differences in linear regression are seldom in a direction perpendicular to the line.

Plots of K_{Ic} versus CVN and RA, regression lines, and correlation coefficients are shown in Fig. 9 for the Process 3 data. It is interesting to note that the correlation coefficients for K_{Ic} versus RA, both muzzle and breech locations, are not significantly different from those for K_{Ic} versus CVN, and yet the abilities of RA and CVN to distinguish between muzzle and breech are very different. Refer again to Table 2, Process 3. Therefore similar correlation of RA and CVN with K_{Ic} does not imply that RA and CVN can similarly distinguish between muzzle and breech material.

The correlation coefficients for all data are shown in Table 3. The coefficient for each separate subgroup of 20 data points was calculated, as was the coefficient for the combination of the two subgroups into one. In one K_{Ic} versus CVN group (Process 1) and in two K_{Ic} versus RA groups (Processes 1 and 3) the combined correlation coefficient was significantly lower than either separate coefficient. This shows that improper combination of data can lead to improper

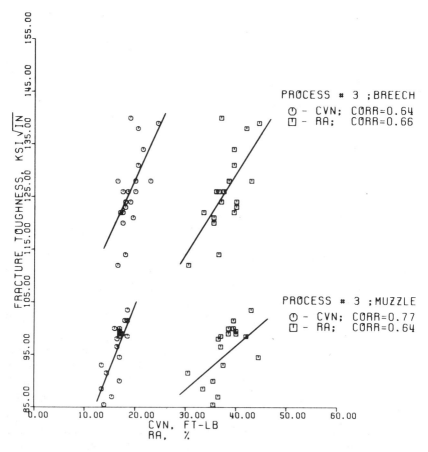

FIG. 9—*Fracture toughness versus Charpy energy (CVN) and reduction in area (RA); muzzle and breech locations; Process 3; ksi · in.$^{1/2}$ × 1.10 = MPa · m$^{1/2}$, ft-lbf × 1.36 = N/m.*

TABLE 3—*Correlation of fracture toughness with Charpy energy and reduction in area.*

Process and Location	Correlation Coefficient for K_{Ic} versus Charpy Energy		Correlation Coefficient for K_{Ic} versus Reduction in Area	
	Separate	Combined	Separate	Combined
(1) Muzzle	0.83	0.43	0.35	0.12
Breech	0.81		0.57	
(2) Muzzle	0.68	0.71	0.47	0.47
Breech	0.70		0.52	
(3) Muzzle	0.77	0.76	0.64	0.17
Breech	0.64		0.66	
(4) Breech	0.88	. . .	0.87	. . .
(5) Muzzle	0.65	0.58	0.05	0.12
Breech	0.44		0.16	
Average	0.71	0.62	0.48	0.22

prediction of one test result from another, K_{Ic} from CVN or RA in this case. The average values of correlation coefficient included in Table 3 show that CVN correlates better with K_{Ic} than does RA, and that the reduction in combined correlation coefficient relative to the separate values is less for CVN than for RA.

Barsom and Rolfe [8] described a correlation between CVN and K_{Ic} which should be considered here. It is

$$K_{Ic} = (5.0\,\text{CVN}\,\sigma_{ys} - 0.25\,\sigma_{ys}^2)^{1/2} \qquad (2)$$

in which K_{Ic} is in ksi $\cdot$ in.$^{1/2}$, σ_{ys} is in ksi, and CVN is in ft-lbf. Since yield strength (σ_{ys}) varies little in the tests here, as shown in Table 2, the most important guidance that can be obtained from the Barsom-Rolfe correlation is that K_{Ic} can be expected to vary with the square root of CVN. This expectation can also be obtained from J integral analysis. For the three-point bend specimen [9].

$$J = 2A/bB \qquad (3)$$

where B is specimen thickness, b is uncracked ligament depth, and A is the total energy under the load versus load-point deflection curve, a quantity quite analogous to Charpy energy. Using the relation between J and K for elastic plane-strain conditions,

$$K = [EJ/(1 - \nu^2)]^{1/2} \qquad (4)$$

it can be easily shown that K_{Ic} can be expected to vary with the square root of CVN. Based on the above, correlation coefficients for K_{Ic} versus (CVN)$^{1/2}$ and

K_{Ic} versus $(RA)^{1/2}$ were calculated for the Process 1 results. The coefficients are 0.83, 0.81, 0.35, and 0.58 for CVN and RA muzzle and breech, respectively. Comparing these values with those in Table 3 for linear correlation with K_{Ic} shows that there is no significant difference between square root and linear correlation of CVN or RA with K_{Ic}.

Scatter Diagram

Plots of the deviation of the K_{Ic} data points from combined linear regression curves of K_{Ic} versus CVN and K_{Ic} versus RA are shown in Fig. 10. The com-

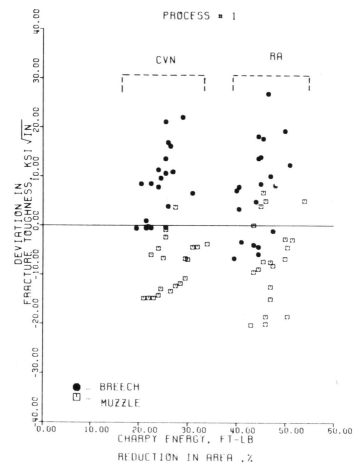

FIG. 10—*Deviation in fracture toughness data from regression line versus Charpy energy (CVN) and reduction in area (RA); breech and muzzle locations; Process 1, ksi · in.$^{1/2}$ × 1.10 = MPa · m$^{1/2}$, ft-lbf × 1.36 = N/m.*

bined regression curves were calculated using both breech and muzzle data and resulted in the combined correlation coefficients of 0.43 and 0.12, shown in Table 3 for CVN and RA, respectively. The plots of deviations, also called *scatter diagrams*, provide a graphic display of both the source and the direction of the deviation of the actual data from the fitted line. In the case here it is clear that nearly all the deviation in K_{Ic} above the K_{Ic} versus CVN and K_{Ic} versus RA regression lines is from breech tests, and the deviation below is from muzzle tests. This result is a further clear indication that there is a significant difference between muzzle and breech fracture toughness and that toughness in these two locations must be analyzed separately, as done here.

Discussion and Conclusions

The statistical distribution and variation of the test results considered here are summarized as: K_{Ic} being the most normal and least variable, RA intermediate in both regards, CVN the least normal and most variable. All test results, including the CVN results, passed the Kolmogorov-Smirnov test for a normal distribution. It is proposed here that the higher variability of the CVN results is related to two aspects of the specimen configuration. First, the volume of material which is critically loaded in the Charpy specimen, that is, the material around the notch tip, is the smallest of the three tests. Second, the Charpy test may be more easily affected by critical configurational variations, such as notch root size and roughness, than the other tests. The critical areas of RA and K_{Ic} specimens, the smooth outer diameter and the fatigue crack tip, respectively, are less subject to critical variations. The solution to the high variation of the CVN test is to increase the minimum required value or to increase the number of CVN tests. An example of the latter is the requirement in ASTM A 723 for three CVN tests, compared with one yield strength test, to test for a given material condition.

The correlation coefficient from linear regression is an important measure of how well one test result can predict another. Regression of K_{Ic} on CVN gives average coefficients of 0.71 and 0.62, for analyses with muzzle and breech separated and combined, respectively. Regression of K_{Ic} on RA gives average coefficients of 0.48 and 0.22, for separate and combined analyses, respectively. The correlation of K_{Ic} with CVN might have been better if the CVN tests had been performed at the temperature of the other tests, $+70°F$ ($+21°C$), rather than at $-40°F$ ($-40°C$). The poorer correlation of K_{Ic} with RA is related to a significant extent to the basic differences in the tests. K_{Ic} is a measure of initial growth of a pre-existing crack with limited plastic deformation, whereas RA is a measure of considerable tensile plastic deformation of a smooth specimen which precedes internal crack initiation and failure. It is concluded that, for the conditions of the tests here, the best prediction of K_{Ic} would be from regression of K_{Ic} on CVN using representative data from a single manufacturing process and a single location within the forging. This procedure will give a good estimate of K_{Ic} from measured values of CVN, provided that enough repeat measurements are performed to account for the variability of the CVN test.

Probability of different means analysis is a valuable complement to correlation analysis for determining how well one test result can predict another. Probability of different means is particularly useful, in the case here, for determining whether or not there is a significant difference in K_{Ic} between two material conditions based on the available CVN or RA data. Comparison of the results in Tables 2 and 3 shows that CVN or RA predicts the same trend as K_{Ic} when (1) the probability of different means of CVN or RA is high, and (2) the combined correlation coefficient of K_{Ic} versus CVN or RA is high. The four highest P values for CVN and RA, in which also the breech value of CVN or RA is higher than the muzzle value, are 96, 34, 22, and 19%, for CVN Processes 3, 2, 5 and RA Process 2, respectively. These subgroups also result in the highest combined correlation coefficients and in the same order, 0.76, 0.71, 0.58, and 0.47, respectively. It is concluded that for the conditions of the tests here, the best determination of a significant difference in K_{Ic} between two material conditions is by the probability of different means of the associated CVN data.

Acknowledgments

We gratefully acknowledge the work of F. A. Heiser, who planned and directed the testing program, and E. E. Coppola, who provided advice on statistical analyses.

References

[1] Underwood, J. H. and Throop, J. F. in *Part-Through Crack Fatigue Life Prediction*, *ASTM STP 687*, J. B. Chang, Ed., American Society for Testing and Materials, 1979, pp. 195–210.
[2] Paris, P. C., Bucci, R. J., Wessel, E. J., Clark, W. R., and Mager, T. R. in *Stress Analysis and Growth of Cracks*, *ASTM STP 513*, American Society for Testing and Materials, 1972, pp. 141–176.
[3] Tauscher, S., "The Correlation of Fracture Toughness with Charpy V-Notch Impact Test Data," USA ARRADCOM Technical Report ARLCB-TR-81012, Benet Weapons Laboratory, Watervliet, N.Y., March 1981.
[4] Kendall, D. P., *Materials Research and Standards*, Vol. 10, No. 12, Dec. 1970.
[5] Rolfe, S. T. and Barsom, J. M., *Fracture and Fatigue Control in Structures—Applications of Fracture Mechanics*, Prentice-Hall, Englewood Cliffs, N.J., 1977.
[6] Bowker, A. H. and Lieberman, G. J., *Engineering Statistics*, Prentice-Hall, Englewood Cliffs, N.J., 1972, pp. 454–458.
[7] Bowker, A. H. and Lieberman, G. J., *Engineering Statistics*, Prentice-Hall, Englewood Cliffs, N.J., 1972, pp. 225–230.
[8] Barsom, J. M. and Rolfe, S. T. in *Impact Testing of Metals, ASTM STP 466*, American Society for Testing and Materials, 1970, pp. 281–302.
[9] Clarke, G. A., Andrews, W. R., Begley, J. A., Donald, J. K., Embley, G. T., Landes, J. D., McCabe, D. E., and Underwood, J. H., *Journal of Testing and Evaluation*, Vol. 7, No. 1, Jan. 1979, pp. 49–56.

William J. Mills[1]

Effect of Fast-Neutron Irradiation on Fracture Toughness of Alloy A-286

REFERENCE: Mills, W. J., **"Effect of Fast-Neutron Irradiation on Fracture Toughness of Alloy A-286,"** *Fracture Mechanics: Fifteenth Symposium, ASTM STP 833.* R. J. Sanford, Ed., American Society for Testing and Materials, Philadelphia, 1984, pp. 499–515.

ABSTRACT: The effect of fast-neutron irradiation on the fracture toughness behavior of Alloy A-286 was characterized at 24 and 427°C using linear-elastic K_{Ic} and elastic-plastic J_{Ic} fracture mechanics techniques. The fracture toughness was found to decrease continuously with increasing irradiation damage at both test temperatures. In the unirradiated and low fluence conditions, specimens displayed appreciable plasticity prior to fracture, and equivalent K_{Ic}-values were determined from J_{Ic} fracture toughness results. At high irradiation exposure levels, specimens exhibited a brittle K_{Ic} fracture mode. The 427°C fracture toughness fell from 129 MPa$\sqrt{m}$ in the unirradiated condition to 35 MPa$\sqrt{m}$ at an exposure of 16.2 dpa (total fluence of 5.2×10^{22} n/cm^2). Room-temperature fracture toughness values were consistently 40 to 60% higher than the 427°C values. Electron fractography revealed that the reduction in fracture resistance was caused by a fracture mechanism transition from ductile microvoid coalescence to channel fracture.

KEY WORDS: fracture (materials), fracture toughness, *J*-integral, neutron irradiation, elevated-temperature tests, fractography, fracture mechanism transition, Alloy A-286

Alloy A-286 (AISI 660) is a heat-resistant Fe-Ni-Cr superalloy strengthened by precipitation of the γ'-phase (Ni$_3$[Al, Ti]). This material is used in a few special reactor structural applications where it is exposed to relatively low levels of neutron irradiation. Knowledge of the fracture toughness response of this material before and after irradiation would aid such applications. Fracture toughness tests on the unirradiated material typically exhibit elastic-plastic behavior, with J_{Ic}-values ranging from 75 to 133 kJ/m^2 [1–3]. Fast-neutron irradiation at intermediate temperatures is likely to result in appreciable degradation in fracture resistance due to substantial displacement-type irradiation damage (dislocation loops and networks, voids, and precipitates). This type of damage frequently produces significant reductions

---

[1] Fellow Engineer, Westinghouse Hanford Company, Richland, Wash. 99352.

in ductility, but quantitative measures of the corresponding changes in fracture resistance are typically not available. This study was undertaken to determine the effect of irradiation on the fracture toughness behavior of Alloy A-286 using both linear-elastic K_{Ic} and elastic-plastic J_{Ic} concepts as applicable. The fracture surface appearance was also characterized by electron fractographic examination to evaluate operative fracture mechanisms at various neutron exposure levels.

Material and Experimental Procedure

The test material, obtained from a hot-rolled 1.27-cm strip of Alloy A-286 Carpenter Steel heat K-58139-2 (electric-arc-melted, consumable-electrode-remelted), was given a standard precipitation heat treatment: solution annealed at 982°C for 30 min, water quenched and aged at 718°C for 16 h. In this condition the alloy displays an austenitic matrix that is strengthened by precipitation of γ'-phase. The typical microstructure (Fig. 1a) reveals many randomly oriented carbide particles coupled with stringers aligned in the rolling direction. These inclusions were identified by wavelength dispersive X-ray analysis as titanium-rich MC-type carbides. Higher magnification (Fig. 1b) reveals smaller unidentified precipitates distributed uniformly throughout the matrix.

The chemical analysis and mechanical properties for this material are listed in Tables 1 and 2 respectively. The unirradiated strength levels are seen to be independent of temperature, but a 40% decrease in ductility was observed at 427°C. Neutron exposures of 2.4 and 7.8 displacements per atom (dpa), corresponding to total fluences of 7.7×10^{21} and 2.5×10^{22} n/cm^2, caused significant irradiation embrittlement; the yield strength was increased by 35% and the irradiated uniform and total elongations were degraded by a factor of approximately two. In a previous study [4], irradiation to 2.4×10^{21} n/cm^2 was found to produce no embrittlement for the same heat of material studied herein. Therefore the threshold level for irradiation damage is between 1 and 2.4 dpa (total fluences of 2.4 and 7.7×10^{21} n/cm^2).

In this study, irradiations were performed in the Experimental Breeder Reactor-II (EBR-II), Position 7A4, in Subassemblies X-267A, X-267B, and X-268. The subassemblies contained weeper canisters that allowed reactor-ambient sodium to circulate around test coupons. Specimens were irradiated at temperatures ranging from 400 to 427°C to total neutron exposures of 2.4 to 16.2 dpa, as calculated using the cross section of Doran and Graves [5]. A summary of neutron irradiation conditions [total fluence, fast fluence ($E > 0.1$ MeV), and dpa] is given in Table 3.

Fracture toughness tests were performed at 24 and 427°C on deeply precracked ($a/W > 0.6$) compact specimens with a width, W, of 29.3 mm and a thickness, B, of 12.7 mm. Specimens were tested on an electrohydraulic

FIG. 1—*Typical microstructure of Alloy A-286.* (a) *Large, randomly oriented carbide particles and stringers aligned in the rolling direction.* (b) *Higher magnification of a stringer.*

TABLE 1—*Chemical composition (weight percent) of Alloy A-286.*

C	Mn	Si	P	S	Cr	Ni	Mo	Cu	V	Ti	Al	B	Fe
0.04	1.27	0.53	0.015	0.002	14.04	25.80	1.30	0.08	0.26	2.17	0.19	0.004	balance

closed-loop machine in stroke control at a stroke rate of 0.25 mm/min. Displacements were measured along the load line by a high temperature LVDT displacement monitoring technique [6]. During each test, the load-line displacement was recorded continuously on an X-Y recorder as a function of load.

In the unirradiated and low neutron exposure (2.4 dpa) conditions, elastic-plastic J_{Ic} concepts were used to characterize the ductile fracture toughness behavior. Values of J_{Ic} were determined by the multiple-specimen R-curve technique in accordance with ASTM E 813. Deeply cracked compact specimens were loaded to various displacements producing different amounts of crack extension, Δa, and then unloaded. After unloading, each specimen was heat tinted to discolor the crack growth region and subsequently broken open so that the amount of crack extension could be measured. The value of J for each specimen was determined from the load versus load-line displacement curve by the following equation from ASTM E 813:

$$J = \frac{2A\,(1 + \alpha)}{Bb\,(1 + \alpha^2)} \tag{1}$$

where

A = area under load versus load-line displacement curve,
b = unbroken ligament size,
$\alpha = [(2a/b)^2 + 2(2a/b) + 2]^{1/2} - (2a/b + 1)$, and
a = crack length.

Single-specimen unloading-compliance J_{Ic}-tests were also performed at 427°C. Periodically during these tests, specimens were unloaded approximately 10% of the maximum load to measure changes in compliance. The analog load-displacement signal was digitized, fed into a minicomputer, and stored on a magnetic tape for future retrieval and analysis. At each unloading, the value of J was calculated by Eq 1, and the unloading compliance was computed from a least-squares regression line fitted through the lower half of the unloading curve. The instantaneous crack length and corresponding crack extension was then determined by the following modification of the Saxena-Hudak equation:

$$a/W = 0.9796 - 3.1350U - 4.4477U^2 + 16.0749U^3 \tag{2}$$

TABLE 2—*Summary of tensile properties of Alloy A-286.*

Temperature, °C	Total Fluence, n/cm²	Fast Fluence (E > 0.1 MeV), n/cm²	Neutron Exposure, dpa	Yield Strength, MPa	Ultimate Strength, MPa	Uniform Elongation, %	Total Elongation, %	Number of Tests
24	0	0	0	769	1058	19	22	1
427	0	0	0	730	999	11	15	3
427	7.7×10^{21}	5.6×10^{21}	2.4	987	1093	4	8	2
427	2.4×10^{22}	1.3×10^{22}	7.8	976	1073	5	6	1

TABLE 3—Irradiation effects on fracture toughness of Alloy A-286.

Temperature, °C	Total Fluence, n/cm²	Fast Fluence $(E > 0.1$ MeV$)$, n/cm²	Neutron Exposure, dpa	J_{Ic}, kJ/m²	K_{Jc}, MPa√m	T	K_Q, MPa√m	$2.5(K_Q/\sigma_{ys})^2$, mm	$W\text{-}a$, mm	K_{Ic}, MPa√m
24	0	0	0	146	180	31	...	...	...	...
24	2.5×10^{22}	1.3×10^{22}	7.8	...	...	...	74	14.4	10.3	...
24	2.5×10^{22}	1.3×10^{22}	7.8	...	...	...	77	15.4	10.3	...
24	5.2×10^{22}	3.7×10^{22}	16.2	...	...	...	57	8.7	9.6	57
427	0	0	0	88	129	16	...	...	...	...
427	7.7×10^{21}	5.6×10^{21}	2.4	42	90	3	...	...	...	...
427	2.5×10^{22}	1.3×10^{22}	7.8	...	...	...	45	5.4	10.5	45
427	2.5×10^{22}	1.3×10^{22}	7.8	...	...	...	49	6.3	9.8	49
427	5.2×10^{22}	3.7×10^{22}	16.2	...	...	...	34	3.0	10.5	34
427	5.2×10^{22}	3.7×10^{22}	16.2	...	...	...	37	3.6	10.5	37

where

$U = 1/(\sqrt{BEC} + 1)$,
E = elastic modulus, and
C = load-line compliance.

Equation 2 takes into account the difference in compliance-crack length calibration at the notch surface, where load-line displacements are frequently monitored, and at the outside surfaces, used in high temperature displacement measurements. This difference is caused by the influence of the pin-loaded holes [7,8]. Compliance values at the two positions were determined as functions of a/W by finite-element techniques [9]. The difference in compliance values was incorporated into the Saxena-Hudak equation [10] to establish the compliance-crack length relationship given in Eq 2. Experimental compliance results for Alloy A-286 and other materials at 427 and 538°C were found to be in agreement with the modified Saxena-Hudak equation.

Single-specimen and multiple-specimen R-curves were constructed by plotting values of J as a function of Δa; J_{Ic} was then taken to be the value of J where a least-squares regression line through the crack extension data points intersected the stretch zone line:

$$J = 2\sigma_f(\Delta a) \tag{3}$$

where σ_f = flow strength = $[\frac{1}{2}(\sigma_{ys} + \sigma_{uts})]$.

The ASTM E 813 specimen size criteria for valid J_{Ic} determination:

$$B, b > 15 \frac{J}{\sigma_f} \tag{4}$$

and

$$B, b > 25 \frac{J_{Ic}}{\sigma_f} \tag{5}$$

were met for all specimens.

Values of the tearing modulus, T, were computed from the equation [11]

$$T = \frac{dJ}{da} \frac{E}{\sigma_f^2} \tag{6}$$

where dJ/da is the R-curve slope. The validity criterion proposed by Hutchinson and Paris [12] for J-controlled crack growth in a fully yielded specimen

$$\omega = \frac{b}{J} \frac{dJ}{da} \gg 1 \tag{7}$$

was met.

At the high irradiation exposures, brittle fracture occurred in the linear-elastic regime; hence the K_{Ic} test procedures outlined in ASTM Test for Plane-Strain Fracture Toughness of Metallic Materials (E 399) were used. The critical load (P_Q), obtained by the 5% secant offset technique, and the average crack length, obtained from the surface, 1/4-thickness, and 1/2-thickness positions, were used to compute the conditional fracture toughness (K_Q) using the standard K-solution given in ASTM E 399:

$$K_Q = (P_Q/BW^{1/2}) \cdot f(a/W) \tag{8}$$

where

$$f(a/W) = \frac{(2 + a/W)}{(1 - a/W)^{3/2}} \cdot (0.886 + 4.64a/W - 13.32a^2/W^2$$
$$+ 14.72a^3/W^3 - 5.6a^4/W^4)$$

The specimen size requirement to assure valid plane strain K_{Ic} fracture toughness determination:

$$a, \ W\text{-}a, \ B \geq 2.5 \ (K_Q/\sigma_{ys})^2 \tag{9}$$

was satisfied at the higher neutron exposures (7.8 and 16.2 dpa at 427°C and 16.2 dpa at 24°C; see Table 3). The K_{Ic} specimens were precracked to an a/W ratio greater than 0.6, which is higher than recommended ratio given in ASTM E 399. This difference is not expected to affect the results, since the minimum specimen dimension, $W\text{-}a$, generally satisfied Eq 9 and catastrophic fracture consistently occurred prior to the 2% increment of effective crack extension allowed at K_Q. The one exception, the 7.8 dpa room temperature results, will be discussed later.

The fracture surface appearance of unirradiated specimens was characterized by direct fractographic examination on a scanning electron microscope (SEM) operated at an accelerating potential of 25 kV. To examine the fracture morphology of irradiated specimens, gold-coated cellulose-acetate replicas were prepared and studied.

Results and Discussion

The unirradiated fracture toughness behavior for Alloy A-286 at 24 and 427°C is summarized in Fig. 2. The J_{Ic}-values differ slightly from those reported in Ref *1* because the tension component correction in Eq 1 was not made originally, and additional data have been included in the current R-curve regressions. At 427°C, the J_{Ic} and tearing modulus (Table 3) decreased by almost a factor of two relative to room-temperature values. This reduction is consistent with the loss in ductility (Table 2) observed at the higher temperature.

Neutron irradiation was found to cause a continuous degradation in fracture

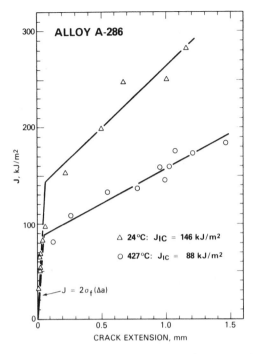

FIG. 2—*R-curves for unirradiated Alloy A-286 at 24 and 427°C.*

resistance (Fig. 3). In the unirradiated condition, typical load-displacement curves exhibited extensive plastic deformation, whereas the low-exposure (2.4 dpa) load-displacement records showed only limited plasticity and a lower maximum load. With higher neutron exposures, specimens failed in the linear-elastic regime at progressively lower loads.

Comparison of the 427°C unirradiated and low fluence R-curves is made in Fig. 4; multiple-specimen and single-specimen results are displayed on the left and right respectively. Both test methods yield comparable fracture toughness responses. Irradiation to 2.4 dpa was found to cause a two-fold reduction in J_{Ic} and a five-fold reduction in tearing modulus. The large degradation in tearing resistance is due to a 70% reduction in the R-curve slope coupled with a 20% increase in flow strength.

The irradiated J_{Ic}-value determined by the multiple-specimen technique (39 kJ/m²) is not technically valid according to ASTM E 813 because only three data points were available to construct the R-curve; ASTM recommends that a minimum of four points be used. The unloading-compliance results met all ASTM criteria; hence the J_{Ic}-value of 42 kJ/m² is considered valid and is reported in Table 3. Both test methods yield the same tearing modulus, $T = 3$.

At the higher neutron exposures, the degradation in fracture toughness be-

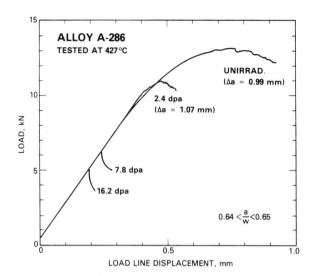

FIG. 3—*Typical load-displacement curves.*

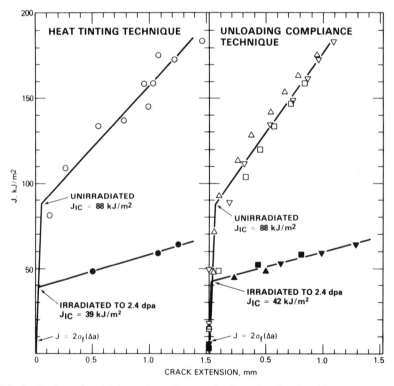

FIG. 4—*Single- and multiple-specimen R-curves for the unirradiated and low neutron exposure conditions.*

havior continued with K_{Ic}-values ranging from 45 to 49 MPa√m at 7.8 dpa and 34 to 37 MPa√m at 16.2 dpa. In Fig. 5, these toughnesses are compared with equivalent plane strain fracture toughness values (K_{Jc}) computed from the unirradiated and 2.4-dpa J_{Ic}-values using the equation [13]

$$K_{Jc} = \sqrt{\frac{E J_{Ic}}{1 - \nu^2}} \qquad (10)$$

where ν represents Poisson's ratio. The plane strain fracture toughness is seen to decrease continuously with increasing irradiation damage, although the degradation rate falls off at the highest exposures. At 16.2 dpa, the K_{Ic}-value is reduced by 70% relative to the unirradiated equivalent toughness.

The room-temperature fracture toughness results are listed in Table 3 and plotted in Fig. 5. The material irradiated to 16.2 dpa fractured in the linear-elastic mode and had a K_{Ic} of 57 MPa√m. Under intermediate exposure conditions (7.8 dpa), the increase in fracture resistance invalidated the K_{Ic}-results since the specimen dimensions no longer met the size criterion given in Eq 9. Furthermore, the values of P_Q for these two tests were equal to

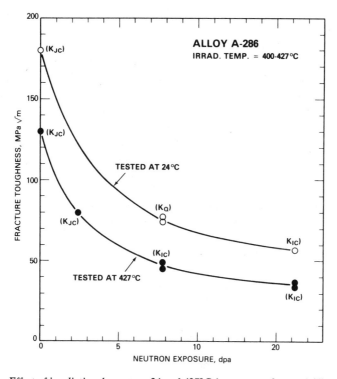

FIG. 5—*Effect of irradiation damage on 24 and 427°C fracture toughness of Alloy A-286.*

P_5, the intersection of the 5% offset secant with the load-displacement record, which means that 2% effective crack extension was allowed at K_Q. The large initial a/W could then result in a slight overestimation of K_Q. Since subsize specimens with long initial crack lengths tend to cause nonconservative toughness values, the intermediate exposure fracture toughnesses are reported as K_Q rather than K_{Ic}. The unirradiated equivalent K_{Jc} fracture toughness was computed using Eq 10.

The room-temperature fracture toughness decreases continuously with increasing irradiation exposure, paralleling the 427°C results. The 24°C values were approximately 40 to 60% higher than elevated temperature results, indicating an overall increase in fracture resistance with decreasing temperature in both the unirradiated and irradiated conditions.

The severe degradation in fracture resistance with increasing irradiation damage demonstrates that brittle fracture would be an important design consideration for highly irradiated Alloy A-286 (that is, for exposures greater than 5 dpa). Critical flaw sizes for design stresses equal to half the unirradiated yield strength are on the order of 5 to 10 mm at 427°C and 10 to 20 mm at 24°C. For exposures less than 2 dpa, the higher toughness levels result in an order of magnitude increase in critical flaw sizes such that brittle fracture would no longer be a primary concern.

Examination of the unirradiated and irradiated fracture surface morphologies revealed that the operative fracture mechanisms were dependent on irradiation exposure. The unirradiated fracture appearance, detailed in Ref 1, consisted of a combination of microvoid coalescence and elongated tear ridges or troughs (Fig. 6). The tear ridges (see upper left corner of Fig. 6) formed around failed stringers that were aligned in the rolling direction. These troughs propagated ahead of the advancing crack front, and they caused a reduction in fracture resistance of the surrounding matrix [1]. The microvoid coalescence mechanism involved a duplex morphology with large primary dimples surrounded by many smaller ones. Early rupture of the large carbide inclusions initiated the primary dimples, but before they could coalesce, many smaller dimples, or void sheets [14,15] were nucleated in the remaining highly strained ligaments. The spacing of the small dimples indicates that they were nucleated by the small, unidentified precipitates shown in Fig. 1b.

The deep, well-defined dimples and tear ridges indicate that their formation involved extensive homogeneous plastic deformation. This is further evidenced by the interwoven patterns observed on the dimple and tear ridge walls. Such deformation markings are indicative of features found on materials strained well into the plastic regime.

In the irradiated condition, the fracture mechanisms were markedly different. The 2.4-dpa fracture surface (Fig. 7a) exhibited a few shallow, ill-defined dimples surrounded by a rather faceted morphology. With higher neutron exposures, the fracture surface took on a highly faceted, crystallographic ap-

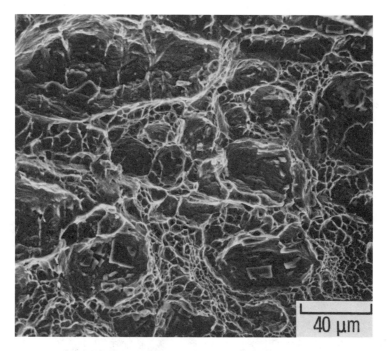

FIG. 6—*Typical fractograph for unirradiated Alloy A-286.*

pearance, reminiscent of channel fracture [16,17]. No evidence of the elongated tear ridges was observed in any irradiated specimen, which indicates that the stringers do not influence the postirradiation fracture behavior.

Channel fracture has been observed in neutron-irradiation Type 304 stainless steel at 371°C [16,17] and in unirradiated Inconel X-750 at 649 and 704°C [18,19]. This fracture mechanism results when all dislocation activity is channeled through narrow deformation zones. The severe dislocation channeling ultimately initiates localized separation along these planar slip bands. In the Type 304 stainless steel, lead dislocations sweep out the fine defects generated by the irradiation displacement damage, creating defect-free channels that are substantially weaker than the surrounding matrix. All subsequent dislocation activity is then channeled through these localized planar regions, and eventually shear cracks nucleate and propagate along the weakened channels. In all likelihood, this heterogeneous deformation model also results in the channel fracture observed in highly irradiated Alloy A-286. Stereo fractography provided additional support for the proposed channel fracture mechanism, because the individual facets were found to be steeply inclined relative to the overall crack plane, a condition that is indicative of shear fracture.

At a neutron exposure of 2.4 dpa, a fracture mechanism transition from

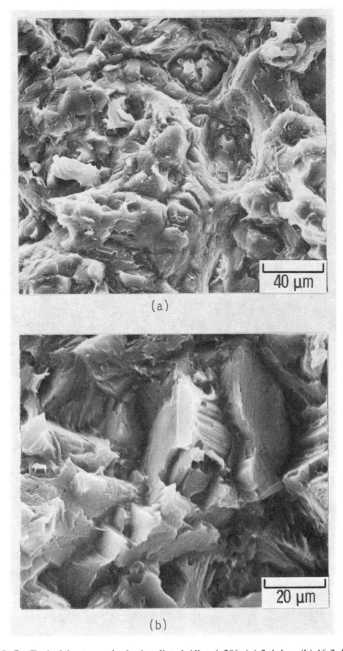

FIG. 7—*Typical fractographs for irradiated Alloy A-286.* (a) *2.4 dpa.* (b) *16.2 dpa.*

ductile microvoid coalescence to channel fracture occurred as evidenced by the ill-defined dimples and facets. The lower irradiation dose results in fewer lattice defects and a lower strength matrix. Straining still produces defect-free channels, but they are only slightly weaker than the surrounding regions, so that limited matrix dislocation mechanisms remain active. As a result, shallow microvoids are initiated but their growth is stunted by the restricted homogeneous slip capabilities.

The severe reduction in fracture toughness with neutron irradiation is attributed to the extensive planar slip and the concomitant channel fracture mechanism. As the channel fracture becomes better defined with increasing exposure, the fracture resistance continues to drop but at a decreasing rate. This suggests that a saturation in toughness degradation occurs when crisp channel fracture dominates the fracture surface. Irradiation embrittlement in Type 304 stainless steel was, in fact, found to saturate at fluences above 3×10^{22} n/cm^2 where the channel fracture mechanism was dominant [16]. Therefore irradiation exposures above 16 dpa are not expected to reduce the K_{Ic} response for Alloy A-286 below 30 to 35 MPa$\sqrt{m}$.

Conclusions

The fracture toughness response of Alloy A-286 irradiated to 2.4, 7.8, and 16.2 dpa (7.7×10^{21}, 2.5×10^{22}, and 5.2×10^{22} n/cm^2 respectively) was characterized at 24 and 427°C using linear-elastic K_{Ic} and elastic-plastic J_{Ic} fracture mechanics techniques. The results of this study are summarized below:

1. The elevated temperature fracture resistance was found to decrease continuously with increasing irradiation damage. In the unirradiated and low neutron exposure conditions, the material failed in an elastic-plastic mode, and the equivalent K_{Jc}, determined from the J_{Ic} fracture toughness, was reduced from 129 to 90 MPa$\sqrt{m}$ due to the irradiation to 2.4 dpa. Brittle fracture occurred at higher exposures, and K_{Ic}-values continued to fall to approximately 35 MPa$\sqrt{m}$ at 16.2 dpa.

2. Room-temperature fracture toughness also decreased with increasing irradiation, paralleling the 427°C results. However, the K_{Jc} or K_{Ic} toughness values at 24°C were consistently 40 to 60% higher than the 427°C values.

3. In the unirradiated condition, the fracture surface exhibited well-defined microvoid coalescence and elongated tear ridges. These morphologies indicate that extensive homogeneous plastic deformation accompanied the tearing process, accounting for the superior fracture resistance.

4. The severe toughness degradation at a neutron exposure of 16.2 dpa was associated with a channel fracture mechanism. In the highly irradiated material, all dislocation activity was localized in planar slip bands, and cracking

eventually initiated and propagated along these channels. The extensive planar slip and accompanying channel fracture resulted in the poor fracture resistance displayed after intense irradiation damage.

5. At 2.4 dpa, a fracture mechanism transition from ductile dimple rupture to channel fracture occurred. Both homogeneous and heterogeneous dislocation mechanisms were active in this regime, which accounts for the modest 30% reduction in fracture toughness observed at this neutron exposure.

Acknowledgments

This paper is based on work performed under U.S. Department of Energy Contract DE-AC06-76FF-02170 with the Westinghouse Hanford Company, a subsidiary of Westinghouse Electric Corporation. The author wishes to gratefully acknowledge Dr. L. D. Blackburn for helpful discussions and Mr. D. J. Criswell for performing the postirradiation fracture toughness tests. Appreciation is also extended to Mr. L. A. James, who planned and prepared the EBR-II fatigue and fracture irradiation experiment.

References

[*1*] Mills, W. J., *Journal of Engineering Materials and Technology*, Vol. 100, 1978, pp. 195–199.
[*2*] Reed, R. P., Tobler, R. L., and Mikesell, R. P., *Advances in Cryogenic Engineering*, K. D. Timmerhaus, R. P. Reed, and A. F. Clark, Eds., Plenum, New York, 1977, pp. 68–79.
[*3*] Wells, J. M., Logsdon, W. A., Kossowsky, R., and Daniel, M. R., "Structural Materials for Cryogenic Application," Research Report 75-904-CRYMT-R2, Westinghouse Research Laboratories, Pittsburgh Pa., Oct. 1975.
[*4*] Steichen, J. M., "The Effect of Strain Rate, Thermal Aging and Irradiation on the Tensile Properties of A-286," HEDL-TME 75-81, Westinghouse Hanford Co., Richland, Wash., March 1976.
[*5*] Doran, D. G. and Graves, N. J. in *Irradiation Effects on the Microstructure and Properties of Metals, ASTM STP 611*, American Society for Testing and Materials, 1976, pp. 463–482.
[*6*] Mills, W. J., James, L. A., and Williams, J. A., *Journal of Testing and Evaluation*, Vol. 5, No. 6, 1977, pp. 446–451.
[*7*] Newman, J. C., Jr., in *Fracture Analysis, ASTM STP 560*, American Society for Testing and Materials, 1974, pp. 105–120.
[*8*] Newman, J. C., Jr., "Stress-Intensity Factors and Crack Opening Displacements for Round Compact Specimens," NASA Technical Memorandum 80174, NASA Langley Research Center, Hampton, Va., Oct. 1979.
[*9*] Ashbaugh, N. E., "Material Evaluation: Part I—Mechanical Property Testing and Materials Evaluation and Modeling," Report AFML-TR-79-4127, Air Force Materials Laboratory, Wright-Patterson Air Force Base, Ohio, Sept. 1979.
[*10*] Saxena, A. and Hudak, S. J., Jr., *International Journal of Fracture*, Vol. 14, No. 5, Oct. 1978.
[*11*] Paris, P. C., Tada, H., Zahoor, Z., and Ernst, H. in *Elastic-Plastic Fracture, ASTM STP 668*, American Society for Testing and Materials, 1979, pp. 5–36.
[*12*] Hutchinson, J. W. and Paris, P. C. in *Elastic-Plastic Fracture, ASTM STP 668*, American Society for Testing and Materials, 1979, pp. 37–64.
[*13*] Begley, J. A. and Landes, J. D. in *Fracture Toughness, ASTM STP 514*, American Society for Testing and Materials, 1974, pp. 1–20.
[*14*] Rogers, H. C., *Ductility*, American Society for Metals, Metals Park, Ohio, 1968, pp. 31–61.

[15] Cox, T. B. and Low, J. R., Jr., *Metallurgical Transactions*, Vol. 5, 1974, pp. 1457–1470.
[16] Hunter, C. W., Fish, R. L., and Holmes, J. J., *Transactions of the American Nuclear Society*, Vol. 15, 1972, pp. 254–255.
[17] Fish, R. L., *Nuclear Technology*, Vol. 31, 1976, pp. 85–95.
[18] Mills, W. J., *Metallurgical Transactions*, Vol. 11A, 1980, pp. 1039–1047.
[19] Mills, W. J. in *Fractography and Materials Science, ASTM STP 733*, American Society for Testing and Materials, 1981, pp. 98–114.

A. Saxena,[1] *T. T. Shih,*[1] *and H. A. Ernst*[1]

Wide Range Creep Crack Growth Rate Behavior of A470 Class 8 (Cr-Mo-V) Steel

REFERENCE: Saxena, A., Shih, T. T., and Ernst, H. A., **"Wide Range Creep Crack Growth Rate Behavior of A470 Class 8 (Cr-Mo-V) Steel,"** *Fracture Mechanics: Fifteenth Symposium, ASTM STP 833*, R. J. Sanford, Ed., American Society for Testing and Materials, Philadelphia, 1984, pp. 516–531.

ABSTRACT: Wide range steady-state creep crack growth rate behavior of an A470 Class 8 steel was characterized at 538°C (1000°F) using constant load and constant deflection rate methods of testing. Both methods yielded mutually consistent results. Crack growth rates in the range of 10^{-4} to 10^{-1} mm/h were successfully correlated with the energy rate line integral, C^*. Results are discussed in light of the recent analytical studies that have appeared in the literature for growing cracks in creeping materials. Various methods of determining C^* for CT specimens are also discussed.

KEY WORDS: creep, fracture, Cr-Mo-V steel, C^*-integral

Creep deformation and crack growth are important design considerations for several components that operate in high temperature environments. Recently, several experimental studies have shown that the energy rate line integral, C^*, first introduced by Landes and Begley and independently by Nikbin et al, is the leading candidate parameter for characterizing steady-state creep crack growth behavior in structural alloys [1–3].

In this study, the steady-state creep crack growth rate behavior of an ASTM A470 Class 8 steel (Cr-Mo-V) is characterized over a wide range of crack growth rates and C^*-values. Two test techniques, the constant displacement rate method and the constant load method, were utilized to obtain creep crack growth rate data over a wide range. The advantages and limitations of the two techniques are discussed. The relevance of C^* for characterizing steady-state creep crack growth behavior is discussed in light of the several analytical

[1]Materials Engineering Department, Westinghouse R&D Center, Pittsburgh, Pa. 15235. Dr. Shih is now with the Steam Turbine-Generator Division.

studies that have recently appeared in the literature.[2] Different methods of calculating C^* for the CT test specimen are also compared and discussed.

Experimental Procedure

Material and Specimens

Standard 25.4-mm (1-in.)-thick compact type (CT) specimens were machined from a large steel forging manufactured in accordance with ASTM Specification for Vacuum-Treated Carbon and Alloy Steel Forgings for Turbine Rotors and Shafts (A470, Class 8). The notches in these specimens were oriented along the radial direction of the forging. The chemical composition and the tensile properties of the material are shown in Tables 1 and 2, respectively.

Limited creep deformation testing was also conducted at 538°C (1000°F). The secondary creep rate as a function of applied stress is shown in Fig. 1 along with some previous data on the same material [4].

Creep Crack Growth Rate Testing

Two types of test techniques were used to obtain creep crack growth rate behavior: (1) constant displacement rate method and (2) constant load method. These are briefly described below.

Constant Displacement Rate Method—This technique was first developed by Landes and Begley [1] for obtaining creep crack growth behavior of diskalloy and was subsequently used by Saxena [3] on 304 stainless steel. In this technique, the tests are conducted under constant ram deflection rate (approximately constant load-line deflection rate) using a servohydraulic test system. The specimens are heated in a three-zone resistance furnace with a temperature control capability of 2°C in the test section of the specimen. A d-c electrical potential system is used to measure crack length. The details of this system are described elsewhere [5]. Typically, three specimens are subjected to constant ram deflection rates between 0.025 to 0.15 mm/h (0.001 to 0.006 in./h). During the test, specimen load, load-line deflection, and crack length are monitored continuously on a strip chart recorder.

In the constant displacement rate tests conducted in this study, a slight variation from this procedure was followed. The deflection rates in the three specimens were changed in steps after crack extensions of approximately 5 mm (0.2 in.) according to a predetermined scheme shown in Fig. 2. This technique optimized the number of data points that were obtained from three tests when the

---

[2]Steady-state creep crack growth rate is defined as follows. Consider a coordinate system with the origin at the crack tip and moving with the growing crack. Ideal steady-state conditions exist when the stresses and strain rates at any distance ahead of the crack tip are independent of time. Quasi-steady-state is defined when the stress and strain rates vary slowly in comparison to the crack growth rate da/dt.

TABLE 1—*Chemical composition (weight percent) of A470 Class 8 steel.*

C	Mn	Si	Cr	Mo	V	P	S	Ni	Cu	Al	Sn	As
0.32	0.78	0.28	1.20	1.18	0.23	0.012	0.011	0.13	0.05	0.005	0.010	0.008

TABLE 2—*Tensile properties of A470 Class 8 steel.*

Test Temperature		0.2% Yield Strength		Ultimate Strength		% Elongation (5 cm gage length)	Reduction of Area, %
°C	°F	MPa	ksi	MPa	ksi		
24	(75)	623	(90.4)	775.6	(112.5)	14.2	39
427	(800)	515.7	(74.8)	624.6	(90.6)	14.2	53
538	(1000)	464	(67.3)	522.6	(75.8)	17.5	75

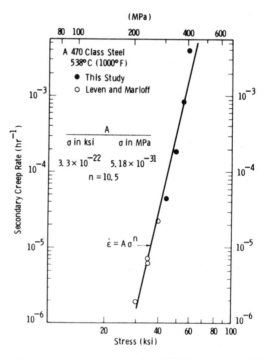

FIG. 1—*Secondary creep rate versus stress for A470 Class 8 steel at 538°C (1000°F).*

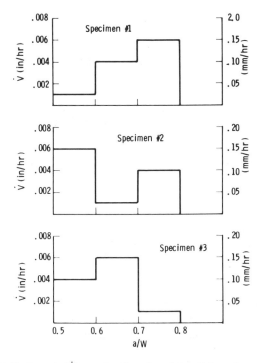

FIG. 2—*Deflection rate ($\dot{V}$) as a function of crack length in various specimens.*

data were reduced by the multiple-specimen graphical technique described later in this section.

Constant Load Tests—These tests were conducted using deadweight-type creep machines. The specimen was heated in a three-zone resistance furnace to the test temperature and then the load was applied by adding weights on the loading pan. The changes in load-line deflection were measured with the help of a dial gage attached to the setup. The deflection from the dial gage was periodically recorded and plotted with time. Typical deflection versus time records for some of the tests are shown in Fig. 3. The deflection versus time behavior exhibited a primary creep region and a steady-state creep region. When substantial amount of deflection had accumulated on the specimen, it was unloaded and subsequently fatigued to failure at ambient temperature. The amount of crack extension due to creep was measured directly on the fractured surface at five points across the thickness of the specimen. Thus a through-the-thickness average crack extension was obtained. See Table 3 for test results.

Data Reduction

The data were plotted as da/dt as a function of C^*. In the deflection rate controlled tests, crack growth rate da/dt was simply obtained by calculating the

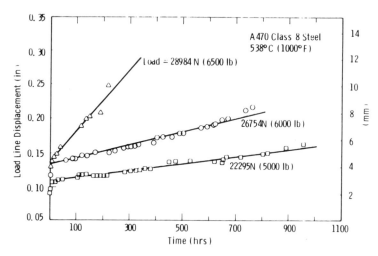

FIG. 3—*Load-line displacement as a function of time for constant-load creep crack growth tests.*

slope of the crack length versus time curve at various crack lengths. Crack growth rate data were obtained excluding crack extensions of approximately 0.5 mm following a step change in deflection rate. This was considered as the transient region during which the specimen response (the load) readjusts to the new deflection rate. In the constant load tests, da/dt was obtained by dividing the crack extension due to creep by the total test time. The transient time which immediately follows loading is also included in calculating the rate. From Fig. 3, it appears that the transition time (time up to the onset of constant deflection rate in the specimen) is negligible compared to the overall test duration; hence inclusion of that in calculating da/dt does not influence the value of da/dt significantly.

Three techniques were used for estimating C^* for compact specimens tested in this study. They were (1) the multiple-specimen graphical method, (2) the fully plastic J-solutions, and (3) a method utilizing the load-deflection rate behavior called the area technique. These methods are described below.

Multiple-Specimen Graphical Method—This method of obtaining C^* is based on the energy rate interpretation [1] of C^* given by

$$C^* = -\frac{dU^*}{da} \tag{1}$$

where U^* is power or the time rate of energy per unit thickness applied at the loading points. A detailed description of this method is given elsewhere [1,3].

Fully Plastic J-Solutions—J-solutions for compact specimens under fully plastic conditions of loading are available from the work of Kumar and Shih [6]. By analogy, these solutions can also be used to calculate C^* when secondary

TABLE 3—Results from constant-load creep crack growth rate tests.[a]

Load, N	Crack Length, mm	Displacement Rate, mm/h	Crack Extension, mm	Test Duration, h	da/dt, mm/h	C^*, $J/m^2/h$
3.56×10^4	16.69	5.08×10^{-5}	0.787	2035	3.86×10^{-4}	4.87
2.67×10^4	26.34	2.38×10^{-3}	1.702	760	2.24×10^{-3}	229.8
2.22×10^4	26.44	1.27×10^{-3}	1.346	950	1.42×10^{-3}	102.1
2.00×10^4	26.42	6.6×10^{-5}	1.727	4203	4.11×10^{-4}	4.78
1.78×10^4	26.54	3.81×10^{-5}	1.499	4847	3.09×10^{-4}	2.45
2.89×10^4	26.00	1.09×10^{-2}	0.965	220	4.38×10^{-3}	1.13×10^3

[a] 1 N = 0.224 lb, 1 mm = 0.03937 in., 1 $J/m^2/h$ = 0.00571 in.-lb/in.^{2}/h, B = 25.4 mm, and W = 50.8 mm.

creep conditions dominate in the entire specimen. The relevant equation for calculating C^* by this technique is

$$C^* = A(W-a)h_1(a/W, n)\,[P/(1.455\alpha(W-a))]^{n+1} \tag{2}$$

where W is specimen width, P is load, a is crack length, and A and n are material constants in Eq 3 valid in the secondary creep regime.

$$\dot{\varepsilon} = A\sigma^n \tag{3}$$

where $\dot{\varepsilon}$ is strain rate, σ is stress, and $h_1(a/W, n)$ is a dimensionless function shown in Fig. 4 for selected values of n. The term $\alpha(a/W)$ is given by

$$\alpha = \left[\left(\frac{2a}{W-a}\right)^2 + 2\left(\frac{2a}{W-a}\right) + 2\right]^{1/2} - \left(\frac{2a}{W-a} + 1\right) \tag{4}$$

Area Technique—Shih [7] and Smith and Webster [8] have independently suggested methods for using the measured loads and deflection rates at the loading points for estimating C^* for CT specimens. This technique is analogous to the area technique commonly used to determine J-integral [9, 10]. Here, the area under the load-deflection rate curve is taken instead of the area under the load-deflection plot used for calculating J. Shih's method, following that of Rice et al [9], models CT specimens as a pure bending case and applies only for deeply cracked specimens ($a/W \rightarrow 1.0$). Smith and Webster's method follows that of Merkle and Corten [10], which considers tension as well as bending and

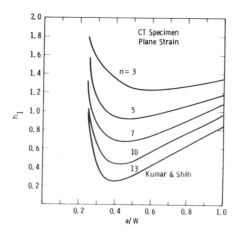

FIG. 4—*Dimensionless function* h_1 *versus a/w for calculating* C^* *in compact specimens.*

thus can be used for $a/W \geq 0.4$. Since the latter is more general, it is adopted in this study. The relevant equations are given below.

$$C^* = \frac{P\dot{V}}{B(W-a)} \frac{n}{n+1} \eta(a/W) \tag{5}$$

where B is the thickness of the specimen, $\dot{V}$ is the deflection rate of the load point, n is a material constant defined in Eq 3, and $\eta(a/W)$ is a simple function of a/W given in Eq 6 for $a/W \geq 0.4$ [11].

$$\eta(a/W) = 2[1.261 - 0.261(a/W)] \tag{6}$$

The term $P\dot{V}(n/n+1)$ in Eq 5 is the area under the load-deflection rate plot.

Results and Discussion

In this section the creep crack growth results are presented and discussed. The various methods of calculating C^* are evaluated and the two techniques used for obtaining the data are compared with each other.

Creep Crack Growth Behavior

The da/dt versus C^* data for A470 Class 8 steel obtained from the constant load and constant deflection rate tests are plotted in Fig. 5. The justification for the use of C^* for plotting da/dt is provided subsequently. Equation 5 has been used to calculate C^* for both types of tests. The C^*-values for the constant deflection rate tests were also calculated using the multiple-specimen technique for comparison. The data from the two data reduction techniques have been plotted using different symbols. All the data appear to follow the same general trend with some scatter.

Good correlation is obtained between da/dt and C^* over a wide range of C^*-values (five orders of magnitude) and crack growth rates (three orders of magnitude). Similar correlation between da/dt and C^*-values was also reported by Shih for 304 stainless steel [7]. The creep crack growth rate behavior in the range of da/dt values of 2.5×10^{-4} mm/h (10^{-5} in./h) reported in this study is of considerable practical significance. It is needed for accurately predicting life of structures designed for long life. The da/dt versus C^* relationship for A470 Class 8 steel at 538°C (1000°F) is given by

$$da/dt = 10^{-4}(C^*)^{0.67} \tag{7}$$

where da/dt is in mm/h and C^* is in J/m²/h. Or,

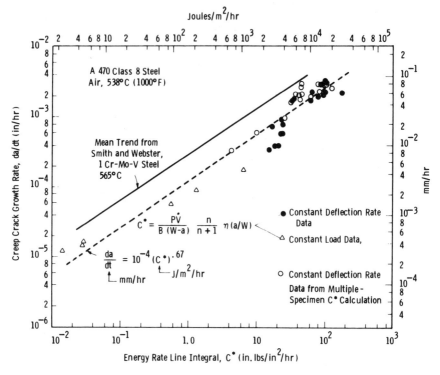

FIG. 5—*Creep growth rate as a function of C* for A470 Class 8 steel.*

$$da/dt = 1.3 \times 10^{-4} (C*)^{0.67} \tag{7a}$$

where da/dt is in in./h and $C*$ is in in.-lb/in.^{2}/h.

The creep crack growth behavior at a higher temperature of 565°C (1050°F) for Cr-Mo-V steel from the work of Smith and Webster [12] is also plotted in Fig. 5 for comparison. There appears to be a significant influence of temperature on the creep crack growth behavior in this material. It should be pointed out, however, that there could be differences in the microstructure between the steels used in Ref 12 and this study and the difference in da/dt versus $C*$ may be caused by that.

Comparison of Methods for Calculating C*

Table 4 lists some representative values of $C*$ obtained from the three procedures described in the previous section of the paper. The discrepancies between the multiple-specimen graphical method and the area technique range between factors of 1.04 to 2. In this range of discrepancies, no quantitative conclusions about the accuracy of the area technique are possible. The $C*$

TABLE 4—Comparison of C*-values calculated from the three methods.[a]

Specimen	Load, N	Crack Length, mm	Displacement Rate, mm/h	C^*, J/m^2/h		
				Multiple-Specimen Graphical Method	Fully Plastic J-solutions	Area Technique
1	2.4×10^4	30.5	0.15	1.63×10^4	39.7	1.57×10^4
	2.23×10^4	31.8	0.15	1.63×10^4	87.6	1.54×10^4
	1.96×10^4	33.0	0.15	1.29×10^4	114.6	1.45×10^4
2	4.24×10^4	27.94	0.10	7.95×10^3	1.4×10^3	1.66×10^4
	2.59×10^4	32.4	0.05	5.2×10^3	9.24×10^2	6.27×10^3
	2.45×10^4	33.02	0.05	7.89×10^3	1.49×10^3	6.13×10^3
3	4.87×10^4	27.94	0.15	1.63×10^4	6.8×10^3	3.01×10^4
	2.59×10^4	34.3	0.10	7.6×10^3	1.75×10^4	1.38×10^4
	2.27×10^4	35.56	0.10	1.09×10^4	2.83×10^4	1.31×10^4

[a] 1 N = 0.224 lb, 1 mm = 0.03937 in., 1 J/m^2/h = 0.00571 in.-lb/in.^{2}/h.

calculations from the fully plastic J-solutions are different by up to two orders of magnitude compared to the C^* values from the multiple-specimen graphical method and the area technique. The exact reason for this discrepancy is unclear at this point and will be the subject of future studies. One reason may be that in order for these solutions to be applicable, the entire specimen should be in secondary stage creep. This condition may not have been met by the tests conducted in this study. Because of these large discrepancies, the data from this method were not used in the plot shown in Fig. 5.

Both the multiple-specimen graphical method and the area technique utilize two measured quantities to calculate C^*, namely the load and deflection rate. The fully plastic J-solutions require only the load and some material constants to calculate C^*. Further, in the use of J-solutions, an assumption has to be made on whether the specimen is under plane strain or under plane stress. The C^*-values calculated assuming plane stress will be 33.9 times larger than those calculated for plane strain. The values reported in Table 4 are for plane strain. The specimens are probably somewhere in between plane stress and plane strain; hence calculations based on either of the two states of stress will be in error.

A power law relationship between load and steady-state deflection rate of the type shown in Eq 8 is assumed in deriving the expression for calculating C^* in the area technique [7,8].

$$\dot{V} = CP^n \qquad (8)$$

where C is a material constant for a given crack length and n is the exponent in Eq 3. When steady-state deflection rate is plotted against the applied load for a series of specimens with nearly identical crack lengths (Fig. 6) the value of the

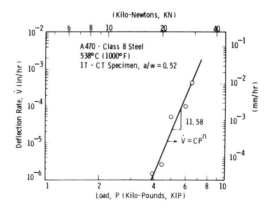

FIG. 6—Load versus steady-state deflection rate ($\dot{V}$) for a 1T-CT specimen at a/w = 0.52; temperature = 538°C.

exponent n was found to be 11.58 as compared to 10.5 which was fitted through the stress versus strain rate data in Fig. 1. Both these values of n are the best estimates derived from linear regression of the respective data sets. No significance can be attached to the small difference between the two values. Hence the limited data available covering only one crack length appear to support the validity of Eq 8.

The value of C^* in the area technique is not sensitive to the value of n. In our calculations we have assumed $n = 11$, which is the average of the two values obtained from the data shown in Figs. 1 and 6. Use of either 10.5 or 11.58 would not have yielded significantly different results.

From comparisons made between the three techniques, it appears that the area technique has the most promise in the future for determining the values of C^*. More work is needed for its further verification.

Comparison of Test Techniques

To obtain creep crack growth rate data over a wide range of growth rates of C^*-values, it is necessary to use both the constant deflection rate and the constant load techniques. The advantages and disadvantages of both techniques are discussed below.

The constant deflection rate method is useful at relatively higher crack growth rates of $da/dt > 5 \times 10^{-3}$ mm/h (2×10^{-4} in./h). It allows data to be obtained over a crack extension of 10 to 15 mm (0.4 to 0.6 in.); thus several data points can be obtained from each test. This technique is not practical for obtaining data at lower crack growth rates, however, because it requires an expensive servohydraulic machine for a long time. Also, under nearly static conditions for long periods of time, the performance of servohydraulic machines can be erratic.

The constant load test is conducted on a relatively inexpensive creep test machine which also provides excellent load stability over a long period of time (the longest test duration in this study was seven months). This setup is therefore very suitable for obtaining data at $da/dt < 5 \times 10^{-3}$ mm/h (2×10^{-4} in./h). The major disadvantage with this technique is that only data over relatively short crack extensions can be obtained. This is because C^* increases rapidly with crack extension in this configuration, causing the crack growth to become unstable. At higher loads or da/dt values, the amount of crack extension prior to instability can be so small that the technique may become impractical [1].

More work is needed in order to optimize test techniques for obtaining creep crack growth rate data. In the constant displacement rate tests, data over relatively large crack extensions can be obtained; however, the data are clustered around the same da/dt and C^* values. Therefore, if a technique for gradually increasing deflection rate (instead of constant deflection rate) with crack extension can be developed, it will significantly reduce test times and the number of specimens required for obtaining data.

Evaluation of C* for Characterizing Creep Crack Growth Rate

For stationary cracks in creeping materials, it has been shown that a C*-controlled HRR-type field (after Hutchinson and Rice and Rosengren [13,14]) develops at the crack tip [15,16]. This has previously been used as a basis for justifying the use of C* for characterizing creep crack growth rate [1,3]. This argument is not strong. We cannot be sure that for growing cracks, the stress and strain rate fields are controlled by C* and that the HRR singularity is valid. Recently, several analytical studies of growing cracks in creeping materials have appeared in the literature [17-19]. It will be of considerable interest to discuss our experimental results in light of the conclusions of these studies. First, the significant conclusions of the analytical studies are briefly reviewed.

Hui and Riedel [17] have shown that for growing cracks in creeping material, a new type of singular stress and strain rate field develops which is considerably different from the HRR field. This new singularity is independent of the load parameter and is dependent only on the current crack growth rate da/dt (or $\dot{a}$) and material parameters (Eq 9). Its derivation is based on the contention that the elastic strain rate for growing cracks cannot be ignored in comparison to the creep strain rate in a region approaching the crack tip ($r \to 0$). Consideration of these elastic strain rate terms results in the new singularity

$$\sigma_{ij} \propto \left[\frac{\dot{a}(t)}{AEr} \right]^{1/(n-1)} \tag{9}$$

where σ_{ij} is the stress tensor at the crack tip, E is the elastic modulus, r_0 is the distance from the crack tip, and A and n are material constants defined in Eq 3.

Far away from the crack tip, it is reasonable to assume that the creep strain rates dominate over the elastic strain rates for cracked bodies undergoing secondary creep. Hence an expression derived by Riedel and Wagner [18] to characterize the stress distribution under steady-state conditions becomes applicable. The equation is

$$\sigma_{ij} = \left(\frac{EC^*}{\dot{a}} \right)^{1/2} \Sigma_{ij}(R, \theta) \tag{10}$$

where R is the distance from the crack tip normalized with respect to the size (r_0) of the region of the dominance of the new singularity, θ is the angular location of the element under consideration, and Σ_{ij} is a function of R and θ. Further, by equating Eqs 9 and 10 at r_0, an approximate value of r_0 has been obtained [18]:

$$r_0 = \frac{1}{A} \left(\frac{\dot{a}}{E} \right)^{(n+1)/2} \left(\frac{1}{C^*} \right)^{(n-1)/2} \tag{11}$$

If r_0 is small compared with the crack length and the other pertinent dimensions of the specimen such as the ligament width, C^* can be expected to characterize the steady-state crack growth behavior because it characterizes the stress and strain rates in the bulk of the region ahead of the crack tip (Eq 10). An order of magnitude calculation of r_0 for one of the tests conducted in this study yielded a value of 10^{-24} mm using a value of 1.69×10^5 MPa (24.5×10^3 ksi) for the E-value. This value is certainly negligible compared with the crack length. It thus appears that the small region in which the new singularity dominates is engulfed in a much larger zone in which C^* characterizes the stress and strain rate behavior. Hence da/dt is expected to correlate with C^*, which is in agreement with the experimental results of this study.

McMeeking and Leckie [19] have used an incremental crack growth model to justify the use of C^* for characterizing creep crack growth rate. In their model they assume that the crack growth occurs in steps and that following each step of crack growth there is a transient period during which C^*-controlled stress and strain rate fields are re-established. Between two steps of crack growth, the crack is considered stationary. They also show that the transient time for re-establishing the C^*-controlled field is negligible in comparison to the total time that elapses between two steps of crack growth. Hence the macroscopic crack growth rate is expected to correlate with C^* because it completely characterizes the steady-state crack tip conditions when the crack is stationary.

From the analytical studies of growing cracks in creeping materials discussed in the preceding paragraphs, it can be concluded that C^* is a good candidate parameter for characterizing steady-state crack growth behavior. Our experimental results over a wide range of C^* and da/dt values provide some experimental support to these conclusions. However, more work on different materials, specimen geometries, and temperatures should be performed in order to further substantiate these conclusions and establish the limitations on the use of C^*. Some specific concerns are discussed below.

When large amounts of crack extensions occur, there are questions about the influence of prior crack growth history on the crack growth rate and on the load-deflection rate behavior of the specimen. There is some experimental evidence in the data obtained in the present study that these history effects may not be important. In our deflection rate controlled tests, the crack extensions were large (10 to 15 mm) and C^* appeared to correlate well with da/dt over the entire crack extension range. Despite these data, a more in-depth experimental study to further investigate the influence of prior growth history is needed. Another important issue is the extent of creep deformation in the specimen required before the use of C^* can be justified for characterizing da/dt. This is particularly important for creep-resistant superalloys or for steels such as the one used in this study at lower temperatures than used here. The transient time required for C^*-controlled field to establish may be large, and significant crack extension may occur under transient conditions. Some quantitative criteria to determine when C^* can be used should be established. These issues will be addressed in detail in forthcoming papers [20, 21].

Summary and Conclusions

Wide range steady-state creep crack growth rate behavior of A470 Class 8 steels was characterized at 538°C (1000°F) using constant load and constant deflection rate methods of testing. The following conclusions can be derived from the results of these tests:

1. The constant load and the constant deflection rate methods for creep crack growth rate testing yield mutually consistent results. The constant load method is more suitable for testing at da/dt values less than 5×10^{-3} mm/h (2×10^{-4} in./h), while the constant deflection rate method is more suitable for testing at da/dt values larger than 5×10^{-3} mm/h (2×10^{-4} in./h).

2. Wide range creep crack growth rate behavior in A470 Class 8 steels correlates well with C^* and can be represented by a simple power law relationship (Eq 7).

3. There is good analytical justification for using C^* for characterizing the steady-state creep crack growth rate behavior.

4. The C^*-values calculated from the area technique were comparable to those calculated by the multiple-specimen graphical technique. The fully plastic J-solution for CT specimen was not found suitable for calculating C^*-values in A470 Class 8 steels at 538°C (1000°F).

Acknowledgments

The authors are indebted to P. J. Barsotti, R. B. Hewlett, and C. Fox for conducting the creep crack growth rate tests. Financial support was provided by the Steam Turbine-Generator Division of Westinghouse Electric Company. The constant encouragement of V. P. Swaminathan and also fruitful discussions with him are gratefully acknowledged.

References

[1] Landes, J. D. and Begley, J. A. in *Mechanics of Crack Growth, ASTM STP 590*, American Society for Testing and Materials, 1976, pp. 128–148.

[2] Nikbin, K. M., Webster, G. A., and Turner, C. E. in *Cracks and Fracture, ASTM STP 601*, American Society for Testing and Materials, 1976, pp. 47–62.

[3] Saxena, A. in *Fracture Mechanics: Twelfth Conference, ASTM STP 700*, American Society for Testing and Materials, 1980, pp. 131–151.

[4] Leven, M. M. and Marloff, R. H., unpublished data, Westinghouse R & D Center, Pittsburgh, March 1977.

[5] Saxena, A., *Engineering Fracture Mechanics*, Vol. 13, 1980, pp. 741–750.

[6] Kumar, V. and Shih, C. F. in *Fracture Mechanics: Twelfth Conference, ASTM STP 700*, American Society for Testing and Materials, 1980, pp. 406–438.

[7] Shih, T. T. in *Fracture Mechanics: Fourteenth Symposium—Volume II: Testing and Ap-

plications, ASTM STP 791, J. C. Lewis and B. Sines eds., American Society for Testing and Materials, 1983, pp. II-232–II-247.

[8] Smith, D. J. and Webster, G. A. in *Elastic-Plastic Fracture: Second Symposium, Volume I—Inelastic Crack Analysis, ASTM STP 803*, American Society for Testing and Materials, 1983, pp. I-654–I-674.

[9] Rice, J. R., Paris, P. C., and Merkle, J. G. in *Flaw Growth and Fracture Toughness Testing, ASTM STP 536*, American Society for Testing and Materials, 1972, pp. 231–245.

[10] Merkle, J. G. and Corten, H. T., *Journal of Pressure Vessel Technology, Transactions of ASME*, Vol. 96, 1974, pp. 286–292.

[11] Clarke, G. A. in *Fracture Mechanics: Thirteenth Conference, ASTM STP 743*, American Society for Testing and Materials, 1981, pp. 553–575.

[12] Smith, D. J. and Webster, G. A., *Journal of Strain Analysis*, Vol. 16, No. 2, 1981, pp. 137–143.

[13] Hutchinson, J. W., *Journal of Mechanics and Physics of Solids*, Vol. 16, 1968, pp. 13–31.

[14] Rice, J. R. and Rosengren, G. F., *Journal of Mechanics and Physics of Solids*, Vol. 16, 1968, pp. 1–12.

[15] Goldman, N. L. and Hutchinson, J. W., *International Journal of Solids and Structure*, Vol. 11, 1975, pp. 575–591.

[16] Riedel, H. and Rice, J. R. in *Fracture Mechanics: Twelfth Conference, ASTM STP 700*, American Society for Testing and Materials, 1980, pp. 112–130.

[17] Hui, C. Y. and Riedel, H., *International Journal of Fracture*, Vol. 17, No. 4, 1981, pp. 409–425.

[18] Riedel, H. and Wagner, W. in *Advances in Fracture Research*, 5th International Conference on Fracture (ICF-5), Cannes, March–April, 1981, pp. 683–690.

[19] McMeeking, R. M. and Leckie, F. A. in *Advances in Fracture Research*, 5th International Conference on Fracture (ICF-5), Cannes, March–April 1981, pp. 699–704.

[20] Saxena, A., Ernst, H. A., and Landes, J. D., *International Journal of Fracture*, Vol. 23, 1983, pp. 245–257.

[21] Saxena, A., "Creep Crack Growth under Non-Steady-State Conditions," paper to be presented at 17th National Symposium on Fracture Mechanics, Albany, N.Y., Aug. 1984, sponsored by ASTM.

Elasto-Plastic Fracture Mechanics

M. G. Vassilaros[1] and E. M. Hackett[1]

J-Integral R-Curve Testing of High Strength Steels Utilizing the Direct-Current Potential Drop Method

REFERENCE: Vassilaros, M. G. and Hackett, E. M., *"J-Integral R-Curve Testing of High Strength Steels Utilizing the Direct-Current Potential Drop Method,"* *Fracture Mechanics: Fifteenth Symposium, ASTM STP 833*, R. J. Sanford, Ed., American Society for Testing and Materials, Philadelphia, 1984, pp. 535–552.

ABSTRACT: J-integral R-curve tests were performed on 1T compact specimens of 3Ni and 5Ni steels under both quasi-static and rapid loading. A direct-current potential drop technique was developed for the crack extension measurements. This technique accounted for specimen plasticity and correlated well with crack extension measurements determined from elastic compliance. Potential drop proved to be a valid technique for the determination of R-curves under rapid loading. Increasing the loading rate elevated the J_I-R curves for both the 3Ni and 5Ni steels with respect to the quasi-static case.

KEY WORDS: direct-current potential drop (DCPD), J-integral, elastic compliance, multispecimen tests, rapid loading, compact specimens, computer data acquisition, high strength steels

Over the past decade the application of the J-integral concept to elastic plastic fracture in metallic materials has gained broad acceptance. Early work by Landes and Begley [1] illustrated the validity and usefulness of the elastic plastic fracture toughness parameter (J_{Ic}) in the evaluation of structural alloys. Subsequent work led to the development of a standard test procedure for the determination of J_{Ic} (ASTM E 813), which was issued in 1981. Recent work by Paris and co-workers [2] has, however, caused increased attention to be focused on the determination of the entire J-integral resistance (J_I-R) curve rather than on J_{Ic} alone. The reason for this shift in focus is that the tearing modulus, which is a function of the slope of the J_I-R curve beyond J_{Ic}, has been used to predict the onset of tearing instability in ductile metals. Further experimental verification of this theory was provided in recent work by Vassilaros and co-workers [3].

[1]David Taylor Naval Ship Research and Development Center, Annapolis, Md. 21402.

Investigation of the tearing instability theory was aided by the prior development of reliable single specimen techniques for the routine determination of J_{Ic} and of the complete J_I-R curve [4,5]. These tests have used elastic compliance techniques to measure crack extension.

A recent review by the National Materials Advisory Board [6] identified a critical need to extend the aforementioned elastic plastic fracture mechanics technology into the dynamic loading regime. Research has been performed in this area [7,8] but primarily to define J_{Ic} and not to determine the entire J_I-R curve. One of the difficulties involved with the determination of dynamic J_I-R curves is that elastic compliance techniques are not readily adaptable to rapid loading. A method for J_I-R curve determination which is readily adaptable to rapid loading is the direct-current potential drop (DCPD) technique.

The object of this investigation was to develop and verify a direct-current potential drop technique for the measurement of crack extension in compact specimens and to employ this technique in the determination of the J_I-R curves for 3Ni and 5Ni steels under both quasi-static and rapid loading.

A four-fold approach was taken for this investigation. The first step involved the generation of a calibration curve for the potential drop technique. Next, the validity of the determination of crack initiation with potential drop was examined using a multispecimen technique performed with 3Ni steel compact specimens. Thirdly, the J_I-R curves generated at quasi-static loading rates using potential drop were correlated with elastic compliance results to establish the reliability and repeatability of the technique. Lastly, the potential drop technique was used to determine the J_I-R curves from rapid tests of 3Ni and 5Ni steel compact specimens.

Direct-Current Potential Drop Theory

The direct current potential drop (DCPD) technique, as applied to compact specimens, assumes that a reasonable correlation can be obtained between potential drop and crack extension during a calibration procedure and that, using this calibration, potential drop can be used to measure crack extension in compact specimens during elastic plastic fracture mechanics testing. Following Wilkowski and Maxey [9], the potential drop technique involves applying a constant current through the compact specimen so that a change in crack length causes a corresponding change in the electric field. This alters the electric potential difference at suitably placed contact points in the vicinity of the crack. Changes in the crack length are then determined by comparing this measured potential difference with a previously developed calibration record for the compact specimen geometry.

The potential drop analysis technique used to obtain the J_I-R curves for this investigation involved several assumptions. Firstly, potential drop was assumed to be, predominantly, a function of specimen geometry. There are numerous studies in the literature which support this assumption. Early work by Johnson [10] on the measurement of cracks in metallic materials using potential drop employed a theoretical calibration which showed the potential to be insensitive to material chemistry, heat treatment, and thickness. The Johnson theoretical expression correlating potential drop with crack length takes into account only specimen depth (W) and the placement of the current inputs and potential outputs on the specimen. More recently, Schwalbe and Hellmann [11] have applied Johnson's expression to three specimen geometries (center cracked tension, compact tension, and single-edge notched beam) with good results. Both Johnson and Schwalbe and Hellmann reported good correlations between the theoretically determined calibration and experimental results. The compact specimens used for this investigation were of a modified geometry, and experimental calibrations were performed for comparison with the results obtained by Schwalbe and Hellmann.

The second assumption made for this analysis was that the potential drop versus crack opening displacement (COD) plot from a specimen will be linear to crack initiation and that deviation from linearity can be used to predict crack initiation in compact specimens. Lowes and Fearnehough [12] first recognized this behavior in using potential drop to test Charpy-sized COD specimens of a structural steel (Fig. 1). More recently, Wilkowski [9] and Schwalbe [13] have verified the validity of this technique for a wide variety of structural alloys. For this investigation, a multispecimen technique was employed on 3Ni steel compact specimens to verify the point of crack initiation.

Lastly, it was assumed that the effect of plasticity on the potential drop after crack initiation was negligible for the 3Ni and 5Ni steels used in this investigation. Wilkowski [9], in testing monotonically loaded compact specimens, has noted that the majority of plasticity is obtained before crack initiation. For the moderately hardening 3Ni and 5Ni steels this assumption should be valid.

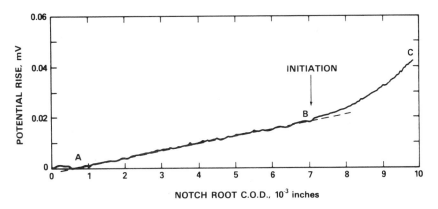

FIG. 1—*Potential rise versus crack opening displacement for C-M structural steel* [12].

J_1-R *Curve Determination*

A schematic representation of the procedure for the determination of the J_1-R curves from potential drop is shown in Fig. 2. At the present state of development, this analysis is conducted after the test is completed, since real time estimates of crack extension are not obtained.

Upon completion of an individual test, the values for the initial crack length and final crack length are determined from the specimen fracture surface. The corresponding normalized potential values related to initial crack length ($\overline{PD}_i$) and final crack length ($\overline{PD}_f$) are determined from the calibration curve (Fig. 2). The values of the measured potentials at crack initiation ($PD_{initial}$) and the final crack length (PD_{final}) are determined from the test data. These parameters ($\overline{PD}_i$, $\overline{PD}_f$, $PD_{initial}$ and PD_{final}) are used to solve the expressions

$$\overline{PD}_i = \frac{PD_{initial} - P_1}{P_2 - P_1} \tag{1}$$

$$\overline{PD}_f = \frac{PD_{final} - P_1}{P_2 - P_1} \tag{2}$$

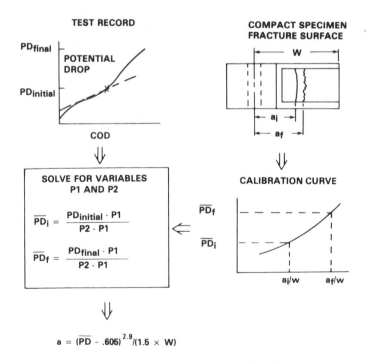

FIG. 2—*Schematic of* J_1-R *curve determination procedure from direct-current potential drop measurement.*

for P_1 and P_2. P_1 is a variable which is a function of the equipment and electrical setup for the test. This variable is designed to account for absolute potential differences due to variations in equipment and electrical setup which produce a constant shift in the measured potential drop. P_2 is a variable designed to account for slight variations in the placement of the electrical leads on the specimen and for specimen resistivity. With P_1 and P_2 determined, the measured potentials from the test data are substituted into the expression

$$\overline{PD} = \frac{PD - P_1}{P_2 - P_1} \tag{3}$$

where P_1 and P_2 are as described above, PD is the measured potential from the test data, and $\overline{PD}$ is the normalized value for potential drop. This normalized value for potential drop is substituted into the power curve fit expression for the calibration curve to determine the crack length. The analysis is therefore "pinned" to the initial and final values of the physically measured crack lengths. This unique aspect of the analysis produces greater reliability and reproducibility for the DCPD technique.

Experimental Approach

Material

A 50.8-mm (2-in.)-thick 5Ni steel plate and a 38.1-mm (1½-in.)-thick 3Ni steel plate were used for all tests. The chemical compositions of the plates are presented in Table 1 and the mechanical properties are presented in Table 2.

Specimens and Test Procedure

Twenty compact specimens of 3Ni steel and five compact specimens of 5Ni steel (Fig. 3) were machined from the aforementioned plates; the notch orientation for all the specimens was T-L. All the specimens were side grooved to a total section reduction of 20% with a standard Charpy V-notch cutter and precracked in accordance with ASTM E 813. All tests were conducted at

TABLE 1—*Chemical composition (weight percent) of steels used for J_1-R curve testing.*

Material	C	Mn	P	S	Cu	Si	Ni	Cr	Mo	V	Ti
3Ni steel (FSU)	0.15	0.33	0.010	0.014	0.25	0.21	2.10	1.25	0.24	0.003	0.002
5Ni steel (FXY)	0.09	0.78	0.010	0.007	0.18	0.26	4.93	0.56	0.43	0.07	0.004

TABLE 2—*Mechanical properties of steels used for J_I-R curve testing.*

Material	Yield Strength (0.2% Offset), MPa (ksi)	Ultimate Tensile Strength, MPa (ksi)	Elongation in 2 in., %	Reduction of Area, %
3Ni steel (FSU)	607 (88)	717 (104)	20	65
5Ni steel (FXY)	958 (139)	985 (143)	21	66

room temperature. The test matrix for this investigation is presented in Table 3.

A Hewlett-Packard 6260B 100-A d-c power supply was used as the current source for the generation of the calibration curve and for all testing. The potential output was conditioned by a Pacific 70A-4 differential amplifier; a gain setting of $\times 2000$ was used.

Quasi-static J_I-R curve testing was performed in a screw-driven tensile machine. Rapid J_I-R curve testing was performed in a fast-acting servohydraulic machine at load-line displacement rates ranging from 5.1 to 30.5 mm/s (0.2 to 1.2 in./s). Nicolet Explorer III digital oscilloscopes were used for data acquisition from the rapid tests. All digital data was processed using Tektronix 4050 minicomputers.

The screw driven and servohydraulic test machines were electrically isolated from the compact specimens by the insertion of Micarta plates between sections of the crosshead components. The COD gage used for the quasi-

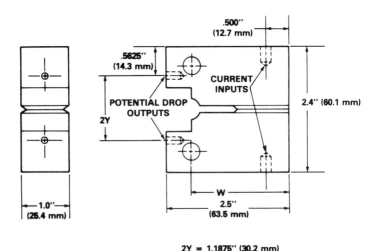

2Y = 1.1875" (30.2 mm)
W = 2.0" (50.8 mm)

FIG. 3—*1T compact specimen used for J_I-R curve testing.*

TABLE 3—J_I-R *curve test matrix for 1T compact specimens;*
a/w = 0.65; 20% side grooved.[a]

| Material | Crack Extension Measurement | | | |
| | E.C. Quasi-Static | E.C. + P.D. Quasi-Static | Potential Drop | |
			Quasi-Static	Rapid
3Ni steel (FSU)	3	8	6	3
5Ni steel (FXY)	1	1	1	2

[a]E.C. = elastic compliance; P.D. = potential drop.

static tests was isolated by the use of a Micarta gage block. The rapidly loaded specimens were isolated from the COD gage by covering the razor blade knife edges with an insulating material.

The following expression developed by Ernst et al [14] was used to determine the value of the *J*-integral for all tests:

$$J_{(i+1)} = \left[J_i + \left(\frac{\eta}{b_i} \right) \frac{A_{i,i+1}}{B_N} \right] \left[1 - \left(\frac{\gamma}{b_i} \right) (a_{i+1} - a_i) \right] \qquad (4)$$

where

$\eta = 2 + (0.522) \, b/W$ for compact specimens,
W = specimen width,
$\gamma = 1 + (0.76) \, b/W$,
b_i = instantaneous length of remaining ligament,
B_N = minimum specimen thickness,
a_i = instantaneous crack length, and
$A_{i,i+1}$ = area under the load versus load-line displacement record between lines of constant displacement at points i and $i + 1$.

Following testing, all specimens were heat tinted at 370°C (700°F) for 30 min to mark the final crack extension. The specimens were then fractured at liquid nitrogen temperature. The fatigue precrack length and final crack extensions were the average of nine equally spaced measurements across the crack surface, excluding the edges.

Results and Discussion

DCPD Calibration Curve

The modified compact specimen used in this investigation is shown in Fig. 3. The current input and potential output locations were chosen to provide a uniform electric field across the path of crack extension and to enhance the reproducibility of the results. With the current input from the top and bottom of the specimen, the potential outputs, as shown in Fig. 3, are located in an electrical "backwater" [15]. Therefore slight variations in the placement of these outputs have a negligible influence on the calibration curve. Locating the outputs nearer to the notch tip would result in increased sensitivity and decreased reproducibility [13]. Reproducibility is decreased due to the fact that the shape of the calibration curve for the notch tip region is more sensitive to changes in the placement of the leads.

The experimental potential drop calibration data generated for the modified compact specimen are shown in Fig. 4. The data were generated by saw cutting four 5Ni steel specimens and one 3Ni specimen to various crack lengths and recording the corresponding potentials. Two of the 5Ni steel specimens were side grooved 20%. The remaining calibration specimens were not side grooved. The recorded potential values were normalized to the potential obtained for the machined notch condition ($a/w = 0.475$). Calibrations were performed at input currents of 20, 40, and 60 A. The data shown in Fig. 4 were generated at an input current of 60 A, which was the current level used for all J_1-R curve testing.

The following expression was fitted to mean values of all data points over the range of crack lengths included for J_1-R curve testing ($0.6 < a/w < 0.8$):

$$P = (1.5 \times a/w)^{2.9} + 0.605 \tag{5}$$

The error between the curve fit and the individual data points over this range was within 3.5%. A power curve fit was chosen to facilitate calculations between the crack length and the normalized potential.

Figure 4 shows the excellent correlation obtained between the power curve fit and the experimental data. This figure also shows the relative insensitivity of the calibration curves to compact specimen side grooving. The curves were also relatively insensitive to input currents of 20, 40, and 60 A.

The error in the measurements of the saw cut notch (a/w) was less than approximately 8% based on an absolute error in measurement of 0.89 mm (0.035 in.) and a remaining ligament of 10.2 mm (0.4 in.).

The experimental calibration performed for this investigation did not correlate well with theoretical calibrations based on Johnson's formula [10]. A comparison of the power curve fit of the experimental data with Johnson's Curve is shown in Fig. 5. Johnson's curve was generated using the expression

$$U/U_0 = \frac{\cosh^{-1}\left[\cosh\left(\pi y/2W\right)/\cos\left(\pi a/2W\right)\right]}{\cosh^{-1}\left[\cosh\left(\pi y/2W\right)/\cos\left(\pi a_0/2W\right)\right]} \tag{6}$$

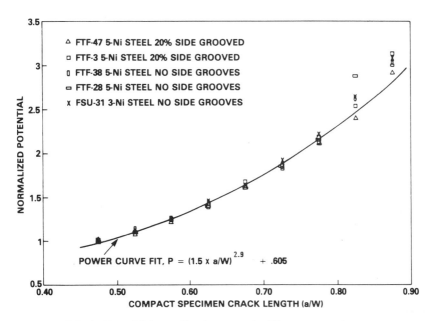

FIG. 4—*Potential drop calibration curves for 1T compact specimens.*

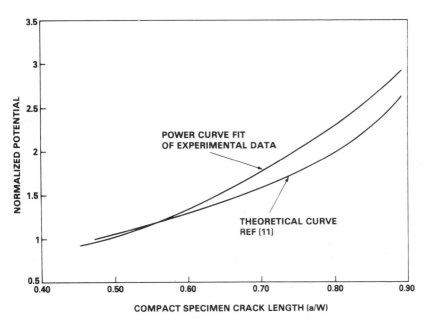

FIG. 5—*Comparison of power curve fit for experimental calibrations and theoretical curve from Schwalbe and Hellmann* [11].

where

U_0 = initial potential,
a_0 = initial crack length,
y = one half the distance between the potential output (see Fig. 3),
U = measured potential,
a = measured crack length, and
W = specimen width (see Fig. 3).

Schwalbe and Hellman [11] and others have shown good correlations between calibrations based on this theoretical expression and those obtained experimentally for the compact specimen geometry. The poor correlations obtained for this investigation could be due to the modified geometry of the compact specimen and slight differences in the placement of the electrical connections. The poor correlations are not totally unexpected, however, since Johnson's expression assumes uniform current density which is probably not achieved in the compact geometry.

Identification of Crack Initiation

As mentioned previously, crack initiation in DCPD tests was determined from deviation from linearity on a plot of potential drop versus load-line crack opening displacement. To check the validity of this criterion, multispecimen tests were performed on six 3Ni steel compact specimens at quasi-static loading rates. Experience with J_I-R curve testing of this steel in IT compact specimens precracked to between 0.65 and 0.75 a/w has shown that crack initiation usually occurs near the maximum load. In the case of the specimens tested, maximum load was attained between load-line COD values of 1.27 and 1.52 mm (0.05 and 0.06 in.). The focus of this testing was to bracket this region to assess the validity of the crack initiation criterion.

The results of three of the multispecimen tests are presented in Figs. 6 to 8. Specimen FSU-16 (Fig. 6) was tested to a load-line COD of approximately 1.02 mm (0.04 in.). No crack extension was expected and there was no evidence of crack extension on the fracture surface. The load-line COD versus DCPD plot for this specimen remained linear. Specimen FSU-19 (Fig. 7) was tested to a load-line COD between 1.27 and 1.52 mm (0.05 and 0.06 in.), the region in which crack initiation was expected. The load-line COD versus DCPD plot for this specimen shows a slight deviation from linearity in this region. The fracture surface shows evidence of crack extension. Figure 8 shows the fracture surface and load-line COD versus DCPD plot for Specimen FSU-21. This specimen was tested to a load-line COD between 1.78 and 2.03 mm (0.07 and 0.08 in.) where a significant amount of crack extension was expected. The fracture surface of this specimen does indeed show a measurable amount of crack extension. The load-line COD versus DCPD

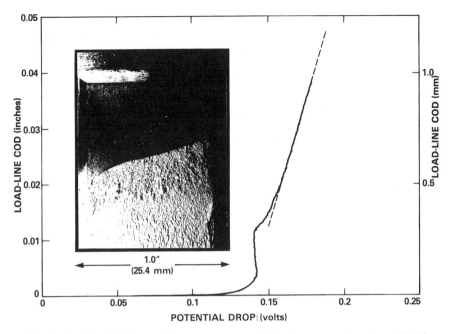

FIG. 6—*Load-line COD versus direct-current potential drop for compact specimen FSU-16.*

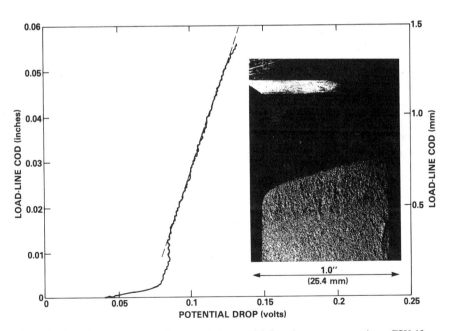

FIG. 7—*Load-line COD versus direct-current potential drop for compact specimen FSU-19.*

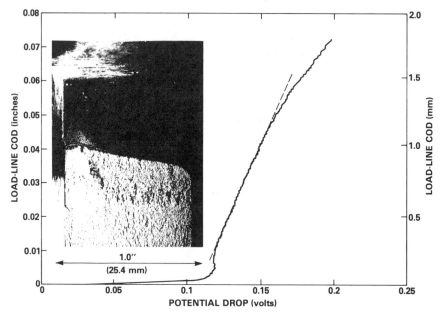

FIG. 8—*Load-line COD versus direct-current potential drop for compact specimen FSU-21.*

plot shows a large deviation from linearity beginning in the region between 1.27 and 1.52 mm (0.05 and 0.06 in.) of load-line COD.

The remaining three specimens were concentrated in the range of load-line COD between 1.27 and 1.52 mm (0.05 and 0.06 in.) and showed results similar to those obtained above.

To summarize, an initial linear relationship between potential drop and load-line crack opening displacement was observed for the 3Ni steel compact specimens. A deviation from this linear region of the curve was shown to correspond closely with crack initiation, thereby verifying the crack initiation criterion first used by Lowes and Fearnehough [12].

J_I–R *Curve Testing*

The J_I–R curve testing was performed in four separate phases:

1. Elastic Compliance Only (Quasi-Static).
2. Elastic Compliance + Potential Drop (Quasi-Static).
3. Potential Drop Only (Quasi-Static).
4. Potential Drop Only (Rapid).

The first two phases employed the single specimen computer interactive procedure of Joyce and Gudas [5]. Phase 1 specimens were used as "controls" and did not employ potential drop. The computer interactive test procedure was modified for Phases 2 and 3 to include potential drop as an additional channel for digital data acquisition. Phase 2 specimens utilized both elastic compliance and potential drop for the crack extension measurements. Phase 3 specimens used only potential drop. The rapidly tested specimens (Phase 4), by necessity, also employed only potential drop for the crack extension measurements.

Results of the J_I-R curve testing are presented in Figs. 9 to 15. Figure 9 illustrates the excellent correlation obtained between the elastic compliance and DCPD techniques for a single 3Ni steel specimen (Elastic compliance + DCPD, Quasi-static). Similar results were obtained for 5Ni steel (Fig. 10). The J_I-R curves for two control specimens from elastic compliance tests (Phase 1) and one specimen of 3Ni steel which was tested with monotonically increasing load monitoring only DCPD are presented in Fig. 11. This figure demonstrates the capability of the potential drop technique to produce a continuous R-curve whereas elastic compliance defined the J_I-R curve at only a relatively few discrete points. Figure 11 also illustrates, again, the excellent correlation obtained between the two techniques. The corresponding plot for 5Ni steel is shown in Fig. 12. The correlation between the two techniques for the specimens in Fig. 12 was not as good as those shown previously. This is attributable to material scatter in this relatively low toughness steel plate.

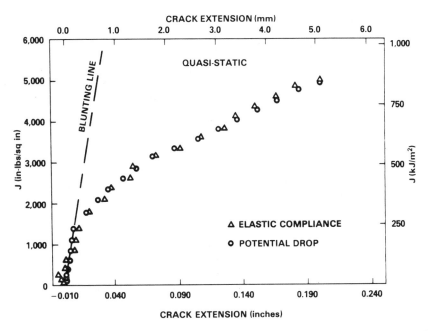

FIG. 9—J_I-R curves for compact specimen FSU-4 (3Ni steel) from elastic compliance and direct-current potential drop measurements at quasi-static loading rates.

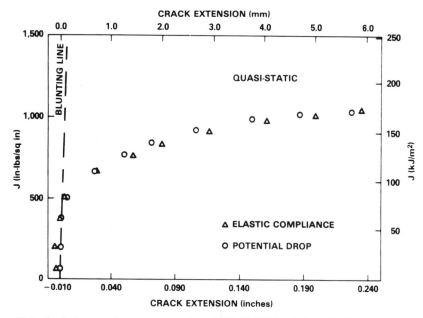

FIG. 10—J_I-R *curves for compact specimen FXY-47 (5Ni Steel) from elastic compliance and direct-current potential drop measurements at quasi-static loading rates.*

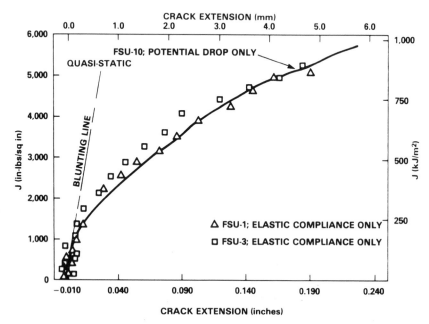

FIG. 11—J_I-R *curves for three 3Ni steel compact specimens showing comparison between direct-current potential drop and elastic compliance measurements at quasi-static loading rates.*

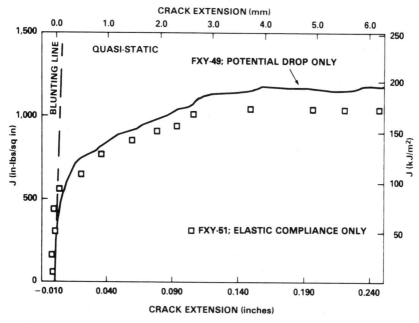

FIG. 12—J_I-R *curves for two 5Ni steel compact specimens showing comparison between elastic compliance and direct-current potential drop measurements at quasi-static loading rates.*

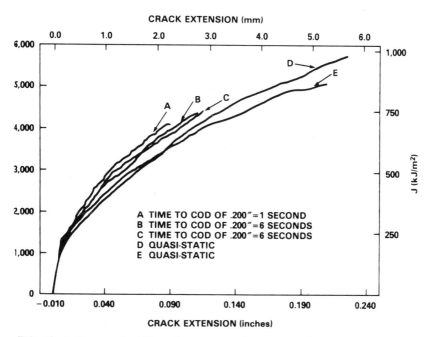

FIG. 13—J_I-R *curves for 3Ni steel compact specimens comparing quasi-static and rapid loading rates.*

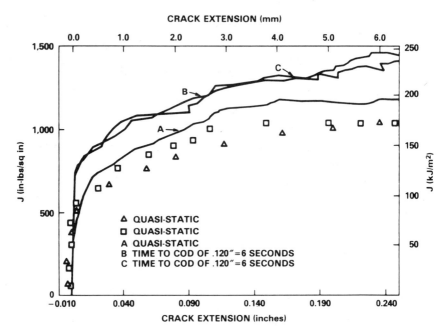

FIG. 14—J_I-R *curves for 5Ni steel compact specimens comparing quasi-static and rapid loading rates.*

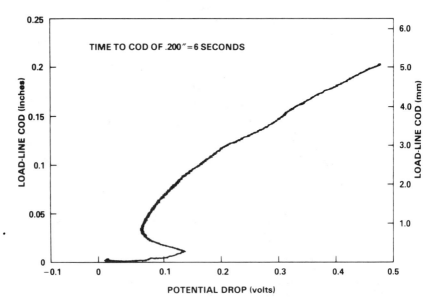

FIG. 15—*Load-line COD versus potential drop for rapidly loaded specimen FSU-27 (3Ni steel).*

Results of rapid loading tests of 3Ni steel specimens are compared with the quasi-static case in Fig. 13. This figure illustrates a general trend toward elevating the J_I-R curve with increasing loading rates. The same general trend was observed for the 5Ni steel (Fig. 14). The rapid loading tests were performed in a manner in which maximum COD for the 3Ni steel was obtained in either 6 or 1 s, and the maximum COD for the 5Ni steel was obtained in 6 s. In order to estimate the rate of J_I application prior to crack initiation, the load-time data from each test were examined and the test time to crack initiation was determined. For the tests shown in Fig. 13, the rate of J_I application was approximately 76 kJ/m^2/s (435 in.-lb./in.^{2}/s) for the 6-s tests and 496 kJ/m^2/s (2834 in.-lb./in.^{2}/s) for the 1-s tests. For the results shown in Fig. 14, the rate of J_I application was approximately 47 kJ/m^2/s (269 in.-lb/in.^{2}/s) for the 6-s tests. Joyce and Czyryca [16] obtained similar results for HY-130 using key curve techniques to indirectly measure crack extension in rapidly loaded compact specimens.

An inversion in the initial linear portion of the load-line COD versus DCPD curves was observed to occur in some of the rapidly loaded specimens (Fig. 15). This effect can produce uncertainty in the determination of the crack initiation point from DCPD, which in turn can cause error in the plasticity correction described earlier. This behavior could be attributable to a capacitance effect occurring across the fatigue precrack during rapid loading.

Summary

A direct-current potential drop technique was developed for the measurement of crack extension in compact specimens. The analysis developed for this technique was directed to account for specimen plasticity.

The potential drop calibration curves obtained for this investigation were shown to be insensitive to compact specimen side grooving and input currents over the range of 20, 40, and 60 A. Over the range of crack lengths obtained during J_I-R curve testing ($0.6 < a/w < 0.8$), the 3Ni and 5Ni steels were found to produce calibration curves which could be fitted with a single equation while producing minimal error. An initial linear relationship between potential drop and load-line crack opening displacement was observed for the specimens tested. A deviation from this linear region was shown to correspond closely with crack initiation in 3Ni steel compact specimens. The use of this criterion to determine crack initiation in J_I-R curve tests was therefore validated.

The J_I-R curves determined for 3Ni and 5Ni steels tested at quasi-static loading rates using the direct-current potential drop technique were found to correlate well with R-curves obtained using the elastic compliance method. The potential drop technique enables the determination of continuous R-curves and provides a means of obtaining the J_I-R curve from a single compact specimen at both static and rapid loading rates. Results from this investigation seem to indicate a general trend towards elevating the J_1-R curve with increasing loading rate.

Acknowledgments

The authors wish to acknowledge the support of Dr. H. H. Vanderveldt of the Naval Sea Systems Command (NAVSEA 05R25). The authors also wish to express thanks to Dr. George Rose of the British Admiralty Marine Technology Establishment for his help and advice during the course of this research and to Mr. John P. Gudas of the David W. Taylor Naval Ship Research and Development Center, who supervised this research program.

References

[1] Landes, J. D. and Begley, J. A. in *Fracture Mechanics, ASTM STP 514*, American Society for Testing and Materials, 1972, pp. 1–20.

[2] Paris, P. C., Tada, H., Zahoor, A., and Ernst, H. in *Elastic-Plastic Fracture, ASTM STP 668*, J. D. Landes, J. A. Begley, and G. A. Clarke, Eds., American Society for Testing and Materials, 1979, pp. 5–36.

[3] Vassilaros, M. G., Joyce, J. A., and Gudas, J. P. in *Fracture Mechanics: Fourteenth Symposium — Volume I: Theory and Analysis, ASTM STP 791*, J. C. Lewis and G. Sines, Eds., American Society for Testing and Materials, 1983, pp. I-65-I-83.

[4] Clarke, G. A., Andrews, W. R., Paris, P. C., and Schmidt, D. W. in *Mechanics of Crack Growth, ASTM STP 590*, American Society for Testing and Materials, 1976, pp. 27–42.

[5] Joyce, J. A. and Gudas, J. P. in *Elastic-Plastic Fracture, ASTM STP 668*, J. D. Landes, J. A. Begley, and G. A. Clarke, Eds., American Society for Testing and Materials, 1979, pp. 451–468.

[6] "Response of Metals and Metallic Structures ft to Dynamic Loading," National Materials Advisory Board Report NMAB-341. Washington, D.C., May 1978.

[7] Hasson, D. F. and Joyce, J. A., *Journal of Engineering Materials and Technology, Transactions of ASME*, Vol. 103, April 1981, pp. 133–141.

[8] Nguyen, P. et al, *Journal of Engineering Materials and Technology, Transactions of ASME*, Vol. 100, July 1978, pp. 253–257.

[9] Wilkowski, G. M. and Maxey, W. A. in *Fracture Mechanics: Fourteenth Symposium — Volume II: Testing and Applications, ASTM STP 791*, J. C. Lewis and G. Sines, Eds., American Society for Testing and Materials, 1983, pp. II-266-II-294.

[10] Johnson, H. H., *Materials Research and Standards*, Vol. 5, 1965, pp. 442–445.

[11] Schwalbe, K. H. and Hellmann, D., *Journal of Testing and Evaluation*, Vol. 9, No. 3, May 1981, pp. 218–221.

[12] Lowes, J. M. and Fearnehough, G. D., *Engineering Fracture Mechanics*, Vol. 3, 1971, pp. 103–108.

[13] Schwalbe, K. H., "Test Techniques for Fracture Mechanics Testing," GKSS 81/E/59, presented at 5th International Conference on Fracture (ICF 5), Cannes, France, 1981.

[14] Ernst, H. A., Paris, P. C., and Landes, J. D. in *Fracture Mechanics: Thirteenth Conference, ASTM STP 743*, R. Roberts, Ed., American Society for Testing and Materials, 1981, pp. 476–502.

[15] Knott, J. F., "The Use of Analogue and Mapping Techniques with Particular References to the Detection of Sharp Cracks," in *The Measurement of Crack Length and Shape during Fracture and Fatigue*, published by Engineering Materials Advisory Service (EMAS), 1980, pp. 113–135.

[16] Joyce, J. A. and Czyryca, E. J. "Dynamic J_1-R Curve Testing of HY-130 Steel", David W. Taylor Naval Ship Research and Development Center, Ship Materials Engineering Department Report SME-81/57, Annapolis, Md., Oct. 1981.

G. M. Wilkowski,[1] *J. O. Wambaugh,*[2] *and K. Prabhat*[1]

Single-Specimen *J*-Resistance Curve Evaluations Using the Direct-Current Electric Potential Method and a Computerized Data Acquisition System

REFERENCE: Wilkowski, G. M., Wambaugh, J. O., and Prabhat, K., **"Single-Specimen *J*-Resistance Curve Evaluations Using the Direct-Current Electric Potential Method and a Computerized Data Acquisition System,"** *Fracture Mechanics: Fifteenth Symposium, ASTM STP 833,* R. J. Sanford, Ed., American Society for Testing and Materials, Philadelphia, 1984, pp. 553–576.

ABSTRACT: This paper describes the direct-current electric potential method of monitoring crack initiation and stable ductile crack growth. In many aspects this method is simpler than the unloading compliance method, and it can be readily incorporated into a computerized data acquisition system. Numerous examples using this technique and a computerized data acquisition system are given for bend and compact specimens, different loading rates, and different materials (Type 304 stainless steel, Zircaloy, and carbon steels).

KEY WORDS: *J–R* curve, stable crack growth, elastic-plastic fracture, electric potential crack growth method

The electric potential (EP) technique, also termed the potential drop or potential difference technique, is being increasingly used for monitoring stable crack growth in monotonic loading during crack growth resistance testing. Wilkowski and Maxey [1] have reviewed this test method and discussed their experience with pressure vessel and pipe fracture experiments. At that time the authors confirmed earlier work by Lowes and Fearnehough [2] that the direct-current (d-c) electric potential method could be used to detect the onset of ductile crack growth not only in typical fracture mechanics speci-

[1] Battelle-Columbus Laboratories, Columbus, Ohio 43201.
[2] Schlumberger Well Services, Houston, Tex.; formerly at Battelle-Columbus Laboratories, Columbus, Ohio 43201.

mens but also in pipe and pressure vessels with surface or through-wall cracks. It was also found that the d-c EP method could be used to document stable ductile crack growth. Although the method is relatively old, recent improvements in d-c amplifiers have made it much more attractive.

The object of this paper is to show how the d-c EP method can be used to detect crack initiation and crack growth in typical laboratory fracture mechanics specimens for single-specimen *J–R* curve testing. The monitoring of crack initiation and crack growth is discussed. Several examples of a computerized data acquisition system are presented. In addition, some of the favorable aspects and points of concern of the d-c EP technique are discussed.

Fundamentals of the Electric Potential Method

The fundamentals of the electric potential method, briefly summarized here for completeness, are given in more detail in Ref *1* or papers cited in Ref *1*. The two basic EP methods are the alternating and direct current methods. Both methods involve applying a remote constant current and monitoring a voltage across the crack. The deeper the crack, the greater the voltage drop across the crack.

For the a-c method, the constant current generally has a frequency in the 2 kHz range and travels along the surface (Fig. 1*a*). The current travels along the crack faces so long as the crack faces are electrically isolated. Generally a very small corrosion product is sufficient to electrically isolate the crack faces for either the a-c or d-c method.

For the d-c EP method, the current penetrates the whole thickness of the specimen. Calibrations will depend on whether the current density is uniform or nonuniform (Figs. 1*b* and 1*c*).

There are many possible locations for the current wires and voltage probes. For example, for a compact tension (CT) specimen, three different current locations have been suggested [3–5]. As shown in Fig. 2*a*, the use of single wires above and below the crack face will result in a nonuniform current distribution. Here standard theoretical calculated calibrations can be erroneous [3]. Using a single-current wire above and below the center of the ligament will yield experimental calibrations closer to the theoretical calibrations. If the current is applied uniformly across the top and bottom of the specimen, however, as shown in the right-hand side of Fig. 2*a*, there is good agreement between experimental and theoretical calibrations [4,5]. Any one of these current locations is acceptable so long as the appropriate calibration is used.

The voltage probe locations can vary in position on the crack face or the side of the specimen. In Fig. 2*b* the voltage probe positions on the side of the specimen can be either at the crack face or across the crack tip. The advantage of having the voltage probes across the crack tip is an increase in the voltage signal [6]. The disadvantage is that although sensitivity is increased, reproducibility is lost since it is difficult to precisely locate these probes rela-

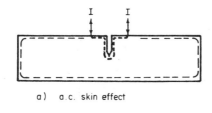

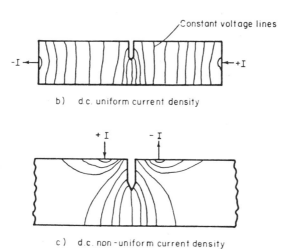

a)   a.c. skin effect

b)   d.c. uniform current density

c)   d.c. non-uniform current density

FIG. 1—*Potential field distribution in single edge notched specimen.*

tive to the crack tip and to keep a consistent fatigue crack length in all specimens. Consequently, it is suggested to keep the voltage probes on the crack face where the reproducibility, especially for crack growth measurements, is less sensitive to location error.

The voltage probes can be located at three positions on the specimen face (Fig. 2c). The most common location is at the midthickness. It has been suggested by some researchers that if crack tunneling occurs, the tunneling could be detected easier if the probes were skewed to opposite corners. At first this seems reasonable; however, if the crack tunnels in the midthickness the differential voltage at the skewed edges is the same as the differential voltage at directly opposite edges. Another possible method of averaging crack tunneling is to place one of the voltage probes at the midthickness and the other at the edge of the specimen. One could postulate that in this case, the differential voltage measured is the average of the midthickness, where the crack tunneling is typically greatest, and the free surface at the edge, where there is little tunneling.

To evaluate if probe position on the crack face has a significant effect on monitoring crack tunneling an experiment using different probe locations

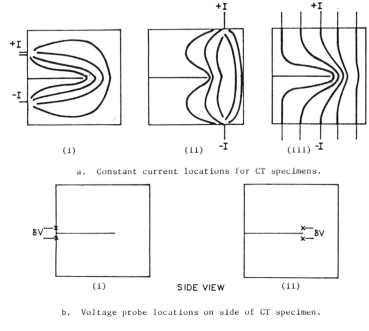

a. Constant current locations for CT specimens.

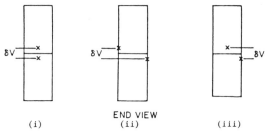

b. Voltage probe locations on side of CT specimen.

SIDE VIEW

END VIEW

c. Voltage probe locations on face of CT specimen.

FIG. 2—*Current and voltage probe locations for CT specimen.*

was conducted. Figure 3 shows an example of d-c EP measurements made using two locations shown in Fig. 2c, (i) and (iii). No difference in the two locations was observed at crack initiation, but for this case with severe tunneling a difference in the EP signals was observed until there was significant crack growth (up to Point A in Fig. 3). In this case, the predicted a/w was 0.616 and 0.630 for probe locations 1-3 and 1-2, respectively. The actual nine-point average a/w was 0.645, which was slightly larger than at the probe locations. For the 1-2 location the error for this severe crack-tunneling case was 2.3%. For cases with less severe crack tunneling (see Figs. 6 and 8) there was little difference between the three locations.

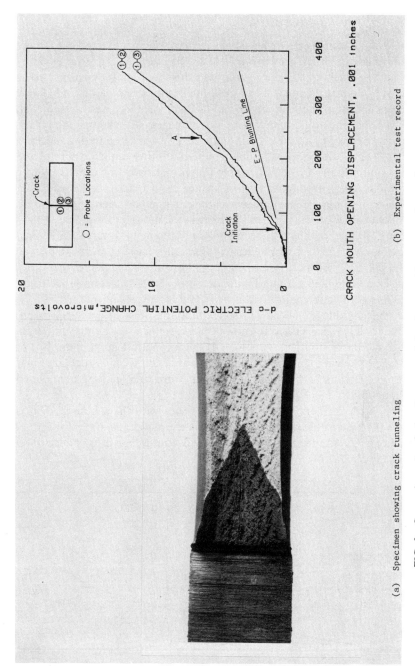

(a) Specimen showing crack tunneling

(b) Experimental test record

FIG. 3—*Comparison of probe locations on crack face for test with severe crack tunneling.*

Determining Crack Initiation

Other than multiple-specimen testing, perhaps the first method used to detect stable ductile crack growth was the d-c EP method. This was done by Lowes and Fearnehough [2], who recorded the crack mouth opening versus d-c electric potential. Figure 4 shows the Lowes data; at low loads, where linear elastic stress existed at the crack tip (up to Point A), there was no change in the EP. As the crack tip blunted, there was a linear change between the crack mouth displacement and the d-c EP up to the point of crack initiation (Point B in Fig. 4). The authors of Ref *1* noted numerous other similar examples and found a similar behavior for extremely tough materials (for example, Type 304 stainless steel), through-wall and surface cracks, and various specimen geometries. The linear behavior between the crack mouth opening and d-c EP up to crack initiation was termed the electric potential blunting line.

The a-c EP method has also been applied to the monitoring of stable ductile crack growth [7,8]. A sample test record is shown in Fig. 5. It can be seen that the a-c EP first increased and reached an inflection point, then after some decrease reached another inflection point, and finally continued to increase. These inflection points in the a-c EP versus displacement records have caused considerable confusion in determining the point of crack initiation. Researchers have determined from multiple-specimen testing that initiation occurs at (1) the first inflection point, (2) the second inflection point [7], or (3) a tangency deviation from the test record after the second inflection point [8] (Fig. 5). Because of this confusion, many investigators have abandoned the a-c method.

Although the authors have limited experience with the a-c method, it is believed that this confusion has occurred because of capacitance or inductance effects across the fatigue crack faces. Because these effects depend on

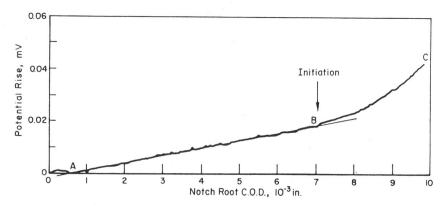

FIG. 4—*Typical d-c potential rise/COD curve for bend specimen testing* [2].

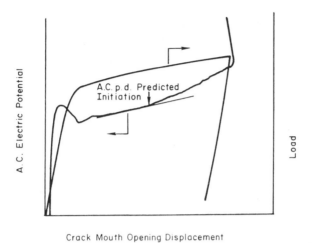

FIG. 5—*Load-line displacement versus a-c EP and load for A533 steel bend specimen* [8].

the area of the fatigue crack faces, this variability in the a-c EP method could be caused by specimen preparation procedures. Further research is needed to clarify these points for the a-c method. Consequently, until these concerns are answered it is suggested that the d-c EP method be used, especially since the electronics are now simpler for the d-c method than for the a-c method.

Crack Growth Measurements and Corrections

For one set of carbon steel data previously reported [1], the amount of stable crack growth was measured by multiple specimens using the nine-point average method and was compared with theoretically predicted crack lengths. These data were from three-point bend specimens of AISI 1020 carbon steel. The crack mouth opening versus d-c EP record is shown in Fig. 6. The EP blunting line was used to determine and calculate *J* at crack initiation J_i. Values of *J* for the amount of crack growth were also calculated using the crack growth from the measured fracture surfaces which exhibited a large amount of crack tunneling (Fig. 7). The tunneling here was slightly less than that shown in Fig. 3. Values of J_i were then determined by extrapolating back to the blunting line. In one case, the test was stopped at the point of first deviation from the EP blunting line. In this case, ductile tearing was determined to just barely start at several locations along the crack tip (Fig. 8). Figure 9 shows that the multiple-specimen J_{Ic} (from extrapolating back to the blunting line, $2J/\sigma_y$) was in close agreement with the range of J_i, which was determined from the EP blunting line technique used from the same seven

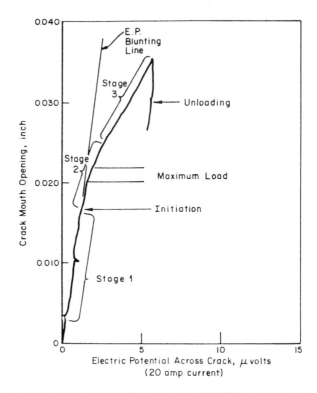

FIG. 6—*Crack mouth opening versus EP data from AISI 1020 carbon steel bend specimen.*

specimens. These results confirm the Lowes and Fearnehough [2] decision to use the d-c EP method to detect the start of ductile crack growth.

For crack growth, the measured a/w was compared with the measured EP and the theoretical calibration. As shown in Fig. 10, the only difference between the experimental EP and the theoretical calibration curve using the method of Johnson [9] was the EP change caused by the crack blunting. This was true even for the specimens exhibiting crack tunneling. Hence for the amount of tunneling in this material, if the EP at the point of crack initiation is used as the initial EP, the stable ductile crack growth agreed with the theoretical calibration curve. This important point has also been found true for ductile crack growth in other specimen geometries [1].

A generalized crack opening versus d-c EP test record [1] is shown in Fig. 11. Stage 1 represents crack blunting and, as previously noted, initiation occurs at the first deviation from the EP blunting line. The second stage is nonlinear and coincides with crack tunneling until the crack growth occurs from the midthickness to the outside edges of the initial crack tip. The third stage is another linear region which corresponds to ductile crack growth where the crack tip opening angle reaches a constant value.

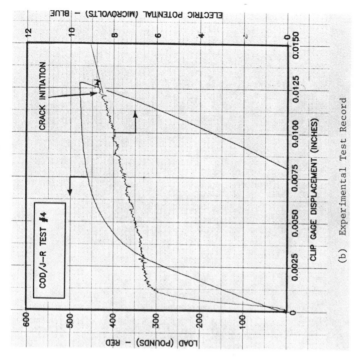

(b) Experimental Test Record

FIG. 7—*Comparison of crack growth on fracture surface just past crack initiation with the d-c EP test record.*

FIG. 8—*Fracture surfaces for AISI 1020 cold-worked carbon steel specimens.*

Some Favorable Aspects and Points of Concern of the d-c EP Method

Table 1 summarizes the favorable aspects and points of concern of the d-c EP method for *J–R* curve testing. Of the favorable points cited, it is worthy of note that the d-c EP method has several advantages over the unloading compliance (UC) method:

1. Full thickness unsided-grooved specimens can be used since the EP can detect tunneling, and average crack lengths equivalent to nine-point measured crack lengths are produced. The UC method frequently requires a correction factor for crack tunneling [10], a factor which may vary for different materials.

2. The d-c EP method can be used at higher strain rates impractical with the UC method; however, at very high rates a-c effects could be induced. This is a point under investigation.

3. The d-c EP method does not require interactive control of the test machine for computerized data acquisition; hence software development and equipment is less expensive.

The selection of the d-c EP or UC method will depend on the researcher's needs, long-range testing plans, and existing equipment.

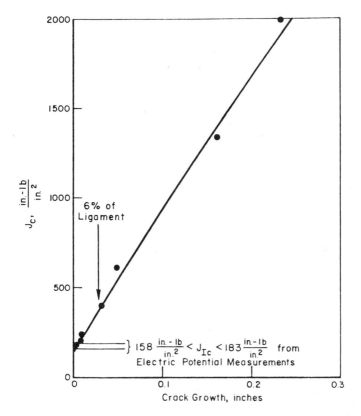

FIG. 9—*J-resistance curve from 11.4-mm (0.45-in.)-thick AISI 1020 carbon steel bend speci-mens.*

None of the points of concern listed for the d-c EP presents any major prob-lem. It is necessary to electrically isolate the specimen from the test machine and isolate any clip gages across the crack mouth. This can be easily accom-plished. Neglecting this isolation will cause current losses to the test machine which will vary with the load on the specimen. Thermal electromotive force (emf) problems with the d-c method can be eliminated by spot-welding wires of the same chemical composition to the specimen, then keeping the connec-tions to copper wires thermally insulated. Temperature changes will also change the specimen's electrical resistivity. There are several ways to account for specimen resistivity changes: (1) keeping the test temperature constant, (2) monitoring the specimen temperature and correcting for resistivity changes, and (3) using reference probes on a dummy specimen or on a posi-tion on the specimen remotely located from the crack location.

Another potential problem is associated with microvolt measurements. Ground loops, radio frequency (RF) induced noise, and earth currents are

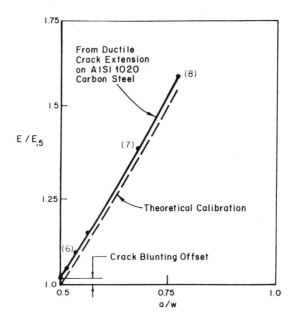

FIG. 10—*Comparison of measured ductile crack growth using nine-point average with theoretical prediction for AISI 1020 carbon steel data. ( ) is specimen number in Fig. 8.*

potential problems, but these can easily be overcome with proper care. It is most important to use a good quality d-c differential amplifier. Within the past five years, the quality of these amplifiers has increased significantly. Commercially available nanovolt meters can be used if testing rates are slow, since their significant filtering improves stability which in turn lowers frequency response.

Computerized Data Acquisition System for J–R Curve Testing

Data from a typical *J–R* curve test can be acquired in several ways. Load, load-line displacement, crack mouth opening, and electric potential across the crack can be recorded on *X–Y* plotters. *J–R* curves can then be obtained by reducing the data manually. Alternatively, the raw data from the *X–Y* plotters can be digitized using a computer and then used to calculate the *J–R* curve. The most efficient and accurate method, however, is to acquire the data in digitized form using a computer-controlled data acquisition unit. Various parameters (for example, EP versus crack mouth opening, or load versus load-line displacement) can then be cross plotted and resistance curves obtained easily.

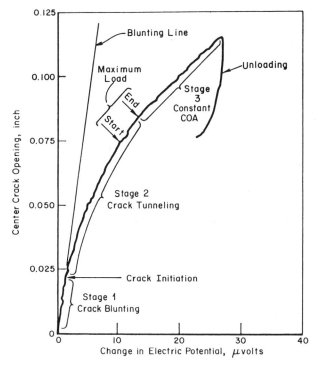

FIG. 11—*Generalized shape of crack displacement versus ʽ-c EP test record.*

TABLE 1—*Favorable aspects and points of concern for d-c EP crack growth monitoring.*

Favorable Aspects

- Single specimen J_i, *J-R* curve
- Detects crack tunneling
- Computer control does not require interactive control of test machine
- High strain-rate testing
- High temperature and environmental testing
- Simple calibrations for standard geometries
- Easily adaptable to computerized data acquisition (does not require high speed data acquisition equipment for standard testing rates)

Points of Concern

- Electrically isolate specimen
- Need to avoid thermal emf
- Temperature fluctuations effect resistivity
- Low voltage problems: noise, ground loops, RF-induced noise, earth currents

Data Acquisition

The experimental setup for data acquisition is shown in Fig. 12. Typical experimental instrumentation consists of a d-c power supply (a 20-A power supply is usually sufficient in most cases) connected to the ends of the test specimen. Wires spot welded across the crack as shown in Fig. 12 are used to measure the electric potential across the crack. If the specimen is to be tested at a controlled temperature, copper wire can be directly connected to the specimen. The thermal emf from these connections cancel each other if they are at the same temperature. If the specimen temperature may vary during the test, long probe wires of the same chemical composition as the specimen should be used. These wires can be connected to copper wires going to the amplifier further from the specimen where the junctions can be thermally insulated. If these wires are long, they should be twisted to avoid any induced earth currents. The wires are then connected to shielded wires (to reduce noise in the signal). Again, care should be taken to thermally insulate this connection to prevent any thermal emf. The EP signal from the shielded wires is then amplified (typically by a factor of 1000) with a differential d-c amplifier and fed into a data acquisition unit. The data acquisition system records load, load-line displacement, and crack mouth opening displacement for

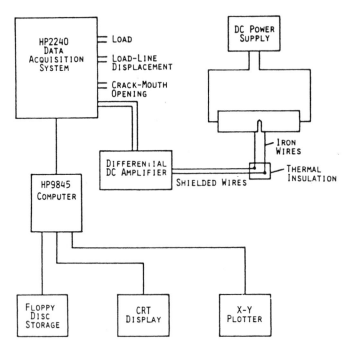

FIG. 12—*Schematic of data acquisition system.*

bend specimens. The data acquisition system is connected to a minicomputer programmed to receive and store data at specified increments in load or load-line displacement or both. These data can then be stored on a flexible disk, displayed on a CRT, or plotted on an *X-Y* recorder. Note that the computer is used to control the data acquisition system and make the desired calculations, but is not used to control the testing machine. This makes software development simpler and the related computer equipment less expensive. Testing can be conducted on standard hydraulic test machines without feedback controls.

Data Reduction

A schematic of the data reduction procedure is shown in Fig. 13. The first step in this procedure consists of obtaining a plot of either the crack mouth

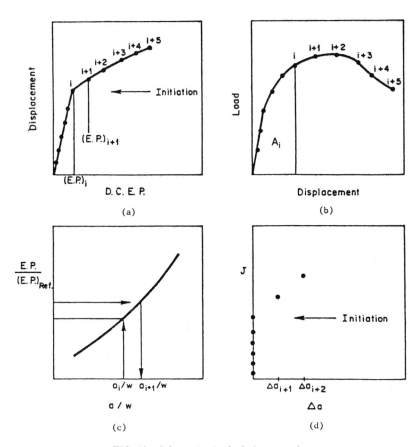

FIG. 13—*Schematic of calculation procedure.*

opening or load-line displacement versus d-c EP across the crack (Fig. 13a). This plot is linear up to crack initiation at which point it deviates from linearity with a distinct change in slope. The point of crack initiation is therefore determined visually from this plot and can be shown on the corresponding load-displacement curve (Fig. 13b). The amount of crack growth beyond initiation can be obtained using a calibration curve of EP/EPreference versus crack depth/specimen width (a/w) (Fig. 13c). Corresponding to the points at which crack growth is calculated, J can be estimated from the area under the load displacement curve as

$$ J = \frac{2A}{Bb} \left[1 - \frac{\left(f\left(\frac{a_0}{w} - 1\right) \Delta a_0 \right)}{b} \right] $$

for a three-point bend specimen, and

$$ J = \frac{A}{Bb} f\left(\frac{a_0}{w}\right) \left[1 - \frac{\left(f\left(\frac{a_0}{w} - 1\right) a_0 \right)}{b} \right] $$

for a CT specimen,

where A is the area under the load displacement curve up to the point under consideration, B is the specimen thickness, b is the remaining ligament, $f(a_0/w)$ is the dimensionless coefficient as defined in Ref 10, and Δa_0 is the amount of crack extension. A plot of the J–R curve can then be obtained very easily (Fig. 13d). The appropriate correction for crack growth was also used in the calculations.

Typical Results Using the Computerized Data Acquisition System

Test data from four different materials (C-Mn-Mo steel, 304 stainless steel, Zircaloy, and an X60 line pipe steel) illustrate typical results that can be obtained using the d-c EP computerized data acquisition system. Table 2 summarizes the composition and tension test data for the materials.

Results from a 0.45T three-point bend specimen of C-Mn-Mo steel tested at 23°C (73°F) are shown in Fig. 14. Figure 14a shows the crack mouth opening versus EP curve and indicates crack initiation. Before initiation, the curve is denoted as the EP blunting line. Crack initiations on a load versus load-line displacement curve and displacement versus crack mouth opening curve are shown in Figs. 14b and 14c, respectively. The J–R curve for the material is given in Fig. 14d, where J at crack initiation is 210 kJ/m² (1200 lb-in./in.²).

Three-point bend ½T and 1T specimens of 304 stainless steel were tested at 20°C (68°F). Crack initiation was detected from a load-line displacement versus EP curve instead of a crack mouth opening versus EP curve. The EP

TABLE 2—*Summary of properties of materials.*

Chemical Composition, wt%										σ_y, MPa (ksi)	σ_u, MPa (ksi)	% Elongation
Mn-Mo-Steel												
C	Mn	P	S	Si	Ni	Cr	Mo	Cu	V	269 (39.1)	460 (66.8)	28.3
0.19	0.62	0.016	0.021	0.39	0.075	0.19	0.057	0.082	0.002			
304 Stainless Steel												
C	Mn	P	S	Si	Ni	Cr	Mo	Cu	V	315 (45.8)	639 (92.8)	69.2
0.55	1.48	0.026	0.000	0.70	8.9	18.1	...	...	...			
Zircaloy												
C	Sn	Fe	Al	Si	Ni	Cr	Hf	N	O	~482 (~70)	...	...
0.012	1.53	0.14	0.0050	0.0069	0.05	0.11	0.0092	0.0037	0.125			
API ×60 Line Pipe Steel												
C	Mn	P	S	Si	Al	V	Cb			445 (64.6)	589 (85.5)	36.0
0.24	1.34	0.012	0.031	0.01	0.003	0.068	<0.01					

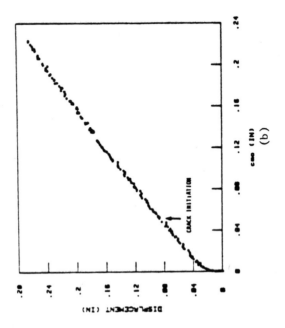

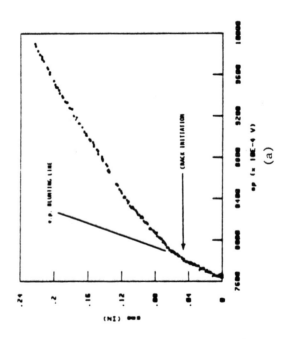

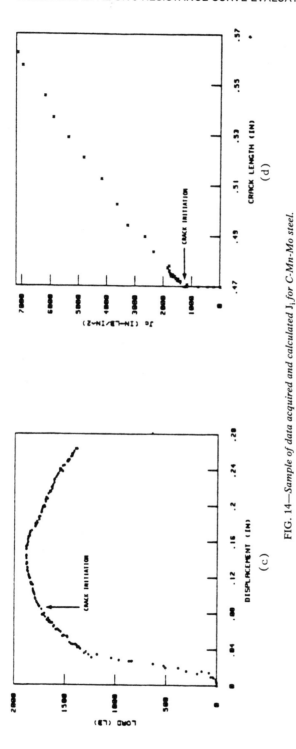

FIG. 14—*Sample of data acquired and calculated* J_i *for C-Mn-Mo steel.*

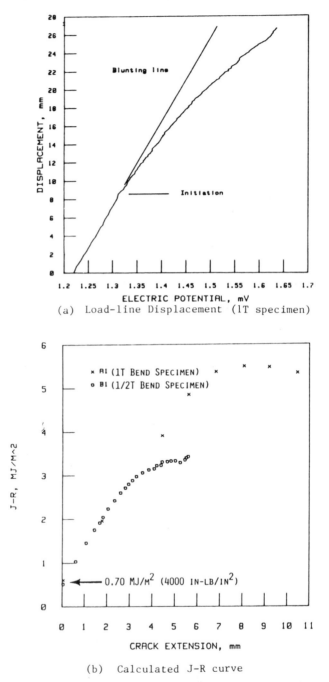

(a) Load-line Displacement (1T specimen)

(b) Calculated J-R curve

FIG. 15—*Data and J-R curve for Type 304 stainless steel bend specimens tested at 20°C (68°F).*

blunting line and point of initiation are indicated in Fig. 15*a*. *J–R* curves from ½T and 1T bend specimens for the same materials are shown in Fig. 15*b*. The J_i value for the material is 700 kJ/m². Beyond a crack extension of 3.8 mm in the ½T specimen and 7.5 mm in the 1T specimen, the crack begins to extend into a region that was previously compressed.

The d-c EP method can also be used to obtain *J–R* curves at elevated temperatures. A 0.2 CT Zircaloy (Zr-2½ Nb) specimen was tested at 149°C (300°F). This was an unusual application because the specimen size was so small (Fig. 16). Figure 17*a* shows the load versus displacement curve from the test. A crack jump towards the end of the test is seen where there is a steep decrease in load. The *J–R* curve for this material is shown in Fig. 17*b*. For this material, J_i was calculated as 66.5 kJ/m² (380 lb-in./in.²). The EP current wire locations were the same as indicated in sketch (i) of Fig. 2*a*; the probe locations were the same as indicated in sketch (i) of Fig. 2*b* and (i) of Fig. 2*c*.

FIG. 16—*Miniature Zircaloy CT specimen used to evaluate J-R curve at 149°C (300°F).*

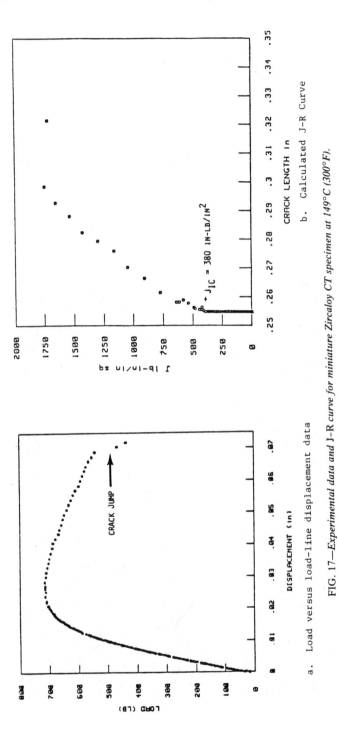

a. Load versus load-line displacement data

b. Calculated J-R Curve

FIG. 17—*Experimental data and J-R curve for miniature Zircaloy CT specimen at 149°C (300°F).*

Another unusual application involved evaluating *J–R* curves at slow and intermediate testing rates. Here 0.3T three-point bend specimens ($a/w =$ 0.57) from X60 line pipe steel were tested at static (1.27 mm/min) and intermediate (12.7 mm/min) rates. The slower rate was such that 0.4 P_{max} was reached in 16 s, which is slower than the 6 s minimum time for *J–R* testing. The intermediate rate corresponded to 0.4 P_{max} reached in 1.6 s. Figure 18 shows *J–R* curves for both rates. The J_i value at intermediate rates was 157.6 kJ/m² (900 lb-in./in.²), while that for the static test was 147.1 kJ/m² (840 lb-in./in.²). The modest increase in J_i was anticipated because of minor dynamic effects on the stress-strain curve beyond initiation; however, such a small increase can be considered as negligible since it is probably within the normal section of J_i at the slow test rate. The *J–R* curves at both rates had the same slope (Fig. 18).

Table 3 summarizes results for the tests conducted on the four materials.

Conclusions

From the work by the authors and others, it has been confirmed that the d-c electric potential (EP) method can be used to accurately monitor crack initiation and crack growth, even if crack tunneling occurs. This method can then easily be adapted for *J–R* curve testing. The d-c EP method has several advan-

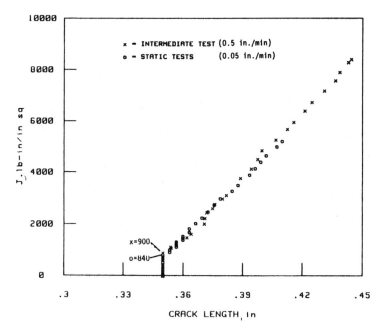

FIG. 18—*Slow and intermediate speed* J-R *curve for* ×60 *line-pipe steel at 20°C (68°F).*

TABLE 3—*Summary of* J-R *curve data.*

	J_i, kJ/m^2 (in.-lb/in.^{2})	dJ/da,^{a} MJ/m^3 (in.-lb/in.^{3})
Mn-Mo steel	210 (1260)	580 (88 300)
$\frac{1}{2}$T 304 stainless steel	700 (4190)	786 (119 700)
1T 304 stainless steel	700 (4190)	786 (119 700)
Zircaloy	67 (398)	275 (41 900)
×60 line pipe steel (intermediate test rate)	158 (944)	602 (91 700)

^{a}Slope of *J-R* curve at initiation.

tages over the unloading compliance method, but care must be taken to electrically isolate the specimen. Adaptation to computerized data acquisition systems is relatively straightforward, which makes J_i or J-R curve testing lower in cost and data reduction less tedious.

References

[1] Wilkowski, G. M. and Maxey, W. A. in *Fracture Mechanics: Fourteenth Symposium— Volume II: Testing and Applications, ASTM STP 791*, American Society for Testing and Materials, 1983, pp. II-266–II-294.

[2] Lowes, J. M. and Fearnehough, G. D., *Engineering Fracture Mechanics*, Vol. 3, 1971, pp. 103–108.

[3] Druce, S. G. and Booth, G. S., "The Effect of Errors in the Geometric and Electric Measurements on Crack Length Monitoring by the Potential Drop Technique," in *The Measurements on Crack Length and Shape During Fracture and Fatigue*, Engineering Materials Advisory Services Ltd., 1980, pp. 136–163.

[4] Schwalbe, K. M. and Mellmann, Dieter, *Journal of Testing and Evaluation*, Vol. 9, No. 3, May 1981, pp. 218–221.

[5] Ritchie, R. O. and Bathe, K. J., *International Journal of Fracture*, Vol. 15, No. 1, Feb. 1979, pp. 47–55.

[6] Srawley, J. E. and Brown, W. F., Jr., in *Fracture Toughness Testing and Its Applications, ASTM STP 381*, American Society for Testing and Materials, 1965, pp. 133–198.

[7] deRoo, P. and Marandet, B., "Application of the A.C. Potential Drop Method to the Prediction of Initiation in Static and Dynamic Testing," in CSNI Workshop on *Ductile Fracture Test Methods*, OECD, Paris, France, Dec. 1982, pp. 75–84.

[8] Ingham, T. and Morland, E., "The Measurement of Ductile Crack Initiation: A Comparison of Data From Multiple and Single Specimen Methods and Some Considerations of Size Effects," Risley Nuclear Power Development Laboratories, UKAEA, Northern Division, England.

[9] Johnson, H. H., *Materials Research and Standards*, Vol. 5, 1965, p. 422.

[10] Ernst, H., Paris, P. C., and Landes, J. D. in *Fracture Mechanics: Thirteenth Conference, ASTM STP 743*, American Society for Testing and Materials, 1981, pp. 476–502.

Dieter Hellmann[1] *and Karl-Heinz Schwalbe*[1]

Geometry and Size Effects on *J-R* and δ-*R* Curves under Plane Stress Conditions

REFERENCE: Hellmann, D. and Schwalbe, K. -H., **"Geometry and Size Effects on *J-R* and δ-*R* Curves under Plane Stress Conditions,"** *Fracture Mechanics: Fifteenth Symposium, ASTM STP 833*, R. J. Sanford, Ed., American Society for Testing and Materials, Philadelphia, 1984, pp. 577–605.

ABSTRACT: *J-R* curves and δ-*R* curves were measured on 5-mm-thick compact, three-point bend, and center cracked tension specimens with different widths and different precrack lengths. The crack tip opening displacement (δ) was measured at the fatigue crack tip. The materials tested were an age-hardened aluminum alloy; the same alloy in a furnace cooled, very soft condition; and an alloy steel. It was found that the *J*-integral correlates three to five times as much crack growth in the tension specimens as in the compact and bend specimens, or, in other words, a valid *R*-curve determined on a center cracked tension specimen covers at least three times more crack growth than a valid *R*-curve determined on a compact or bend specimen of the same width and crack length. Furthermore, δ correlates even more crack growth: a valid δ-*R* curve determined on a compact or bend specimen covers about five times more crack growth than a valid *J-R* curve. For center cracked tension specimens the amount of crack growth measured was not sufficient to establish limits for a valid δ-*R* curve; however, at least 60% of the original ligament width can be expended by crack growth for a valid δ-*R* curve.

It is worth noting that no geometry effect on the *J-R* curve and δ-*R* curve was found as long as the *R*-curve data could be considered "valid."

KEY WORDS: nonlinear fracture mechanics, geometry effect, size effect, *J-R* curve, δ-*R* curve, crack tip opening displacement, plane stress

Nomenclature

a Crack length
a_0 Fatigue precrack length
Δa Amount of crack growth
b Ligament width

[1]Research Engineer and Head of Materials Technology Department, respectively, GKSS-Research Center Geesthacht, Geesthacht, Federal Republic of Germany.

B Thickness
E Young's modulus
J J-integral
J_0 J at onset of crack growth
K_0 Stress intensity at onset of crack growth
n Strain-hardening exponent
W Specimen width or half width (Fig. 1)
ϵ Strain
ϵ_Y Yield strain
δ Crack tip opening displacement
δ^0 Crack mouth opening displacement of CCT specimens
σ Applied stress
σ_F Flow stress
σ_n Net section stress
σ_n^{limit} Limiting net section stress
σ_Y Yield strength
$\sigma_{0.2}$ 0.2% offset yield strength

The R-curve concept claims the ability to predict instability of a cracked structural part. For the prediction of instability two conditions must be satisfied:

1. The R-curve must be a material property which is to a certain degree independent of geometrical parameters.
2. For the structural part in question the driving force must be known.

Although in the field of linear elastic fracture mechanics a considerable proportion of problems seems to be solved satisfactorily, in the case of nonlinear material behavior both conditions given above are far less certain owing to the complexity of the problems.

Size Effects

Theoretical considerations [1-4] lead to certain size requirements for J-controlled crack growth which have to be met for crack growth to be independent of the size of the cracked body:

$$\text{uncracked ligament length} \quad \begin{cases} b > \omega J/(dJ/da) & (1) \\ b > \rho J/\sigma_F & (2) \end{cases}$$

$$\text{allowable amount of crack growth} \quad \Delta a < \alpha b_0 = \alpha(W - a_0) \quad (3)$$

where σ_F is the flow stress and ω, ρ, and α are constants to be determined.
The ability of J to control the crack tip field depends on the loading config-

uration and on the strain-hardening properties of the material. Thus no universal values for the constants ω, ρ, and α are to be expected. From the calculations made in Refs *1* to *4* it can be concluded that high work hardening and bending promote a J-controlled crack tip field; in other words, smaller specimens are acceptable in that case than for low work hardening and tension configurations.

Since nonlinear fracture mechanics is particularly important for nuclear reactor pressure vessel steels, the major part of the research effort has been aimed at the evaluation of these steels. Thus the constants ω, ρ, and α in Eqs 1 to 3 have been evaluated for A533-B steel and plane strain conditions [*1-4*]:

$$\text{bending} \begin{cases} \omega \approx 10 & (4) \\ \rho = 25 \ldots 50 & (5) \\ \alpha \approx 0.06 & (6) \end{cases}$$

$$\text{center cracked tension} \begin{cases} \omega \approx 80 & (7) \\ \rho \approx 200 & (8) \\ \alpha \approx 0.01 & (9) \end{cases}$$

As pointed out by Shih et al [*2*] a COD-based resistance curve may be less restricted with respect to the crack growth than the J-R curve.

To the knowledge of the present authors, systematic experimental studies confirming or contradicting these size requirements are still lacking, although some geometry and size effects have been studied [*5-7*].

Driving Force

The driving force is not only a function of geometry but of the material's flow properties. Some proposals facilitating the computation of J and δ as a driving force have already been made [*8-10*].

The present paper describes part of work in progress which is aimed at finding evidence for:

1. Influence of size and geometry on R-curves.
2. Empirical criteria for conditions beyond which the parameters J and δ no longer correlate crack growth.

In earlier work [*11,12*] it was demonstrated that K-R curves of high-strength materials for different specimen sizes and crack lengths coincide in their initial parts but split off after varying amounts of crack growth. The splitting points (that is, the points beyond which the stress intensity factor K no longer correlates crack growth) satisfy the condition $\sigma_n \approx 0.9\sigma_{0.2}$. It was hoped that criteria for crack growth in the nonlinear regime could be established in a similar way:

1. Ability of the d-c potential drop method and compliance method to measure crack length.

2. Simple analytical expressions for the driving force to predict instability.

The experimental work is being done at GKSS Research Center, Geesthacht, Federal Republic of Germany. An accompanying theoretical investigation is being carried out at the University of Darmstadt.

The first part of the experimental program described in the present paper consisted of in-plane geometry variation used to avoid additional effects of thickness. The thickness was 5 mm and led to full shear fracture, with some exceptions noted in the paper.

Materials and Specimens

The major part of testing was conducted on 2024 aluminum alloy in the age-hardened T351 condition and in an annealed condition which includes furnace cooling to avoid any ageing effects. The material properties are given in Table 1. This material was chosen as a model material since in the furnace-cooled (FC) condition it has a high work-hardening capacity; also, in the FC condition its tensile strength is three times its yield strength ($\sigma_{0.2}$). The high work-hardening capacity is also reflected by the high work-hardening exponent (n), which is defined by the Ramberg-Osgood equation

$$\epsilon/\epsilon_Y = \sigma/\sigma_Y + \beta(\sigma/\sigma_Y)^{1/n} \tag{10}$$

Furthermore, owing to the low strength of the material, testing machines of moderate size can be used even for the center cracked tension specimens. Finally, the as-received T351 condition represents a high-strength aluminum alloy with relatively little work hardening.

Some additional tests were conducted on 35NiCrMo16 steel (Table 1).

Three specimen geometries were investigated: single edge notched three-point bend (SENB), compact (CT), and center cracked tension (CCT)

TABLE 1—*Tensile properties.*[a]

Material	$\sigma_{0.2}$, MPa	Ultimate Tensile Strength, MPa	Work-Hardening Exponent (n)	β
2024-T351 aluminum	317	440	0.091	0.8
2024-FC aluminum	75	217	0.294	1.5
35NiCrMo16 steel	510	726	0.161	2.5

[a]σ_Y in Eq 10 was set equal to $\sigma_{0.2}$.

(Fig. 1). Because of symmetry, the specimen width of the CCTs was designated as $2W$.

For the width and half width (W) the values of 50 and 100 mm were chosen; the steel, however, was tested only with $W = 50$ mm. A single specimen with $W = 25$ was tested. In all cases, thickness (B) was 5 mm. The precrack length ratio a_0/W varied between 0.2 and 0.9; Table 2 shows the test matrix.

Test Procedure

The testing machines employed were 40, 160, and 1000 kN Schenck servohydraulic machines. To keep the growing crack stable as long as possible, in the case of the CCT specimens the testing machines were controlled by the crack mouth opening displacement (COD) clip gage. For the compact and bend tests the testing machines were stroke controlled. In this manner all

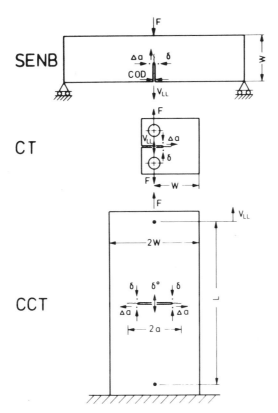

FIG. 1—*Specimen types used for the present investigation with indication of the quantities measured.*

tests could be extended far beyond maximum load. The design of the clip gage was described in Ref *12*.

The compact and bend tests were conducted in accordance with ASTM Test for J_{Ic}, a Measure of Fracture Toughness (E 813), except that requirements for the validity of J_{Ic} measurements were not followed. An outline of the test procedure for the CCT specimens is given in Refs *13* and *14*.

The following quantities were measured:

CCT Specimens:

• Specimen elongation (v_{LL}) between gage points on the specimen close to the grips; gage span $= L$ (Fig. 1).
• Crack mouth opening displacement (δ^0) in the specimen's centerline [*12–14*] for purposes other than *R*-curve determination (not covered by the present paper).

CT Specimens:

• Crack opening displacement along the load line in accordance with ASTM E 813.

SENB Specimens:

• Deflection of specimen by a special mechanism to exclude the roller indentations from the measurement [*15*].

TABLE 2—*Test matrix.*

Material		2024-FC			2024-T351			35NiCrMo16		
W, mm	a_0/W	CT	SENB	CCT	CT	SENB	CCT	CT	SENB	CCT
25	0.7	...	...	X	...	...	...	...	...	...
	0.2	...	X	...	...	...	...	...	X	...
	0.3	X	X	...	...	...	...	X	X	...
	0.4	...	...	X	...	...	...	...	...	X
50	0.5	XX	XX	X	X	X	X	X	X	...
	0.6	X	...	...	...	...	...	...	X	...
	0.7	XX	X	X	...	...	...	X	X	...
	0.8	...	...	X	X	...	...	...	...	X
	0.9	...	...	...	...	...	...	...	...	X
	0.2	...	...	X	...	...	...	...	...	...
	0.3	X	...	...	...	...	...	...	...	...
	0.4	...	...	...	...	...	...	...	...	...
100	0.5	XX[a]	X[a]	X	...	X[a]	X	...	...	...
	0.6	...	...	...	...	...	...	...	...	...
	0.7	XX[a]	...	X	...	...	...	...	...	...
	0.8	...	...	...	...	...	...	...	...	...

[a] Specimen tested with antibuckling guides.

All Specimen Types:

- Load.
- Crack tip opening displacement (δ) measured at the original fatigue crack tip using a special displacement gage (Fig. 2) attached to the specimen by a thin blade made of spring steel (gage span = 5 mm).
  - Crack length determined by the d-c potential drop method [16].

Some CT specimens were equipped with antibuckling guides (Table 2).

Data Evaluation

Crack length was calculated from the potential drop signal using Johnson's formula [16]. The *J*-integral was evaluated as follows:

CCT Specimens:

J at initiation of growth [17]:

$$J_0 = \frac{K_0^2}{E} + \frac{U_0^*}{B(W - a_0)} \tag{11}$$

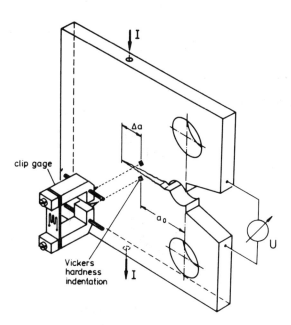

FIG. 2—*Measurement of crack tip opening displacement at fatigue crack tip.*

J during growth, modified for growth in analogy to the procedure reported in Ref *18*:

$$J_i = J_{i-1} + \frac{2\,\Delta U^*}{B(b_{i-1} + b_i)} + \frac{2}{E(b_{i-1} + b_i)}[K_i^2 b_i - K_{i-1}^2 b_{i-1}] \quad (12)$$

where i and $i - 1$ indicate two consecutive points on the test record. For U_0^* and ΔU^* see Fig. 3, and for the other symbols see Nomenclature.

CT Specimens:

J at initiation of growth [19]:

$$J_0 = \frac{U_0}{B(W - a_0)}\,\eta\left(\frac{a_0}{W}\right) \quad (13)$$

J during growth [20]:

$$J_i = \left[J_{i-1} + \eta_{i-1}\frac{\Delta U}{B(W - a_{i-1})}\right]\left[1 - \frac{\gamma_{i-1}}{W - a_{i-1}}(a_i - a_{i-1})\right]$$

$$\eta_i = 2 + 0.522\left(1 - \frac{a_i}{W}\right) \quad (14)$$

$$\gamma_i = 1 + 0.76\left(1 - \frac{a_i}{W}\right)$$

SENB Specimens:

J at initiation of growth [21]:

$$J_0 = \frac{2U_0}{B(W - a_0)} \quad (15)$$

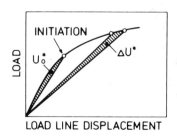

FIG. 3—*Evaluation of J-integral for CCT specimens in accordance with Ref 17.*

J during growth [*18*]:

$$J_i = J_{i-1} \frac{W - a_i}{W - a_{i-1}} + \frac{2\Delta U}{B(W - a_{i-1})} \tag{16}$$

where U_0 is the deformation energy at initiation of growth determined from the area under the load-deflection curve, and ΔU is the area bounded by the actual test record trace and the lines of constant displacement $v_{LL_{i-1}}$ and v_{LL_i}.

Finite-element calculations [*22*] demonstrated that the *J*-formulas used for the test evaluation work very well even for the smallest *a/W* ratios investigated.

Results

The results obtained on SENB and CT specimens for varying width and crack length are very similar, so that it is only necessary to show some typical

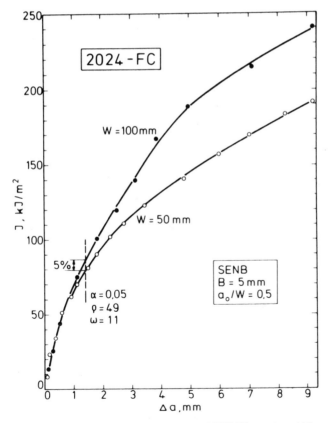

FIG. 4a—*J-R-curves for bend specimens of 2024-FC, varying widths.*

examples. Furthermore, owing to limited space, mainly the *R*-curves of 2024-FC will be shown; the other two materials behaved in a very similar way. Figures 4 and 5 show the effect of width and crack length for SENB and CT specimens made of 2024-FC. At the beginning, the *J–R* curves and *δ–R* curves are independent of specimen width and precrack length. After certain amounts of crack growth that depend on *W* and a_0/W, however, curve splitting occurs. The most interesting detail is that the crack tip opening displacement (δ) can correlate much larger amounts of crack growth than *J*.

In Figs. 4*a* and 4*b* it is indicated that the points of curve splitting were defined as 5% deviation in J or δ of the curve with the smaller ligament from the curves with the larger ligaments. For these points the parameters α, ρ, and ω from Eqs 1 to 3 were evaluated (see Discussion).

One of the *R*-curves plotted in Fig. 5*b* is considered to be invalid, since the crack path was oblique with respect to the ligament. Some specimens were equipped with antibuckling guides to see whether buckling affected the results. Since no large effect could be detected (Fig. 5*b*), the remaining specimens were tested without antibuckling guides.

Apart from the specimen with $a_0/W = 0.2$, the CCT specimens also ex-

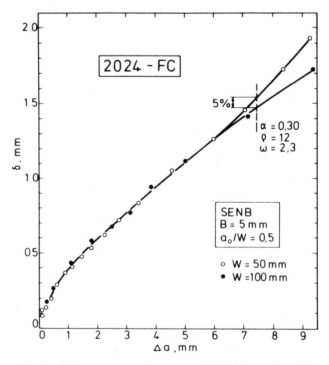

FIG. 4*b*—*δ-R curves for bend specimens of 2024-FC, varying widths.*

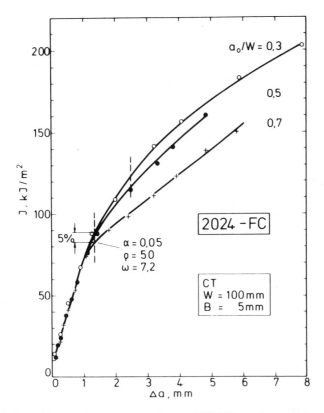

FIG. 5a—*J-R-curves for compact specimens of 2024-FC, varying crack lengths.*

hibited an initial *R*-curve section independent of a_0/W and W (Fig. 6*a*). Curve splitting, depending on W and a_0/W, occurs after much larger amounts of crack growth than on CT specimens (and SENB specimens not shown here). The specimen with $a_0/W = 0.2$ exhibited a higher crack growth resistance in terms of the *J*-integral than the other configurations of Fig. 6*a*.

With one exception, no curve splitting of the *δ-R* curves of the CCT specimens can be observed within the measured range of crack growth (Fig. 6*b*). It is remarkable that even the specimen with $W = 25$ mm and $a_0/W = 0.7$ lies in the scatterband although almost the complete original ligament was consumed by crack growth. The only curve deviating from the common scatterband is that of the specimen with $W = 50$ mm and $a_0/W = 0.7$. Since this specimen deviates from the general trend, it is assumed that the test is invalid although neither the specimen nor the test records exhibit anomalous behavior. This test will be duplicated as soon as possible to see whether the result will be reproduced or follow the general trend. It is of par-

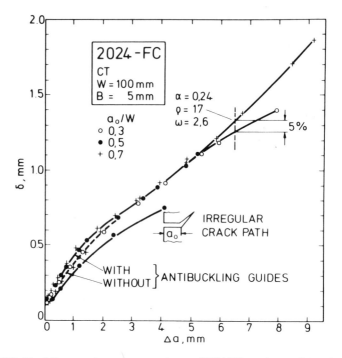

FIG. 5b—δ-R *curves for compact specimens of 2024-FC, varying crack lengths.*

ticular interest that the data points of $a_0/W = 0.2$ lie in the common scatter-band in spite of the large deviation of the *J-R* curve (Fig. 6*a*). Similar trends were observed on the other two materials, but with different absolute values of the parameters α, ρ, and ω.

Discussion

J-R *Curves*

For a given specimen geometry the most important quantity with respect to curve splitting is the width of the ligament $b = W - a_0$ because curve splitting occurs earlier in specimens which are narrower or which have longer fatigue precracks (Fig. 7). The symbols α, ρ, and ω, the values of which were determined at the splitting points, will now be used with the index *J* to indicate that they refer to *J-R* curves. The values for the α_J, ρ_J, and ω_J criteria (Eqs 4 to 9) exhibit the ranges shown in Table 3 (the individual values are plotted in Fig. 8). Table 3 contains Eqs 17 to 22.

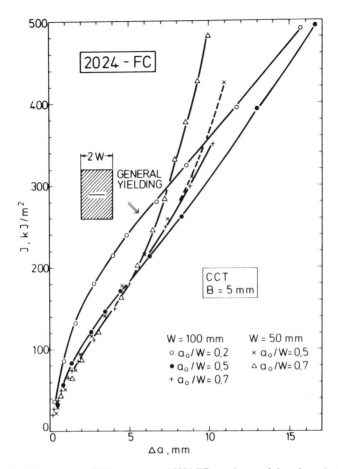

FIG.6*a*—J-R *curves for CCT specimens of 2024-FC, varying crack lengths and widths.*

All 2024-FC and 35NiCrMo16 specimens exhibited full shear fractures. Thus the results may be attributed to a plane stress situation as opposed to the results in Eqs 4 to 9 which have been derived for plane strain situations. The age-hardened 2024-T351 alloy, however, exhibited a certain degree of normal stress fracture, amounting to about 60% of the thickness.

The exact values for α_J, ρ_J, and ω_J may depend on:

1. The way in which they were derived: comparison of experimental *R*-curves as in the present paper, comparison of theoretical near-field and far-field *J* as in Ref 2, or comparison of the near tip field in fully plastic condition with the small-scale yielding solution as in Ref 3.

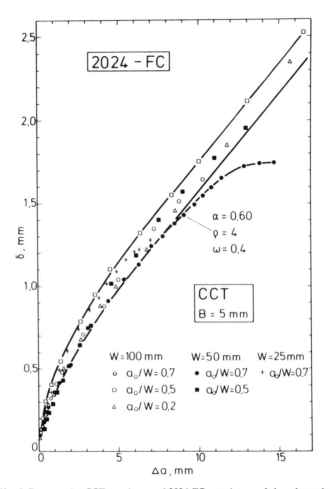

FIG. 6b—δ-R *curves for CCT specimens of 2024-FC, varying crack lengths and widths.*

2. The amount of deviation which is taken to establish the size require-
ments α_J, ρ_J, and ω_J. This deviation is 5% in J and δ in the present investigation, but is less well-defined in other investigations.

3. The strain-hardening properties of the material.

4. The state of stress.

Thus a comparison of our results with those given in the literature is difficult.
In spite of this fact, however, and although there is much scatter in values for
α_J, ρ_J, and ω_J some provisional conclusions can be drawn:

1. In spite of the different methods of verifying the criteria for J-controlled
crack growth, our data for α_J, ρ_J, and ω_J obtained for the bending configurations are in remarkably good agreement with the data given by Eqs 4 to 6.

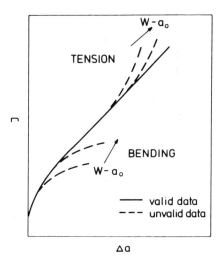

FIG. 7—*Schematic of splitting behavior of* J-R *curves as a function of ligament width and loading mode.*

2. Future theoretical and experimental work aimed at establishing reliable minimum size requirements should be based on comparable definitions of a breakdown of *J*-control in order to obtain results from different sources that really can be compared to each other. For example, if one looks at Fig. 29 in Ref *2*, one recognizes that 6% crack extension is characterized by a 20% difference between the near-field J ($\hat{=}J_2$) and the far-field J. If the difference were reduced to 5%, the allowable amount of crack growth would shrink to about 3% and consequently ρ and ω would increase.

3. There does not seem to be a significant effect of the crack tip constraint in the bending configurations on the minimum size requirements for *J*-controlled crack growth, since for the whole range from plane strain to plane stress similar results are obtained.

TABLE 3—*Values of* α, ρ, *and* ω *at the splitting points of* J-R *curves.*

		2024-FC	2024-T351	35NiCrMo16	Eq
Bending (CT + SENB)	$\omega_J = 4$ to 11		4.5 to 7	12 to 14	(17)
	$\rho_J = 27$ to 60		50 to 85	34 to 40	(18)
	$\alpha_J = 0.04$ to 0.1		0.06 to 0.07	0.04	(19)
Tension (CCT)	$\omega_J = 2$ to 3		5.6	5.5 to 7	(20)
	$\rho_J = 7$ to 13		47	9 to 10	(21)
	$\alpha_J = 0.25$ to 0.35		0.1	0.12 to 0.17	(22)

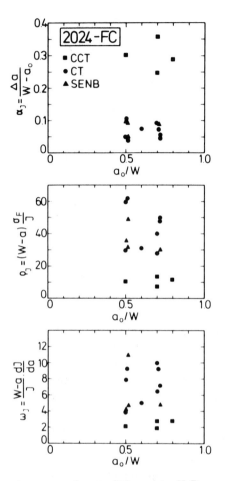

FIG. 8—*Values for α_J, ρ_J, and ω_J at splitting points of* J-R *curves for 2024-FC.*

4. From theoretical considerations (see, for example, Ref *4*) one would expect that higher strain-hardening exponents would lead to less strict size requirements. Considering the ρ_J-values in Eqs 18 and 21, less hardening materials (2024-T351) do indeed seem to require larger minimum specimen sizes. However, the data obtained on 2024-T351 are not sufficient for a convincing conclusion.

5. An interesting result is that in contrast to plane strain the CCT specimens in plane stress have a much higher measurement capacity than the bending configurations (compare Eqs 7 and 8 with Eqs 20 to 22). In terms of the allowable amount of crack growth, tension is by a factor of at least three superior to bending. For 2024-T351 the difference between bending and tension is smaller since there is some fraction of plane strain fracture mode.

6. Using only *R*-curve data qualified by the splitting point criterion (that is, no data of a curve were used that were beyond the curve's splitting point as defined by Fig. 4*a*), a geometry-independent *R*-curve is obtained for each material (Fig. 9). These *R*-curves are geometry-independent in the sense that, within the range of variables investigated, a variation of the in-plane parameters crack length, specimen width, and loading mode yields *R*-curve data that fall into a common scatter band. As pointed out previously, however, in a specific test the termination of a valid *R*-curve depends on the ligament length and loading mode (Fig. 8). The *R*-curves shown in Fig. 9 are of course valid only for specimen thickness equal to 5 mm.

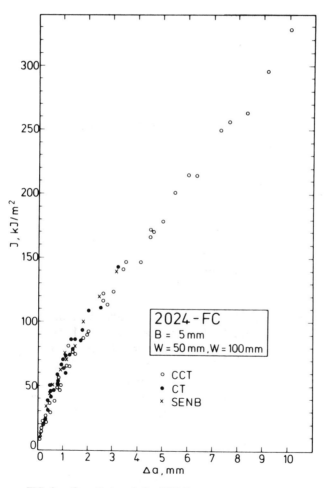

FIG. 9*a*—*Compilation of all valid J–R curve data for 2024-FC.*

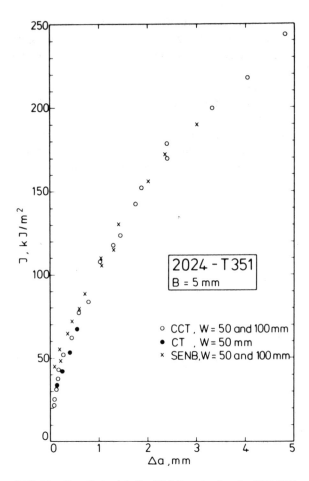

FIG. 9b—*Compilation of all valid J-R curve data for 2024-T351.*

When determining a *J-R* curve remote yielding must be carefully avoided, since it contributes to the load line displacement and hence to the nominal *J* although it has nothing to do with the crack. Figure 6a shows clearly that *J* can be considerably overestimated when remote yielding occurs. It can be shown for the *R*-curve of the specimen with $a_0/W = 0.2$ that *J* becomes bigger than that of the other specimens exactly at that point where general yield is reached, that is, where the gross section stress equals $\sigma_{0.2}$.

Significance of δ

The crack tip opening displacement (CTOD) generally depends on gage length, state of stress, and strain-hardening properties [4,23]. For the ad-

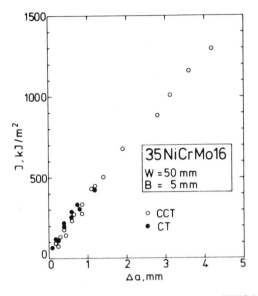

FIG. 9c—*Compilation of all valid J-R curve data for 35NiCrMo16.*

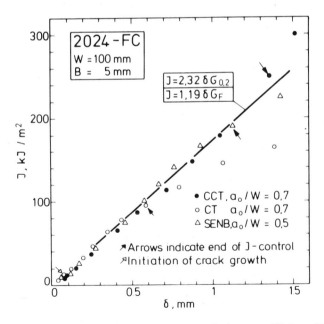

FIG. 10—*J-integral as a function of crack tip opening displacement (δ) for a CT, a SENB, and a CCT specimen of 2024-FC. Only qualified J-data are used for establishing the J-δ relationship; nonvalid J-data are plotted to show deviations from trend of valid data.*

vancing crack the arrangement shown in **Fig. 2** measures the relative displacement of the crack faces, since the material between the gage points and the crack faces is essentially stress free. Before initiation of growth a certain contribution of the material elements between the gage points is to be expected; this is, negligible, however, in the case of ideally plastic materials [24]. A detailed finite-element study is presently being conducted to gain further insight into the effect of a finite gage length on pre-initiation CTOD.

The influence of strain hardening on CTOD can be studied by relating the J- and δ-values for each material. Figure 10 shows data for three specimens of 2024-FC. These data and those of the other materials investigated yield

$$2024\text{-FC} \qquad J = 2.32\delta\sigma_{0.2} \qquad\qquad (23)$$

$$2024\text{-T351} \qquad J = 1\delta\sigma_{0.2} \qquad\qquad (24)$$

$$35\text{NiCrMo16} \qquad J = 1.33\delta\sigma_{0.2} \qquad\qquad (25)$$

These results show that for a given J the CTOD decreases with increasing strain hardening of the material.

The influence of strain hardening on CTOD can be treated explicitly; that is, the CTOD is obtained numerically [4] or analytically [23]. A very simple approach (ASTM E 813) is to start with the ideally plastic relationship

$$J \approx \sigma_Y \cdot \delta \qquad\qquad (26)$$

and to replace the yield strength (σ_Y) by the flow stress

$$\sigma_F = 0.5(\sigma_{0.2} + \text{UTS}) \qquad\qquad (27)$$

One then obtains

$$J \approx \sigma_F \cdot \delta \qquad\qquad (28)$$

For the three materials tested the following results are obtained (Fig. 10):

$$2024\text{-FC} \qquad\qquad J = 1.19\delta\sigma_F \qquad\qquad (29)$$

$$2024\text{-T351} \qquad\qquad J = 0.84\delta\sigma_F \qquad\qquad (30)$$

$$35\text{NiCrMo16} \qquad\qquad J = 1.1\delta\sigma_F \qquad\qquad (31)$$

The factor on the right-hand side of these equations is not exactly equal to unity; however, the relatively close data grouping about Eq 28 shows that much of the strain-hardening effect is described by that equation.

It is worth noting (Fig. 10) that there is a unique relationship between J and δ for all specimen types. This is a prerequisite for geometry-independent *J-R* and δ-*R* curves.

δ-R *Curves versus* J-R *Curves*

Although it was expected that the CTOD would yield more consistent results than *J*, it was surprising to see the large degree of difference between the ability of δ and *J* to correlate stable crack growth. Figure 11 shows the values of α_δ, ρ_δ, and ω_δ for the splitting points of the δ-*R* curves so far as they could be detected; since in the case of the CCT specimens only one splitting

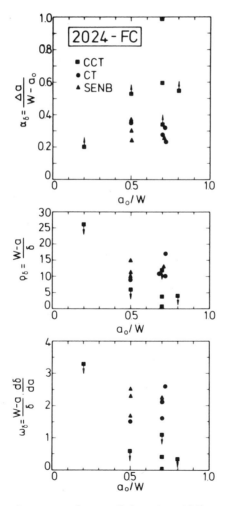

FIG. 11—*Values for α_δ, ρ_δ, and ω_δ at splitting points of δ-R curves for 2024-FC.*

TABLE 4—*Values of α, ρ, and ω at the splitting points of δ-R curves.*

				2024-FC	2024-T351	35NiCrMo16
Bending (CT + BEND)	ω_δ =	1.5 to 2.6			1.9 to 2.3	2.8
	ρ_δ =	9 to 17			12 to 17	8
	α_δ =	0.24 to 0.37			0.28 to 0.30	0.26
Tension (CCT)	ω_δ <	0.4 to 3.3			<1.1	<1.6 to 2.2
	ρ_δ <	4 to 26			<13	<5 to 7
	α_δ >	0.2 to 0.6			>0.37	>0.25 to 0.31

point could be observed, the values given in Fig. 11 for the CCTs are minimum values for α_δ and maximum values for ρ_δ and ω_δ, taken at the points of final crack extension. The values obtained are shown in Table 4.

Figure 12 presents the ranges of α_J and α_δ obtained for the three materials tested together with the plane strain values of Eqs 6 and 9. The graph shows that for *J-R* curves the measurement capacity of the bending configurations with respect to the amount of crack growth is independent of the material and state of stress. The measurement capacity of the CCT specimens, however, can be about five times bigger for a high work-hardening material in plane stress, decreasing with decreasing work hardening, and reaching a very low level under plane strain conditions. Concerning δ-R curves, the measurement capacity of CT and SENB specimens again remains unaffected by the material but is roughly five times as high as that for *J-R* curves. The capacity of the

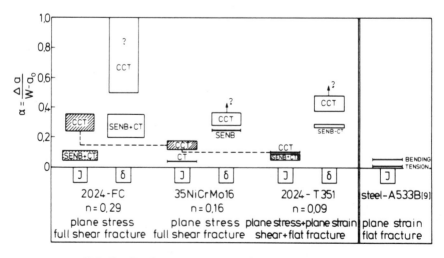

FIG. 12—*Graphical display of scatterbands of α_J and α_δ values.*

CCT specimens is still unknown but seems to be at least 50% of the original ligament. Additional tests will be done to clarify this point. If experimental data are only taken when they are qualified according to the above conditions, a geometry-independent *R*-curve is obtained. Figure 13 shows the data for 2024-FC. This geometry-dependent *R*-curve was found for *J*-control as well (Fig. 9), but now it has much larger amounts of crack growth.

The superiority of δ versus *J* with respect to correlation of crack growth may be explained qualitatively. It has been shown by theoretical considerations [1–4] that beyond a certain stage of deformation or a certain amount of crack growth the *J*-integral no longer represents a one-parameter characterization

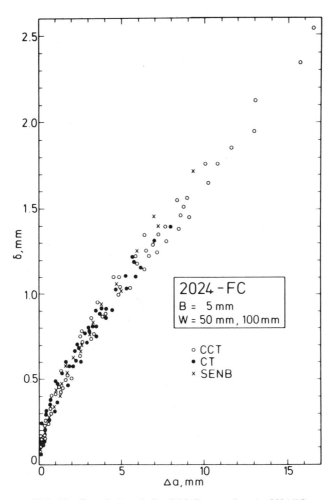

FIG. 13—*Compilation of all valid δ-R curve data for 2024-FC.*

of the events occurring at the crack tip. In addition, a formally determined J becomes meaningless if the load line deflection is affected by remove plasticity (Fig. 6a, specimen with $a_0/W = 0.2$).

Both of these cases illustrate the limits that are set to a global parameter (determined by applied load and load point displacement) to describe the crack tip state. None of these limitations seem to apply for δ since this is a quantity determined locally at the site of interest. Its magnitude is a direct measure of the events occurring at the crack tip irrespective of the conditions remote from the crack tip. The problem with J (in the experiments the far-field J is determined) is the transfer function that has to determine uniquely

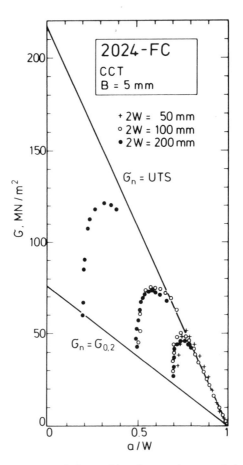

FIG. 14a—*Development of applied stress (σ) and net section stress (σ_n) during crack growth in CCT 2024-FC specimens.*

the local behavior from the global behavior whereas by δ the local behavior is determined directly.

These considerations and the experimental evidence presented in this paper suggest that in plane stress the CTOD is a much better correlation parameter for crack growth than the *J*-integral as determined experimentally by present practice.

Ligament Conditions During Crack Growth

In a CCT specimen the nominal net section stress (σ_n) is a good indicator for the degree of plasticity the ligament underwent during the test. The graphs in Fig. 14 show that in case of the very tough materials 2024-FC and 35NiCr-Mo16 initiation of crack growth coincides with the attainment of net section yield conditions; that is, all the crack growth occurs under fully plastic conditions.

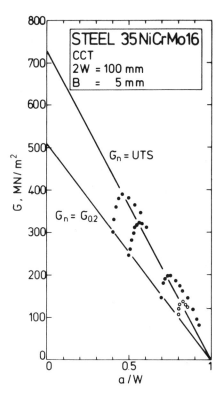

FIG. 14*b*—*Development of applied stress (σ) and net section stress (σₙ) during crack growth in CCT 35NiCrMo16 specimens.*

The following conclusions can be drawn from Figs. 14a and 14b:

1. The maximum load carrying capacity of the actual net section is at or somewhat beyond the locus given by the tensile strength (UTS) of the material:

$$\sigma_n^{\text{limit}} = \text{UTS} \tag{32}$$

2. This condition is reached after maximum load. Thus the maximum load is not a measure of limiting net section conditions in terms of net section stress as given by Eq 32.

3. Neither the maximum net section stress nor the net section stress at maximum load coincides with the often used limit load formula

$$\sigma_n^{\text{limit}} = 0.5(\sigma_{0.2} + \text{UTS}) \tag{33}$$

4. An assessment of the real net section limit conditions (as given by Eq 32) based on the precrack length a_0 is nonconservative; that is, the limiting net section stress $\sigma_n^{\text{limit}} = \text{UTS}$ will be reached during the course of loading. This will occur only after appreciable crack growth whereby the gross stress (or applied load) can be reduced significantly (Fig. 15). On the other hand, in cases

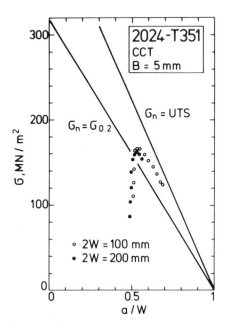

FIG. 14c—*Development of applied stress (σ) and net section stress (σ_n) during crack growth in CCT 2024-T351 specimens.*

where a rough estimate for a quick assessment is needed, a_0 can be used to calculate a net section stress and equate it to 0.5 $(\sigma_{0.2}$ + UTS) in order to estimate the maximum load (Fig. 15). However, the real conditions can only be obtained by an R-curve analysis. According to Fig. 14c, in 2024-T351 crack growth starts well below net section yield. Within the range of measured Δa the condition given by Eq 32 was not attained in this material.

Conclusions

The investigation of stable crack growth under conditions of plane stress and fully plastic behavior resulted in the following conclusions:

1. In the initial part of the J-R curve there is no effect of specimen width, precrack length, and specimen type.

2. Curve splitting occurs depending on the width of the original ligament $(W - a_0)$ and on the specimen type.

3. This curve splitting was used to establish minimum size criteria for the specimen. These criteria show good coincidence with those published in the literature for plane strain in the case of the CT and SENB specimens. CCT specimens, however, have a bigger measurement capacity than the bending configurations.

4. When establishing limitation criteria for J to correlate cracl growth theo-

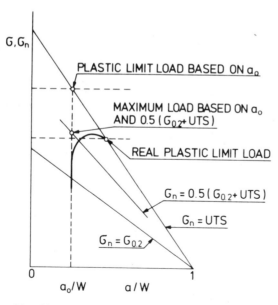

FIG. 15—*Same as Figs. 14, schematically, together with indication for limit load assessments.*

retically or experimentally, more accurate definitions of this limitation are needed (for example, in terms of allowable errors).

5. The ability of the crack tip opening displacement (δ) to correlate crack growth is much better than that of the J-integral; for a given specimen configuration with bend-type loading δ can correlate about five times as much crack growth as J.

6. While in some circumstances J may be overestimated due to remote plasticity, no such error is possible for δ since δ measures the local behavior whereas J is determined by the global response of the specimen.

Acknowledgments

The authors gratefully acknowledge the financial support of this work by the Deutsche Forschungsgemeinschaft. Fruitful discussions with Prof. T. Seeger, Prof. D. Gross, H. Amstutz, and A. Cornec are appreciated.

References

[1] Hutchinson, J. W. and Paris, P. C. in *Elastic-Plastic Fracture, ASTM STP 668*, American Society for Testing and Materials, 1979, pp. 37–64.

[2] Shih, C. F., de Lorenzi, H. G. and Andrews, W. R. in *Elastic-Plastic Fracture, ASTM STP 668*, American Society for Testing and Materials, 1979, pp. 65–120.

[3] McMeeking, R. M. and Parks, D. M. in *Elastic-Plastic Fracture, ASTM STP 668*, American Society for Testing and Materials, 1979, pp. 175–194.

[4] Shih, C. F., *Journal of the Mechanics and Physics of Solids*, Vol. 29, 1981, pp. 305–326.

[5] Vassilaros, M. G., Joyce, J. A., and Gudas, J. P. in *Fracture Mechanics: Twelfth Conference, ASTM STP 700*, American Society for Testing and Materials, 1980, pp. 251–270.

[6] Garwood, S. J. in *Fracture Mechanics: Twelfth Conference, ASTM STP 700*, American Society for Testing and Materials, 1980, pp. 271–295.

[7] Tanaka, K. and Harrison, J. D., *International Journal of Pressure Vessels and Piping*, Vol. 6, 1978, pp. 177–201.

[8] Bucci, R. J., Paris, P. C., Landes, J. D., and Rice, J. R. in *Fracture Toughness, Proceedings of the 1971 National Symposium on Fracture Mechanics, Part II, ASTM STP 514*, American Society for Testing and Materials, 1972, pp. 40–69.

[9] Shih, C. F., German, M. D., and Kumar, V., *International Journal of Pressure Vessels and Piping*, Vol. 9, 1981, pp. 159–196.

[10] Paris, P. C. and Johnson, R. E., "A Method of Application of Elastic-Plastic Fracture Mechanics to Nuclear Vessel Analysis," undated manuscript.

[11] Schwalbe, K.-H. and Setz, W. in *Proceedings*, Third European Colloquium on Fracture, J. C. Radon, Ed., Pergamon Press, London, 1980, pp. 277–285.

[12] Schwalbe, K.-H. and Setz, W., *Journal of Testing and Evaluation*, Vol. 9, 1981, pp. 182–194.

[13] Schwalbe, K.-H., Setz, W., Schwarmann, L., Geier, W., Wheeler, C., Rooke, D., de Koning, A. U., and Easterbrook, J., "Anwendung der Bruchmechanik auf Querschnitte geringer Dicke (Application of Fracture Mechanics to Thin-Walled Structures)," German text with English summary, *Fortsch.-Ber. VDI-Z*, Vol. 18, No. 9, 1980, 114 pp.

[14] Wheeler, C., Eastabrook, J. N., Rooke, D. P., Schwalbe, K.-H., Setz, W., and de Koning, A. U., *Journal of Strain Analysis*, Vol. 17, 1982, pp. 205–213.

[15] Schwalbe, K.-H., "Test Techniques," in *Advances in Fracture Research*, Vol. 4, Proceedings of the 5th International Conference on Fracture (ICF5), Cannes, France, 1981, published by Pergamon Press, London, 1982.

[*16*] Schwalbe, K.-H. and Hellmann, D., *Journal of Testing and Evaluation*, Vol. 9, 1981, pp. 218-221.

[*17*] Landes, J. D., Walker, H., and Clarke, G. A. in *Elastic-Plastic Fracture, ASTM STP 668*, American Society for Testing and Materials, 1979, pp. 266-287.

[*18*] Garwood, S. J., Robinson, J. N., and Turner, C. E., *International Journal of Fracture*, Vol. 11, 1975, p. 528.

[*19*] Clarke, G. A. and Landes, J. D., *Journal of Testing and Evaluation*, Vol. 7, 1979, p. 264.

[*20*] Ernst, H. A., Paris, P. C., and Landes, J. D. in *Fracture Mechanics: Thirteenth Conference, ASTM STP 743*, American Society for Testing and Materials, 1981, pp. 476-502.

[*21*] Rice, J. R., Paris, P. C., and Merkle, J. G. in *Progress in Flaw Growth and Fracture Toughness Testing, ASTM STP 536*, American Society for Testing and Materials, 1973, pp. 231-245.

[22] Amstutz, H., report in preparation.

[*23*] Schwalbe, K.-H., *International Journal of Fracture*, Vol. 9, 1973, p. 381.

[*24*] Cornec, A., report in preparation.

P. De Roo,[1] *B. Marandet,*[1] *G. Phelippeau,*[1]
and G. Rousselier[2]

Effect of Specimen Dimensions on Critical *J*-Value at the Onset of Crack Extension

REFERENCE: De Roo, P., Marandet, B., Phelippeau, G., and Rousselier, G., "**Effect of Specimen Dimensions on Critical *J*-Value at the Onset of Crack Extension,**" *Fracture Mechanics: Fifteenth Symposium, ASTM STP 833*, R. J. Sanford, Ed., American Society for Testing and Materials, Philadelphia, 1984, pp. 606–621.

ABSTRACT: This paper presents the experimental results of a test program designed to evaluate the effect of specimen dimensions on the critical *J* transition curve. These tests were conducted, along with K_{Ic} measurements, on four quenched-and-tempered medium strength steels with 15 to 200 mm thick specimens. In the cleavage fracture domain, J_c yielded toughness values that overestimated K_{Ic}, especially with the smallest specimens. Although ASTM E 813 requirements were met, the cleavage to ductile tearing transition temperature increased with increasing the *J*-specimen size. The level of the ductile plateau was found to be strongly dependent on the uncracked ligament length. It is pointed out that J_{Ic} measurements may overestimate the plane-strain fracture toughness, the ASTM E 813 size requirements being not restrictive enough.

KEY WORDS: fracture criterion, size effect, cleavage, slow crack growth, *J*-integral, ferritic steels, ductile brittle transition, elastic-plastic

A considerable effort has been devoted for the past ten years to characterizing the fracture behavior of structural materials in the elastic-plastic regime by a one-parameter criterion such as J_{Ic}. Many workers think that under some circumstances the critical value J_{Ic} measured on a subsized specimen can be used as a means of determining the plane-strain fracture toughness K_{Ic}. According to the ASTM E 813 procedure for the experimental determination of

---

[1] Institut de Recherches de la Sidérurgie Française (IRSID), Saint Germain en Laye, France.
[2] Electricité de France (EDF) Les Renardières, Moret Sur Loing, France.

J_{Ic}, valid results are obtained if the initial ligament length $(W - a)$ and the thickness B meet the requirements

$$(W - a), \qquad B \geq 25 \, \frac{J_{Ic}}{\sigma_y} \qquad \qquad (1)$$

where σ_y is the effective yield strength of the material at the test temperature.

To aid comparison with fracture toughness K_{Ic}, an equivalent stress intensity factor is usually derived from J_{Ic} through the equation

$$K_{Jc} = \sqrt{\frac{E \, J_{Ic}}{(1 - \nu^2)}}$$

The same formula is used for calculating K_{Jc} from the critical J-value, J_c, measured at cleavage initiation. Much experimental evidence [1,2] has supported the argument that, in the temperature range where fracture occurs suddenly by cleavage, there is quite good agreement between K_{Ic} and K_{Jc}. Recent results [3-7] show, however, that tests on specimens which meet the thickness and the ligament length requirements (Eq 1) can produce K_{Jc} values that overestimate K_{Ic} by a factor of two or more. A statistical model based on the Weibull distribution has been proposed to explain these differences [3,8].

In the temperature range where slow crack growth and ductile fracture take place, the independence of J_{Ic} values with respect to the specimen dimensions B and $(W - a)$ has been inferred from a limited amount of data concerning a particular class of steels [1,9]. Other results show, on the contrary, that the measured J_{Ic} values increase continuously with increasing specimen size although the conditions on minimum dimensions are satisfied [10-12]. In spite of these contradictions, it is usually considered that J_{Ic} measurements yield conservative estimates of K_{Ic} when fracture occurs by ductile tearing instabilities [1,7,13].

In this paper we present the experimental results of a systematic test program designed to evaluate the influence of specimen size on the cleavage to ductile fracture transition and to determine the effect of thickness and uncracked ligament length on the critical value J_{Ic} measured at the onset of stable crack growth.

Materials and Specimens

The materials used in this investigation were three grades of Ni-Cr-Mo (AFNOR 28NCD 8-5), Ni-Cr-Mo-V (AFNOR 26NCDV 11-6), and Cr-Ni-Mo (AFNOR 2ONCD 8) rotor forging steels, plus a Ni-Cr-Mo (AFNOR 2ONCD 14) steel plate for reactor coolant pump fly-wheels. The chemical compositions, heat treatments, and mechanical properties are presented in Tables 1 to 3 respectively.

These quenched-and-tempered medium strength steels exhibited a bainitic microstructure. Great care was taken in studying the microstructure gradients in the plate, particularly in the forgings where segregations and a non-uniform inclusion distribution might lead to toughness gradients through the disks. Compact tension specimens were removed from between the 1/4 and 3/4 thickness after Charpy V tests had shown the best toughness homogeneity at this location.

Preliminary tests on standard compact tension specimens of different sizes (25 to 200 mm thick) were carried out to determine the plane-strain fracture toughness (K_{Ic}) of each material at various temperatures in the transition regime. The J-tests were subsequently performed with compact specimens (hereafter referred to as CTJ specimens) modified in order to permit the mea-

TABLE 1—*Chemical composition (weight percent) of investigated steels.*

Steel	C	Mn	Si	S	P	Ni	Cr	Mo	V
28NCD 8-5	0.295	0.63	0.275	0.007	0.009	1.95	1.35	0.425	...
20CND 8	0.225	0.65	0.175	0.010	0.010	1.00	1.73	0.625	0.023
26NCDV 11-6	0.275	0.36	0.070	0.007	0.008	2.77	1.59	0.42	0.096
20NCD 14	0.167	0.343	0.24	0.007	0.006	3.23	1.71	0.549	...

TABLE 2—*Heat treatment of investigated steels.*

Steel	Temperature, °C	Time at Temperature, h	Type of Cooling
28NCD 8-5	850	6.5	water quenching
	620	7	furnace cooling
20CND 8	875	13	water quenching
	640	16	air cooling
26NCDV 11-6	845	21	water quenching
	630	21	furnace cooling
20NCD 14	885	8.5	water quenching
	630	6.5	air cooling

TABLE 3—*Mechanical properties of investigated steels.*

Steel	σ_{ys}, MPa	σ_{uts}, MPa	Elongation, %	Reduction of Area, %	FATT, °C	TK28[a]	Ductile Level, J
28NCD 8-5	675	831	19	67	−20	−100	145
20CND 8	675	790	18	67	+35	−10	140
26NCDV 11-6	730	850	21	69	−80	−110	145
20NCD 14	680	800	21	72	−85	−110	160

[a]Temperature at which the Charpy V energy equals 28 J.

surement of load-line deflection (Fig. 1). As shown in Table 4, CTJ specimens had homothetic dimensions ($W = 2B$) except for the 28NCD 8-5 steel that was used to investigate separately the influence of thickness B and uncracked ligament length ($W - a$) on J_{Ic} values. The notch depth was such that after fatigue precracking, the crack length ratio a/W was in the 0.55 to 0.65 range.

J_{Ic} Test Procedure

A continuous plot of load versus load-point displacement was obtained for each specimen, and J was calculated by the equation

$$J = \frac{2A}{B(W - a)} f(a/W)$$

where A is the area under the load, load-point displacement record (in energy units) up to the point of interest, B is the specimen thickness, ($W - a$) is the initial uncracked ligament length, a is the original crack size, and $f(a/W)$ is a dimensionless coefficient that corrects for the tensile component of loading [14].

The tests were generally conducted at different temperatures from the transition region to the upper shelf. In the case of a ductile initiation of failure, the critical value J_{Ic} was determined by two methods:

1. The interrupted loading and heat tinting method recommended by ASTM E 813. This procedure defines J_{Ic} in the J-Δa diagram as the intersection of the best straight line that fits the experimental data with a blunting line. This method is also called the multispecimen technique.

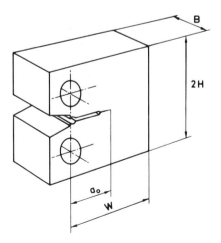

FIG. 1—*CTJ specimen used for J test.*

TABLE 4—*Specimens used for J_{Ic} determination.*

Steel	Type of Specimen	Width, mm	Thickness, mm
28NCD 8-5	CTJ15	30	5-10-15
	CTJ25	50	5-10-15-20-25
	CTJ40	80	15-20-25-40
	CTJ75	150	25-50-75
20CND 8	CTJ15	30	15
	CTJ25	50	25
	CTJ40	80	40
20NCD 14	CTJ25	50	25
	CTJ50	100	50
	CTJ75	150	75
	CTJ150	300	150
26NCDV 11-6	CTJ50	100	50
	CTJ75	150	75
	CTJ150	300	150

2. The a-c potential drop method developed some years ago at IRSID [2]. The first stages of stable crack growth are associated during the loading with the minimum of the output voltage. The critical value J_{Ic} can thus be derived from a single test.

Results and Discussion

Fracture Toughness Transition Curves

The specific experimental results used in the present study to establish the basic fracture toughness K_{Ic} versus temperature T transition curves for the 28NCD 8-5, 20CND 8, and 20NCD 14 steels are presented in Table 5. As can be seen from this table, a few K_Q values do not meet the ASTM E 399 requirement of $B \geq 2.5(K_Q/\sigma_y)^2$. Nevertheless, it is widely recognized that this proportional factor may vary among materials [16,17]; furthermore, modified criteria have been proposed [18]. In the case of the medium strength, quenched-and-tempered steels used in this investigation, the ASTM condition appeared to be too stringent, since the measured value of K_Q was independent of thickness for $B \geq (K_Q/\sigma_y)^2$ and $(P_{max}/P_Q) \leq 1.1$. These K_Q values are therefore considered as valid K_{Ic} and are plotted in the diagrams K_{Ic} versus T (Figs. 2 to 4). The fracture toughness transition curve was also estimated for each steel from the Charpy V energy transition curve by the correlation proposed by Marandet and Sanz [15]. As shown in Figs. 2 to 4, this curve fits very well the experimental data of the 28NCD 8-5 and 20NCD 14 steels and lies slightly above those from the 20CND 8 steel.

TABLE 5—*Results of plane-strain fracture toughness measurements.*

Steel	Temperature (T), °C	Type of Specimen	Thickness (B), mm	a/W	K_Q, MPa$\sqrt{m}$	P_{max}/P_Q	2.5 $(K_Q/\sigma_{ys})^2$, mm
28NCD 8-5	−8	CT150	150	0.484	256	1.13	345
	−40	CT150	150	0.484	209	1.0	216
	−40	CT150	150	0.474	242	1.0	296
	−40	CT100	100	0.523	226	1.0	257
	−40	CT150	100	0.475	210	1.0	220
	−70	CT150	150	0.485	158	1.0	116
	−70	CT100	100	0.522	174	1.0	139
	−70	CT100	100	0.521	166	1.0	128
	−70	CT75	75	0.500	200	1.0	185
	−70	CT50	50	0.514	162	1.0	121
	−97	CT75	75	0.498	132	1.0	47
	−97	CT75	75	0.496	124	1.0	42
	−155	CT25	25	0.658	77	1.0	20
20CND 8	+40	CT200	200	0.502	174	1.0	171
	+40	CT200	200	0.492	125	1.0	88
	+20	CT150	150	0.478	135	1.0	100
	+20	CT150	150	0.481	86pop	1.47	41
	+20	CT150	150	0.482	122	1.0	82
	+20	CT150	150	0.482	117	1.0	76
	+10	CT75	75	0.489	123	1.0	81
	0	CT75	75	0.495	84	1.0	38
	0	CT75	75	0.494	95	1.0	48
	−20	CT75	75	0.491	81	1.0	33
	−40	CT50	50	0.548	52	1.0	14
20NCD 14	−70	CT150	150	0.516	207	1.0	190
	−70	CT150	150	0.516	220	1.0	215
	−150	CT150	150	0.518	92	1.0	35
	−140	CT25	25	0.600	87	1.0	24
	−140	CT25	25	0.600	98	1.0	25
	−196	CT25	25	0.512	35	1.0	26

These three basic K_{Ic} versus T curves are plotted in Figs. 5 to 7 as the true fracture toughness behavior of the materials to be compared with K_{Jc} values derived from J_c and J_{Ic} measurements on smaller specimens. As shown in Figs. 5 to 7, the K_{Jc} transition curves obtained with different specimen sizes exhibit the same trends for the three steels. Hollow points denote CTJ specimens which demonstrated cleavage instability, and solid data points represent specimens that exhibited ductile crack initiation. It can be seen that the plane-strain fracture toughness K_{Ic} values measured in small-scale yielding conditions with large specimens provide a lower bound of the K_{Jc} values at initiation of cleavage fracture of smaller specimens loaded in the elastic-plastic regime. As found by other authors [5,6] the K_{Jc} versus T curve is shifted towards lower temperatures as specimen size decreases.

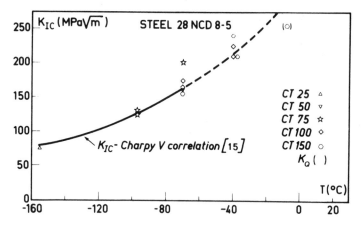

FIG. 2—K_{Ic} fracture toughness transition curves for 28NCD 8-5 steel.

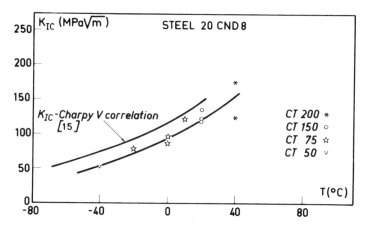

FIG. 3—K_{Ic} fracture toughness transition curves for 20CND 8 steel.

Although the overestimation factor is slight at the lowest temperatures, it becomes important in the transition region. Milne and Chell [4] proposed a qualitative explanation based on the cleavage fracture model of Ritchie et al [19]. The former authors argue that in smaller specimens, where large-scale plasticity occurs, the competition between the loss of stress triaxiality and the increase of the applied load produces a K_{Jc} value which exceeds K_{Ic} and will eventually lead to another fracture mode, the ductile tearing.

As shown in Figs. 5 to 7, high K_{Jc} values were found for smaller specimens which failed by ductile mechanism and met ASTM E 813 size requirements, whereas the valid K_{Ic} specimen failed by cleavage. Clearly, a thin J-integral

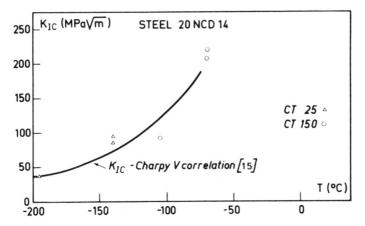

FIG. 4—K_{Ic} *fracture toughness transition curves for 20NCD 14 steel.*

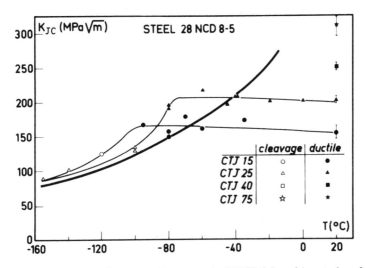

FIG. 5—K_{Jc} *fracture toughness transition curves for 28NCD 8-5 steel (onset of crack growth detected by a-c potential drop method).*

specimen cannot provide the same toughness value as a thicker K_{Ic} specimen if the mechanism of fracture is different. The fact that J-measurement exhibits a plateau must be considered a limitation of the method, since K_{Ic} values continually increase with temperature so long as no metallurgical parameter induces ductile tearing.

Moreover, the temperature at which the fracture initiation mode changes, as well as the level of the upper shelf, depend on the specimen size (Figs. 5 to

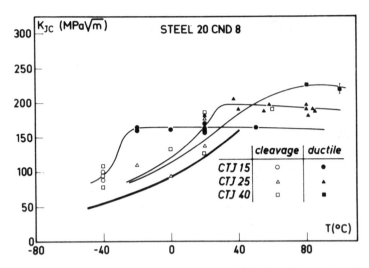

FIG. 6—K_{Jc} *fracture toughness transition curves for 20CND 8 steel (onset of crack growth detected by a-c potential drop method).*

7). The larger the specimen, the higher the brittle to ductile transition temperature and the toughness plateau. This phenomenon makes it a puzzling problem to determine K_{Ic} from a J_{Ic} value measured in the ductile regime. For example, let us consider the 28NCD 8-5 steel (Fig. 5). Tests carried out with 25-mm-thick specimens at temperatures below $-40°C$ overestimated K_{Ic}

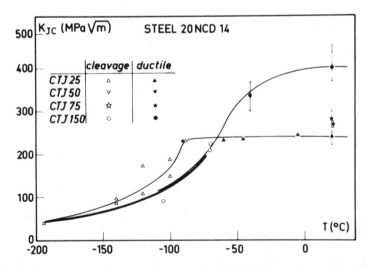

FIG. 7—K_{Jc} *fracture toughness transition curves for 20NCD 14 steel (onset of crack growth detected by a-c potential drop method).*

whereas tests above this temperature gave the opposite result. For 15-mm-thick specimens, this crossover temperature is about $-70°C$. Similar observations can be made on the two other transition curves.

It should be noted that the onset of stable crack growth was detected by means of the a-c potential drop method. The ductile crack extension before failure was checked by examining the fracture surfaces of the specimens. The K_{Jc} values plotted in Figs. 5 to 7 were derived from J_{Ic} measured at the minimum of the output voltage. Nevertheless, the specimen size dependence of the K_{Jc} plateau cannot be attributed to this method, since it was also found when using the ASTM E 813 multispecimen technique, as will be seen later on.

Ductile Crack Growth Initiation

Since a specimen size dependence of J_{Ic} was found, it was worth studying whether it was the thickness or the uncracked ligament length, or both, which

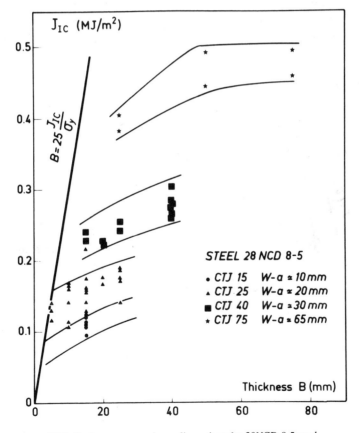

FIG. 8—J_{Ic} versus specimen dimensions for 28NCD 8-5 steel.

influenced the J_{Ic} measurements in the ductile regime. For this purpose we used CTJ specimens with W ranging from 30 to 150 mm and with reduced thicknesses (Table 4). Because of the number of specimens that such an investigation required, only 28NCD 8-5 steel was studied.

Figure 8 shows the J_{Ic} values calculated at the onset of stable crack growth detected by the a-c potential drop method versus the specimen thickness. The different symbols are related to the CTJ specimen types and therefore to different uncracked ligament lengths. As can be seen in this figure, the effect of specimen thickness on J_{Ic} at a given ligament length seems rather weak. The J_{Ic} values slightly increased with thickness, but this rise was not significant compared to the scatterband of the results. On the other hand, it is clear that at a given thickness, J_{Ic} increases with ligament length. For example, for a specimen thickness equal to 25 mm, the mean J_{Ic} value increased from 0.17 to 0.40 MJ/m² as the ligament length increased from 20 to 65 mm.

The independence of J_{Ic} versus specimen thickness in the ductile regime is also shown on a J versus Δa plot (Fig. 9). This J-R curve was established with CTJ 40 specimens ($W - a \simeq 30$ mm) 15 to 40 mm thick by the usual multi-specimen technique. It can be observed that all the data points lie within the same scatterband, irrespective of the specimen thickness. We can also notice

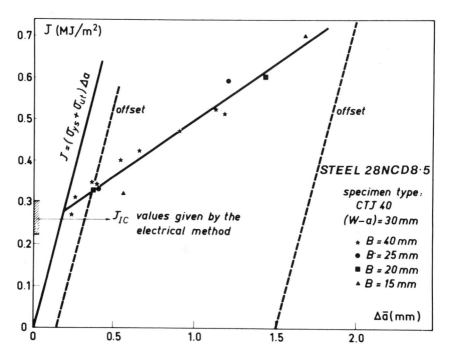

FIG. 9—J *versus* Δa *R-curve for 28NCD 8-5 steel obtained with CTJ 40 specimens of various thicknesses and constant ligament length.*

the very good agreement between the J_{Ic} value calculated in accordance with ASTM E 813 and the mean value given by the a-c potential drop method.

It can thus be considered that the observed specimen size effect on J_{Ic} measurements at the ductile plateau is mainly related to the uncracked ligament length. As mentioned previously, this effect is also evident when evaluating J_{Ic} with ASTM E 813. Figure 10 shows the J versus Δa curves obtained with four different types of CTJ specimens ($W - a = 10, 20, 30$, and 65 mm) for the 28NCD 8-5 steel at room temperature. It appears clearly that the intersection point of the blunting line and the best straight line that fits each series of data points rises with increasing ligament length. The slope dJ/da does not change from one specimen size to the other except for the smallest specimens which exhibit a steeper J-R curve. This may be because of the tunneling found in these thin specimens. Although there were less data available for the 20NCD 14 steel, similar J-R curves obtained with CTJ 25, 50, and 150 are shown in Fig. 11 and lead to the same conclusion. It should be noted that for the investigated steels the ASTM procedure required Δa values that could only be obtained after the maximum load (Figs. 10 and 11).

All the J_{Ic} values measured at the ductile plateau of the four steels are plotted on Fig. 12 as function of the uncracked ligament length. In spite of the

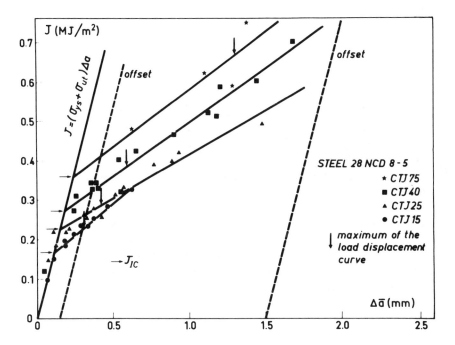

FIG. 10—*Plot of* J *versus crack extension for 28NCD 8-5 steel as related to specimen size* (W = 2 B).

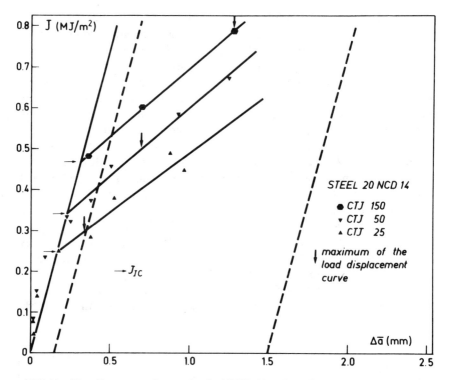

FIG. 11—*Plot of J versus crack extension for 20NCD 14 steel as related to specimen size* (W = 2 B).

differences between the J_{Ic} values calculated by the two methods, especially for the larger specimens, Fig. 12 shows the general trend of J_{Ic} increasing with increasing $(W - a)$.

These results indicate that at the onset of stable crack growth for these specimen dimensions the crack tip field is no longer J-dominated. The observed influence of the uncracked ligament length suggests that the occurrence of general yield may promote the initiation of ductile tearing. Indeed, an elastic-plastic finite element calculation [20] has shown that for a CT 50 specimen of A 508 C1.3 steel the measured J_{Ic} value corresponds to a loading point at which plastic deformation spreads over the whole ligament. An intrinsic measure of the plane-strain ductile initiation toughness would thus require a specimen size which would enable initiation of tearing to be obtained before general yield.

Conclusions

Four quenched-and-tempered medium strength steels were investigated in order to compare the plane-strain fracture toughness K_{Ic} to J_{Ic} measurements

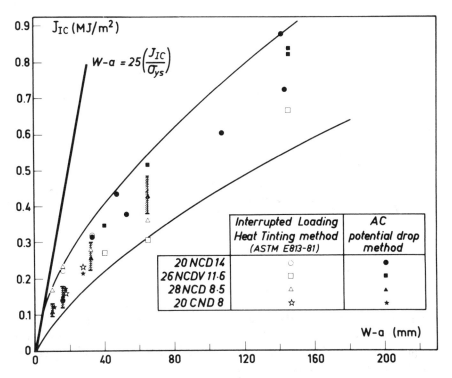

FIG. 12—J_{Ic} *versus uncracked ligament length for the four steels investigated.*

performed with different specimen sizes. The following conclusions can be drawn from the experimental results as shown in Fig. 13:

1. In the lowest temperature range K_{Jc} slightly exceeds K_{Ic}, but the *J*-integral method remains a very attractive way to evaluate K_{Ic} with subsized specimens. In the transition region where fracture also occurs by cleavage, however, the overestimation factor may become larger, depending on the specimen thickness.

2. *J* specimens exhibit a change of the fracture mode from cleavage to ductile tearing, whereas K_{Ic} specimens fail by cleavage. Although ASTM E 813 requirements are met, this transition temperature as well as the level of the ductile plateau increase with increasing specimen dimensions.

3. ASTM E 813 procedure and an a-c potential drop method were used for calculating J_{Ic} at the upper shelf. Both techniques show the trend of J_{Ic} plateau rising with increasing uncracked ligament length, irrespective of thickness.

4. J_{Ic} measurements do not necessarily provide conservative estimates of K_{Ic}, depending on the temperature, unless it can be proved that the full section material exhibits a ductile behavior.

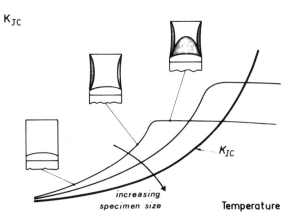

FIG. 13—*Schematic variation of* K_{Jc} *with temperature for different specimen sizes.*

5. In order to obtain reliable toughness values for the steels considered, the *J*-specimen size should be much larger than required by ASTM E 813. Further work is needed to explain the influence of the ligament length on J_{Ic} and to take the tensile properties of materials into account in a more precise size requirement. Experimental programs and elastic-plastic finite element calculations should be conducted for that purpose.

References

[1] Landes, J. D. and Begley, J. A. in *Developments in Fracture Mechanics Test Methods Standardization, ASTM STP 632*, American Society for Testing and Materials, 1977, pp. 57–81.

[2] Marandet, B. and Sanz, G. in *Flaw Growth and Fracture, ASTM STP 631*, American Society for Testing and Materials, 1977, pp. 462–476.

[3] Landes, J. D. and Shaffer, D. H. in *Fracture Mechanics: Proceedings of the 12th National Symposium, ASTM STP 700*, American Society for Testing and Materials, 1980, pp. 368–382.

[4] Milne, I. and Chell, G. G. in *Elastic-Plastic Fracture, ASTM STP 668*, American Society for Testing and Materials, 1979, pp. 358–377.

[5] Pisarski, H. G., "Influence of Thickness on Crack Opening Displacement (COD) and *J* Values," WI Report 90/1979, The Welding Institute, Abington Hall, U.K., 1979.

[6] Hagiwara, Y. and Mimura, H., *Tetsu to Hagane*, Vol. 65, No. 2, Feb. 1979, pp. 58–66.

[7] Scarlin, R. B. and Shakeshaft, M., *Metals Technology*, Jan. 1981, pp. 1–9.

[8] Andrews, W. R., Kumar, V., and Little, M. N. in *Fracture Mechanics: Proceedings of the 13th National Symposium, ASTM STP 743*, American Society for Testing and Materials, 1981, pp. 576–598.

[9] Griffis, C. A., "Elastic-Plastic Fracture Toughness: A Comparison of *J*-Integral and Crack Opening Displacement Characterizations," paper presented at the 2nd National Congress on Pressure Vessels and Piping, ASME, San Francisco, 1975.

[10] Blauel, J. G. and Hollstein, T., "On the Determination of Material Fracture Parameters in Yielding Fracture Mechanics," paper presented at the 2nd European Colloquium on Fracture, ECF2, Darmstadt, West Germany, 1978.

[11] Williams, J. A., "Ductile Fracture Toughness of Heavy Section Pressure Vessel Steel Plate:

A Specimen Size Study of ASTM A 533 Steels," Technical Report, Handford Engineering Development Laboratory, Richland, Wash., 1979.

[12] Carlson, K. W. and Williams, J. A. in *Fracture Mechanics: Proceedings of the 13th National Symposium, ASTM STP 743*, American Society for Testing and Materials, 1981, pp. 503-524.

[13] Logsdon, W. A. in *Mechanics of Crack Growth, ASTM STP 590*, American Society for Testing and Materials, 1976, pp. 43-60.

[14] Clarke, G. A. and Landes, J. D., *Journal of Testing and Evaluation*, Vol. 7, No. 5, Sept. 1979, pp. 264-269.

[15] Marandet, B. and Sanz, G. in *Flaw Growth and Fracture, ASTM STP 631*, American Society for Testing and Materials, 1977, pp. 72-95.

[16] Srawley, J. E., Jones, M. H., and Brown, W. F., *Materials Research and Standards*, Vol. 7, No. 6, June 1967, pp. 262-266.

[17] Wessel, E. T., *Engineering Fracture Mechanics*, Vol. 1, 1968, pp. 77-103.

[18] Ritter, J. C., *Engineering Fracture Mechanics*, Vol. 9, 1977, pp. 529-540.

[19] Ritchie, R. O., Knott, J. F., and Rice, J. R., *Journal of the Mechanics and Physics of Solids*, Vol. 21, 1973, pp. 395-410.

[20] Marandet, B., Devaux, J. C., and Pellissier-Tanon, A., "Correlation between General Yielding and Tear Initiation in CT Specimens," paper presented at CNSI Specialist Meeting on Tear Instabilities, St. Louis, Mo., 1979.

B. Marandet,[1] *G. Phelippeau,*[1] *and G. Sanz*[1]

Influence of Loading Rate on the Fracture Toughness of Some Structural Steels in the Transition Regime

REFERENCE: Marandet, B., Phelippeau, G., and Sanz, G., "**Influence of Loading Rate on the Fracture Toughness of Some Structural Steels in the Transition Regime,**" *Fracture Mechanics: Fifteenth Symposium, ASTM STP 833,* R. J. Sanford, Ed., American Society for Testing and Materials, Philadelphia, 1984, pp. 622-647.

ABSTRACT: The fracture toughness of five structural steels was determined as a function of temperature in static and dynamic conditions ($\dot{K} \approx 2 \times 10^4$ MPa $\sqrt{m}$/s). The most significant effect of increasing the strain rate was to shift the K_{Ic}–temperature curve towards higher temperatures. The amplitude of the temperature shift between the transition temperatures obtained in static (TK_{Ic}) and dynamic (TK_{Id}) conditions were compared with the predictions given by different models. It appeared that the variations of yield strength cannot always account for the change of fracture toughness with temperature and strain rate.

A correlation between Charpy V-notch (CVN) transition temperature (TK 28) and dynamic toughness transition temperature (TK_{Id}) was established. This correlation is an extension of the one proposed previously between TK 28 and TK_{Ic}. Assuming that fracture toughness transition is a thermally activated process, it was possible to calculate theoretically the slopes of the correlations; these were found to be in good agreement with the experimental results.

KEY WORDS: fracture properties, toughness, dynamic tests, Charpy V-notch, steels, correlations

The reliability of steel structures is usually based on the use of the critical stress intensity factor (K_{Ic}) determined in quasi-static conditions ($0.55 < \dot{K}$, MPa $\sqrt{m}$/s < 2.75) in accordance with ASTM Test for Plane-Strain Fracture Toughness of Metallic Materials (E 399). A number of investigations have shown that fracture toughness of low and medium strength steels is generally

[1]Institut de Recherches de la Sidérurgie Française (IRSID), St. Germain en Laye, Cedex, France.

strain rate sensitive. The fracture parameters (critical flaw size or fracture load) can be greatly overestimated if the structure is subject to a rapid loading due to an accident (collision, sudden pressure rise, thermal shock, etc.). It is therefore essential to evaluate the dynamic fracture toughness of steels in order to assess the safety of some components and structures.

The most significant effect of increasing loading rate is to shift the quasi-static K_{Ic} transition curve towards higher temperatures. Barsom [1] proposed an empirical relation to calculate this temperature shift (ΔT) from the yield strength of the material. Many authors [2-7] have tried to quantify the fracture toughness variations with temperature (T) and strain rate ($\dot{\epsilon}$) by means of the parameter $T \log (A/\dot{\epsilon})$ and to find a relationship between the fracture toughness and the yield strength measured at the same temperature and strain rate.

The object of this study was to evaluate the effect of high-speed loading on the fracture toughness of structural steels. To this end, the static and dynamic ($\dot{K} = 2 \times 10^4$ MPa $\sqrt{m}$/s) fracture toughness of five low and medium strength steels was determined as a function of temperature by using the J-integral concept. Different models for predicting the temperature shift between static and dynamic fracture toughness transition curves were studied. Another purpose of this work was to propose a correlation between Charpy V-notch (CVN) and K_{Id} transition temperatures similar to the one established previously between CVN and K_{Ic} transition temperatures for the same type of steels [8].

Materials

Tests were conducted on five medium-strength structural steels:

1. *C-Mn steel microalloyed with niobium (AFNOR E 36)*—This steel was studied in the as-rolled conditions in order to obtain a relatively high CVN transition temperature. Specimens were taken from a 160-mm-thick plate, with an L-T orientation.

2. *Quenched and tempered bainitic Ni-Cr-Mo rotor forging steel (AFNOR 28 NCD 8-5)*—The test ring came from an ingot forged to the following dimensions: outer diameter 1500 mm, inner diameter 580 mm, and thickness 400 mm. Specimens were taken from the test ring so that the notches were radial and their end was located on a circle of 1180 mm diameter (ASTM C-R orientation).

3. *Weldable cast steel (AFNOR 15 MDV 04-03)*—Blocks capable of 100-mm-thick CT specimens were cast to those dimensions and then normalized and tempered. All specimens were cut out with the notch in the same region of the blocks.

4. *Cr-Mo steel (AFNOR 10 CD 9-10) used for pressure vessels*—This steel was melted in an electrical furnace, vacuum degassed, and cast into ingots

from which 110-mm-thick plates were obtained. Blocks measuring 1500 by 300 by 100 mm^3 were then quenched, tempered, and stress-relieved. The specimens were taken from the blocks with an L-T orientation.

5. *C-Mn normalized structural steel (AFNOR E 36)*—Specimens were removed from a 26-mm-thick plate in the T-L orientation

The chemical compositions and ASTM designations of these steels are given in Table 1. The heat treatment conditions are reported in Table 2. As indicated in Table 3, the steels cover a wide range of mechanical properties: the yield strength at room temperature varies from 300 to 675 MPa and the CVN transition temperature defined at 28 J varies from −100 to 0°C.

Experimental Technique

Determination of Quasi-Static Fracture Toughness

Specimens of various dimensions were used to determine the fracture toughness of the steels in quasi-static conditions. The critical stress intensity factor K_{Ic} was measured directly in accordance with ASTM E 399 on large specimens up to 150 mm thick or on smaller specimens at lower temperatures.

In the case of elastic-plastic behavior, the fracture toughness was evaluated by means of the *J*-integral method in accordance with ASTM Test for J_{Ic}, a Measure of Fracture Toughness (E 813) recommended procedure. When fracture initiated by ductile tearing, J_{Ic} was obtained with a single specimen by using the potential drop method developed by Marandet and Sanz [9]. This technique makes it possible to detect the initiation of crack growth during the test. The toughness is then estimated by

$$K_{Jc} = \sqrt{\frac{J_{Ic}E}{(1 - \nu^2)}}$$

where E is the Young's modulus and ν is the Poisson's ratio.

Determination of Dynamic Fracture Toughness

An original technique has been developed recently at IRSID [10] for determining the fracture toughness of steels at K rates of about 2×10^4 MPa $\sqrt{m}/s$ for small specimens by the *J*-integral method.

Specimens and Loading System—All dynamic fracture toughness tests were performed on 25-mm-thick CT specimens modified in order to clip an extensometer between two knives machined on the very axis of the loading pin holes (Fig. 1). Specimens were fatigue precracked in accordance with the ASTM procedure with a crack length to specimen width ratio (a/W) of about 0.55 and tested on a closed-loop servohydraulic machine with load frame and

TABLE 1—*Chemical composition (weight percent) of test steels.*

Reference Steel	AFNOR Steel Grade	ASTM Steel Grade	C	Mn	Si	P	S	Al	Ni	Cr	Mo	Nb	V
A	E 36 Nb-V	A 440	0.16	1.25	0.32	0.19	0.018	0.011	...	...	...	0.020	0.038
B	28 NCD 8-5	A 471	0.295	0.63	0.275	0.009	0.007	...	1.95	1.35	0.425	...	...
E	15 MDV 04-03	Mn-Mo-V cast steel	0.140	1.50	0.50	0.02	0.02	...	...	...	0.3	...	0.06
F	10 CD 9-10	A 387 Gr. 22	0.110	0.455	0.275	0.014	0.025	...	0.115	2.10	1.05	...	...
FT	E-36	A 633 Gr. C	0.170	1.41	0.32	0.022	0.019	0.059	0.031	0.026	0.005	0.028	...

TABLE 2—*Heat treatments of test steels.*

Steel	Heat Treatment[a]
A (150 mm thickness)	as rolled
B (400 mm thickness)	850°C at 6.5 h/WQ + 620°C at 7 h/AC
E (100 mm thickness)	940°C at 4 h/AC + 650°C at 3 h/AC
F (110 mm thickness)	940°C at 1 h/WQ + 700°C at 1 h/AC + 675°C at 4 h/AC
FT (26 mm thickness)	normalized

[a]WQ = water quenched, AC = air cooled.

TABLE 3—*Mechanical properties of test steels.*

	Tension Tests (Room Temperature)				CVN		
Steel	$\sigma_{\text{ys 0.002}}$, MPa	σ_{uts}, MPa	Elongation, %	Reduction in Area, %	TK at 28 J, °C	FATT, °C	Ductile Level, J
A	300	515	29	68	0	...	120
B	675	831	20	67	−100	−20	145
E	377	523	24	44	−5	+55	125
F	404	624	24	76	−75	−20	160
FT	360	530	33	71	−65	−30	100

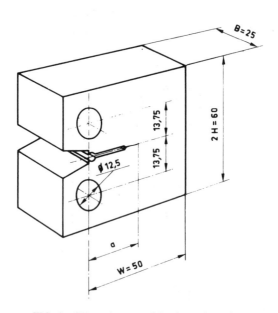

FIG. 1—*CT specimen used for dynamic testing.*

load cell capacities of 250 kN. The hydraulic system of this machine has a sufficient delivery (170 L/min) to produce a ram speed of about 500 mm/s over a calibrated stroke of 100 mm. This results in a loading rate $\dot{K} = \Delta K/\Delta t$ of about 2×10^4 MPa $\sqrt{m}$/s in the elastic range.

During the test, the shock generated by the instantaneous deformation of the specimen is damped by means of a mechanical device consisting of a stack of truncated-cone washers. The washers are grouped in three sets of different thicknesses in a cylindrical sleeve screwed onto the load cell (Fig. 2). The washers are flattened when the ram moves. This variable-rigidity device makes it possible to increase the loading rate of the specimen in successive steps over a more-or-less large portion of the elastic range. The maximum loading rate is reached after the complete crushing of all the washers; the specimen is then connected to the ram moving at full speed.

Detection of Initiation—The a-c potential drop method originally used to determine the onset of ductile crack growth in quasi-static conditions was modified to operate in dynamic conditions. The principle of the apparatus is

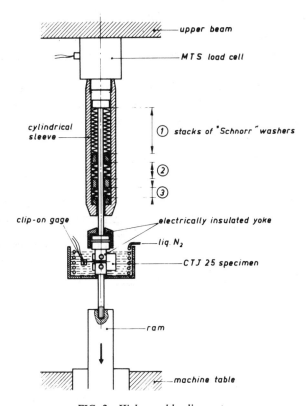

FIG. 2—*High-speed loading system.*

presented in Fig. 3. A high-frequency low-intensity current (10 kHz, 3 A) passes through a reference and a test specimen. The potential drop is measured on either side of the mechanical notch of each specimen. The sum of those two voltages in opposite phase is then recorded.

As stated previously [9], the initiation of stable crack growth is associated with a minimum of the electrical potential. This was confirmed by SEM examinations of fracture surfaces.

Recording and Processing of Signals—A schematic diagram of the signals processing system is given in Fig. 4. During the test, electrical signals proportional to the load, displacement, potential drop, and ram stroke were recorded with a frequency of 125 kHz by a minicomputer. The data were then sent to a central computer which can automatically plot the diagrams of load and potential drop as a function of load point displacement and calculate the K_{Id}, J_{Id}, or K_{Jd} values that characterize the fracture toughness of the material.

Figures 5a and 5b show typical variations of the load, displacement, and potential drop versus time for brittle and ductile behaviors. The corresponding load versus displacement curves are also shown.

Determination of Yield Strength at High Strain Rate

Dynamic tension tests were carried out over the same temperature range at a strain rate $\dot{\epsilon}$ equal to the calculated crack tip strain rate during a dynamic fracture toughness test. Cylindrical tension specimens ($\phi = 5$ mm) were loaded on the servohydraulic machine also used for the high-speed toughness tests. The damping system placed in the loading line was similar to the one previously described and made it possible to avoid the mechanical vibrations that may occur during the shock.

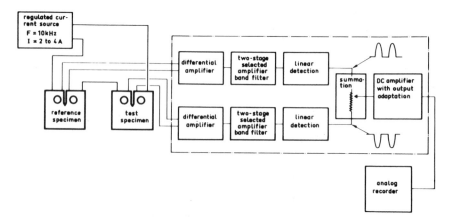

FIG. 3—*Diagram of circuit used to detect crack initiation under dynamic loading.*

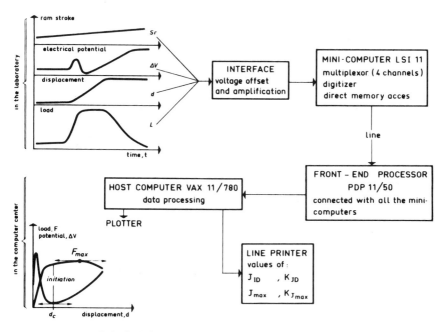

FIG. 4—*Schematic diagram of signal processing system.*

The test data (load and ram displacement) were recorded by the minicomputer. The curve giving the variations of load as a function of time was then automatically restored from these data. As no displacement gage could be fixed on the specimen, only the lower yield strength (σ_{ys}) and the ultimate tensile stress (σ_{uts}) were measured.

Results and Interpretation

Effect of Loading Rate on Fracture Toughness

The static and dynamic fracture toughness are plotted as a function of test temperature in Fig. 6 for the five steels considered in this study. The symbol K_{Ic} (or K_{Id}) refers to linear load-displacement recordings. The ASTM condition $P_{max}/P_Q \leq 1.1$ was always fulfilled, although in a few cases the specimen dimensions were not large enough to meet the size requirement B and $a \geq 2.5$ $(K_{Ic}/\sigma_{ys})^2$. Nevertheless, we verified on several materials that these K_{Ic}-values were also in good agreement with K_{Jc} (or K_{Id}) deduced from the J-integral. It can thus be assumed that they yield a correct evaluation of the fracture toughness, the condition on specimen dimensions being usually too restrictive. The critical stress intensity factor derived from the J-integral is noted K_{JCf} (or K_{JDf})

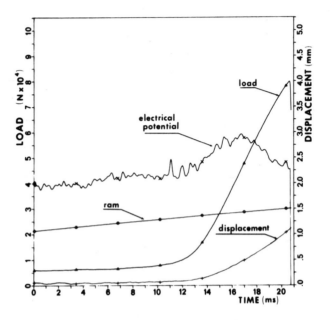

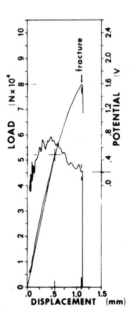

FIG. 5a—*Typical brittle fracture records during dynamic loading.*

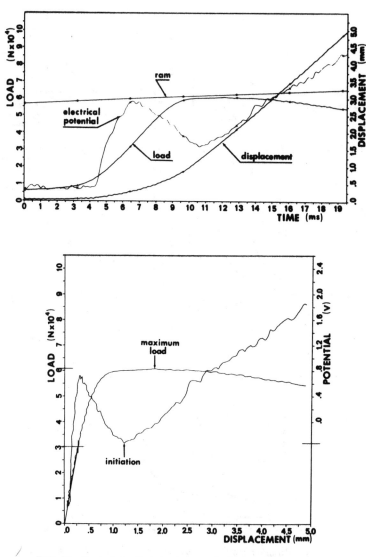

FIG. 5b—*Typical ductile behavior records during dynamic loading.*

when fracture occurs by cleavage without subcritical crack extension and $K_{J\text{Ci}}$ (or $K_{J\text{Di}}$) when fracture is first initiated by ductile tearing.

For the steels investigated, an increase in the loading rate results in a decrease in the cleavage fracture toughness. The magnitude of this decrease may be very important in the transition zone where fracture toughness varies considerably with temperature. Thus the dynamic fracture toughness transition curve can be deduced from the quasi-static curve by a temperature shift

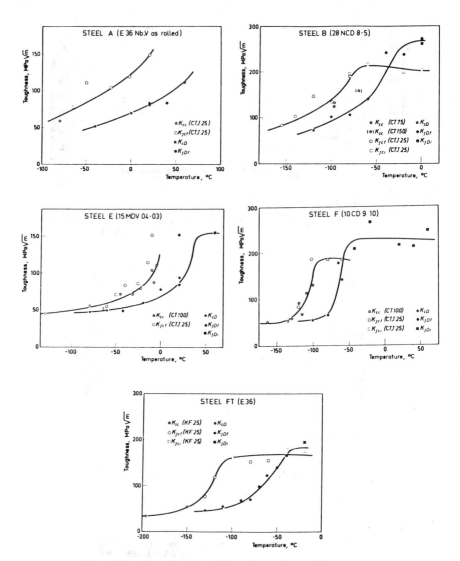

FIG. 6—*Variation of static and dynamic toughness with temperature.*

ΔT towards higher temperatures. An increase in the loading rate seems to be equivalent to a decrease in the test temperature. It is possible to define the temperature shift by $\Delta T = TK_{\text{Id}} - TK_{\text{Ic}}$, where TK_{Ic} and TK_{Id} are the temperatures at which K_{Ic} and K_{Id} respectively are equal to 100 MPa $\sqrt{\text{m}}$. Values of ΔT were found to vary from 40 to 70°C depending on the steel considered.

For Steels B and F, the upper shelf plateau associated with ductile fractures was determined in static and dynamic conditions with CT 25 specimens. In this region, a substantial increase in the initiation toughness K_{JCi} with increasing loading rate was observed. As mentioned by other authors [11], this rise in the ductility is related to the increase in the yield strength of the materials with strain rate.

The ability of different methods to predict these variations of fracture toughness with the loading rate and the temperature shift ΔT was examined.

Application of Relation Proposed by Barsom

Barsom [1] has proposed a relation to estimate the temperature shift between static and dynamic fracture toughness from the material yield strength σ_{ys} and crack tip strain rate $\dot\epsilon$. For strain rates in the range $10^{-3}\,s^{-1} < \dot\epsilon < 10$ s^{-1} and for steels having yield strengths less than about 965 MPa, the relationship between ΔT, $\dot\epsilon$, and σ_{ys} is

$$\Delta T = (83 - 0.08\,\sigma_{ys})\,\dot\epsilon^{0.17} \qquad (1)$$

with ΔT in °C, σ_{ys} in MPa, and $\dot\epsilon$ in s^{-1}. One should bear in mind that this empirical relation is just an approximation, since the only material property that goes in the formula is σ_{ys} at room temperature.

The value of $\dot\epsilon$ at a point on the elastic-plastic boundary can be calculated from the stress intensity rate $\dot K$ by the equation [12]

$$\dot\epsilon = \frac{2\,\sigma_{ys}}{E} \times \frac{\dot K}{K}$$

where σ_{ys} is the yield strength for the test temperature and strain rate, and E is the Young's modulus of the material.

A mean value of $\dot\epsilon$ was calculated in the case of dynamic fracture testing and used as a constant in this study:

$$\dot\epsilon = \frac{2 \times 500}{200\,000} \times \frac{2 \times 10^4}{100} = 1\,s^{-1}$$

This approximate value provides the correct order of magnitude for $\dot\epsilon$.

Figure 7 shows the temperature shift ΔT measured at a fracture toughness level of 100 MPa $\sqrt{m}$ as a function of yield strength for the five steels investigated. The straight line that corresponds to the relation proposed by Barsom is indicated on the same diagram. In spite of a rather large scatter, which may be due to the uncertainty in the determination of ΔT, there is rather good agreement between this relation and the experimental data. Nevertheless, ΔT is not as strongly dependent on the material yield strength as estimated by the empirical relation proposed by Barsom.

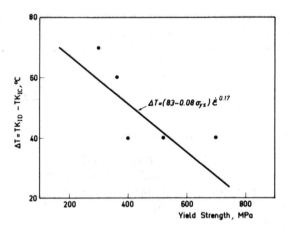

FIG. 7—*Temperature shift* $\Delta T = TK_{Id} - TK_{Ic}$ *as a function of yield strength.*

The Zener-Hollomon Parameter

The Zener-Hollomon parameter was originally proposed to account for the yield strength variations of steels with temperature and strain rate. A convenient form of this parameter is

$$P = T \log (A/\dot{\epsilon})$$

where T is the absolute temperature (degrees Kelvin), A is a frequency factor (s^{-1}), and $\dot{\epsilon}$ is the strain rate (s^{-1}).

Several authors [2–7] have since used this parameter to gather on a unique curve fracture toughness data obtained at different crack tip strain rates ($\dot{\epsilon}$). The value of the frequency factor A usually taken is 10^8 s^{-1} but experimental work done by Priest [13] and Vlach et al [14] has clearly shown that this factor is strongly material dependent.

Static and dynamic cleavage fracture toughness values were plotted in Fig. 8 as a function of the temperature rate parameter P for the different steels. In each case, the frequency factor was chosen to obtain the best fit between static and dynamic toughness data. The factor A was found to be 10^{13} for Steel A, 10^{14} for Steel B, 10^{26} for Steel E, 10^{12} for Steel F, and 10^{11} for Steel FT. These results confirm that the frequency factor depends very much on the steel considered.

This type of approach implies that the temperature shift ΔT varies with the toughness level. The relationship between absolute temperature T_1 and T_2 at which $K_{Ic} = K_{Id}$ can be written as

$$\frac{T_1}{T_2} = \frac{\log(A/\dot{\epsilon}_2)}{\log(A/\dot{\epsilon}_1)} = C$$

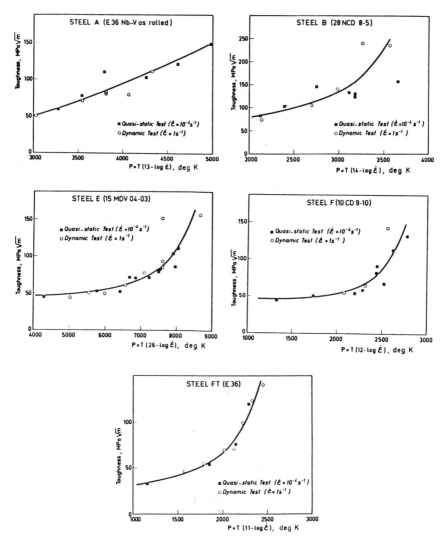

FIG. 8—*Fracture toughness as a function of the Zener-Hollomon parameter.*

where $\dot{\epsilon}_1$ and $\dot{\epsilon}_2$ are the strain rates in static and dynamic conditions respectively, and C is a constant which is a function of the material and the toughness level. Therefore the temperature shift $\Delta T = T_2 - T_1$ becomes larger as the toughness increases. This is in contradiction to the relation proposed by Barsom.

Although it accounts rather well for the strain rate effect, the Zener-Hollomon parameter is not very useful in predicting dynamic toughness. In fact, one should first establish the complete K_{Ic} temperature curve of a steel and carry out at least one test in dynamic conditions to determine the frequency factor A before using the parameter P to estimate the fracture toughness at any strain rate.

Relation Between Yield Strength and Fracture Toughness of Steels

Much work [15-17] has been devoted to relating the cleavage fracture toughness of steels to the yield strength determined at the same temperature and strain rate. Such a relationship postulates a critical cleavage stress (σ_c) which is independent of temperature and strain rate. It has been shown that σ_c depends on metallurgical factors such as the ferrite grain size and carbide thickness.

The most widely accepted local criterion for cleavage fracture was developed by Ritchie, Knott, and Rice (RKR) [18]. According to their model, cleavage fracture initiates when the maximum principal stress ahead of the crack tip exceeds σ_c over a characteristic distance X_0 which can also be related to microstructural parameters such as ferrite grain size and carbide distribution. The RKR model originally proposed for mild steels has been successfully applied by Curry [19], Parks [20], and Ritchie et al [6] to other ferritic steels. Furthermore, Beremin [21] has used this model to explain the apparent increase in the toughness due to warm prestress on an A 508 Cl.3 steel.

Hutchinson [22] and Rice and Rosengren [23] (HRR) have calculated the stress field ahead of a sharp crack in elastic-plastic conditions. This solution was then modified in the plastic zone by Rice and Johnson [24] to account for blunting. For a material with a stress-strain law of the form $\epsilon_p = \alpha(\sigma/\sigma_{ys})^{N-1}$ (σ/E), the HRR solution gives the expression of the maximum stress ahead of the crack:

$$\sigma_{YY}/\sigma_{ys} = f(\alpha, N)r^{-[1/(N+1)]} (K/\sigma_{ys})^{2/(N+2)} \qquad (2)$$

where r is the distance from the crack tip.

The application of the RKR criterion leads to

$$K_{Ic} \cdot \sigma_{ys}^{(N-1)/2} = \sigma_c^{(N+1)/2} f(\alpha, N)^{[(N+1)/2]} X_0^{1/2} \qquad (3)$$

This equation enables one to calculate the variation of K_{Ic} with temperature and strain rate knowing σ_{ys}, α, and N. Assuming that α and N are almost temperature and strain rate independent, Eq 3 becomes

$$K_{Ic}(\dot{\epsilon}, T) \cdot \sigma_{ys}(\dot{\epsilon}, T)^{(N-1)/2} = \sigma_c^{(N+1)/2} f(\alpha, N)^{[(N+1)/2]} X_0^{1/2} = C \qquad (4)$$

where C is constant for a given material. This provides a unique relationship between K_{Ic} and σ_{ys}. Moreover, the slope of the log K_{Ic}-log σ_{ys} curve at different T and $\dot{\epsilon}$ is directly related to the strain-hardening exponent. This assumption was verified by Pineau [25] for A 508 Cl.3 and A 533 B steels.

In order to examine whether the RKR model can be applied to the steels investigated, the cleavage fracture toughness was plotted in Fig. 9 as a function of the yield strength determined at the same temperature and loading rate. In the case of Steels A, E, F, and FT the experimental data from dynamic tests do not coincide very well with those from quasi-static tests. This indicates that the strain-hardening exponent N of those materials is probably sensitive to temperature or strain rate or both. On the other hand, all the data obtained on Steel B lie on the same curve. On a logarithmic plot (Fig. 9) a linear relationship is observed. The value of the slope enables us to calculate $N = 8.5$ by means of Eq 4.

Tension tests performed in quasi-static conditions at $+20°$C and $-110°$C gave a value $N = 8$. This value is in good agreement with the value $N = 8.5$ derived from the fracture toughness versus yield strength plot.

The evolution of fracture toughness with temperature and strain rate can be easily predicted by applying the RKR model if the work-hardening exponent is not temperature and strain rate dependent. Unfortunately, as we have pointed out in this study, this is not the case for most of the structural steels.

Correlation Between Fracture Toughness and CVN

Existing Correlation

Various empirical methods have been proposed to evaluate the fracture toughness of steels from CVN in the transition region. The method developed by Marandet and Sanz [26] is based on results obtained on numerous grades of structural steels with yield strengths varying from 300 to 1000 MPa. These authors have established a good correlation between fracture toughness K_{Ic} and CVN energy transition temperatures. By defining TK_{Ic} as the temperature at which $K_{Ic} = 100$ MPa $\sqrt{m}$ and TK 28 as the temperature corresponding to an impact energy $KV = 28$ J, the relationship is $TK_{Ic} = 1.4\ TK$ 28 (°C). This relation was based on results from 58 steel grades yielding a correlation coefficient of 0.96.

According to Marandet and Sanz, the K_{Ic} transition curve can be deduced from the CVN transition curve as follows:

1. A K_{Ic}-temperature curve is derived from the CVN transition curve using the experimental relation $K_{Ic} = 19\ (KV)^{1/2}$.
2. This curve is then shifted along the temperature axis so that $TK_{Ic} = 1.4$ TK 28.

This method is only valid for impact energies less than 80 J and a crystallinity of the Charpy specimen greater than 80 to 85% at the temperature TK 28.

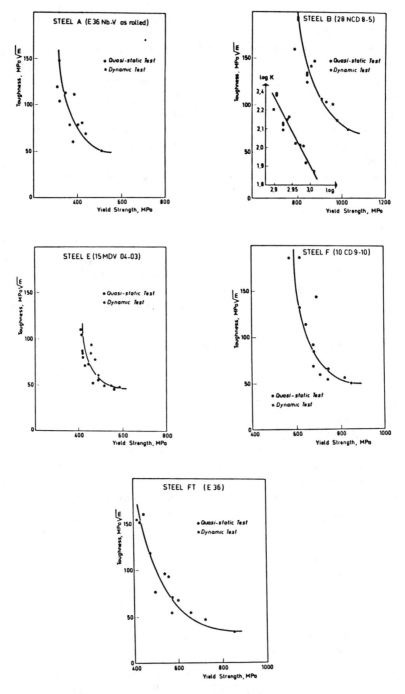

FIG. 9—*Fracture toughness as a function of yield strength.*

In a study on the comparison of various brittle fracture tests, Cheviet et al [27] have shown that the reciprocal of the transition temperature is degrees. Kelvin determined by different tests (Robertson, Battelle, Pellini, Schnadt, Charpy, impact tension) can be correlated by a linear relationship. Figure 10 plots $1/TK_{Ic}$ in a similar way as a function of $1/TK\ 28$ for the results collected by Marandet and Sanz [26]. The points were fitted by a straight line, and the slope given by normal linear regression was found to be 1.63.

The linear relationship observed between the transition temperatures on the one hand and between the reciprocal of these temperatures on the other are not contradictory. The two relations are plotted on the same diagram in Fig. 11. In view of this diagram it is clear that, in the temperature range investigated, the branch of hyperbola corresponding to the relation between the reciprocal of the transition temperatures can be approximated by a straight line with a slope of 1.4.

Theoretical Determination of the Slope of the Correlations Between Transition Temperatures

It is known that the transition temperature determined by the standard K_{Ic} fracture toughness test or any other brittle fracture test increases when the strain rate increases. It has been shown in the present work that this effect may be described by means of the Zener-Hollomon parameter $T \log (A/\dot{\varepsilon})$. For a given steel, one can write $T_c \log (A/\dot{\varepsilon}) = C$ where T_c is the transition temperature and the constant C is a function of the steel considered and of the

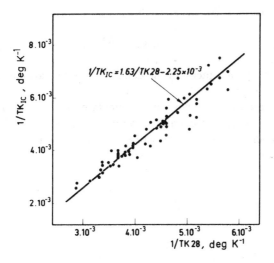

FIG. 10—*Correlation between* $1/TK_{Ic}$ *and* $1/TK\ 28$.

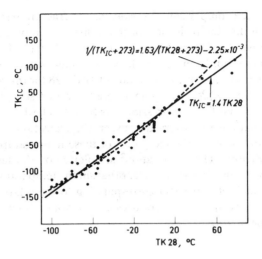

FIG. 11—*Correlation between* TK$_{Ic}$ *and* TK 28.

toughness level at which the transition is defined. This expression can be re-written as

$$\frac{1}{T_c} = a - b \log \dot{\epsilon}.$$

where a and b are constant for a given material.

These relations indicate that fracture transition is a thermally activated process and that the following Arrhenius equation can be applied:

$$\dot{\epsilon} = Ae^{-Q/RT}$$

where A is the frequency factor of the Zener-Hollomon parameter and R is the Boltzmann constant.

From these equations it is possible to calculate the slope of the correlations between CVN, static, and dynamic fracture toughness transition temperatures by considering only the effect of strain rate. We first assume that the correlations can be expressed by

$$y = \alpha + \beta x$$

where

$$x = \frac{1}{T_c}$$

and

$$y = \frac{1}{T_c'}$$

where T_c and T_c' are the transition temperatures (degrees Kelvin) for two different brittle fracture tests with strain rates $\dot{\varepsilon}$ and $\dot{\varepsilon}'$ respectively.

As stated previously, the effect of strain rate is given by

$$x = a - b \log \dot{\varepsilon}$$

$$y = a - b \log \dot{\varepsilon}'$$

By considering two steels (1 and 2), one can write

$$y_1 = \alpha + \beta x_1$$

$$y_2 = \alpha + \beta x_2$$

$$x_1 = a_1 - b_1 \log \dot{\varepsilon}$$

$$y_1 = a_1 - b_1 \log \dot{\varepsilon}'$$

$$x_2 = a_2 - b_2 \log \dot{\varepsilon}$$

$$y_2 = a_2 - b_2 \log \dot{\varepsilon}'$$

Thus the slope of the correlation is given by

$$\beta = \frac{1 - k \log \dot{\varepsilon}'}{1 - k \log \dot{\varepsilon}}$$

where

$$k = \frac{b_2 - b_1}{a_2 - a_1}$$

The values of k and $\dot{\varepsilon}$ in the case of the impact Charpy test are calculated in the Appendix from the results of a study by Cheviet et al. These values were found to be approximately $k = 0.1$ and $\dot{\varepsilon} = 15 \text{ s}^{-1}$. As the crack tip strain rate is about 10^{-4} s^{-1} for quasi-static fracture toughness tests, the calculation of the slope of correlation between $1/TK_{Ic}$ and $1/TK$ 28 gives B = 1.57. This value is in very good agreement with the experimental value of 1.63.

The observed agreement tends to prove the validity of the theoretical analysis. Thus the results of this study should enable us to verify the existence of a correlation between TK_{Ic} and TK_{Id} and also between $TK\,28$ and TK_{Id} and to predict the slope of the linear relations between the inverse of those transition temperatures.

Application to Dynamic Fracture Toughness

Within the framework of previous studies, the transition temperatures for CVN, quasi-static, and dynamic ($\dot{K} \simeq 2 \times 10^4$ MPa $\sqrt{m}$) fracture toughness tests were determined on various steels. The results obtained on structural steels, including A 508 C1.3, A 533 B, and a low-alloy Mn-Ni-Mo steel, are summarized with those of this study in Table 4.

Figure 12 plots $1/TK_{Ic}$ as a function of $1/TK_{Id}$ for all the results available. The experimental points can be fitted by a straight line. A linear regression gives the relation

$$1/TK_{Ic} = 1.51/TK_{Id} - 9.75 \times 10^{-4}$$

where TK_{Ic} and TK_{Id} are in degrees Kelvin. The value of the slope, 1.51, is close to the calculated value of 1.39.

In the same way, $1/TK_{Id}$ is plotted as a function of $1/TK\,28$ in Fig. 13. Again a good correlation is observed between those two parameters. A linear regression leads to the relation

$$1/TK_{Id} = 1.15/TK\,28 - 1.13 \times 10^{-3}$$

As expected, the value of the slope is close to one because the crack tip strain rates are not very much different in the case of a CVN test ($\dot{\varepsilon} \approx 15\ \mathrm{s}^{-1}$) and in

TABLE 4—*Transition temperatures of various structural steels studied at IRSID.*

Steel	σ_{ys} at 20°C, MPa	$TK\,28,$ °C	$TK_{Ic},$ °C	$TK_{Id},$ °C
A	300	0	−25	+45
B	700	−100	−130	−90
E	400	−5	−10	+30
F	520	−75	−110	−70
FT	365	−65	−125	−65
(1) A 508 C1.3	475	−20		+15
(2) A 508 C1.3	460	−55	−85	−40
(3) A 508 C1.3	475	−55	−55	−5
(4) A 508 C1.3	470	−55	−70	−20
(5) A 508 C1.3	475	−80	−100	−25
(6) A 508 C1.3	460	−65	−90	−35
Mn-Ni-Mo	610	−45	−80	−40
A 533 B C1.3	730	−65	...	−45

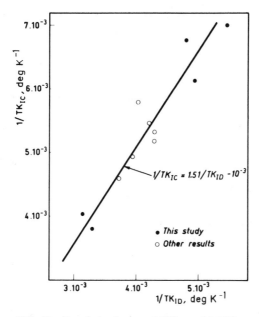

FIG. 12—*Correlation between* $1/TK_{Ic}$ *and* A/TK_{Id}.

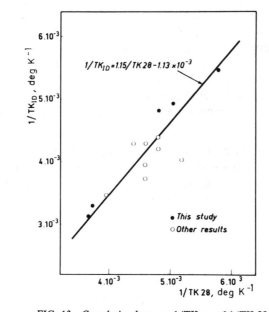

FIG. 13—*Correlation between* $1/TK_{Id}$ *and* $1/TK\ 28$.

the case of a dynamic fracture toughness test with $\dot{K} \approx 2 \times 10^4$ MPa $\sqrt{m}/s$ ($\dot{\varepsilon} \approx 1 \ s^{-1}$). The calculated value of the slope is 1.13.

It can be seen that the agreement between the calculated and experimental values of the slope is excellent. This tends to support the view that strain rate is the main factor that accounts for the variation in the slope of the correlations between the transition temperatures of various brittle fracture tests.

Conclusion

For the five structural steels investigated, it was established that an increase in the loading rate results in a decrease in the toughness when fracture occurs by cleavage. Indeed, for ductile fractures we have observed a certain increase in the measured toughness which may be related to the increase in the yield strength with strain rate.

The amplitude of the temperature shift between static and dynamic toughness transition curves was compared with the predictions given by different models:

1. It appears that the temperature shift ΔT is not as strongly dependent on the material yield strength as estimated by the empirical relation proposed by Barsom. Nevertheless, this relation gives a rather good approximation of ΔT.

2. For each steel, it is possible to relate the effect of temperature to the effect of strain rate on fracture toughness by using the Zener-Hollomon parameter $T \log (A/\dot{\varepsilon})$. However, the frequency factor A depends very much on the material. Therefore this factor has to be determined for each steel to predict the fracture toughness at any loading rate.

3. The variations of yield strength are not always sufficient to account for the evolution of fracture toughness with temperature and strain rate in the case of the steels investigated. In order to apply a cleavage fracture criterion such as the one proposed by Ritchie et al [8], one must also take into consideration the influence of temperature and strain rate on the strain-hardening exponent.

By considering dynamic fracture toughness measurements from previous studies on medium strength steels, a correlation between CVN transition temperatures and dynamic toughness transition temperatures was established. Defining TK_{Id} as the temperature at which $K_{Id} = 100$ MPa $\sqrt{m}$ with a loading rate $\dot{K} \simeq 2 \times 10^4$ MPa $\sqrt{m}/s$ and $TK\ 28$ as the temperature at which the CVN energy $= 28$ J, the relationship is

$$\frac{1}{TK_{Id}} = \frac{1.15}{TK\ 28} - 1.13 \times 10^{-3}$$

where TK_{Id} and $TK\ 28$ are in degrees Kelvin.

A similar relation exists between TK 28 and TK_{Ic}, where TK_{Ic} is the temperature at which $K_{Ic} = 100$ MPa $\sqrt{m}$. Marandet and Sanz [9] have shown in a previous study that this correlation can be approximated by $TK_{Ic} = 1.4\ TK$ 28 (°C).

Assuming that fracture toughness transition is a thermally activated process, we can calculate the slopes of the correlations. These slopes are in good agreement with the experimental results.

Acknowledgments

This study was supported by the French Délégation Générale à la Recherche Scientifique et Technique. The authors also wish to acknowledge C. Bertoletti and P. Boisgontier for their contribution to the experimental work.

APPENDIX

The strain rate $\dot{\varepsilon}$ at the crack tip of a Charpy specimen can be derived from the results obtained by Cheviet et al [28]. Defining TB 3 as the temperature at which the fracture energy is 30 J for a Schnadt test at low speed (0.1 m/s) and TK 3 as the temperature at which the fracture energy is 30 J for a Schnadt test at high speed (5 m/s), the following correlations were found:

$$\frac{1}{TB\ 3} = \frac{1.26}{TK\ 3} - 0.34 \quad \text{(correlation coefficient } r = 0.96) \tag{5}$$

$$\frac{1}{TK\ 3} = \frac{0.72}{TK\ 28} + 0.66 \quad (r = 0.94) \tag{6}$$

$$\frac{1}{TB\ 3} = \frac{0.91}{TK\ 28} + 0.37 \quad (r = 0.88) \tag{7}$$

where the transition temperatures are in degrees Kelvin.

These correlations were established from results obtained on 55 steel grades. As $\dot{\varepsilon} \approx 100$ s^{-1} in the case of the low-speed Schnadt test and $\dot{\varepsilon} \approx 5000$ s^{-1} in the case of the high-speed Schnadt test [29], Eq 5 enables us to calculate the coefficient k as

$$k \approx \frac{1.26 - 1}{1.26 \log 5000 - \log 100} = 0.1$$

References

[1] Barsom, J. M., in *Proceedings*, International Conference on Dynamic Fracture Toughness, Welding Institute and American Society for Metals, London, July 1976, pp. 281-302.

[2] Corten, H. T. and Shoemaker, A. K., *Journal of Basic Engineering, Transactions of ASME*, Series D, Vol. 89, No. 1, March 1967, pp. 86-92.

[3] Shoemaker, A. K. and Rolfe, S. T., *Journal of Basic Engineering, Transactions of ASME*, Series D, Vol. 91, No. 3, Sept. 1969, pp. 512-518.

[4] Shoemaker, A. K., *Journal of Basic Engineering, Transactions of ASME*, Series D, Vol. 91, No. 3, Sept. 1969, pp. 506-511.

[5] James, L. A., *Welding Research Supplement*, Vol. 51, Oct. 1972, pp. 506-507.

[6] Ritchie, R. O., Server, W. L., and Wullaert, R. A., *Metallurgical Transactions*, Vol. 10A, Oct. 1979, pp. 1557-1570.

[7] Krabiell, A. and Dahl, W., "Influence of Strain Rate and Temperature on the Tensile and Fracture Properties of Structural Steels," in *Proceedings*, Fifth International Conference on Fracture, Cannes, 1981, pp. 393-400.

[8] Marandet, B. and Sanz, G. in *Flow Growth and Fracture, ASTM STP 631*, American Society for Testing and Materials, 1977, pp. 72-95.

[9] Marandet, B. and Sanz, G. in *Flow Growth and Fracture, ASTM STP 631*, American Society for Testing and Materials, 1977, pp. 462-476.

[10] Marandet, B., Phelippeau, G., and Sanz, G., "Experimental Determination of Dynamic Fracture Toughness by J-Integral Method," in *Proceedings*, Fifth International Conference on Fracture, Cannes, 1981, pp. 375-383.

[11] Tsukada, H., Iwadate, T., Tanaka, Y., and Ono S. in *Proceedings*, Fourth International Conference on Pressure Vessel Technology, The Institution of Mechanical Engineers, London, May 1980, Vol. I, pp. 369-373.

[12] Irwin, G. R., *Journal of Engineering for Power, Transactions of ASME*, Vol. 86A, No. 4, 1964, pp. 444-450.

[13] Priest, A. H., "Influence of Strain Rate and Temperature on the Fracture and Tensile Properties of Several Metallic Materials," in *Proceedings*, International Conference on Dynamic Fracture Toughness, Welding Institute and American Society for Metals, London, July 1976, pp. 281-302.

[14] Vlach, B., Man, J., Holzmann, M., and Bilek, Z., "The Relation Between Static and Dynamic Fracture Toughness of Structural Steels," in *Proceedings*, Fourth International Conference on Fracture, Waterloo, June 1977, pp. 531-539.

[15] Hahn, G. T. and Rosenfield, A. R., *Transactions of ASM*, Vol. 59, 1966, p. 909.

[16] Hahn, G. T., Hoagland, R. G., and Rosenfield, A. R., *Metallurgical Transactions*, Vol. 2, Feb. 1971, pp. 537-541.

[17] Wilshaw, R. T., Rau, C. A., and Tetelman, A. S., *Engineering Fracture Mechanics*, Vol. 1, 1968, pp. 191-211.

[18] Ritchie, R. O., Knott, J. F., and Rice, J. R., *Journal of the Mechanics and Physics of Solids*, Vol. 21, 1973, pp. 395-410.

[19] Curry, D. A., *Material Science and Engineering*, Vol. 43, 1980, pp. 135-144.

[20] Parks, D. M., *Journal of Engineering Materials and Technology, Transactions of ASME*, Vol. 98, Jan. 1976, pp. 30-36.

[21] Beremin, F., "Study of Instability of Growing Cracks Using Damage Functions: Application to Warm Prestress Effect," presented at the Specialist Meeting on Instability, St. Louis, Mo., 25-27 Sept. 1979.

[22] Hutchinson, J. W., *Journal of the Mechanics and Physics of Solids*, Vol. 16, 1968, pp. 13-31.

[23] Rice, J. R. and Rosengren, G. F., *Journal of the Mechanics and Physics of Solids*, Vol. 16, 1968, pp. 1-12.

[24] Rice, J. R. and Johnson, M. A. in *Inelastic Behaviour of Solids*, McGraw-Hill, New York, 1970, pp. 641-672.

[25] Pineau, A., "Review of Fracture Micromechanisms and Local Approach to Predicting Crack Resistance in Low Strength Steels," in *Proceedings*, Fifth International Conference on Fracture, Cannes, 1981, pp. 553-577.

[26] Marandet, B. and Sanz, G., *Mécanique Matériaux Electricité*, No. 328-329, April-May 1976, pp. 77-84.
[27] Cheviet, A., Grumbach, M., Prudhomme, M., and Sanz, G., *Revue de Métallurgie*, March 1970, pp. 217-236.
[28] Cheviet, A., Grumbach, M., Prudhomme, M., and Sanz, G., *Revue de Métallurgie*, Vol. 67, March 1970, pp. 217-236.
[29] Schnadt, M., *Soudage et Techniques Connexes*, Vol. 18, 1964, pp. 288-295.

D. E. McCabe[1] *and H. A. Ernst*[1]

Crack Growth Resistance Measurement by Crack Opening Displacement Methods

REFERENCE: McCabe, D. E. and Ernst, H. A., **"Crack Growth Resistance Measurement by Crack Opening Displacement Methods,"** *Fracture Mechanics: Fifteenth Symposium, ASTM STP 833*, R. J. Sanford, Ed., American Society for Testing and Materials, Philadelphia, 1984, pp. 648–665.

ABSTRACT: In a previous investigation, tests had been conducted on compact specimens made of A508 Class 2A steel to study specimen geometry effects on J_R-Curve behavior. No particular geometry effects were shown in specimens ranging in size from ½T to 10T when conditions for *J*-controlled growth were satisfied. These same specimens had been instrumented to develop crack opening displacement (COD) toughness values using a plastic hinge model. The COD-type *R*-Curve behavior was examined herein for specimen geometry effects and crack opening angle (COA) behavior.

Plastic hinge (the position of a center of plastic rotation), COD, and COA were determined using a frame-type fixture that had been originally introduced by Andrews. The smallest specimen included herein was 1T and the largest was 10T. COD inferred from the plastic hinge model was determined at the original crack tip position (δ_0) and at the advancing crack tip position (δ_t).

Experimental COD and COA data and some theoretical evaluation schemes indicated that there is geometry-dependent δ_t versus Δa_p type *R*-Curve behavior. It was therefore concluded that J_R and experimentally measured COD are incompatible with respect to geometry independence of *R*-Curve behavior. It was shown that the plastic hinge ratio is strongly influenced by ligament dimensions and that this hinge ratio ultimately controls the δ_t and δ_0 values of COD once the plastic displacement becomes significant with respect to total displacement.

KEY WORDS: cracks, crack opening displacement, crack opening angle, plastic hinge, J_R-Curve, compact specimens

The two elastic-plastic fracture toughness criteria most widely used to evaluate structural materials are *J*-integral and crack opening displacement (COD). It is now well demonstrated that structural materials usually display stable crack growth prior to instability events when tests are conducted on the upper shelf of the transition temperature behavior [1,2]. Elastic-plastic *R*-Curves are

[1] Westinghouse R&D Center, Pittsburgh, Pa. 15235

used to characterize this behavior, and the material toughness has for the most part been expressed in terms of J_R. This is so even though COD methods have been widely available and in fact represent the more senior of the two elastic-plastic disciplines. Under COD practices, crack initiation toughness and toughness at maximum load have been specifically identified as design parameters [3]. The objective of the present work, therefore, was to examine the viability of COD and COD related concepts for R-Curve toughness descriptions. There have been several suggestions on how COD can be measured and some background discussion at this point would be useful.

Both of the aforementioned elastic-plastic toughness parameters are known to have definite limitations when used under conditions of stable crack growth [4,5]. If critical toughness indications are restricted to the safe limitation of crack growth initiation, such as for J_{Ic} or δ_i (COD), the analytical and experimental treatments tend to be uncomplicated. In the case of δ_i, a rather simple link to linear elastic fracture mechanics analysis methods is suggested through the following equation [6]:

$$G = m \delta_i \sigma_0 \qquad (1)$$

Here, G represents a pseudo linear-elastic parameter used in the same sense as elastic-plastic J. COD displacement separation is measured at the starting crack tip position, normal to the crack plane, and is directly associated with crack tip blunting. The δ_i crack surface separation up to the onset of growth is therefore related to the material ductility in the crack growth resistance development process. Equivalency with G or J is achieved through the introduction of material flow strength (σ_0) and m (a constant of proportionality). The magnitude of m is influenced by conditions of constraint, and the values vary from 1 for plane stress to 3.0 for plane strain.

When crack growth develops, the measurement and interpretation of COD becomes a little more complex. As an example, a stably propagating crack tip becomes relatively sharp, quite unlike the blunted parabolic shape established before the onset of stable growth or as indicated in finite-element calculated crack tip contours [7]. See Fig. 1a. Therefore, even if extremely refined measurement techniques were available for measuring the advanced local crack tip displacement, the values obtained would not be directly representative of the increased crack growth resistance development indicated by other toughness criteria such as J_R or K_R. It is necessary, therefore, to redefine COD measurement to fit practical experimental realities. Two choices are (1) measurement of hinge-implied crack separation at a fixed location, such as the original crack tip position, henceforth designated δ_0; or (2) measurement of implied COD at the advancing crack tip position, henceforth designated δ_t. The usual practice in experimental procedures is to use hinge point modeling of specimen displacement and COD measured at the original crack tip position (British Standard BS 5762, Methods of Crack Opening Displacement (COD) Testing).

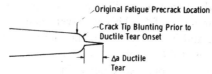

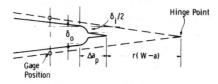

A) Schematic of Crack Shape at Midthickness After Stable Crack Growth.

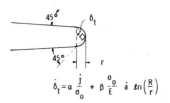

B) Inferred Crack Opening Displacement From Hinge Rotation Assumption.

$$\delta_t = \alpha \frac{J}{\sigma_0} + \beta \frac{\sigma_0}{E} \; \dot{a} \; \ell n \left(\frac{R}{r} \right)$$

C) Crack Tip Opening Displacement Described by 45° Projection

FIG. 1—*Crack opening displacement models under ductile tear conditions.*

Plastic hinge behavior develops in bend-type specimens and is evidenced after general yield when all displacements along the crack plane become a linear function of the distance from a hinge point. See Fig. 1*b*. Hence displacements measured at any two locations along the crack plane could then be used to infer a displacement at any other chosen alternative location. It is evident that after some appreciable stable growth, a significant part of the inferred δ_0 may be due to physical separation of the crack surfaces. Consequently, one might expect the equivalency between δ_0 and G values (Eq 1) to become untenable.

If one assumes that plastic deformation dominates, the computation of displacement, δ_0 is obtained very simply from similar triangle construction as

$$\delta_0 = \left[\frac{H_a - a_0}{H_a} \right] \delta_{LL} \tag{2}$$

where

H_a = experimentally determined hinge length from load line to hinge point.
a_0 = initial crack size, and
δ_{LL} = total load line displacement.

Present computational procedures for determining crack tip opening displacements (δ_t) call for the separation of elastic and plastic components of measured displacements. Elastic K_I is used to infer δ_t elastic, and the plastic component is used with plastic hinge approximation to obtain δ_t plastic. The following equation can be used with a clip gage mounted on the load-line of compact specimens:

$$\delta_t = K_I^2/2\sigma_0 E + (\delta_{LL})_P \left[\frac{r(1 - a/W)}{r(1 - a/W) + a/W} \right] \qquad (3)$$

where

$(\delta_{LL})_P$ = plastic component of displacement (δ_{LL} total − δ_{LL} elastic),
$\quad r = (H_a - a)/(W - a)$,
$\quad \sigma_0$ = material yield strength,
$\quad W$ = specimen width,
$\quad a_p$ = updated crack size, and
$\quad E$ = elastic modulus.

A recent interest is measurement of crack opening angle (COA) [8]. This fracture toughness parameter has been identified from numerical analyses [9], and the characteristic noted is that it initially starts high and then rapidly settles down to almost constant values with continued stable growth. COA, unlike plastic hinge inferred values of COD, does in fact physically exist and can be calculated from δ_0 as

$$\delta^* = \delta_0 - \delta_i \qquad (4a)$$

$$\text{COA} = \delta^*/\Delta a_p \qquad (4b)$$

As was implied earlier, δ_i is δ_0 at the onset of stable crack growth, and is experimentally obtained from plastic hinge modeling.

Computational practices in standard methods for COD (British Standard BS 5762 and ASTM Draft Test Method for Crack Tip Opening Displacement (CTOD) Testing) use an assumption that the hinge behavior for most materials is essentially fixed, having a rotation point at $0.4(W - a)$ forward of the crack tip location. If this is acceptable, then it is necessary to measure displacement at only one location along the crack plane, and $r = 0.4$ is used in Eq 3.

Recently, numerical methods have been applied rather extensively to produce new COD-type crack growth criteria [9,10]. In general, such studies tend to focus attention on displacement behavior very local to the crack tip, where experimental methods are notoriously weak. A constant crack tip opening angle (CTOA) appears to be the preferred growth model and, as with COA, calculated CTOA rapidly subsides to somewhat constant values during simulated crack growth. To be sure, CTOA is slightly mesh size and finite-element technique sensitive. Nevertheless, it is reputed to behave as a sufficiently geometry-independent property to serve as a controlling parameter to predict load-displacement behavior for untested geometries. A finite element based local crack tip δ_t measurement location has been suggested by Rice and Sorensen [11]. Forty-five degree lines emanating from the crack tip as shown in Fig. 1c intersect the crack surface profile where δ_t is measured. The δ_t defined thusly was shown to be related to the specimen deformation behavior, and from this, a simple δ_t-J equivalency similar to Eq 1 was shown to exist:

$$J = \delta_t \sigma_0 / d_n \qquad (5)$$

The coefficient d_n that relates J and δ_t is obtained through careful numerical analysis. It is strongly dependent on work-hardening exponent n and to a lesser extent on yield strain σ_0/E. Values of d_n can range from 1 to 0.3, depending upon material work hardening and conditions of constraint [12]. As a consequence of Eq 5, along with the observation that CTOA tends to constant values, Shih et al have noted a rather confining implication about the shape of elastic-plastic J_R-Curves. Rearranging the terms and differentiating with respect to crack size results in

$$\frac{d\delta_t}{da} = \left[\frac{n}{n+1} \frac{d_n}{\sigma_0} \right] \frac{dJ_R}{da} \qquad (6)$$

Since $d\delta_t/da$ or CTOA is identified as fixed, Eq 6 suggests that the slope of J_R-Curves should be essentially constant.

The Eq 5 definition for δ_t has been introduced into this discussion to acknowledge its existence among the various crack tip displacement models. The present investigation addresses COD related properties of δ_0, δ_t and COA that are experimentally determined, and the point is made that these results do not necessarily support or discredit the numerical method based definitions of crack tip behavior.

Experimental Procedure

Data had previously been developed on A508 Class 2A steam generator tube plate steel using a test matrix designed to evaluate specimen plan view dimensional effects on J_R-Curve behavior [13]. The specimens were 20% side-grooved compacts, ranging in size from 1T to 10T (Table 1). The test material was not always of uniform properties throughout the sampling plan, but representative properties of the majority are given in Table 2. For the critical COD comparisons made herein, the specimens were carefully chosen for uniformity of J_R-Curve and strength behavior previously identified in the aforementioned investigation (Fig. 2). The more critical of the COD size effect comparisons were made on specimens appearing in this figure.

TABLE 1—*Test matrix of elastic-plastic methodology to establish* δ_t-Δa_p *R-Curves.*[a]

Thickness, in. (gross/net)	Plan View Size	Ligament Size, in.				
		1/2	1	2	4	10
1.0/0.8	1T	$w = 2$	$w = 2$	...	...	...
	2T	...	...	...	...	...
	4T	...	$w = 8$	$w = 8$	$w = 8$	$w = 20$
2.0/1.6	2T	$w = 4$	...	$w = 4$	...	...
	4T	...	$w = 8$	$w = 8$	...	...
4.0/3.2	4T	...	$w = 8$	...	$w = 8$	...
10/8	10T	...	...	...	...	$w = 20$

[a]1 in. = 25.4 mm.

TABLE 2—*Tensile and Charpy properties of A508-2A at the mid-slab location where specimens were taken.*

Yield Strength		Tensile Strength		Elonga-tion, %	Reduction of area, %	R_B	CVN		FATT	
ksi	MPa	ksi	MPa				ft-lb	J	°F	K
55	380	80	550	30	63	87	115	157	110	396

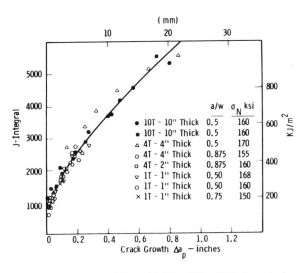

FIG. 2—J_R-*Curves for material in the 1030 to 1170 MPa (150 to 170 ksi) nominal strength category.*

Specimens were instrumented as shown in Fig. 3a; displacement measurements were made at five locations. An exception to this is that there were no V_0 gages on the smallest (1T size) specimens because of practical physical limitations. Hinge point was experimentally determined from locations V_{LL} and V_2 or from the externally mounted frame using V_{LL} and V_a. The more precise of the two data sets came from the external frame, which was designed after a method by Andrews [14]. As shown in Fig. 3b, the fulcrum point is effectively maintained adjacent to the crack tips of the specimens for all a/W (inside a drilled hole). It is always slightly ahead of and above the starting crack tip position. The reference measurement locations are V_{LL} and V_a and these in combination point to the hinge position where specimen displacement is zero. Hinge length from the load line to the hinge point H_a is indicated.

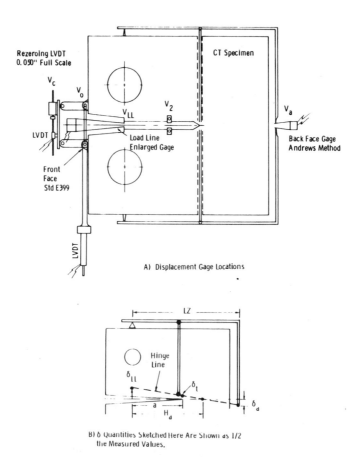

FIG. 3—(a) *Instrumentation of specimens with five measurement points.* (b) *Schematic of imaginary hinge line.*

Crack growth was followed using unloading compliance methods. The data used in this report came from EPRI-sponsored work [15].

Results

Crack Tip COD (δ_t)

Figure 4 plots δ_t obtained by the ASTM equation in the ASTM CTOD draft method where r is fixed at 0.4 against δ_t obtained using the experimental plastic hinge. The comparison covers the full range of specimen sizes from 1T to 10T, but relative crack size (a/W) in this case was fixed at 0.5 throughout. This is excellent agreement. Figure 5 illustrates what happens when the relative crack size is varied. In this example, initial ligament sizes are purposefully small, comparing 1T size specimens, and this had the effect of accentuating any variability caused by plastic hinge ratio effects. Fixing hinge ratio at 0.4, as in the ASTM equation, forces a less crack size dependent δ_t-type R-Curve, shown on the left. However, the actual experimental hinge ratio, r was strongly affected by crack size or ligament size or both, and the result was that the R-Curve behavior became strongly geometry dependent.

To broaden the specimen size effects evaluation, Fig. 6 compares δ_t-Δa_p

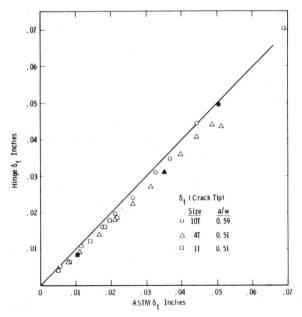

FIG. 4—*Crack tip opening displacement (CTOD) comparing experimental hinge ratio equation and ASTM equation.*

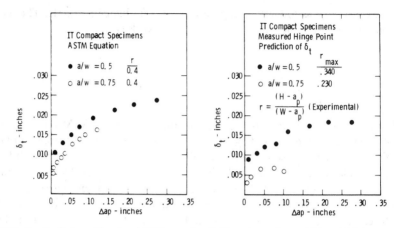

FIG. 5—*Prediction of δ using ASTM equation* (left) *and experimental hinge* (right). *1T compact specimens were used.*

R-Curves for 1T, 4T, and 10T specimens directly with a/W fixed at about 0.5 and experimentally determined hinge ratios. The upper plateau plastic hinge ratios developed can be seen with each curve. The solid datum point that appears with each data set represents where the elastic and plastic components of displacement were of about equal magnitude. Before this level, K_1 (or J elastic) tends to dominate calculated δ_t and the *R*-Curve behavior is geometry indepen-

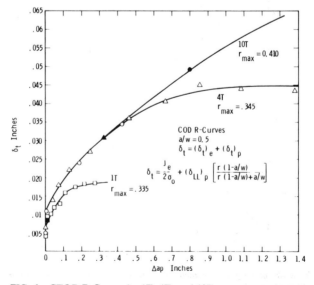

FIG. 6—*CTOD R-Curves for 1T, 4T, and 10T compact specimens.*

dent. Beyond this level, plastic hinge effects begin to exert control and specimen geometry dependence with variable experimental hinge ratio develops.

Theoretical δ_t

A recently published handbook of elastic-plastic solutions [16] can be used
to generate theoretical load-displacement test record behavior. Both calibration J and δ_{LL} can be determined at assumed applied loads, requiring only input information on material flow strength properties. The calculated behavior
can be used to predict δ_t as if the data source had been from an actual test.
Such test condition simulations have the advantage of modeling the specimen
behavior for a completely homogeneous material, and this eliminates any questions associated with experimental measurement variabilities. An added bene-
fit is that the simulation is made for a nongrowing crack situation.

The analysis uses tensile property information in the form of the Ramberg-
Osgood work-hardening law

$$\frac{\epsilon}{\epsilon_0} = \frac{\sigma}{\sigma_0} + \beta \left(\frac{\sigma}{\sigma_0} \right)^n \tag{7}$$

where

$\epsilon_0 =$ true yield strain,
$\sigma_0 =$ true yield strength,
$\beta =$ work-hardening coefficient, and
$n =$ work-hardening exponent.

Example data were generated from typical A508 Class 2A tensile properties
having $n = 7$ and $\beta = 0.4$. The matrix of specimen geometries for which these
test record simulations were made is listed in Table 3. As can be seen, the specimen plan view size simulations ranged from 1T to 10T and crack size ratio
(a/W) from 0.5 to 0.95. Crack tip position δ_t is predicted using an assumed
plastic hinge ratio of $r = 0.4$. Since there is no stable crack growth, the rela-

TABLE 3—Specimen dimensions assumed in handbook
simulation solutions.[a]

Plan View Size	W, in.	$a/W = 0.5$	$(W - a)$, 1 in.
1T	2.0	X	X
2T	4	X	
4T	8	X	
10T	20	X	X

[a]1 in. = 25.4 mm.

tionship between δ_t and calculated load line displacement δ_{LL} will be linear at a fixed crack size ratio.

Figure 7 shows calibration J versus δ_t values for specimens of varied size using no separation of δ_{LL} elastic versus δ_{LL} plastic and a/W fixed at 0.4. The separation between the various specimen sizes is clearly due to the geometry effects in δ_t. In Ref 13, the model material had displayed relatively geometry-independent J_{Ic} at about 1100 in.-lb/in.^{2} (192.5 kJ/m^2), and Fig. 7 suggests that δ_t, when determined by pure plastic hinge equations, is different by a factor of about three at this J level. If Eq 3 is used, with elastic δ_{LL} versus plastic δ_{LL} displacement separation, this drastic difference with specimen size can be reduced (Fig. 8). At J_{Ic} of 1100 in.-lb/in.^{2}, the difference in δ_t is reduced to about 20 to 30%. Again, elastic J reduced the size dependency of δ_t, but as soon as the hinge began to exert significant control, the size dependency expanded.

Since the size dependency of δ_t–Δa_p R-Curves comes from an apparent weakness of the plastic hinge approximation, ligament dimensions should attract more attention. In fact, if ligament dimensions control, specimens of equal remaining ligament size should display size independent δ_t–J_R behavior. Again using Eq 3, theoretical J, and δ_{LL}, we calculated δ_t for ligaments fixed at 25.4 mm (1-in.) with plan view size varied from 1T to 10T (Fig. 9). As we suggest, the geometry dependence vanishes. To further support this assertion of the dominance of ligament dimensions, experimental data in Fig. 10 show δ_t

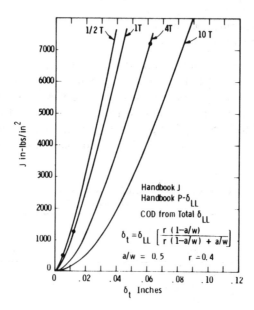

FIG. 7—*Handbook-predicted J and CTOD using plastic hinge (total hinge without J elastic removed).*

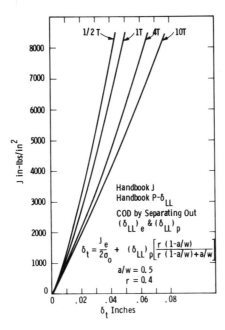

FIG. 8—*Handbook-predicted J and CTOD (separate J elastic and plastic hinge contributions).*

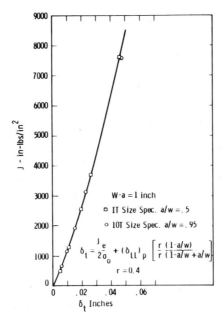

FIG. 9—*Handbook-predicted J and CTOD (separate J elastic and plastic hinge contributions; fixed ligament size).*

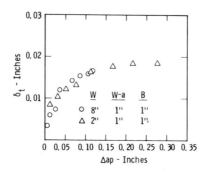

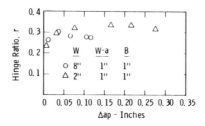

FIG. 10—*Trend in CTOD and hinge ratio (r) with crack growth (4T and 1T plan view size: fixed ligament size).*

versus Δa_p for 4T and 1T plan view size specimens where the thickness and remaining ligament length were essentially the same. Note that the hinge ratio r was about the same for both specimens. This lead to identical δ_t–Δa_p R-Curve behavior. The fact that ligament dimensions tend to control plastic hinge ratio was detected to be endemic throughout the Table 1 test matrix. Figure 11 summarizes the trends seen in r with the various ligament sizes, W–a, and specimen thicknesses. Hinge ratio appears to be a function of both the remaining ligament dimensions, W–a and thickness B. This figure contains specimens of slightly varied strength; a fact that had been discovered in the earlier J_R-Curve work [13]. Nevertheless, the trends in ratio r were sufficiently strong to overpower these strength variations. This figure makes it clear that current experimental methods of inferring δ_t from plastic hinge will result in geometry-dependent fracture toughness indications. The usual size of fracture toughness specimens (1T compact) will suffer the greatest sensitivity to geometry variations.

Experimental Crack Opening Angle (COA)

As was discussed earlier, COA is determined with crack mouth displacement measurement δ_0 using Eqs 2 and 4. By plotting δ_0 against crack growth Δa_p, as in Figs. 12 and 13, COA is obtained directly from the straight line

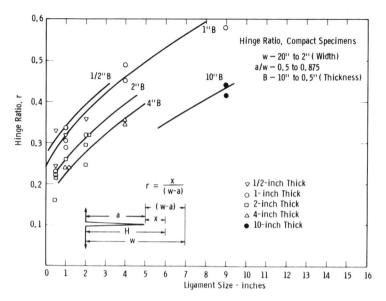

FIG. 11—*Hinge ratio versus thickness* (B) *and ligament size.*

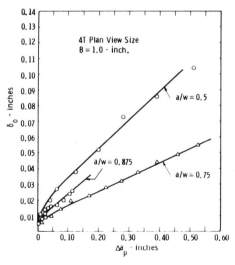

FIG. 12—*Displacement at the initial crack tip position versus stable crack growth (1-in.-thick 4T specimens; initial crack size varied).*

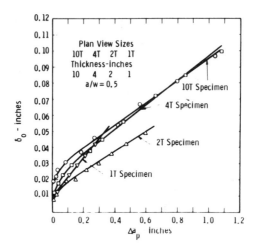

FIG. 13—*Displacement at the initial crack tip position versus stable crack growth (overall specimen size varied; same relative crack ratio).*

slopes. Intercepts with the coordinate axes correspond to rough estimates of δ_i. In Fig. 12, specimen plan view size (4T) and thickness (1-in.) is maintained constant, and relative crack size a/W is varied from 0.5 to 0.875. In Fig. 13, specimen dimensions are scaled consistently between 1T and 10T size and a/W is fixed at about 0.5. In all cases, the specimens display COA-type behavior with linear slopes, but it is readily apparent that both the COA and δ_0 behavior are strongly geometry-dependent properties.

Figure 14 shows how the COD R-curve is affected by use of Eq 2 versus Eq 3. The initial agreement for the 1T specimen at small Δa_p is due to the fact that the specimen was in full plastic hinge at onset of stable growth. The rapid divergence with continued growth is due to an increased component of δ_0 being from the physical separation of crack surfaces as described in Fig. 1. The 10T specimen suffers a poor crack growth initiation δ_0 versus δ_t comparison because the specimen is dominant elastic at this point, and Eq 2 is inaccurate from the pure plastic hinge assumption. For the range of crack growth shown (10T specimen), there is only about 10% of ligament change as opposed to about 30% in the 1T specimen. The separation between δ_0 and δ_t is far more gradual with continued stable growth because of this.

Comparison of Figs. 12 to 14 show linearity of δ_0 with Δa_p once a full plastic hinge behavior is set up; this is reminiscent of the constant COA behavior predicted in numerical methods. On the other hand, δ_t versus Δa_p type R-Curves have the same general shape of J_R-Curves, and in some cases an Eq 1 type coefficient of m can be defined that would relate J_R and δ_t. Unfortunately, as was pointed out before, δ_t is geometry dependent and a fitted m value would not apply to data from other specimens.

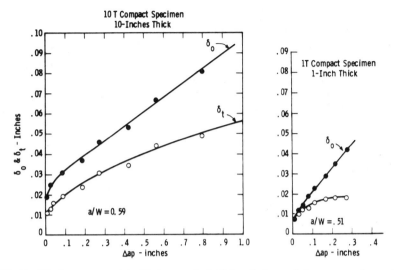

FIG. 14—*Displacement at original fatigue crack tip* (δ_o) *and advanced tip versus crack growth.*

Conclusions

The COD test method presently under development within ASTM Committee E-24 on Fracture Testing uses an equation that assumes a universal plastic hinge rotation ratio of 0.4 for all materials, specimen sizes, and remaining ligament sizes. Comparisons of δ_t calculated by ASTM versus the experimentally determined hinge method were good so long as the experimental hinge ratio r was reasonably close to 0.4. Experimental hinge ratios often varied consistently with varied specimen ligament dimensions, however, and agreement was not good over all specimen geometries.

A handbook on elastic-plastic solutions [16] was used to develop simulated specimen load-displacement behavior and J calibration using typical tensile properties for A508 Class 2A. A fixed hinge ratio of 0.4 was used in conjunction with predicted δ_{LL} values to predict δ_t. These data showed that specimen geometry affects the relationship between δ_t and J_R. The geometry effects observed develop when plastic hinge effects begin to exert significant control. If ligament dimensions are fixed and plan view size varied, the specimen size influence on the relationship between J_R and δ_t vanishes. Experimental trend evidence was presented in Figs. 10 and 11 to support this observation.

Plastic hinge behavior is controlled by (1) remaining ligament dimensions, (2) conditions of constraint (thickness B), and (3) material ductile flow properties. Because of this, COD determined by inference using experimental hinge methods is geometry dependent, and the information obtained from one geometry cannot be used to predict the crack growth behavior of another geometry. Hence the principal value of such COD data can only be for relative comparisons of given grades of materials. Strict specimen geometry control should be a stipulated experimental requirement.

Numerical analysis has indicated that there is constant crack opening angle behavior during slow stable crack growth. The reported behavior is that COA initially starts high and then rapidly reduces to constant values. Experimental COA was determined herein and all specimens confirmed that there is in fact constant COA from δ_0 measurements. However, experimental COA was specimen geometry dependent.

Acknowledgments

The work presented in this paper was supported by the Electric Power Research Institute under Contract RP 1238-2. The authors would like to acknowledge Dr. D. M. Norris, Jr., EPRI project manager for this program, who contributed to the planning of the work.

The authors also acknowledge the assistance of their colleagues at Westinghouse. W. H. Pryle assisted with material handling and specimen preparations; F. X. Gradich assisted with the equipment development; L. W. Burtner, T. R. Fabis, and W. T. Broush helped with the testing; and R. R. Hovan and J. Selchan assisted with the preparation of the manuscript.

References

[1] Merkle, J. G. in *Elastic-Plastic Fracture, ASTM STP 668*, American Society for Testing and Materials, 1979, pp. 674–702.

[2] Fearnehough, G. D., Lees, G. M., Lowes, J. M., and Weiner, R. T., "The Role of Stable Ductile Crack Growth in Failure of Structures," presented at Applied Mechanics Group Conference, Dec. 1970.

[3] Dawes, M. G., "The COD Design Curve Approach" in *Advances in Elastic-Plastic Fracture Mechanics*, Applied Science, New York, 1979.

[4] Hutchinson, J. W. and Paris, P. C. in *Elastic-Plastic Fracture, ASTM STP 668*, American Society for Testing and Materials, 1979, pp. 37–64.

[5] Tanaka, K. and Harrison, J. P., *International Journal of Pressure Vessels and Piping*, Vol. 6, 1978, pp. 177–202.

[6] Written discussion prepared by A. A. Wells, F. M. Burdekin, and D. E. Stone, *Fracture Toughness Testing and Its Applications, ASTM STP 381*, American Society for Testing and Materials, 1964, pp. 400–405.

[7] Garwood, S. J. and Turner, C. E., *International Journal of Fracture*, Vol. 14, 1978, pp. 195–198.

[8] DeKoning, A. U., "A Contribution to the Analysis of Slow Crack Growth," Report NLR MP 75035U, The Netherlands National Aerospace Laboratory, 1975.

[9] Kanninen, M. F. et al in *Elastic-Plastic Fracture, ASTM STP 668*, American Society for Testing and Materials, 1979, pp. 121–150.

[10] Shih, C. F., *Journal of the Mechanics and Physics of Solids*, Vol. 29, No. 4, 1981, pp. 305–326.

[11] Sorensen, E. P. in *Elastic-Plastic Fracture, ASTM STP 668*, American Society for Testing and Materials, 1979, pp. 151–174.

[12] Shih, C. F., German, M. D., and Kumar, V., "An Engineering Approach for Examining Crack Growth and Stability in Flawed Structures," Report 80CRD205, General Electric Company, Sept. 1980.

[*13*] McCabe, D. E., Landes, J. D., and Ernst, H. A. in *Elastic-Plastic Fracture: Second Symposium, Volume II—Fracture Resistance Curves and Engineering Applications, ASTM STP 803*, American Society for Testing and Materials, 1983, pp. II-562–II-581.

[*14*] Andrews, W. R. and Shih, C. F. in *Elastic-Plastic Fracture, ASTM STP 668*, American Society for Testing and Materials, 1979, pp. 426–450.

[*15*] McCabe, D. E. and Landes, J. D., "Elastic-Plastic Methodology to Establish *R*-Curves and Instability Criteria," Topical Report on Compilation of Data, EPRI Contract RP 1238-2, Electric Power Research Institute, 11 Dec. 1982.

[*16*] Kumar, V., German, M. D., and Shih, C. F., "An Engineering Approach for Elastic-Plastic Fracture Analysis," EPRI Topical Report NP-1931, Electric Power Research Institute, July 1981.

Y. W. Cheng, [1] *R. B. King,* [1] *D. T. Read,* [1] *and H. I. McHenry* [1]

Post-Yield Crack Opening Displacement of Surface Cracks in Steel Weldments

REFERENCE: Cheng, Y. W., King, R. B., Read, D. T., and McHenry, H. I., **"Post-Yield Crack Opening Displacement of Surface Cracks in Steel Weldments,"** *Fracture Mechanics: Fifteenth Symposium, ASTM STP 833*, R. J. Sanford, Ed., American Society for Testing and Materials, Philadelphia, 1984, pp. 666–681.

ABSTRACT: Crack-mouth-opening displacements (CMOD) of surface cracks are measured as functions of stress and strain in tensile panels of API 5LX-70 steel plates and welded pipe segments. The experimental results are compared with analytical predictions. For CMOD versus stress, a previously developed model provides good agreement between experiment and analysis for the base metal and the welds. At stresses above net-section yielding, it is observed in seven of nine base metal tests that all the remote displacement is transferred to the crack tip through slip bands extending from the crack tip to the plate edges at 45 deg angles; the two exceptions are in specimens with small (less than 5% of the cross-sectional area) cracks where yielding occurred in the gross section. A model based on this observation is used to calculate CMOD versus strain for net-section yielding; analysis and experiment agree in the intended range, that is, net-section yielding. In the welded specimens, the yield strength of the weld was higher than that of the base metal. Consequently, the base metal began yielding before the weld, and the CMOD in the weld was lower than predicted by the analytical model. A strain-partitioning model partially accounts for the observed behavior, provided the crack is large enough to permit net-section yielding in the weld.

KEY WORDS: crack opening displacement, elastic-plastic fracture mechanics, pipeline steel, surface cracks, welds, yielding

The application of elastic-plastic fracture mechanics (EPFM) to surface cracks requires a relationship among applied stress (strain), crack size, and crack driving force. When the crack driving force equals the toughness of the material, crack growth starts. For stress below yield, an EPFM relationship based on the crack-tip-opening displacement (CTOD) concept has previously been derived and verified [1] for the case of steel plates. The present study ex-

---

[1] Fracture and Deformation Division, National Bureau of Standards, Boulder, Colo. 80303.

tended the investigation to include tests of surface cracks in steel welds, which are present in a large number of engineering structures such as pipelines.

The approach used in this study was to measure the crack-mouth-opening displacement (CMOD) as a function of applied stress or applied strain in tensile panels for various surface-crack sizes. Previous work [1] has shown that CMOD is simply related to CTOD, a parameter closely related to crack driving force. The experimental results for post-yield CMOD versus applied strain are presented, and the various yielding patterns are discussed.

Test Materials

The base metal was an API 5LX-70 pipeline steel in the form of 16-mm-thick plate. The chemical composition is given in Table 1. The longitudinal tensile properties were measured at room temperature using standard 6.35-mm-diameter, 25-mm-gage-length specimens. The results are summarized in Table 2.

Welds were made in 1220-mm-diameter, 16-mm-thick-wall API 5LX-70 pipe, using manual and automatic processes with positioning, procedures, and consumables simulating field practice. The manual welds were made by the shielded metal-arc process using AWS E8010G electrodes. Ten weld passes at 24 to 30 V and 150 to 170 A were used to fill the joint, a 60 deg included angle V-joint. The automatic welds were made by the gas metal-arc process using AWS E70S-6 filler wire and a 50/50 mixture of CO_2 and argon for gas coverage (1.27 m^3/h flow rate). Six weld passes at 23 V and 180 to 190 A were used to fill the joint, a duplex V-joint with a 60 deg included angle for 6 mm at the root and a 14 deg included angle for the upper 11 mm. Chemical analyses of the weld metals are shown in Table 1. Radiographic inspection showed that all the welds tested were sound and met the weld quality standards of API 1104.

The tensile properties of the welds were measured at room temperature using standard 6.35-mm-diameter, 25-mm-gage-length specimens. The specimens were oriented so that the reduced section was totally within the weld metal and the specimen axis was parallel to the weld axis. The results are summarized in Table 2.

Experimental Procedures

Three series of specimens with different surface crack lengths and depths were tested: one series each for the base metal, the manual welds, and the automatic welds. The test matrix is shown in Table 3. The base metal specimens were tensile panels (Fig. 1) that were notched in three different ways: (1) a saw cut of 0.4 mm width, (2) a saw cut followed by fatigue sharpening, or (3) an electrical-discharge-machined (EDM) notch of 0.4 mm width. The weld metal specimens were tensile panels taken from the 1220-mm-diameter pipe

TABLE 1—*Chemical analyses (weight percent).*

	C	Si	Mn	P	S	Al	Cu	Cr	Ni	Mo	V	Nb	N
Base metal	0.08	0.3	1.45	0.015	0.003	0.039	0.08	0.09	0.03	0.10	0.07	0.036	0.009
Manual weld	0.12	0.17	0.67	0.005	0.010	...	...	0.13	1.6	0.031	...	...	...
Automatic weld	0.11	0.43	1.3	0.011	0.005	...	...	0.073	0.92	0.053	...	...	...

TABLE 2—*Tensile properties of test materials at room temperature.*

Material	Yield Strength (0.2% offset), MPa	Ultimate Tensile Strength, MPa	Elongation in 25 mm, %	Reduction of Area, %	Notes
Base metal	491[a]	549	26	69	average of three tests
Manual weld (AWS E8010G)	465	555	28	66	...
Automatic weld (AWS E70S-6)	725	810	21	66	...

[a]The upper yield strength of the base metal was 532 MPa.

TABLE 3—*Test matrix.*

Specimen[a]	Plate Width (W), mm	Crack Depth (a), mm	Crack Length (ℓ), mm
P1[b]	102	6.65	40.9
P2[b]	102	9.07	43.5
P3[b]	102	5.64	31.2
P4[b]	102	6.88	48.3
P5[b]	76	10.54	31.2
P6	76	4.57	13.5
P7	76	3.56	31.8
P8	76	5.08	14.7
P9	381	3.81	114.3
A1	76	8.13	29.7
A2	102	5.72	47.5
A3	76	2.49	17.3
A4	76	4.57	11.8
M1	76	8.13	29.2
M2	102	5.84	47.5
M3	76	4.32	11.8
M4	76	2.16	15.0

[a] P = base metal, A = automatic weld, M = manual weld.
[b] Data presented in Ref *1*.

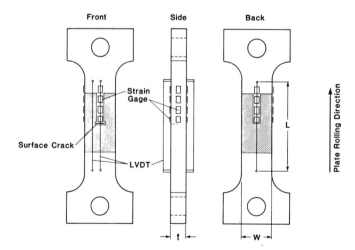

FIG. 1—*Configuration of test specimen. Shaded areas are coated with photoelastic material or brittle lacquer.*

with the weld transverse to the tensile axis. The pipe curvature was retained in the specimens, and the weld crown was removed to obtain a uniform cross section. The surface notches, all prepared by EDM, were located on the concave side of the pipe at the root of the V-shaped weld.

Each of the specimens in Table 3 was loaded in tension, and the CMOD was measured as a function of nominal stress (σ) and gage-length strain (ϵ_L). In the base-metal specimens, ϵ_L was taken as the average strain over a gage length of 305 mm, except for the 381-mm-wide specimen that had a gage length of 610 mm. The experimental procedures are described in a previous paper [1]. Briefly, the specimens were pulled in tension under displacement control with a closed-loop servo-controlled hydraulic testing machine. The values of CMOD were measured with a clip-on displacement gage, mounted directly onto the crack mouth. The values for ϵ_L were obtained from displacements measured over the chosen gage length with linear variable differential transducers (LVDTs). For observation of the deformation patterns near the cracks, the specimens were coated with photoelastic material and instrumented with electrical-resistance strain gages.

Results and Discussion

For the test matrix shown in Table 3, CMOD has been measured as a function of applied stress (σ) and of gage-length strain (ϵ_L). In Ref 1, it was shown that CMOD is equal to CTOD after ligament yielding, and thus CMOD can be regarded as a measure of the driving force for fracture. In this section, the experimental results of CMOD versus σ are compared with the analytical predictions of the modified critical COD model [1]:

$$\text{CMOD} = \text{CMOD (elastic)}, \quad \text{for } \sigma < 1 - (a/t)\bar{\sigma} \tag{1a}$$

and

$$\text{CMOD} = \frac{2(\ell + 2r_y)}{E}[\sigma - (1 - a/t)\bar{\sigma}] + \text{CMOD}|_{\text{LY}}, \quad \text{for } \sigma \geq 1 - (a/t)\bar{\sigma} \tag{1b}$$

where

  ℓ = crack length,
  r_y = the Dugdale plastic zone size solution with a finite-width correction [2],
  E = Young's modulus of elasticity,
  a = crack depth,
  t = plate thickness,
  $\bar{\sigma}$ = flow stress of the material (average of the yield and the ultimate strengths), and
  $\text{CMOD}|_{\text{LY}}$ = CMOD at the onset of ligament yielding; that is, $\sigma = 1 - (a/t)\bar{\sigma}$.

The derivation of Eq 1b is described in the Appendix. Equation 1a is evaluated using the elasticity solution of Kobayashi [3] for cracks with aspect ratios (a/ℓ) greater than 0.1. For long cracks ($a/\ell < 0.1$), the Kobayashi solution is not applicable and the elastic CMOD is calculated using King's [4] simplified version of the line-spring model. Also, for long cracks the plastic zone development through the thickness becomes more important, and is accounted for by adding a plastic zone correction (r_{yd}) to the crack depth. The Rice [2] solution for a strip-yield plastic zone in a center-cracked finite-width panel is used. In using this solution for an edge crack, as required by the line-spring model, bending is neglected; the underestimation of r_{yd} is considered negligible for the present purposes.

Equation 1b is used from the onset of ligament yielding, $\sigma = 1 - (a/t)\bar{\sigma}$, until net-section yielding occurs. After net-section yielding, CMOD increases without bound at a constant applied stress and is shown as a vertical line in the CMOD-versus-σ curve at $\sigma = \bar{\sigma}[1 - (a\ell/tW)]$.

Base Metal Results

As shown in Table 3, nine base metal tests were conducted. The results of the first five tests were presented in terms of CMOD versus σ in Ref 1. Good agreement was obtained between the experimental data and the analytical predictions of Eq 1. Four additional tests have been conducted to evaluate Eq 1 for smaller and larger crack sizes. The results, summarized in Fig. 2, show good agreement between experiment and analysis. Note that the shape of the CMOD-versus-σ curve for the long crack ($\ell = 114.3$ mm) test differs from that of the shorter crack tests; owing to the plastic zone correction, nonlinearity oc-

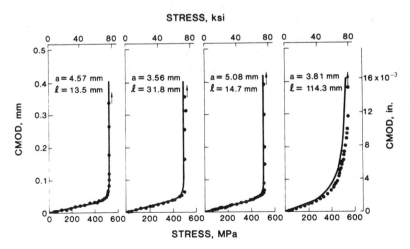

FIG. 2.—*CMOD versus nominal stress for base metal specimens. Symbols are experimental data; solid lines are analytical predictions based on the modified critical COD model.*

curs at a lower stress and there is a more gradual transition to net-section yielding. Thus the modified critical COD model appears to be valid for a broad range of crack sizes.

The results in terms of CMOD versus ϵ_L are shown in Figs. 3 and 4. Two distinct behaviors are observed. Seven of the specimens exhibited a nearly bilinear relationship between CMOD and ϵ_L, with the slope sharply increasing at strain levels slightly below the yield strain (Fig. 3). Specimens 6 and 8 exhibited similar behavior at low strains, but at high strains the CMOD reached a plateau value, increasing only slightly with strain. The bilinear behavior results when the strains that exceed yield only occur in the net section, that is, net-section yielding (NSY). Photoelastic observations on the front and the back surfaces of the specimen and strain gage measurements indicate that, for NSY, yielding is confined to slip bands, which extend at approximately 45 deg angles from the crack tip to the specimen edges. The plateau develops when strain hardening elevates the flow strength in the slip bands and strains above yield occur in the gross section, that is, gross-section yielding (GSY). Once GSY occurs, the remote displacements are absorbed along the entire test section and CMOD increases slowly with ϵ_L.

The bilinear relationship between CMOD and ϵ_L can be modeled at stress levels below NSY using Eq 1, provided the appropriate relationship between ϵ_L and σ has been established. At strains above NSY, the slopes of the CMOD-versus-ϵ_L curves are equal to the gage length; this occurs because all the remote displacement is transmitted to the crack tip through the slip bands. Thus, after NSY,

$$\text{CMOD} = \text{CMOD}\big|_{\text{NSY}} + L\,(\epsilon_L - \epsilon_{\text{NSY}}) \tag{2}$$

where L is the gage length and the subscript NSY refers to the value of the quantity at the onset of NSY.

A relationship between ϵ_L and σ is necessary to model CMOD versus ϵ_L at stresses below NSY and necessary to compute $\text{CMOD}\big|_{\text{NSY}}$ and ϵ_{NSY} for Eq 2. The value of ϵ_L is the sum of the elastic strain (σ/E) plus the increment of ϵ due to the presence of the crack. The remote displacements (Δ) in a center-cracked panel are given by Ref 5:

$$\Delta = \frac{2\sigma\ell}{E}\,V\!\left(\frac{\ell}{W}\right) \tag{3}$$

where $V(\ell/W)$ is as given in Ref 5. For a surface crack, Irwin's equivalent through-thickness crack concept [6] used to derive Eq 1 can be used to modify Eq 3; that is, σ is reduced by the closure stress (σ_c) caused by the uncracked ligament:

$$\sigma_c = \bar{\sigma}\left(1 - \frac{a}{t}\right) \tag{4}$$

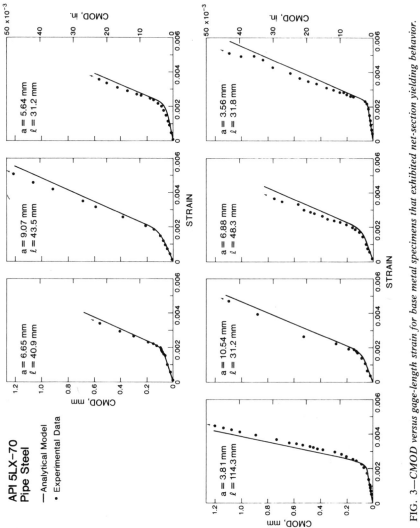

FIG. 3—*CMOD versus gage-length strain for base metal specimens that exhibited net-section yielding behavior.*

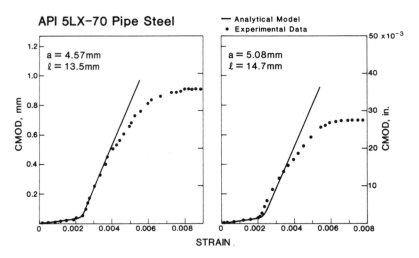

FIG. 4—*CMOD versus gage-length strain for base metal specimens that exhibited mixed behavior.*

To account for the crack tip plasticity, the effective crack length (ℓ_{eff}), which includes the plastic zone size correction r_y, is used in place of ℓ [1,2]. The resultant expression becomes

$$\epsilon_L = \frac{\sigma}{E} + \frac{2(\sigma - \sigma_c)}{E} \frac{\ell_{eff}}{L} V\left(\frac{\ell_{eff}}{W}\right) \qquad (5)$$

Given Eqs 1, 2, and 5, it is possible to calculate CMOD versus ϵ_L for the NSY case. For stresses below and equal to NSY, Eq 1 is used to calculate CMOD and σ is converted to ϵ_L using Eq 5. For stresses above NSY, Eq 2 is used. Comparison of the experimental and analytical results of CMOD versus ϵ_L shown in Figs. 3 and 4 indicates good agreement.

For Specimens 6 and 8, NSY is followed by GSY at high strains. The transition from NSY to GSY and CMOD versus ϵ_L for GSY have not been modeled. In the present study, the transition occurred at a ratio of crack area to cross-sectional area of 5%. The area ratio at which the transition occurs should be higher for materials with more strain hardening.

Weld Metal Results

As shown in Table 3, four automatic-weld specimens and four manual-weld specimens were tested. As shown in Table 2, the yield and ultimate strengths of the welds were significantly different for both types of weld and for the base metal. These differences were far less in transverse-weld tension tests on 76-mm-wide, full-thickness tensile panels with a 305-mm gage length. The stress-strain curves obtained using strain gages mounted in the center of the weld at midthickness on both edges are shown in Fig. 5. For the CMOD-versus-σ model, only the flow stress values are of importance: 596 MPa for the automatic weld and 557 MPa for the manual weld. For CMOD-versus-ϵ modeling using the two-tensile-bar model (discussed below), however, the shape of the stress-strain curve is also important.

The experimental results of CMOD versus σ for the automatic and manual welds are compared with the analytical predictions in Figs. 6 and 7. The general behavior trends observed in the experiment are reasonably modeled; however, the agreement between experiment and analysis is not as good as it was for the base metal tests (see Fig. 2 and Ref 1). The correlations are best at low stresses, but the CMOD is underestimated (Specimens A2 and M4 are exceptions) for stresses between the stress required for ligament yielding and the stress required for NSY. At still higher stresses, CMOD is overestimated. To model the observed behavior with Eq 1, it would be necessary to use a variable flow stress. This option is physically sensible, because the stress-strain curves for the welds exhibit more work hardening than the base metal curves. The quality of the correlations is also influenced by an increase in experimental

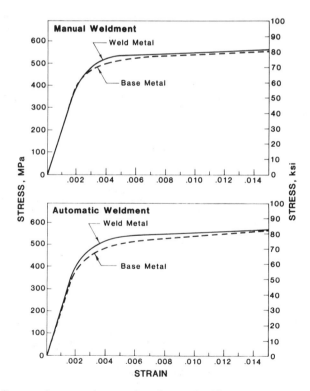

FIG. 5—*Stress-strain curves of automatic and manual weldments as determined in transverse weld tension tests.*

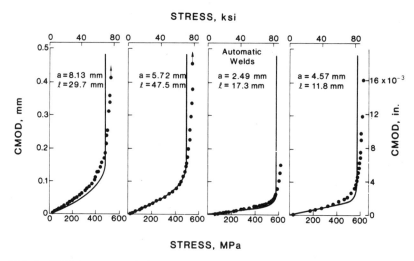

FIG. 6—*CMOD versus nominal stress for automatic weld specimens. Symbols are experimental data; solid lines are analytical predictions based on the modified critical COD model.*

uncertainty for testing the welded specimens over that encountered in base metal tests. The principal sources of error are: (1) bending caused by distortion, misalignment, and the curved test section, (2) the presence of residual stresses in the welded specimen, and (3) thickness variations caused by hand grinding the crown and root of the weld.

The experimental results of CMOD versus ϵ_L for the automatic and manual welds are shown in Figs. 8 and 9. The two characteristic trends observed in the

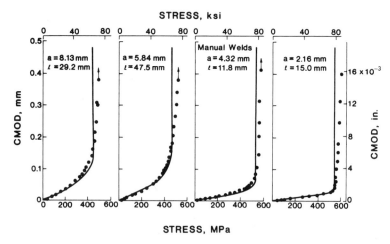

FIG. 7—*CMOD versus nominal stress for manual weld specimens. Symbols are experimental data; solid lines are analytical predictions based on the modified critical COD model.*

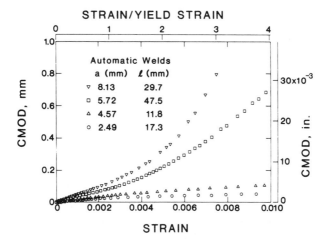

FIG. 8—*CMOD versus gage-length strain for automatic weld specimens.*

base metal tests, a bilinear curve for NSY and a plateau when GSY occurs, are not apparent in the weld metal results; however, the NSY and GSY designations are still appropriate. The CMOD-versus-ϵ_L curves for NSY behavior in welded specimens are more parabolic than bilinear; the four specimens with the largest cracks (A1, A2, M1, M2) exhibited NSY. The CMOD-versus-ϵ_L curves for GSY behavior in welded specimens show a very small increase in CMOD with ϵ_L; the four specimens with the smallest cracks (A3, A4, M3, M4) exhibited GSY.

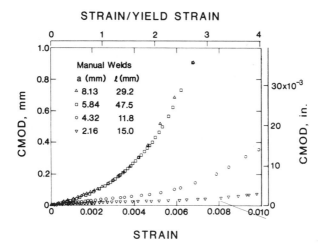

FIG. 9—*CMOD versus gage length strain for manual weld specimens.*

The differences in the CMOD-versus-ϵ_L curves for the base metal and the welds can be explained by consideration of the higher yield strength and increased work-hardening rates in the weld. For small cracks, the base metal starts yielding while the weld is still elastic. Subsequent increases in stress applied to the weld depend on the work hardening in the base metal. Simply, the base metal stretches and the weld does not, and consequently CMOD increases slowly with strain. For the larger cracks (Specimens A1, A2, M1, M2), 16 to 20% of the cross-sectional area of the weld is cracked. Consequently, the crack plane yields first and NSY can occur. Even at the 20% crack area to cross-sectional area ratio, the weld work hardens sufficiently to cause yielding in the base metal.

For the NSY case, the apportionment of strain in the base metal and the weld can be modeled with the two-tensile-bar analogy shown in Fig. 10.

For a given value of remote imposed displacement, $\Delta = \epsilon_L L$, at the ends of two tensile bars in series, the compatibility condition is

$$\epsilon_1 L_1 + \epsilon_2 L_2 = \epsilon_L L \tag{6}$$

where L is the gage length, L_2 is the width of the weld, and $L_1 = L - L_2$ (Fig. 10). The force equilibrium is

$$[\sigma_1(\epsilon_1)] \cdot A_1 = [\sigma_2(\epsilon_2)] \cdot A_2 \tag{7}$$

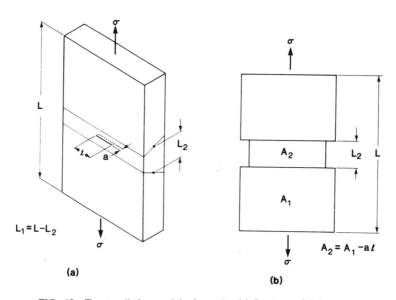

FIG. 10—*Two-tensile-bar model schematics. (a) Specimen. (b) Analogy.*

where $\sigma_1(\epsilon_1)$ and $\sigma_2(\epsilon_2)$ are the stress-strain curves for Materials 1 and 2, and A_1 is the gross-section area and A_2 is the net-section area in the plane of the crack (Fig. 10). Since the stress-strain curves are nonlinear, Eqs 6 and 7 are solved numerically for ϵ_1 and ϵ_2. Assuming all the strain in the weld (ϵ_2) goes into the crack, then CMOD $= \epsilon_2 L_2$.

A comparison of the measured CMOD-versus-ϵ_L data with the values calculated using the two-tensile-bar analogy is shown in Fig. 11. The results indicate that strain partitioning due to the difference in the stress-strain curves of the base metal and the weld is a usable physical model of weldment behavior.

Conclusions

Crack-mouth-opening displacements (CMOD) of surface cracks have been measured as functions of stress and strain in tensile panels of API 5LX-70 steel plates and welded pipe segments. The conclusions drawn from the investigation are:

1. For CMOD versus stress, predictions from the modified critical COD model agree well with the experimental results for the base metal and the welds.

2. For strains above net-section yielding (NSY), the relationship between CMOD and strain is complex depending upon the specimen type (plate or weld) and crack size.

A. For large cracks in base metal, all the remote displacement goes into the crack tip through the slip bands extending from the crack tip to the plate

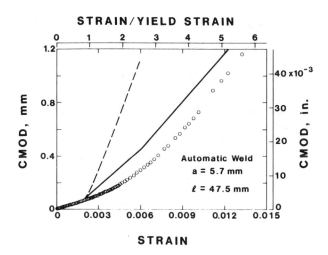

FIG. 11—*CMOD versus gage-length strain for an automatic weld metal specimen exhibiting NSY. Symbols are experimental data; solid lines are predictions from two-tensile-bar model; dashed lines are predictions from simple NSY models.*

edges at 45 deg, that is, NSY. This behavior is modeled and predictions from the model agree well with the experiment.

B. For cracks with crack area less than 5% of the cross-sectional area, NSY is followed by gross-section yielding (GSY) in which the remote displacement is distributed evenly along the length of the specimen.

C. For large cracks in steel welds, NSY is observed. The observed CMOD is lower than predicted by the model because the yield strength of the weld is higher than that of the base metal. A model based on strain partitioning accounts for the observed behavior and provides better predictions.

D. For small cracks in steel welds, GSY is observed throughout the strain range studied.

Acknowledgments

The authors express their appreciation to Mr. David McColskey for his technical assistance. Dr. James Early, Jr., and Mr. Ted Anderson are acknowledged for the tensile properties of weldments and base metal. This work was supported by the U.S. Department of Transportation, Office of Pipeline Safety Regulation.

APPENDIX

Modified Critical COD Model

The critical COD model [6] treats the surface crack in a plate as a through-thickness center crack after the ligament is yielded. The opening (δ) at the middle of the center crack of length (ℓ) in an infinite plate under a remote stress is

$$\delta = \frac{2\ell\sigma}{E} \qquad (8)$$

For a surface crack, the value of δ in Eq 8 is reduced by the remaining ligament. The effect of ligament depth can be estimated by considering a closing force distributed over the crack-face area. Assuming the ligament is yielded, the total closing force (F_c) is

$$F_c = \ell(t - a)\bar{\sigma} \qquad (9)$$

where $\bar{\sigma}$ is flow strength and is estimated as the average of the yield strength and the ultimate tensile strength, t is the plate thickness, and a is the crack depth. Distributing this closing force over the area ℓt gives a closing stress (σ_c) on the equivalent through-thickness crack of

$$\sigma_c = (1 - a/t)\bar{\sigma} \qquad (10)$$

This closing force opposes the remote stress (σ) and the resultant opening of the surface crack is then

$$\delta = \frac{2\ell(\sigma - \sigma_c)}{E} \qquad (11)$$

To account for the crack opening due to the crack-tip plasticity, the effective crack length, which includes the plastic zone size correction (γ_y), is used in place of ℓ. The resulting expression when Eq 10 is substituted into Eq 11 becomes

$$\delta = \frac{2(\ell + 2\gamma_y)}{E} [\sigma - (1 - a/t)\bar{\sigma}] \qquad (12)$$

where γ_y is a Dugdale plastic zone size solution [8] with a finite-width correction and is evaluated as

$$\frac{\sin\left(\dfrac{\pi}{2}\dfrac{\ell}{W}\right)}{\sin\left(\dfrac{\pi}{2}\dfrac{\ell_p}{W}\right)} = \cos\left(\dfrac{\pi}{2}\dfrac{\sigma'}{\bar{\sigma}'}\right) \qquad (13)$$

where $\ell_p = \ell + 4\gamma_y$, $\sigma'/\bar{\sigma}' = [1 - (t/a)][(1 - (\sigma/\bar{\sigma}'))]$, and W is the plate width. Equation 13 was derived [2,9] based on solutions of periodic array of colinear cracks in an infinite plate [10]. Equation 12 applies when the ligament is yielded. Before yielding, it is assumed that $\delta = 0$ when $\sigma < \sigma_c$.

To account for the elastic component of CMOD, the results of Kobayashi [3] were used. Thus the analytical predictions of CMOD of a surface crack in the elastic-plastic region consist of the sum of CMOD calculated from the modified critical COD model plus the linear elastic CMOD up to ligament yielding.

References

[1] Cheng, Y. W., McHenry, H. I., and Read, D. T. in *Fracture Mechanics: Fourteenth Symposium—Vol. II: Testing and Applications, ASTM STP 791*, American Society for Testing and Materials, 1983, pp. II-214–II-231.

[2] Rice, J. R. in *Fatigue Crack Propagation, ASTM STP 415*, American Society for Testing and Materials, 1967, pp. 247–311.

[3] Kobayashi, A. S. in *Proceedings, Second International Conference on Mechanical Behavior of Materials*, Boston, 16-20 Aug. 1976, American Society for Metals, Metals Park, Ohio, 1976, pp. 1073–1077.

[4] King, R. B., *Engineering Fracture Mechanics*, Vol. 18, No. 1, 1983, pp. 217–231.

[5] Tada, H., Paris, P. C., and Irwin, G. R., *The Stress Analysis of Cracks Handbook*, Del Research Corporation, Hellertown, Pa., 1973.

[6] Irwin, G. R. in *The Surface Crack: Physical Problems and Computational Solutions*, J. L. Swedlow, Ed., ASME, New York, 1972, pp. 1–10.

[7] McHenry, H. I., Read, D. T., and Begley, J. A. in *Elastic-Plastic Fracture Mechanics, ASTM STP 668*, American Society for Testing and Materials, 1979, pp. 632–642.

[8] Dugdale, D. S., *Journal of the Mechanics and Physics of Solids*, Vol. 8, 1960, pp. 100–108.

[9] King, R. B., unpublished work, the National Bureau of Standards, Boulder, Colo.

[10] Irwin, G. R., *Journal of Applied Mechanics, Transactions of ASME*, Vol. 24, 1957, p. 361.

T. Nicholas,[1] N. E. Ashbaugh,[2] and T. Weerasooriya[2]

On the Use of Compliance for Determining Crack Length in the Inelastic Range

REFERENCE: Nicholas, T., Ashbaugh, N. E., and Weerasooriya, T., **"On the Use of Compliance for Determining Crack Length in the Inelastic Range,"** *Fracture Mechanics: Fifteenth Symposium, ASTM STP 833*, R. J. Sanford, Ed., American Society for Testing and Materials, Philadelphia, 1984, pp. 682–698.

ABSTRACT: Compliance measurements were made during creep and fatigue crack growth to establish crack length. During various tests some of the data from several nickel-base superalloys demonstrated a lack of a one-to-one correspondence between crack length and compliance. In particular, decreases in compliance were observed when going from fatigue to creep and when going from low frequency to higher frequency fatigue cycling. On the other hand, no anomalous behavior was observed in ductile copper when loaded to high K values and subsequently unloaded. It is concluded that a probable cause of the anomalous compliance values during transient loading conditions is due to a complex three-dimensional stress state which may be additionally influenced by environmental factors.

KEY WORDS: compliance, cracks, fracture mechanics, inelastic material behavior, creep, fatigue

One of the most reliable and widely used techniques for determining crack length in fracture and fatigue experiments is the compliance method. From the slope of the load-displacement curve in a cracked body, the crack length can be determined from calculations based on linear elastic fracture mechanics (LEFM) or reference experimental calibrations. If an accurate displacement measurement technique is available, accurate compliance measurements can be made which, in turn, can provide accurate crack length determinations in many of the commonly used fracture and fatigue specimen geometries. For fatigue cracks where LEFM analysis is valid—that is, where plastic zone sizes are small in comparison to the crack length—the compliance technique provides highly reliable data. In cases where crack tip plasticity may not be

[1]Materials Research Engineer, AFWAL Materials Laboratory, (MLLN) Wright-Patterson Air Force Base, Ohio 45433.
[2]Research Engineers, University of Dayton Research Institute, Dayton, Ohio 45469.

negligible, the compliance technique should still be valid if the measurements are made during the unloading portion of a fatigue cycle, thereby assuring elastic behavior throughout most of the specimen. One precaution which must be taken in this case, however, is to ensure that crack closure does not occur in the region of the unload cycle where the compliance data are obtained. Data obtained during unloading above the start of closure usually provide accurate values of compliance for crack length calculations.

Several recent investigations have extended the use of the compliance technique to creep crack growth studies [1,2]. Since no fatigue cycling is present in creep crack growth experiments (constant load), the experiments are periodically interrupted for a very short time period to partially unload and then quickly reload the specimen. Compliance measurements during this unloading cycle are used to calculate the crack length. As in the case of fatigue cycling, the unloading behavior is assumed to be entirely elastic. A rapid rate of unloading is used to ensure that no time-dependent strains occur due to relaxation.

Some of the experimental data from creep crack growth investigations have revealed an anomalous behavior when using compliance to determine crack length [1,2]. In the early stages of creep crack growth following fatigue precracking, the compliance has been observed to decrease. This implies that either the crack is shortening, which is physically impossible provided closure effects are not present, or that the compliance of a crack during creep crack growth is less than the compliance for a crack during fatigue for the same length of crack. In a separate investigation, compliance values at the beginning of fatigue cycling following creep crack growth provided crack lengths which were shorter than those measured on the fracture surface [3].

These results demonstrate a lack of a one-to-one correspondence between crack length and compliance, particularly when changing from one type of loading to another. To help resolve this apparent inconsistency, additional data were taken from on-going investigations, and several experiments were conducted using other materials and test specimen geometries. Materials exhibiting time-dependent and time-independent material behavior were included as were several different displacement measurement techniques. Additionally, data from the literature were collected to supplement the experimental results. This paper presents a compilation of results from various experimental investigations in which the authors participated and from tests specifically designed to produce compliance data during transient loading conditions. Some possible explanations for the observed behavior and some general guidance for the use of compliance are provided.

Theory

Compliance or the reciprocal of stiffness is the rate of change of the displacement between two points on a body with respect to a load applied to the body.

Compliance is a function of the materials response, the location of the applied load, the location of the two points between which the displacement is determined, and the geometry of the body. For the ideal linear elastic material, compliance for any given geometry, applied load, and two points is constant; that is, a load-displacement curve is linear, assuming that the deformation of the body is small. Compliance values can be obtained from solutions to an elasticity problem or from experimental measurements.

For a body containing a crack, compliance is generally a function of crack length. If the two points where displacement is measured are carefully chosen, a one-to-one relationship between compliance and crack length can be obtained for a given geometry, including the crack length, assuming linear elastic material behavior. For two-dimensional elasticity problems where the body containing a crack is deformed in either of the three modes of deformation under a simple system of applied forces, the stress intensity factor (K) can be related to the compliance (C_{LP}) for the load-point displacement by the formula [4].

$$K^2 = \lambda \frac{\partial C_{\mathrm{LP}}}{\partial a} \tag{1}$$

where λ is a positive proportionality factor and a is a measure of crack length. Compliance measurements for bodies containing narrow slots to simulate cracks are discussed by Bubsey et al [5]. They show that the strain energy release rate (G) or the stress intensity factor (K) can be obtained from derivatives with respect to crack length of the measured compliance values.

Compliance can be determined from relative displacements of any two points in a body. For common types of laboratory specimens having a crack or cracks, compliance is often associated with crack-opening displacement (COD) and appears as a monotonic increasing function of crack length. The load-point compliance is also a monotonic increasing function of crack length for a nonzero stress intensity factor (Eq 1). Hence for common laboratory specimens, compliances for crack-opening and load-point displacements are uniquely related to crack length. Compliance can be calculated and measured at any other convenient points in a specimen. For some of the results reported, displacement measurements were made at the top and bottom of the CT specimen along the load line. The numerical results for these compliance values were obtained from finite-element analysis and are shown as δ_4 as a function of $\alpha = a/W$ in Fig. 1. Also shown in that figure are the compliance values for load-point displacements (δ_3), load-line displacements across the notch surface (δ_2), and COD (δ_1). The values of δ_1 and δ_2 agree closely with those obtained by Newman [6]. Note that the values of compliance for displacements along the load line taken at the top and bottom of the specimen (δ_4) differ from the values at other points along the load line, especially for shorter crack lengths. This points out the importance of using the appropriate numerical

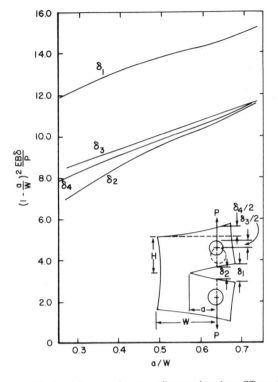

FIG. 1—*Analytical results for various compliance values for a CT specimen.*

solution for the specific location on a specimen where displacements are measured. In calculations of compliance, the nondimensional term $E'BC$ is used where E' is the appropriate modulus for plane stress or plane strain. As pointed out by Bubsey et al [5], there is a minor degree of ambiguity about the relationship between experimentally measured compliance and that obtained by numerical analysis. This is because the analysis assumes either plane stress ($E' = E$) or plane strain [$E' = E/(1 - \nu^2)$], while experimentally the stress field is generally three dimensional and the effective E' lies somewhere between the plane stress and plane strain values.

In the experimental determination of compliance, nonlinearity in the observed load-displacement curve can be obtained for many reasons. Included in these are misalignment in the load train, friction between specimen and load train, closure along the crack plane which is equivalent to a change in geometry or crack length, and inelastic or nonlinear material response. The last must be avoided if an unbiased determination of compliance is to be made. Clarke et al [7] used partial unloading in J-integral testing to ensure elastic response for their compliance measurements. Thus inelastic material behavior in the form of plasticity or creep which may exist during loading or sustained load on a specimen can be minimized by unloading so that the stress state throughout the specimen is always moving away from a local yield surface.

Whatever the cause, nonlinear behavior in the load-displacement curve must be avoided if a valid determination of compliance is to be made. In many observations of load-displacement records, no consistency in results relating compliance to crack length for a given material and specimen configuration could be obtained by the authors when the load-displacement records were nonlinear.

Results and Observations

The following series of results on compliance anomalies have been obtained from various experiments conducted by the authors and co-workers. The first results in which the authors observed an apparent anomaly in the crack length based on compliance measurements were from an investigation reported by Donath et al [1]. Some typical examples shown in Figs. 2 and 3 are from creep crack growth tests on CT specimens of IN 100 at 732°C and were obtained using linear variable differential transformer (LVDT) transducers at the ends of long rods attached to plates on the top and bottom of the specimen as described in Ref 2.

Figure 2 shows results for a specimen initially precracked at room temperature and then loaded to the maximum load at 732°C corresponding to a stress intensity of 38.5 MPa · m$^{1/2}$. The initial compliance value at zero time is obtained from the loading curve which was linear up to maximum load. Subsequent compliance values were obtained by partially unloading and reloading

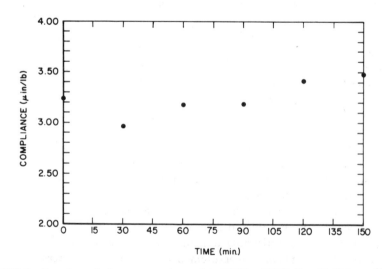

FIG. 2—*Compliance during creep crack growth in IN 100 at 732°C; K = 38.5 MPa · m$^{1/2}$.*

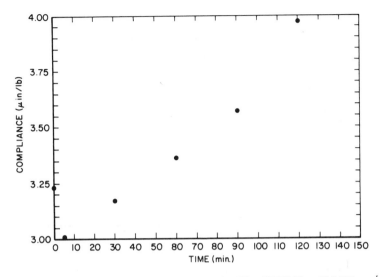

FIG. 3—*Compliance during creep crack growth in IN 100 at 732°C; K = 49.5 MPa · m^{1/2}.*

the specimen at various time intervals. The amount of unloading ranged from 15 to 25% of maximum load. The compliance values were taken from the unloading curves, although the difference between unloading and reloading values was negligible in all cases. The first value of compliance was taken 30 min after the start of the test and is nearly 9% below the initial loading value. Figure 3 shows results at a higher K of 49.5 MPa · m$^{1/2}$ where the first compliance was taken after only 5 min. A decrease of nearly 7% is noted although, as in the first test, all subsequent values increased corresponding to crack extension under constant load. This decrease in compliance was noted in all 23 tests performed under constant load corresponding to initial K values ranging from 25.3 to 49.4 MPa · m$^{1/2}$ and covering a range of specimen thickness from 5.4 to 18.2 mm.

A series of similar tests [8] was subsequently performed using the same material, temperature, and apparatus, with side-grooved specimens. The side grooves amounted to 20% of the total thickness of the specimen which ranged from 5.6 to 15.1 mm in total thickness. The side grooves impose a state of plane strain along most of the crack front [9]. The compliance results from one of these tests is shown in Fig. 4. It can be seen that the decrease in compliance is very small. In a total of nine tests on side-grooved specimens, the results showed little or no decrease in compliance after the initial loading.

The same material (IN 100) at the same temperature (732°C) was used in a series of tests described by Sharpe [10]. In these tests, ungrooved center cracked panels were subjected to crack growth under sustained load while displacement was monitored between two indents on either side of the crack using a highly precise laser interferometric displacement measuring system. In addi-

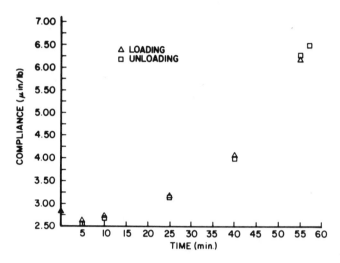

FIG. 4—*Compliance for CT specimens with side grooves (IN 100, 732°C).*

tion to displacement-time data, compliance values were also obtained by periodic unloading and reloading as in the previous tests. Here, the amount of unloading was 20% of the maximum load. The first compliance value was taken anywhere from 20 min after initial loading to 3 s in several tests. In all cases, a decrease in compliance was observed after the initial loading to maximum load. As in the previous tests, the initial compliance was measured during loading while subsequent values were based on unloading. However, as before, the unloading and reloading compliances were identical within experimental accuracy. The initial loading compliance was taken between half and maximum load but was linear except near maximum load (Fig. 5). Also shown in that figure is the first unloading compliance taken after 3 s. A decrease in compliance upon unloading was observed in all cases, and it was noted that the amount of decrease was greater as the K value for the sustained load was increased. Another observation was that except for the initial decrease in compliance, all subsequent values correlated well with increases in crack length.

A hybrid experimental-numerical (HEN) procedure [11] was used to model the deformation observed in Sharpe's tests. The procedure involved the use of a finite-element computer code with a viscoplastic material model and a node-popping scheme to simulate crack extension. The crack extensions were obtained by matching the total displacements calculated with the code with those measured using the interferometric technique. Figure 6, taken from Ref 11, shows the crack extension from the HEN procedure (dashed line) compared with the compliance-based crack length determined at 20-min intervals. After the decrease to the first point at 20 min, the two curves are essentially parallel. The total crack extension measured on the fracture surface agreed well with

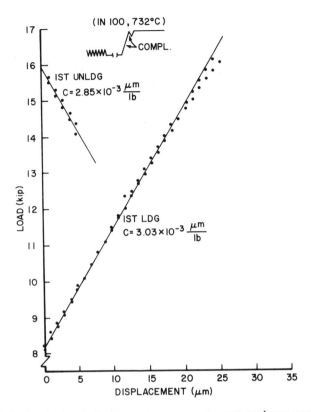

FIG. 5—*Interferometric displacement measurements across crack near crack tip.*

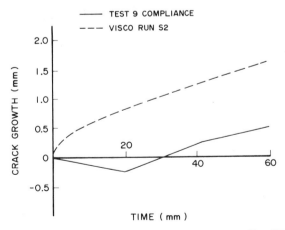

FIG. 6—*Computed and measured compliance for CT specimen of IN 100 at 732°C under creep crack growth* [11].

the HEN calculation (dashed line in Fig. 6), while the compliance technique grossly underpredicted the observed value. The compliance curves at 0, 20, 40, and 60 min are shown in Fig. 6 in Ref *10* and clearly demonstrate that the compliance value at 20 min is less than that at zero time. Reference *10* also provides details of the experimental procedures for obtaining the displacement data and compliance.

In all the foregoing experiments, the initial compliance values were taken during the loading portion of a cycle. A series of tests was conducted in which compliance was measured during the unloading portion of the fatigue cycle prior to the subsequent sustained load portion of the test. Using IN 100 at 732°C again, CT specimens instrumented with a highly sensitive clip gage attached across the crack mouth were subjected to the loading pattern depicted in Fig. 7. In this case, the fatigue cycling was conducted at the same temperature as the sustained load, whereas fatigue cycling was conducted at room temperature in the previous tests. The numbers 35/35 in the figure refer to the maximum value of K during fatigue followed by the value of initial K during the sustained load portion in units of MPa · m$^{1/2}$. Compliance was obtained on the unloading portion of the last fatigue cycle as well as on the unloading portion of each excursion during the sustained load portion of the test by fitting a straight line using least-square error minimization. The unloading-reloading was done under computer control at the same load rate as in the fatigue cycling and occurred 10 s after loading to maximum load and every 30 s thereafter. In the two cases where the sustained load exceeded the maximum load in fatigue,

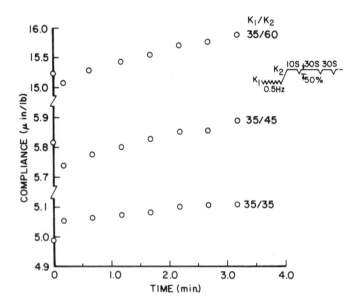

FIG. 7—*Unloading compliance during fatigue/creep interaction.*

designated as 35/60 and 35/45 in Fig. 7, one can see a decrease in the value of compliance after reaching maximum load. In two tests where the prior cycling was at the same maximum load level, one of which is shown and designated 35/35 in the figure, there was an increase in one case (shown) and essentially little change in the other.

Further evidence of an apparent reduction in compliance when going from fatigue to creep was reported by Larsen and Nicholas in other tests on IN 100 at 732°C [3]. CT specimens were subjected to continuous fatigue cycling interrupted by occasional hold times ranging in duration from 1 to 20 min. Compliance values were obtained from mouth displacement data using a clip gage extensometer during the unloading portion of the fatigue cycling. At an R-ratio of 0.1 for the fatigue cycles, displacement-load data were acquired between 90 and 40% of maximum load to avoid possible hysteresis and closure effects in the calculations. All the load-displacement curves obtained in this manner were linear and were fitted with a straight line using a least-squares error minimization technique. The amount of crack extension during the hold time was determined from compliance values taken on the fatigue cycles immediately before and after the hold time. These crack extensions were also determined after the test by measurements on the fracture surface using a nine-point averaging technique. In all cases, the crack extension determined using the compliance technique was less than the measured value from the fracture surface. A typical fracture surface is depicted in Fig. 8 which shows the nine-

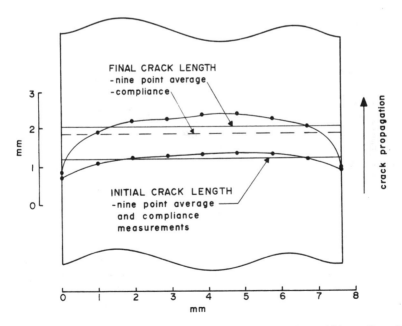

FIG. 8—*Crack front profiles and effective crack length from* (a) *averaging and* (b) *compliance* [3].

point average crack and that determined from compliance measurements. The compliance data for the fatigue cycling are shown as circles in Fig. 9. Shown also are the data points (squares) representing the optically measured value from the fracture surface. The crack extension from these latter measurements is designated as Δa_c. As in the previous cases, there appears to be a decrease in compliance or apparently less crack extension when going from fatigue cycling to sustained loading.

A similar series of tests was performed on Inconel 718 CT specimens at 649°C using a high-resolution extensometer. Blocks of fatigue cycles at a constant value of maximum K of 38.5 MPa · $m^{1/2}$ and an R-ratio of 0.1 were interrupted with hold times of 3 min at maximum load. The frequency of the fatigue cycles was 0.5 Hz and a single block consisted of 900 cycles. Crack extension during the 3-min hold time was determined from compliance values immediately before and after the hold time and also from fracture surface measurements using a seven-point averaging technique. The results, presented in Table 1, show an average crack extension of 0.16 measured on the fracture surface, over double that determined from compliance measurements. The data illustrate the consistently low predictions of crack extension from the compliance measurements compared with those measured on the fracture surface in creep crack growth following fatigue.

To evaluate the effect of large amounts of plasticity at the crack tip on the compliance, several tests were conducted on oxygen-free, high-conductivity

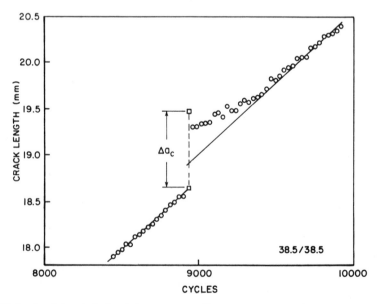

FIG. 9—*Crack length during fatigue from compliance measurements. Squares indicate crack lengths measured from fracture surface for beginning and end of 20-min hold time.*

TABLE 1—*Measurements of crack extension due to 3-min hold time following fatigue.*

	Initial Crack Length, mm		Crack Extension During 3-min Hold Time, mm	
Sequence	Compliance	Fracture Surface	Compliance	Fracture Surface
1	5.3061	5.3111	0.0533	0.1295
2	6.3322	6.3525	0.0533	0.1600
3	7.5692	7.5565	0.0813	0.1778
4	8.6411	8.6157	0.1143	0.1676
5	9.7180	9.7003	0.0864	0.1727
		Average	0.0777	0.1615

(OFHC) copper CT specimens. The specimens were precracked at a maximum value of K of approximately 10 MPa · m$^{1/2}$ and then loaded to several increasingly higher K levels ranging from approximately 10 to 40 MPa · m$^{1/2}$. After each loading, the specimen was unloaded to half the maximum load. Three sets of experiments were conducted using different displacement measurement techniques. The laser interferometric displacement measurement technique was used in two of the tests; a clip gage and an LVDT extensometer were used in the other tests. Load-displacement records were obtained for both the loading and unloading at each progressively higher maximum load. The crack did not appear to extend at any of the load levels, but significant amounts of plastic deformation could be seen at the highest load levels on the surface of the specimen ahead of the crack tip. Additionally, nonlinearity was observed near maximum load on the loading portion of the load-displacement curve cycle at the highest load levels.

Results of the loading and unloading compliances from the tests using the three measurement techniques are presented in Fig. 10. Compliance measurements are all in arbitrary units since only changes are of interest here. The LVDT data show an apparent increase in compliance as higher loads are obtained, although much of this trend can be attributed to scatter or experimental accuracy. The clip gage and interferometry data, on the other hand, provided results with very little scatter because of the high resolution. These data show no apparent change in compliance over the entire range of loads. The maximum value of K reached was 38.5 MPa · m$^{1/2}$ which is felt to be close to the fracture toughness of the material.

Some of the typical loading and unloading curves obtained by the two techniques are shown in Fig. 11. The LVDT extensometer results showed some hysteresis when going from loading to unloading, which accounts for scatter in the data on compliance. The data obtained using this technique showed a lack of linearity during unloading at the high K levels, typically nearly a bilinear behavior. This is not a closure effect because compliance goes in the wrong direction (increases) and the load is reduced only to half the previous maxi-

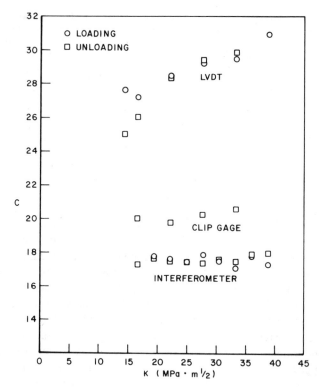

FIG. 10—*Loading and unloading compliances on copper CT specimens loaded to progressively higher K levels.*

mum. The data for unloading were obtained from the lower linear portion of the curve because it appeared parallel to the loading curve. Taking the upper portion gave extremely low values of compliance. Overall, little confidence is placed in the data obtained with the LVDT extensometer, because of the large amount of hysteresis observed in a mechanical system that involves the use of long extender rods and tubes [2]. It is important to point out, however, that the displacements involved at the highest K levels amounted to approximately 0.1-mm full scale, which requires extensometry and instrumentation capable of resolving nearly 0.1 μm for accurate compliance calibrations.

The clip gage data showed less scatter than the LVDT data and no apparent change in compliance with K level (Fig. 10). There was some amount of hysteresis in the load-displacement curves and the results are accurate to no better than ±5%. The data from the interferometry measurements, taken at points 0.15 mm behind the crack tip, show essentially linear behavior on loading and unloading except at the highest load levels. The data, additionally, are very reproducible and show no change in compliance over the entire load range for both loading and unloading. Figure 11, which shows the curves at the

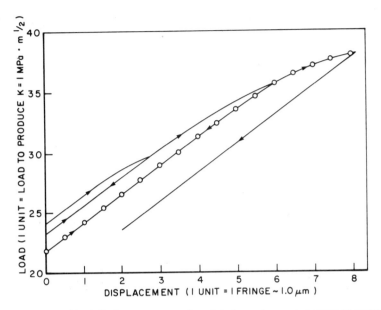

FIG. 11—*Typical load-displacement curves from interferometry measurements on copper CT specimens. Data points are shown for only one loading curve.*

higher load levels, illustrates the nonlinear behavior upon subsequent reloading as loads beyond those previously encountered are applied. The unloading curves, however, are essentially linear. Note that the resolution for the displacement measurements as shown on one curve in Fig. 11 is one half of a fringe corresponding to approximately 0.5 µm.

Other anomalous results obtained in our laboratory involve numerous measurements of compliance in constant K tests under computer control over a range of frequencies in IN 100 and Inconel 718 at temperatures where both cycle-dependent and time-dependent behavior are observed. In going from blocks of fatigue cycling at one frequency to blocks at a higher frequency, we consistently note a decrease in compliance or crack length between the end of the block of low frequency cycles and the beginning of the next block of higher frequency cycles. Conversely, if the frequency is decreased, there appears to be an increase in compliance. In constant K testing this requires slight adjustments to the modulus or calibration constant relating compliance to crack length.

Lower frequencies or time-dependent behavior require a slightly lower modulus or calibration constant than higher frequencies or cycle-dependent behavior in order to avoid apparent jumps in crack length. In fact, in almost all cases observed, a fatigue cycle following either slower frequency cycling or hold times resulted in a decrease in compliance. The decrease in compliance in a single cycle during a hold time following fatigue can then be thought of as a result of two competing mechanisms, thus explaining the apparently inconsistent results of Fig. 7: (1) after fatigue, the compliance apparently increases when creep takes place, and (2) after some creep, a fatigue cycle shows a decrease. For creep at the same level as fatigue in Fig. 7 (the 35/35 case), the net effect is an apparent increase. For the other two cases shown having higher K in creep than fatigue, the fatigue cycle after 10 s of creep shows a decreasing compliance.

In addition to the results obtained in our laboratory, several instances of anomalous compliance data have been reported elsewhere. Tobler [12] has reported lower apparent crack lengths in AISI 310S stainless steel in single specimen unloading-compliance J-integral tests than those measured directly from the fracture surface. Asaro has observed an "apparent" negative crack growth in J-integral tests in 4140 steel using a highly accurate differential compliance method for crack growth determination.[3] At loads where blunting was first observed, compliance measurements indicated an apparent decrease in crack length. This apparent shortening of the crack before crack extension occurs has also been mentioned by a number of colleagues engaged in J-integral testing and is usually handled by a slight adjustment of modulus in the computations.

Discussion

Several possible explanations can be proposed to explain the apparently anomalous results from compliance measurements. The decrease in compliance using assumed linear elastic material behavior implies a shortening of the crack or decrease in effective crack length. One would immediately suspect a closure effect of some type, yet this behavior was observed in creep crack growth and in instances where displacement values were obtained only near maximum load. Unless some type of closure effect occurs over the entire load range, this explanation has to be ruled out.

Another explanation that could account for some of the earlier observations of a decrease in compliance is the nonlinear material behavior during the loading portion of the cycle. In fact, in the tests on copper specimens using a clip gage, it was noted that the load-displacement curves were rather linear in the loading portion of the cycle when going from a previously achieved load level to a higher load level. This was rather surprising because it was apparent that large plastic deformation was occurring around the crack tip. The slope of the additional loading curve was less than the subsequent unloading curve and could explain an apparent decrease in compliance after reaching maximum load. However, a number of tests were run where compliances were obtained during the unloading portion of a cycle only and the decrease in compliance was still obtained. It seems reasonable to assume linear elastic material behav-

---

[3] Asaro, R. J., Brown University, Providence, R.I., private communication, 1981.

ior throughout the specimen when unloading from a maximum load. Furthermore, all the load-displacement records during unloading were highly linear. The obvious conclusion that the crack was shortened does not, however, make sense physically.

An alternative explanation involves changes in the three-dimensional stress and strain state after a cracked body has been loaded so that significant inelastic deformation occurs near the crack tip. To explain an apparent decrease in compliance, one would have to propose going from a state of essentially plane stress, where the compliance is governed by modulus E, to a state of plane strain, where the effective modulus is higher and given by $E/(1 - \nu^2)$. In the extreme case, the material ahead of the crack tip could be totally confined in two directions (thickness and direction of crack growth); in this case a state of uniaxial strain would exist which is governed by an effective modulus given by $K + 4G/3$. It would be necessary to demonstrate that during loading the region around the crack tip was in a state of essentially plane stress—that is, no lateral constraint exists—but that after maximum load was achieved and the crack tip region was deformed inelastically sufficient constraint was built up such that the effective modulus of the material in the direction normal to the crack plane was that of plane strain or uniaxial strain. Since no three-dimensional elastic solutions are available for cracked bodies, this explanation is highly speculative and not physically or intuitively obvious.

Acknowledgments

This research was conducted at the Materials Laboratory, Air Force Wright Aeronautical Laboratories, and supported under Air Force Project 2307. The authors would like to express their appreciation to Lt. Ron Sincavage and Mr. Jay Jira for providing the interferometry measurements, to Mr. George Ahrens for his assistance in performing experiments, and to Dr. Robert Donath for providing creep crack growth and compliance data.

References

[1] Donath, R. C., Nicholas, T., and Fu, L. S. in *Fracture Mechanics: Thirteenth Conference, ASTM STP 743*, Richard Roberts, Ed., American Society for Testing and Materials, 1981, pp. 186–206.
[2] Donath, R. C., "Crack Growth Behavior of Alloy IN100 under Sustained Load at 732°C (1350°F)," AFWAL-TR-80-4131, Wright-Patterson Air Force Base, Ohio, April 81.
[3] Larsen, J. M. and Nicholas, T. in *Fracture Mechanics: Fourteenth Symposium—Volume II: Testing and Applications, ASTM STP 791*, J. C. Lewis and G. Sines, Eds., American Society for Testing and Materials, 1983, pp. II-536–II-552.
[4] Irwin, G. R. in *Structural Mechanics, Proceedings of First Naval Symposium*, J. N. Goodier and N. J. Hoff, Eds., Pergamon Press, New York, 1960, p. 567.
[5] Bubsey, R. T. et al, " Compliance Measurements," *Experimental Techniques in Fracture Mechanics*, SESA Monograph No. 1, Society for Experimental Stress Analysis, 1973, pp. 76–95.
[6] Newman, J. C., Jr., "Crack-Opening Displacements in Center-Crack, Compact and Crack-

Line Wedge-Loaded Specimens," NASA TN D-8268, National Aeronautics and Space Administration, Washington, D.C., July 1976.

[7] Clarke, G. et al in *Mechanics of Crack Growth, ASTM STP 590*, American Society for Testing and Materials, 1976, pp. 27-42.

[8] Ashbaugh, N. E. in *Fracture Mechanics: Fourteenth Symposium—Volume II: Testing and Applications, ASTM STP 791*, J. C. Lewis and G. Sines, Eds. American Society for Testing and Materials, 1983, pp. II-517-II-535.

[9] Shih, F., deLorenzi, H. G., and Andrews, W. R., *International Journal of Fracture*, Vol. 13, 1977, pp. 544-548.

[10] Sharpe, W. N., Jr., in *Fracture Mechanics: Fourteenth Symposium—Volume II: Testing and Applications, ASTM STP 791*, J. C. Lewis and G. Sines, Eds., American Society for Testing and Materials, 1983, pp. II-157-II-165.

[11] Hinnerichs, T., Nicholas, T., and Palazotto, A., *Engineering Fracture Mechanics*, Vol. 16, 1982, pp. 265-277.

[12] Tobler, R. L. in *Elastic-Plastic Fracture: Second Symposium, Volume II—Fracture Resistance Curves and Engineering Applications, ASTM STP 803*, C. F. Shih and J. P. Gudas, Eds., American Society for Testing and Materials, 1983, pp. II-763-II-776.

C. H. Popelar,[1] J. Pan,[2] and M. F. Kanninen[3]

A Tearing Instability Analysis for Strain-Hardening Materials

---

REFERENCE: Popelar, C. H., Pan, J., and Kanninen, M. F., "**A Tearing Instability Analysis for Strain-Hardening Materials,**" *Fracture Mechanics: Fifteenth Symposium, ASTM STP 833*, R. J. Sanford, Ed., American Society for Testing and Materials, Philadelphia, 1984, pp. 699–720.

ABSTRACT: The usefulness of the tearing instability analysis procedure depends upon the availability of J and dJ/da estimates for load/structure conditions of practical interest. Most of the J-estimation solutions obtained so far have assumed that a limit load condition exists. To broaden the applicability of these approaches, the role of material hardening in stable crack growth and fracture instability is addressed in this paper. The approach employs Turner's η-factor which links J to the strain energy and the remaining ligament in a cracked component. The resulting analysis is quite general and includes all the presently known J-estimation analyses as special cases.

KEY WORDS: tearing instability, J-resistance curve, J-estimation method, pipe fracture, fracture mechanics

While many different nonlinear approaches have been found to have merit, most current efforts are based upon the use of the J-integral [1]. In this approach, J measures the propensity for crack advance for a given crack structure geometry and applied loading while a J-resistance curve represents the material fracture toughness as a function of crack growth. A further step introduces the tearing modulus parameter, $T = (E/\sigma_0^2)\, dJ/da$, to determine the onset of fracture instability. Here a denotes the crack length while E and σ_0 are the elastic modulus and flow stress, respectively [2]. Equating the applied J and T values to the material values associated with the J-resistance curve to determine the point of fracture instability is then known as a tearing instability analysis.

[1]Engineering Mechanics Department, The Ohio State University, Columbus, Ohio 43210.
[2]Stress Analysis and Fracture Section, Battelle Columbus Laboratories, Columbus, Ohio 43201.
[3]Engineering and Materials Sciences Division, Southwest Research Institute, San Antonio, Tex. 78284.

While limitations on the tearing instability approach exist, mainly due to the geometry dependence of J-resistance curves, they can be partly neutralized through the use of J-resistance curves taken from compact tension or bend specimen experiments. Because these specimens provide a high degree of triaxial constraint, the resulting J-resistance curves will be lower-bound values that should provide conservative predictions. The analysis effort then narrows to the determination of the applied J and dJ/da for the conditions of interest. This can be done with elastic-plastic finite-element analyses. Such approaches, however, are usually far too cumbersome for practical engineering assessments. Approximate methods, generally known as J-estimation schemes, are therefore desirable.

The J-estimation methods that have already been developed have employed the assumption of limit load conditions [3–6]. While these may be intuitively acceptable for Type 304 stainless steel components and others in which considerable plastic deformation generally precedes fracture instability, they are certainly not generally valid. Accordingly, the research described in this paper addresses the role of material hardening in stable crack growth and fracture instability. The objectives of the work are to both broaden the applicability of J-estimation methods and to assess the role of material hardening in those tearing instability analyses that are based upon limit load conditions.

Basis of Tearing Instability Analysis

The Tearing Modulus

Consider a body with a through-thickness crack as illustrated in Fig. 1. Let P be a generalized load per crack tip and per unit thickness of the body. When more than one crack tip exists, assume the flawed body and its loading are such that each tip experiences the same crack driving force. Take Δ to be the

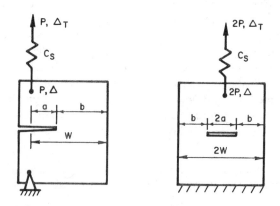

FIG. 1—*Typical flawed structures with one and two crack tips.*

generalized load-point displacement through which P acts to do work on the body. The linear spring in Fig. 1 models the compliance (C_s), of the testing machine and/or the elastic compliance of the associated structure through which the body is loaded. The prescribed total displacement (Δ_T), can then be written as

$$\Delta_T = C_M P + \Delta \tag{1}$$

Here $C_M = mBC_s$ where m is the number of crack tips and B is the thickness of the body. The following arguments are based on the supposition that the loading is controlled by manipulating Δ_T, which will be either held fixed or will be monotonically increasing at an infinitesimally slow rate.

At impending (or during) crack extension, equilibrium between the crack-driving force, $J = J(P, a)$, and the material's resistance to ductile fracture and tearing, $J_R = J_R(\Delta a)$, requires that

$$J(P, a) = J_R(\Delta a) = J_R(a - a_0) \tag{2}$$

where $a = a_0 + \Delta a$ is the current crack length and a_0 is its initial length. The crack extension is said to be stable if an arbitrarily small increase, $\delta a > 0$, in the current crack length with the total displacement held fixed does not produce a driving force in excess of the material's fracture resistance. That is, the equilibrium state defined by Eq 2 is stable if

$$J(P, a + \delta a) < J_R(a + \delta a - a_0) \tag{3}$$

for fixed Δ_T and $\delta a > 0$.

The expansion of this inequality about the current crack length gives

$$\left(\frac{dJ[P, a]}{da} \right)_{\Delta_T} \delta a + \frac{1}{2} \left(\frac{d^2J[P, a]}{da^2} \right)_{\Delta_T} (\delta a)^2 + \cdots < \frac{dJ_R(a - a_0)}{da} \delta a$$

$$+ \frac{1}{2} \frac{d^2J_R(a - a_0)}{da} (\delta a)^2 + \cdots \tag{4}$$

in which Eq 2 has been used together with obvious assumptions on the existence of the differentials of the functions involved.

Inequality 4 can be rewritten as

$$\left[\frac{dJ(P, a)}{da} \right]_{\Delta_T} - \left[\frac{dJ_R(a - a_0)}{da} \right] + \frac{\delta a}{2} \left\{ \left[\frac{d^2J(P, a)}{da^2} \right]_{\Delta_T} \right.$$

$$\left. - \frac{d^2J_R(a - a_0)}{da^2} \right\} + \cdots < 0$$

where the omitted terms are of higher order in δa. Since this inequality must be satisfied for vanishingly small $\delta a > 0$, then for a stable equilibrium state

$$\left(\frac{dJ[P, a]}{da} \right)_{\Delta_T} < \frac{dJ_R[a - a_0]}{da} \tag{5}$$

The equilibrium is unstable if

$$\left(\frac{dJ[P, a]}{da} \right)_{\Delta_T} > \frac{dJ_R[a - a_0]}{da} \tag{6}$$

The demarcation between stable and unstable (neutral) equilibrium is then

$$\left(\frac{dJ[P, a]}{da} \right)_{\Delta_T} = \frac{dJ_R[a - a_0]}{da} \tag{7}$$

Although it will not be used in the following sections, it might be noted that Paris et al [4] have arrived at an equivalent statement by defining the dimensionless parameters

$$T = \frac{E}{\sigma_0^2} \left(\frac{dJ}{da} \right)_{\Delta_T} \quad \text{and} \quad T_R = \frac{E}{\sigma_0^2} \frac{dJ_R}{da} \tag{8}$$

where σ_0 is an appropriate flow stress. In terms of these parameters, Eqs 5 and 6 become

$$T < T_R \quad \text{for stability and} \quad T > T_R \quad \text{for instability} \tag{9}$$

where T is formally identified as the tearing modulus and T_R is its material property counterpart.

Determination of Fracture Instability

Equations 2 and 7 provide two relations for determining the load P and the crack length a at the limit of stable crack growth. As is usually done in a resistance curve analysis for dead loading ($C_s = \infty$), if J_R and J with P as a parameter are plotted versus a, then the limit of stable growth defined by Eqs 2 and 7 is identified with the point of tangency between the resistance curve J_R and the parametric driving force curve J. This is illustrated in Fig. 2.

A general expression for $(dJ/da)_{\Delta_T}$ can be developed by assuming that J and Δ are functions only of P and a. Hence for arbitrary dP and da

$$dJ = \left(\frac{\partial J}{\partial a} \right)_P da + \left(\frac{\partial J}{\partial P} \right)_a dP \tag{10}$$

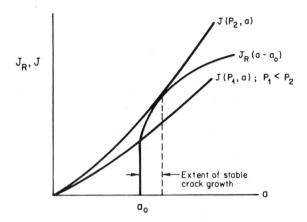

FIG. 2—*Schematic of typical resistance curve analysis.*

With Δ_T held fixed it follows from Eq 1 that

$$d\Delta_T = C_M dP + \left(\frac{\partial \Delta}{\partial a} \right)_P da + \left(\frac{\partial \Delta}{\partial P} \right)_a dP = 0 \tag{11}$$

The combination of Eqs 10 and 11 to eliminate dP then gives

$$\left(\frac{dJ}{da} \right)_{\Delta_T} = \left(\frac{\partial J}{\partial a} \right)_P - \left(\frac{\partial J}{\partial P} \right)_a \left(\frac{\partial \Delta}{\partial a} \right)_P \left[C_M + \left(\frac{\partial \Delta}{\partial P} \right)_a \right]^{-1} \tag{12}$$

Now, using the energy release rate interpretation of J, the crack-driving force can be expressed as

$$J = - \left(\frac{\partial U}{\partial a} \right)_\Delta = \left(\frac{\partial U^*}{\partial a} \right)_P \tag{13}$$

where

$$U = \int_0^\Delta P \, d\Delta \quad \text{and} \quad U^* = \int_0^P \Delta \, dP \tag{14}$$

are, respectively, the strain energy and the complementary energy of the cracked body per crack tip and per unit thickness of the body.

Equivalently, again assuming that the necessary continuity considerations are satisfied, Eqs 13 and 14 can be combined to give

$$J = - \int_0^\Delta \left(\frac{\partial P}{\partial a} \right)_\Delta d\Delta = \int_0^P \left(\frac{\partial \Delta}{\partial a} \right)_P dP \tag{15}$$

According to Castigliano's theorem

$$\Delta = \left(\frac{\partial U^*}{\partial P} \right)_a$$

and, hence, by differentiating this expression

$$\left(\frac{\partial \Delta}{\partial a} \right)_P = \frac{\partial^2 U^*}{\partial a \partial P} = \frac{\partial}{\partial P} \left(\frac{\partial U^*}{\partial a} \right)_P = \left(\frac{\partial J}{\partial P} \right)_a \tag{16}$$

The introduction of Eq 16 into Eq 12 leads to

$$\left(\frac{dJ}{da} \right)_{\Delta_T} = \left(\frac{\partial J}{\partial a} \right)_P - \left(\frac{\partial J}{\partial P} \right)_a^2 \left[C_M + \left(\frac{\partial \Delta}{\partial P} \right)_a \right]^{-1} \tag{17}$$

This is a key result for what follows. To illustrate, it can be seen that for dead loading ($C_M \rightarrow \infty$) Eq 17 reduces to

$$\left(\frac{dJ}{da} \right)_{\Delta_T} = \left(\frac{\partial J}{\partial a} \right)_P \tag{18}$$

Since in general $(dJ/da)_{\Delta_T} \leq (\partial J/\partial a)_P$, then clearly dead loading provides the most adverse condition for stable crack growth. At the other extreme, it is clear that $(dJ/da)_{\Delta_T}$ is an absolute minimum when $C_M = 0$. This corresponds to fixed grip loading.

Basis of the Approach

The η-Factor

To determine $(dJ/da)_{\Delta_T}$ for an arbitrary compliant loading, it is necessary to evaluate the partial derivatives $(\partial J/\partial a)_P$ and $(\partial J/\partial P)_a$ in Eq 18. To accomplish this it is convenient to decompose the load point displacement according to

$$\Delta = \Delta_{nc} + \Delta_c \tag{19}$$

where Δ_{nc} is the load point displacement in the absence of a crack and Δ_c is the remainder due to the presence of the crack. The displacement component due to the presence of the crack can be further decomposed as

$$\Delta_c = \Delta_{ce} + \Delta_{cp} \tag{20}$$

where the additional subscripts "e" and "p" are used to identify, here and in the sequel, the linear elastic and plastic components, respectively.

Equation 20 permits rewriting Eq 15 as

$$J = J_e + J_p = \int_0^P \left(\frac{\partial \Delta_{ce}}{\partial a} \right)_P dP + \int_0^P \left(\frac{\partial \Delta_{cp}}{\partial a} \right)_P dP \tag{21}$$

since Δ_{nc} is independent of a. It is then permissible to write

$$\Delta_{ce} = C_c(a) P \quad \text{and} \quad C_c(0) = 0 \tag{22}$$

where $C_c(a)/m$ is the contribution to the elastic compliance of the flawed body due to the presence of each crack tip. It follows from Eqs 21 and 22 that

$$J_e = \frac{dC_c}{da} \int_0^P P \, dP = \frac{1}{C_c} \frac{dC_c}{da} \int_0^{\Delta_{ce}} P \, d\Delta_{ce} \tag{23}$$

Alternatively,

$$J_e = \frac{\eta_e}{b} \int_0^{\Delta_{ce}} P \, d\Delta_{ce} = \frac{\eta_e}{b} U_{ce} \tag{24}$$

where U_{ce} is the elastic contribution to the strain energy due to the crack, b is the remaining ligament length (see Fig. 1), and by definition

$$\eta_e = \frac{b}{C_c} \frac{dC_c}{da} = -\frac{b}{P} \left(\frac{\partial P}{\partial a} \right)_{\Delta_{ce}} \tag{25}$$

is a dimensionless geometric factor. The η_e-factor defined in Eq 25 is the reciprocal of the η-factor originally introduced by Turner [7].

Based upon the developments of Rice et al [3], Sumpter and Turner [8] proposed writing

$$J_p = \frac{\eta_p}{b} \int_0^{\Delta_{cp}} P \, d\Delta_{cp} = \frac{\eta_p}{b} U_{cp} \tag{26}$$

where U_{cp} is the contribution to the plastic strain energy due to the crack. The dimensionless parameter η_p is similar to η_e in that it is assumed to be a function of the flawed configuration and independent of the deformation. It is necessary and sufficient for the existence of such an η_p that P and Δ_{cp} be related by the separable form

$$P = f(a) g(\Delta_{cp}) \tag{27}$$

in which $f(a)$ is a function of geometry only and $g(\Delta_{cp})$ is a function of Δ_{cp} but independent of a. This form exists at limit load for a deeply cracked body when the remaining ligament experiences primarily bending and for a body exhibiting power law hardening that is subjected to a single monotonically increasing load parameter.

The use of the η_p-factor simplifies the task of determining J. It allows the stability of crack growth to be assessed rigorously when η_p exists and approximately when it does not. It also permits the stability of J-controlled crack growth to be formulated generally. To date the assumed existence of η_p appears not to be any more severe than are the assumptions regarding the form of the load-displacement function in alternative approaches. Turner [9] and Paris et al [10] argued that η_p does not rigorously exist when the plasticity in the remaining ligament changes substantially as it develops from small-scale yielding to the fully plastic state. In this case the separable form of Eq 27 does not exist. However, any other approach that relies on a relationship of this type will also suffer the same shortcoming.

Use of the η-Factor

The combination of Eq 26,

$$J_p = \int_0^P \left(\frac{\partial \Delta_{cp}}{\partial a} \right)_P dP \tag{28}$$

and

$$d\Delta_{cp} = \left(\frac{\partial \Delta_{cp}}{\partial P} \right)_a dP \tag{29}$$

for a fixed crack length yields

$$\eta_p = \frac{b}{P} \frac{(\partial \Delta_{cp}/\partial a)_P}{(\partial \Delta_{cp}/\partial P)_a} \tag{30}$$

or, equivalently,

$$\eta_p = -\frac{b}{P} \left(\frac{\partial P}{\partial a} \right)_{\Delta_{cp}} \tag{31}$$

Hence Eq 31 is the plastic counterpart of Eq 25.

For fixed crack length, Eqs 24 and 30 yield

$$dJ = \frac{\eta_e}{b} P d\Delta_{ce} + \frac{\eta_p}{b} P d\Delta_{cp}$$

and

$$\left(\frac{\partial J}{\partial P}\right)_a = \left(\frac{\partial J_e}{\partial P}\right)_a + \left(\frac{\partial J_p}{\partial P}\right)_a = P\left[\frac{\eta_e}{b}\left(\frac{\partial \Delta_{ce}}{\partial P}\right)_a + \frac{\eta_p}{b}\left(\frac{\partial \Delta_{cp}}{\partial P}\right)_a\right]$$

(32)

Since Δ_{nc} is independent of a, Eqs 16 and 32 imply that

$$\left(\frac{\partial J_e}{\partial P}\right)_a = \left(\frac{\partial \Delta_{ce}}{\partial a}\right)_P = P\frac{\eta_e}{b}\left(\frac{\partial \Delta_{ce}}{\partial P}\right)_a$$

(33)

and

$$\left(\frac{\partial J_p}{\partial P}\right)_a = \left(\frac{\partial \Delta_{cp}}{\partial a}\right)_P = P\frac{\eta_p}{b}\left(\frac{\partial \Delta_{cp}}{\partial P}\right)_a$$

(34)

Now, noting that $U_{cp} = P\Delta_{cp} - U_{cp}^*$, it follows from Eq 26 that

$$J_p = \frac{\eta_p}{b}[P\Delta_{cp} - U_{cp}^*]$$

Differentiating this expression gives

$$\left(\frac{\partial J_p}{\partial a}\right)_P = \frac{\eta_p}{b}\left[P\left(\frac{\partial \Delta_{cp}}{\partial a}\right)_P - \left(\frac{\partial U_{cp}^*}{\partial a}\right)_P\right]$$

$$+ \frac{\eta_p}{b^2}\left[1 - \frac{b}{\eta_p}\frac{\partial \eta_p}{\partial b}\right]U_{cp} = \left(\frac{\eta_p P}{b}\right)^2\left(\frac{\partial \Delta_{cp}}{\partial P}\right)_a$$

$$+ \left[1 - \eta_p - \frac{b}{\eta_p}\frac{\partial \eta_p}{\partial b}\right]\frac{J_p}{b}$$

(35)

In arriving at Eq 35, Eqs 13, 30, 31, and 34 and $\partial/\partial a = -\partial/\partial b$ were employed. In a similar way

$$\left(\frac{\partial J_e}{\partial a}\right)_P = \left(\frac{\eta_e P}{b}\right)^2\left(\frac{\partial \Delta_{ce}}{\partial P}\right)_a + \left[1 - \eta_e - \frac{b}{\eta_e}\frac{\partial \eta_e}{\partial b}\right]\frac{J_e}{b}$$

(36)

and therefore

$$\left(\frac{\partial J}{\partial a}\right)_P = -\frac{J}{b} + \left[2 - \eta_e - \frac{b}{\eta_e}\frac{\partial \eta_e}{\partial b}\right]\frac{J_e}{b}$$

$$+ \left[2 - \eta_p - \frac{b}{\eta_p}\frac{\partial \eta_p}{\partial b}\right]\frac{J_p}{b} + \left(\frac{P}{b}\right)^2\left[\eta_e^2\left(\frac{\partial \Delta_{ce}}{\partial P}\right)_a + \eta_p^2\left(\frac{\partial \Delta_{cp}}{\partial P}\right)_a\right] \quad (37)$$

Finally, the introduction of Eqs 32 and 37 into Eq 17 produces

$$\left(\frac{\partial J}{da}\right)_{\Delta_T} = -\frac{J}{b} + \left[2 - \eta_e - \frac{b}{\eta_e}\frac{\partial \eta_e}{\partial b}\right]\frac{J_e}{b}$$

$$+ \left[2 - \eta_p - \frac{b}{\eta_p}\frac{\partial \eta_p}{\partial b}\right]\frac{J_p}{b} + \left(\frac{P}{b}\right)^2\left\{\left[\eta_e^2\left(\frac{\partial \Delta_{ce}}{\partial P}\right)_a\right.\right.$$

$$\left.+ \eta_p^2\left(\frac{\partial \Delta_{cp}}{\partial P}\right)_a\right]\left[C_M + \left(\frac{\partial \Delta}{\partial P}\right)_a\right] - \left[\eta_e\left(\frac{\partial \Delta_{ce}}{\partial P}\right)_a\right.$$

$$\left.\left.+ \eta_p\left(\frac{\partial \Delta_{cp}}{\partial P}\right)_a\right]^2\right\}\left\{C_M + \left(\frac{\partial \Delta}{\partial P}\right)_a\right\}^{-1} \quad (38)$$

If the relation between the load and the load-point displacement is known, either experimentally or analytically, then the tearing modulus can be computed from Eq 38. The stability of crack growth can then be examined via Eq 9.

Relations for an Extending Crack

Equations 24 and 26 are strictly valid only for a nonextending crack even though they are frequently used to determine J-resistance curves. Since J is based upon deformation theory, it is independent of the path leading to the current values of a and Δ_c, provided the conditions for J-controlled crack growth are satisfied. Thus, for arbitrary increments of a and Δ_{cp}, Eq 26 gives

$$dJ_p = \frac{\eta_p}{b}\left[\left(\frac{\partial U_{cp}}{\partial \Delta_{cp}}\right)_a d\Delta_{cp} + \left(\frac{\partial U_{cp}}{\partial a}\right)_{\Delta_{cp}} da\right] = \frac{\eta_p U_{cp}}{b^2}\left[1 - \frac{b}{\eta_p}\frac{\partial \eta_p}{\partial b}\right]da$$

Since $J_p = -(\partial U_{cp}/\partial a)_{\Delta_{cp}}$ and $P = (\partial U_{cp}/\partial \Delta_{cp})_a$, then the preceding equation can be rewritten as

$$dJ_p = \frac{\eta_p}{b}Pd\Delta_{cp} + \frac{J_p}{b}\left[1 - \eta_p - \frac{b}{\eta_p}\frac{\partial \eta_p}{\partial b}\right]da$$

Because dJ_p is an exact differential, then

$$J_p = \int_0^{\Delta_{cp}} \frac{\eta_p P}{b} d\Delta_{cp} + \int_{a_0}^a \frac{J_p}{b} \left[1 - \eta_p - \frac{b}{\eta_p} \frac{\partial \eta_p}{\partial b} \right] da \qquad (39)$$

This holds for any path followed by a and Δ_{cp} to their current values. The analogous expression for J_e is

$$J_e = \int_0^{\Delta_{ce}} \frac{\eta_e P}{b} d\Delta_{ce} + \int_{a_0}^a \frac{J_e}{b} \left[1 - \eta_e - \frac{b}{\eta_e} \frac{\partial \eta_e}{\partial b} \right] da \qquad (40)$$

When there is no crack growth, Eqs 39 and 40 reduce to Eqs 26 and 24, respectively. The second term of each of these equations represents a correction to account for crack extension. The use of the minicomputer in automated data acquisition and reduction simplifies the task of properly accounting for the second term. In this manner the simultaneous measurement of load, load-point displacement, and crack extension permits, until instability intercedes, the determination of the J-resistance curve from a single test.

Some Illustrative Examples

Bend Specimens

For compliant loading of a deeply cracked bend specimen (for example, see Fig. 3a the generalized load and displacement are $P = M$ and $\Delta = \theta$, re-

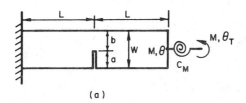

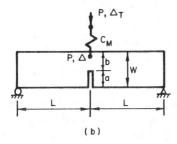

FIG. 3—*Compliant loading of* (a) *bend specimen and* (b) *three-point bend specimen.*

spectively. It follows from dimensional analysis that $P = M = b^2 F(\theta_c)$ [3]. Consequently, Eqs 25 and 31 lead to $\eta_e = \eta_p = 2$ for this specimen. In this case Eq 38 reduces to

$$\left(\frac{dJ}{da}\right)_{\theta_T} = -\frac{J}{b} + \frac{4M^2}{b^2}\left[\frac{C}{1 + C(\partial M/\partial \theta_c)_a}\right] \tag{41}$$

where

$$C = C_M + C_{nc} = C_M + d\,\theta_{nc}/dM \tag{42}$$

is the combined compliance. Equation 41 agrees with the development of Hutchinson and Paris [11] who used an alternative approach.

This analysis may also be applied to the compliant loading of the three-point bend specimen of Fig. 3b. In this case

$$\left(\frac{dJ}{da}\right)_{\Delta_T} = -\frac{J}{b} + \frac{4P^2}{b^2}\left[\frac{C}{1 + C(\partial P/\partial \Delta_c)_a}\right] \tag{43}$$

in which

$$C = C_M + C_{nc} = C_M + d\Delta_{nc}/dP \tag{44}$$

With $\eta_e = \eta_p = 2$, Eqs 39 and 40 combine to give

$$J = 2\int_0^{\Delta_c} \frac{P}{b}\,d\Delta_c - \int_{a_0}^a \frac{J}{b}\,da \tag{45}$$

for a growing crack.

Center Cracked Tension Panel

From Paris et al [10] it can be seen for a deeply cracked center cracked panel that

$$\eta_p = 2 - P\Delta_{cp}\left\{\int_0^{\Delta_{cp}} P\,d\Delta_{cp}\right\}^{-1} \tag{46}$$

With this η_p, Eq 38 can be shown to give the same tearing modulus obtained by Hutchinson and Paris [11]. In general, η_p given by Eq 46 is a function of the deformation. This, of course, is contrary to Turner's original assumption and the one assumed here that η_p is a function of configuration only and independent of the deformation. This dependence of η_p on the deformation, however, is likely to be rather weak. If P is a homogeneous function of Δ_{cp} (for ex-

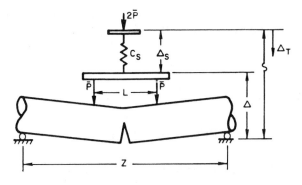

(a)   Four-Point Bend Loading System

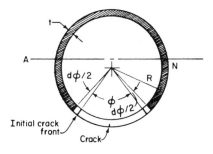

(b)   Cross Section of a Through-Wall Cracked Pipe

FIG. 4—(a) *Four-point bend loading system.* (b) *Cross section of a through-wall cracked pipe.*

ample, P and Δ_{cp} are related by a power law or a fully yielded limit state exists), then η_p will be independent of the deformation. If the dependence of η_p upon deformation is weak, then to a first approximation η_p can be taken outside the integral as in Eq 26. To the extent that the latter is an appropriate approximation, then Eq 38 is general and its development as such does not depend upon the restriction that η_p be independent of the deformation.

Circumferentially Cracked Pipe

For an example of a more realistic problem, which is included as a special case of the present development, consider the compliant four-point bend test of a circumferentially cracked pipe depicted in Fig. 4a. This problem was first treated by Zahoor and Kanninen [5,6]. The limit load for this configuration is

$$P_0 = \frac{8\sigma_0 R^2 t}{Z - L} \, h(\phi) \tag{47}$$

where ϕ is the arc subtended by the crack, t and R are the pipe's thickness and mean radius, respectively, σ_0 is the flow stress, and

$$h(\phi) = \cos(\phi/4) - \tfrac{1}{2}\sin(\phi/2) \tag{48}$$

For this configuration

$$b = R(2\pi - \phi)/2 \quad \text{and} \quad P = \bar{P}/t \tag{49}$$

It can be argued from dimensional considerations that Δ_{cp} must be a function of $\bar{P}/P_0$ or, equivalently,

$$P = h(\phi)\,g(\Delta_{cp}) \tag{50}$$

Note that $\partial/\partial a = (2/R)(\partial/\partial\phi)$ and introducing Eq 50 into Eq 31 gives

$$\eta_p = -(2\pi - \phi)\,h'/h \equiv (2\pi - \phi)\,Rt\beta \tag{51}$$

where the prime denotes a differentiation with respect to ϕ. With the aid of Eqs 49 and 51, Eq 26 yields[4]

$$J_p = 2\beta \int_0^{\Delta_{cp}} \bar{P}\,d\Delta_{cp} \tag{52}$$

for a nongrowing crack. For an extending crack, Eqs 39 and 51 combine to give

$$J_p = 2\beta \int_0^{\Delta_{cp}} \bar{P}\,d\Delta_{cp} + \int_{\phi_0}^{\phi} \gamma J_p\,d\phi \tag{53}$$

where $\gamma = h''/h'$ and β is evaluated for the initial crack angle ϕ_0.

When the elastic contributions to J and Δ can be neglected compared with their plastic counterparts, Eq 38 reduces to

$$\left(\frac{dJ}{da}\right)_{\Delta_T} = \left[1 - \eta_p - \frac{b}{\eta_p}\frac{d\eta_p}{db}\right]\frac{J}{b} + \left(\frac{\eta_p P}{b}\right)^2\left[\frac{C}{1 + C(\partial P/\partial\Delta_{cp})_a}\right] \tag{54}$$

where

$$C = C_M + d\Delta_{nc}/dP \tag{55}$$

---

[4]The factor of two in Eq 52 was inadvertently dropped in Ref 5.

Substitution of Eq 51 and $C_M = 2C_s t$ into Eq 54 leads to

$$\left(\frac{dJ_p}{da}\right)_{\Delta_T} = \frac{4t\,(\beta\bar{P})^2(2C_s + C_e)}{1 + (2C_s + C_e)(2\bar{P}/\partial\Delta_{cp})_a} + \frac{2\gamma}{R}J_p \qquad (56)$$

where

$$C_e = d\Delta_{nc}/d\bar{P} = (Z - L)^2(Z + 2L)/24EI \qquad (57)$$

is the elastic compliance of the uncracked pipe and EI is its flexural rigidity. Aside from the missing factor of two, Eqs 52, 53, and 57 agree with those of Ref 5.

Equations 38 to 40 also contain the development of Ernst et al [12] as a special case. Those authors assumed that $\eta_e = \eta_p = \eta$, which holds in particular for deeply cracked predominately bend specimens, but not in general.

Tearing Instability Analysis for Power Law Hardening

Basic Approach

The form of Eqs 26 and 38 is particularly well suited for use with the GE/EPRI elastic-plastic handbook [13] for power law hardening materials of the type

$$\frac{\epsilon}{\epsilon_y} = \alpha\left(\frac{\sigma}{\sigma_y}\right)^n \qquad (58)$$

where σ_y and ϵ_y are the yield stress and strain, respectively, and α and n are material properties. Ilyushin's theorem permits writing

$$J_p = \alpha\,\epsilon_y\,\sigma_y\,b\,g_1\,(a/W)\,h_1\,(a/W, n)(P/P_0)^{n+1} \qquad (59)$$

$$\Delta_{cp} = \alpha\,\epsilon_y\,a\,g_3\,(a/W)h_3\,(a/W, n)(P/P_0)^n \qquad (60)$$

where P_0 is a reference limit load based upon σ_y. The known dimensionless functions g_1 and g_3 are selected for convenience of presentation. The dimensionless functions h_1 and h_3 depend upon a/W and n but are independent of P. These functions can be found tabulated in Ref 13 for selected fracture specimens and flawed pressure vessels.

For a power law hardening material

$$U_{cp} = \frac{n}{n + 1}P\Delta_{cp} \qquad (61)$$

which upon introduction into Eq 26 gives

$$J_p = \frac{n}{n+1} \frac{\eta_p}{b} P\Delta_{cp}$$ (62)

Substituting Eqs 59 and 60 into Eq 62 then gives

$$\eta_p = \frac{n+1}{n} \frac{b^2\sigma_y}{P_0 a} \frac{g_1(a/W)}{g_3(a/W)} \frac{h_1(a/W, n)}{h_3(a/W, n)}$$ (63)

which can be evaluated for the specimens and structures included in the GE/EPRI handbook [13]. Computations demonstrate that η_p can depend rather strongly upon the crack length and the hardening exponent for tension loadings. A typical result is shown in Fig. 5. However, as shown in Fig. 6, for bend specimens, η_p is virtually independent of the hardening exponent for a range of crack lengths. For deeply cracked bodies η_p also appears to be independent of the hardening exponent and approaches a constant value that is specimen dependent.

Having established η_p, $\partial\eta_p/\partial b$ can be determined numerically. Since it is the quantity

$$\frac{b}{\eta_p} \frac{\partial\eta_p}{\partial b} = \frac{d(\ln \eta_p)}{d(\ln b)}$$

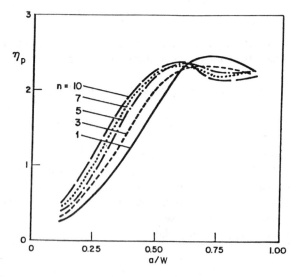

FIG. 5—*The η_p factor deduced from the GE/EPRI plastic fracture handbook [13] as a function of crack length and hardening index for a single-edge cracked tension specimen.*

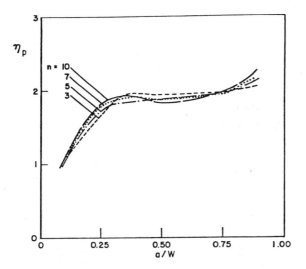

FIG. 6—*The η_p factor deduced from the GE/EPRI plastic fracture handbook [13] as a function of crack length and hardening index for a bend specimen.*

that appears in Eq 38, it may be more convenient to form the numerical derivative of $d(\ell n\ \eta_p)/d(\ell n\ b)$. It also follows from Eq 60 that

$$\left(\frac{\partial \Delta_{cp}}{\partial P}\right)_a = \frac{n\alpha\sigma_y a}{EP_0} g_3(a/W) h_3(a/W, n)\left(\frac{P}{P_0}\right)^{n-1} \qquad (64)$$

This completes the determination of the plastic contributions in Eq 38 for $(dJ/da)_{\Delta_T}$. The elastic contributions to this quantity can be evaluated in a similar manner using LEFM handbooks; see, for example, that by Tada et al [14].

The principal advantage of this approach over that suggested in Ref 13 is that only a single numerical differentiation (compared to four) is required. Furthermore, as suggested by Parks et al [15], a judicious choice of g_1 and g_3 can even simplify this single numerical differentiation. To the extent that the tabulated functions h_1 and h_3 exist in Ref 13 for the configuration of interest, this approach can be used to assess the stability of crack growth in power law hardening materials.

For a flawed configuration not included in the GE/EPRI handbook [13] one can generate h_1 and h_3 functions for the configuration of interest or adopt an alternative approach. A rather efficient approximate procedure is the following one. Suppose that an approximation for $\eta_p = \eta_p(b)$ can be developed; say, through a combination of dimensional analysis and Eq 31 as was done in the previous examples. Since

$$J_p = \frac{\eta_p}{b} U_{cp} = -\left(\frac{dU_{cp}}{da}\right)_{\Delta_{cp}} = \left(\frac{dU_{cp}}{db}\right)_{\Delta_{cp}} \qquad (65)$$

then

$$\frac{dU_{cp}}{U_{cp}} = \frac{\eta_p}{b} db \tag{66}$$

for $\Delta_{cp} = \Delta_{cp}^0 = $ constant.

Integrating Eq 66 produces

$$\frac{U_{cp}(b, \Delta_{cp}^0)}{U_{cp}(b_0, \Delta_{cp}^0)} = \left(\frac{b}{b_0}\right)^{\eta_p(b)} \Gamma(b/b_0) \tag{67}$$

where

$$\Gamma(b/b_0) = \exp -\left\{\int_{b_0}^{b} \ell n\,(b/b_0)\,\frac{d\eta_p}{db}\,db\right\} \tag{68}$$

A representative reference length of the remaining ligament of the flawed configuration is denoted by b_0 while Δ_{cp}^0 is an associated prescribed load-point displacement. From Ilyushin's theorem it follows that

$$\frac{U_{cp}(b, \Delta_{cp})}{U_{cp}(b, \Delta_{cp}^0)} = \left(\frac{\Delta_{cp}}{\Delta_{cp}^0}\right)^{(n+1)/n} \tag{69}$$

The combination of Eqs 67 and 69 leads to

$$U_{cp}(b, \Delta_{cp}) = U_{cp}(b_0, \Delta_{cp}^0)\left(\frac{b}{b_0}\right)^{\eta_p(b)} \Gamma\left(\frac{b}{b_0}\right)\left(\frac{\Delta_{cp}}{\Delta_{cp}^0}\right)^{(n+1)/n} \tag{70}$$

Therefore

$$J_p = \frac{\eta_p}{b} U_{cp}(b_0, \Delta_{cp}^0)\left(\frac{b}{b_0}\right)^{\eta_p(b)} \Gamma\left(\frac{b}{b_0}\right)\left(\frac{\Delta_{cp}}{\Delta_{cp}^0}\right)^{(n+1)/n} \tag{71}$$

From Castigliano's theorem

$$\left(\frac{\partial U_{cp}}{\partial \Delta_{cp}}\right)_a = P = \frac{(n+1)}{n}\frac{U_{cp}(b_0, \Delta_{cp}^0)}{\Delta_{cp}^0}\left(\frac{b}{b_0}\right)^{\eta_p(b)}$$

$$\times \Gamma\left(\frac{b}{b_0}\right)\left(\frac{\Delta_{cp}}{\Delta_{cp}^0}\right)^{1/n} \tag{72}$$

from which it follows that

$$\left(\frac{\partial \Delta_{cp}}{\partial P}\right)_a = \left\{ \frac{n+1}{n^2} \frac{U_{cp}(b_0, \Delta_{cp}^0)}{(\Delta_{cp}^0)^2} \left(\frac{b}{b_0}\right)^{\eta_p(b)} \right.$$

$$\left. \times \Gamma(b/b_0) \left(\frac{\Delta_{cp}}{\Delta_{cp}^0}\right)^{(1-n)/n} \right\}^{-1} \quad (73)$$

For a given power law hardening material and a flawed configuration, a single reference computation for a prescribed load-point displacement Δ_{cp}^0 and remaining ligament length b_0 is sufficient to determine $U_{cp}(b_0, \Delta_{cp}^0)$. Finite-element methods can efficiently perform this type of computation. Having established $U_{cp}(b_0, \Delta_{cp}^0)$, it is clear from Eqs 71 to 73 that J_p, P, and $(\partial \Delta_{cp}/\partial P)_a$ can be determined for any other crack length and load-point displacement of this configuration. When these quantities are combined with their elastic counterparts, everything is in place for performing the tearing instability analysis.

Application of the Theory

A comparison of this approach and the known solution from the GE/EPRI handbook for bend specimens is made in Tables 1 and 2 for plane stress and plane strain, respectively. Based upon the previous dimensional analysis, $\eta_p = 2$ is used for the bend specimens. In these tables $b_0 = W/2$ is the reference length of the remaining ligament, and J_0 is the corresponding value of J_p for a fixed Δ_{cp}. If there were perfect agreement, then the ratio $(J_p/J_0)_h$ from the handbook [13] to J/J_0 determined from Eq 71 would be unity. For the most part the agreement is fairly good. Significant differences appear for short cracks ($a/W < 0.25$); smaller departures occur for the larger values of n and $a/W > 0.75$. The large variances for $n = 1$ are of little practical consequence

TABLE 1—*Comparison of handbook solution and approximate solution for plane stress three-point bend specimen* $(J/J_0)_h/(J/J_0)$.

					$n =$				
a/W	1	2	3	5	7	10	13	16	20
1/8	25.78	0.574	0.707	0.640	0.699	0.713	0.729	0.742	0.764
1/4	4.484	0.853	0.828	0.906	0.948	0.946	0.950	0.957	0.977
3/8	1.818	1.018	0.979	0.991	0.945	0.984	0.985	0.990	0.992
1/2	1.000	1.000	1.000	1.000	1.000	1.000	1.000	1.000	1.000
5/8	0.653	0.955	0.977	0.999	0.995	0.999	1.000	1.003	1.011
3/4	0.512	0.884	0.926	0.945	0.956	0.961	0.960	0.962	0.969
7/8	0.444	0.862	0.922	0.962	0.985	1.003	1.018	0.923	1.025

TABLE 2—*Comparison of handbook solution and approximate solution for plane strain three-point bend specimen* $(J/J_0)_h/(J/J_0)$.

					$n =$				
a/W	1	2	3	5	7	10	13	16	20
1/8	25.96	0.553	0.541	0.581	0.613	0.636	0.638	0.637	0.641
1/4	4.498	0.798	0.793	0.866	0.910	0.953	0.941	0.952	0.957
3/8	1.813	0.993	0.997	1.006	0.984	1.004	0.984	0.980	0.963
1/2	1.000	1.000	1.000	1.000	1.000	1.000	1.000	1.000	1.000
5/8	0.657	0.971	1.016	1.040	1.049	1.050	1.028	1.041	1.031
3/4	0.509	0.930	1.014	1.032	1.050	1.078	1.070	1.080	1.074
7/8	0.444	0.892	0.989	1.056	1.098	1.150	1.136	1.144	1.097

since linear elastic solutions are readily available. The range of a/W and n for which the differences are significant is, as expected, the same range where the plots in Fig. 6 differ appreciably from $\eta_p = 2$.

It is also clear from Tables 1 and 2 that, except for $n = 1$, better agreement occurs for crack lengths close to the reference crack length. Thus it is advisable to perform the reference computation for the anticipated flaw size. By doing so, the error introduced by the approximation inherent to this approach will be minimized when neighboring flaw sizes are considered. In addition, for reasons of precision it is better to work with Δ_{cp} than P. If Eq 72 is used to eliminate Δ_{cp} in Eq 71, then the term $(b/b_0)^{-\eta_p n}$ appears in J_p. Because of this term a small error in η_p, when multiplied by a large value of n, can produce a substantial error in J_p if b/b_0 differs appreciably from unity.

With this approach a tearing instability analysis can be performed in the following manner. For an assumed value of the crack extension Δa, $J_R(\Delta a)$ is determined from the resistance curve. A value of Δ_{cp} is selected and J_p and P are computed from Eqs 71 and 72. The value of P can be used to compute J_e using LEFM methods. The sum $J = J_e + J_p$ is compared with J_R. If $J \neq J_R$, then Δ_{cp} is adjusted appropriately and the procedure repeated until $J = J_R$. Next, $(dJ/da)_{\Delta_T}$ is computed using Eq 38, and the stability of the crack growth is assessed by means of Eq 9. Alternatively, the same procedure can be followed to determine $(dJ/da)_{\Delta_T}$ versus $J = J_R$. The value of J at instability is identified with the point of intersection (if it exists) of this curve and dJ_R/da versus J_R. The resistance curve can be used to determine the limit of stable crack growth. When J_e can be neglected compared to J_p, then $J_p = J_R$ so that Δ_{cp} can be determined without iteration from Eq 71, whereupon P follows from Eq 72.

Discussion

The *J*-estimation methods developed previously have been based upon the assumption that a limit load condition exists. Because the restriction to near limit load conditions could be a grave oversimplification that may lead to un-reliable estimates of the fracture instability point, the role of material harden-ing on stable crack growth and tearing instability must be assessed. In this paper the role of material hardening on stable crack growth and tearing insta-bility was examined through the use of a tearing instability analysis employing Turner's deformation-independent η-factor. The η-factor links J to the strain energy and the remaining ligament in a cracked component. An important as-sumption that is conventionally made is to split the deformation into additive elastic and plastic components, the former often being negligible in compari-son to the latter. The assumed existence of the plastic component of η appears to be no more severe than the usual assumptions found in the literature re-garding the form of the load-displacement function.

Despite the usefulness of the η-factor approach, the analysis difficulties have not been entirely overcome. Rather, they have simply been concentrated onto the determination of the η-factor. In limit load conditions, η is readily identified. In conditions where it is more difficult to obtain this parameter, reasonable procedures are available. For example, for power hardening mate-rials, η can be deduced for the specimens and structures included in the GE/EPRI elastic-plastic fracture handbook [13]. These computations demon-strate that η can depend rather strongly upon the crack length and hardening exponent for tension loadings. In bend specimens, however, η is virtually in-dependent of the hardening exponent for a range of crack lengths. For deeply cracked bodies η also appears to be independent of the hardening exponent and approaches a constant value that is specimen dependent.

Ilyshin's theorem for a power hardening material and the energy interpreta-tion of the J-integral permits writing the strain energy and the load in terms of η, the remaining ligament, the load point-displacement, and an undetermined coefficient. This coefficient is ultimately determined by performing a single reference computation for a specific hardening exponent. Having performed such a computation, J and dJ/da can be readily determined for other crack lengths and load intensities. These parameters and a J-resistance curve are sufficient for performing a tearing instability analysis to determine the fracture instability point and, hence, the margin of safety for a cracked structure.

Conclusions

The basis for tearing instability analyses incorporating material hardening (analyses that can be applied to cracked nuclear piping and other systems re-quiring an elastic-plastic fracture mechanics approach) has been provided in this paper. The generality of the approach was shown by specializing the results obtained herein to limit load conditions, whereupon all previous J-estimation results are recovered as special cases. A specific application of the approach is given in a companion paper in this volume [16].

Acknowledgments

This research was supported by the U.S. Nuclear Regulatory Commission, Metallurgy and Materials Research Branch, under Contract NRC-04-81-178. The authors would like to express their appreciation to Messrs. Jack Strosnider and Milton Vagins of the NRC for their encouragement of this work.

References

[1] Kanninen, M. F., Popelar, C. H., and Broek, D., *Nuclear Engineering and Design*, Vol. 67, 1981, pp. 27–55.
[2] Paris, P. C., "Analyses and Applications with Plastic Tearing Instability," presented at Second International Symposium on Elastic-Plastic Fracture Mechanics, ASTM, Philadelphia, Oct. 1981.
[3] Rice, J. R., Paris, P. C., and Merkle, J. G. in *Progress in Flaw Growth and Fracture Toughness Testing, ASTM STP 536*, American Society for Testing and Materials, 1973, pp. 231–245.
[4] Paris, P. C., Tada, H., Zahoor, A., and Ernst, H. in *Elastic-Plastic Fracture, ASTM STP 668*, American Society for Testing and Materials, 1979, pp. 5–36.
[5] Zahoor, A. and Kanninen, M. F., *Journal of Pressure Vessel Technology*, Vol. 103, 1981, pp. 352–358.
[6] Zahoor, A. and Kanninen, M. F. in *Elastic-Plastic Fracture: Second Symposium, ASTM STP 803, Volume II—Fracture Resistance Curves and Engineering Applications*, American Society for Testing and Materials, 1983, pp. II-291–II-308.
[7] Turner, C. E., *Material Science and Engineering*, Vol. 11, 1973, pp. 275–282.
[8] Sumpter, J. D. G. and Turner, C. E. in *Cracks and Fracture, ASTM STP 601*, American Society for Testing and Materials, 1976, pp. 3–15.
[9] Turner, C. E. in *Fracture Mechanics: Twelfth Conference, ASTM STP 700*, American Society for Testing and Materials, 1980, pp. 314–337.
[10] Paris, P. C., Ernst, H., and Turner, C. E. in *Fracture Mechanics: Twelfth Conference, ASTM STP 700*, American Society for Testing and Materials, 1980, pp. 338–351.
[11] Hutchinson, J. W. and Paris, P. C. in *Elastic-Plastic Fracture, ASTM STP 668*, American Society for Testing and Materials, 1979, pp. 37–64.
[12] Ernst, H. A., Paris, P. C., and Landes, J. D. in *Fracture Mechanics: Thirteenth Conference, ASTM STP 743*, American Society for Testing and Materials, 1981, pp. 476–502.
[13] Kumar, V., German, M. D., and Shih, C. F., "An Engineering Approach for Elastic-Plastic Fracture Analysis," EPRI NP-1931, Project 1287-1, Topical Report, Electric Power Research Institute, Palo Alto, Calif., July 1981.
[14] Tada, H., Paris, P. C., and Irwin, G. R., *Stress Analysis of Cracks Handbook*, Del Research Corp., Hellertown, Pa., 1973.
[15] Parks, D. M., Kumar, V., and Shih, C. F. in *Elastic-Plastic Fracture: Second Symposium. Volume I—Inelastic Crack Analysis. ASTM STP 803*, American Society for Testing and Materials, 1983, pp. I-370–I-383.
[16] Pan, J., Ahmad, J., Kanninen, M. F., and Popelar, C. H., this publication, pp. 721–745.

J. Pan,[1] *J. Ahmad,*[2] *M. F. Kanninen,*[2] *and C. H. Popelar*[3]

Application of a Tearing Instability Analysis for Strain-Hardening Materials to a Circumferentially Cracked Pipe in Bending

REFERENCE: Pan, J., Ahmad, J., Kanninen, M. F., and Popelar, C. H., **"Application of a Tearing Instability Analysis for Strain-Hardening Materials to a Circumferentially Cracked Pipe in Bending,"** *Fracture Mechanics: Fifteenth Symposium, ASTM STP 833*, R. J. Sanford, Ed., American Society for Testing and Materials, Philadelphia, 1984, pp. 721–745.

ABSTRACT: All previous tearing instability analyses for circumferentially cracked pipes require the use of a limit-load assumption. In this paper a more general approach is developed for power law hardening materials. Specifically, a simple J and $(\partial J/\partial a)_{\Delta_T}$ estimation scheme for a circumferentially cracked pipe subjected to four-point bending is developed using a power hardening law for the plastic part of deformation. The predictions of crack initiation and tearing instability are compared with experimental data. Comparisons with tearing instability predictions using the limit-load approach are also provided.

KEY WORDS: tearing instability, J-resistance curve, J-estimation method, pipe fracture, fracture mechanics

Nuclear plant piping can exhibit considerable strength even when containing large cracks. Significant plastic deformation and crack tip blunting generally occur prior to the initiation of crack growth. In addition, crack growth initiation is usually followed by stable crack growth before the onset of fracture is reached. Because linear elastic fracture mechanics (LEFM) treatments cannot properly account for these effects, nonlinear fracture mechanics procedures must be used to assess the integrity of nuclear piping. As found by Kanninen et al [1], most such approaches are based upon the use of the *J*-

[1]Mechanics Section, Battelle-Columbus Laboratories, Columbus, Ohio 43201.
[2]Engineering and Materials Sciences Division, Southwest Research Institute, San Antonio, Tex. 78284.
[3]Engineering Mechanics Department, The Ohio State University, Columbus, Ohio 43201

integral/tearing modulus concept, a methodology now commonly referred to as tearing instability analysis [2].

The J-estimation methods that have been developed for cracked nuclear piping have been based upon the assumption that a limit-load condition exists; for examples, see Refs 3 to 6. Even so, these analyses require that at least one experimental load-displacement curve be available for the load/structure condition and material of concern. The assumption of near limit-load conditions could be a grave oversimplification and, in some instances, could lead to unreliable estimates of the fracture instability point. Accordingly, the role of material hardening on stable crack growth and tearing instability must be addressed to minimize this possibility. In this paper a simple method that accounts for material hardening [7] is applied to predict tearing instability for a circumferentially cracked pipe subjected to bending loads.

Basis for Instability of a Cracked Body

A two-dimensional cracked body subjected to displacement-controlled conditions is considered (Fig. 1). Let P denote the generalized load, and Δ the work-conjugate load-point displacement of the cracked body. Within the context of small-strain deformation theory of plasticity, Δ can be expressed as a function of the load P and the straight crack length a. The cracked body is

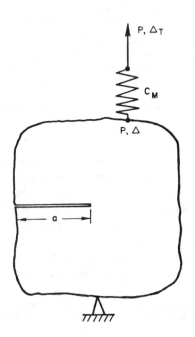

FIG. 1—*Two-dimensional cracked body.*

loaded through a linear elastic spring having a compliance C_M. The total load-point displacement, (Δ_T) can then be written as

$$\Delta_T = C_M P + \Delta(P, a) \tag{1}$$

In deformation theory of plasticity the J-integral can be expressed as [8, 9]

$$J = J(P, a) \tag{2}$$

For a small amount of crack growth, a material resistance curve J_R can be proposed such that J_R is only a function of crack extension [9]. Hence, for a stably growing crack,

$$J(P, a) = J_R(a - a_0) \tag{3}$$

where a represents the current crack length and a_0 the initial crack length. A universal J-resistance curve for a material is questionable [10]. Nevertheless, there can be a resistance curve for a given cracked geometry and material within the condition of J-controlled growth [9].

The functional relations of Eqs 1 to 3 lead to

$$\Delta_T = \Delta_T(J, a_0) \tag{4}$$

The differential form of Eq 4 can be easily obtained as

$$\frac{d\Delta_T}{dJ} = \left[\left(\frac{dJ_R}{da}\right)\left(\frac{\partial J}{\partial P}\right)_a\right]^{-1}\left(C_M + \left(\frac{\partial \Delta}{\partial P}\right)_a\right)\left(\frac{dJ_R}{da} - \left(\frac{\partial J}{\partial a}\right)_{\Delta_T}\right) \tag{5}$$

where

$$\left(\frac{\partial J}{\partial a}\right)_{\Delta_T} = \left(\frac{\partial J}{\partial a}\right)_P - \left(\frac{\partial \Delta}{\partial a}\right)_P\left(\frac{\partial J}{\partial P}\right)_a\left(C_M + \left(\frac{\partial \Delta}{\partial P}\right)_a\right)^{-1} \tag{6}$$

Available experimental data show that J_R is a monotonically increasing function of crack extension $(a - a_0)$ and that dJ_R/da decreases as J_R increases. For a hardening material, the term

$$\left[\left(\frac{dJ_R}{da}\right)\left(\frac{\partial J}{\partial P}\right)_a\right]^{-1}\left(C_M + \left(\frac{\partial \Delta}{\partial P}\right)_a\right)$$

in Eq 5 is positive. Therefore, $d\Delta_T/dJ = 0$ when $dJ_R/da = (\partial J/\partial a)_{\Delta_T}$. This point can be determined graphically (Fig. 2) [9,11]. The instability point is determined at the intersection of the material curve $(dJ_R/da), J_R)$ and the

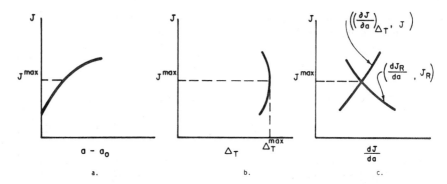

FIG. 2—*Tearing instability point at the maximum total displacement of the system.*

applied curve $((\partial J/\partial a)_{\Delta_T}, J)$. However, the J-resistance curve shown in Fig. 2a together with Eqs 1 and 2 is sufficient to determine the J-Δ_T curve (Fig. 2b). The maximum value of Δ_T corresponds to the instability point determined from the intersection of the material curve and the applied curve (Figs 2b and 2c). Nevertheless, the special case shown in Fig. 3 is possible such that, throughout the loading history, $dJ_R/da \geq (\partial J/\partial a)_{\Delta_T}$. In this case, the point at $dJ_R/da = (\partial J/\partial a)_{\Delta_T}$ cannot be identified as an instability point under an increasing total displacement condition. Therefore, when $(\partial J/\partial a)_{\Delta_T} > dJ_R/da$, the system becomes unstable under the increasing total displacement condition.

For a rigid–perfectly plastic material, the functional forms of Eqs 1 and 2 become

$$\Delta_T = C_M P + \Delta \tag{7}$$

and

$$J = J(\Delta, a) \tag{8}$$

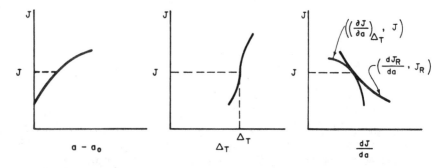

FIG. 3—*No tearing instability associated with the point at* $dJ_R/da = (\partial J/\partial a)_{\Delta_T}$.

whereupon the load P can be expressed as a function of a only when the limit-load condition is reached. Thus

$$P = P(a) \tag{9}$$

Note that for a rigid–perfectly plastic material the independent variables Δ and a (instead of P and a for hardening materials) are chosen. The functional form of Eq 4 can still be constructed; however, the differential form then becomes

$$\frac{d\Delta_{\mathrm{T}}}{dJ} = \left[\left(\frac{dJ_{\mathrm{R}}}{da}\right)\left(\frac{\partial J}{\partial \Delta}\right)_a\right]^{-1}\left(\frac{dJ_{\mathrm{R}}}{da} - \left(\frac{\partial J}{\partial a}\right)_{\Delta_{\mathrm{T}}}\right) \tag{10}$$

where

$$\left(\frac{\partial J}{\partial a}\right)_{\Delta_{\mathrm{T}}} = \left(\frac{\partial J}{\partial a}\right)_\Delta - C_{\mathrm{M}}\left(\frac{\partial J}{\partial \Delta}\right)_a \frac{dP}{da} \tag{11}$$

Therefore the instability relations for materials characterized by the deformation theory of plasticity are also valid for rigid–perfectly plastic materials.

Basis of Tearing Instability Analysis for a Circumferentially Cracked Pipe Subjected to Four-Point Bending

Engineering Estimation Scheme

Circumferential cracks initiated at the inner surface of the pipe wall due to stress corrosion have been observed in a number of boiling water reactor piping systems. To determine the margin of safety of these cracked piping systems, laboratory tests were designed to gain some insight into the fracture behavior of circumferentially cracked pipes subjected to bending loads [5,12]. Figure 4 represents a circumferentially cracked pipe subjected to four-point bending. In the figure, R denotes the mean radius of the pipe, t the thickness of the pipe, $2a$ the length of the circumferential crack subtending an angle 2θ, and Z and L the length dimensions shown.

An elastic spring with compliance C_s acts between the load point and the cracked pipe. The displacement of the cracked pipe is denoted as Δ, and the total displacement is denoted as Δ_{T}. The total load applied to the system is denoted as P_t. The total displacement Δ_{T} is

$$\Delta_{\mathrm{T}} = \Delta_s + \Delta \tag{12}$$

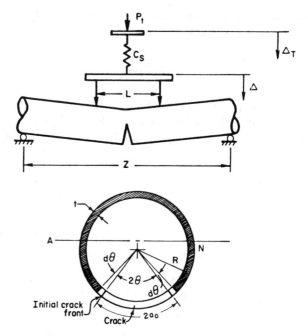

FIG. 4—*Circumferentially cracked pipe subjected to four-point bending.*

where Δ_s represents the displacement due to the elastic spring. Hence

$$\Delta_s = C_s P_t \tag{13}$$

The displacement of the cracked pipe can be decomposed into

$$\Delta = \Delta_{nc} + \Delta_c \tag{14}$$

where Δ_{nc} represents the displacement of the pipe without a crack, and Δ_c represents the displacement due to the presence of the crack. Since the presence of a crack decreases the load-carrying capacity, the load carried by the cracked pipe will usually be in the elastic range of a pipe with no crack. It is therefore assumed that

$$\Delta_{nc} = C_{nc} P_t \tag{15}$$

According to the engineering estimation scheme proposed in Refs *13* and *14* for Ramberg-Osgood materials, Δ_c can be interpolated between the elastic solution Δ_{ce} and the fully plastic solution Δ_{cp}. Here it is assumed that

$$\Delta_c = \Delta_{ce} + \Delta_{cp} \tag{16}$$

The elastic part Δ_{ce} is given by

$$\Delta_{ce} = C(\theta)P_t \tag{17}$$

where $C(\theta)$ represents the elastic compliance as a function of the crack angle θ. The plastic part Δ_{cp} is given by

$$\Delta_{cp} = \overline{f}(\theta, n)\left(\frac{P_t}{P_0}\right)^n \tag{18}$$

where n represents the hardening exponent of the power law relation [13,14], P_0 is a reference load, and $\overline{f}$ is a function of θ, n, and other geometric dimensions and material properties.

The studies of a single-edge cracked plate in bending for power hardening materials show that the load-displacement curve can be well represented by a limit-load solution obtained in Merkle-Corten fashion [15] and a single coefficient which is a function of the hardening exponent only [16]. For the cracked pipe in bending, it is therefore assumed that

$$\Delta_{cp} = f(n)\left(\frac{P_t}{P_0}\right)^n \tag{19}$$

where f is a function of the hardening exponent n, and P_0 is a limit-load solution. Alternatively, Eq 19 can be rewritten as

$$\Delta_{cp} = \left(\frac{P_t}{c(n)P_0}\right)^n \tag{20}$$

where c is a function of the hardening exponent n.

The needed lower-bound limit-load solution [4,5] can be written as

$$P_0 = \frac{16\sigma_0 R^2 t}{Z - L} F(\theta) \tag{21}$$

where σ_0 is the yield stress, and $F(\theta)$ is

$$F(\theta) = \cos\frac{\theta}{2} - \frac{1}{2}\sin\theta \tag{22}$$

The J-integral of the deformation theory of plasticity is defined as [8]

$$J = \frac{1}{2Rt}\frac{\partial}{\partial\theta}\int_0^{P_t}\Delta dP_t \tag{23}$$

Substituting Eqs 16, 17, 20, and 21 into Eq 23 gives

$$J = J_e + J_p \qquad (24)$$

where

$$J_e = \frac{1}{4Rt} C'(\theta) P_t^2 \qquad (25)$$

and

$$J_p = \left(-\frac{n}{n+1}\right) c(n) \frac{8\sigma_0 R}{Z-L} F'(\theta) \left(\frac{P_t(Z-L)}{c(n)16\sigma_0 R^2 t F(\theta)}\right)^{n+1} \qquad (26)$$

or, in terms of Δ_{ce} and Δ_{cp},

$$J_e = \frac{1}{4Rt} \frac{C'(\theta)}{C^2(\theta)} \Delta_{ce}^2 \qquad (27)$$

and

$$J_p = \left(-\frac{n}{n+1}\right) c(n) \frac{8\sigma_0 R}{Z-L} F'(\theta) \Delta_{cp}^{(n+1)/n} \qquad (28)$$

where $C'(\theta)$ is the derivative of $C(\theta)$ with respect to θ, and $F'(\theta)$ is the derivative of $F(\theta)$ with respect to θ. From the basic discussion of the last section, the instability point of the system can be readily determined graphically in the J-Δ_T plot which can be obtained from a J-resistance curve and the estimation formulae for the load-displacement curve and the J integral. However, the characteristics of the system can be shown more vividly in the plot of J as a function of $(\partial J/\partial a)_{\Delta_T}$, as suggested in Ref 11.

Determination of $(\partial J/\partial a)_{\Delta_T}$

Within the context of the deformation theory of plasticity, J can be expressed as

$$J = J(P_t, \theta) \qquad (29)$$

The total displacement Δ_T is

$$\Delta_T = C_s P_t + \Delta(P_t, \theta) \qquad (30)$$

where Δ is expressed as a function of P_t and θ. Using Eqs 29 and 30 and noting that $(\partial J/\partial a)_{\Delta_T} = (\partial J/R\partial\theta)_{\Delta_T}$ gives

$$\left(\frac{\partial J}{\partial a}\right)_{\Delta_T} = \left(\frac{\partial J}{R\partial\theta}\right)_{P_t} - \left(\frac{\partial\Delta}{R\partial\theta}\right)_{P_t}\left(\frac{\partial J}{\partial P_t}\right)_\theta\left(C_s + \left(\frac{\partial\Delta}{\partial P_t}\right)_\theta\right)^{-1} \tag{31}$$

Since

$$\Delta = \left(\frac{\partial\left(\int_0^{P_t}\Delta dP_t\right)}{\partial P_t}\right)_\theta \tag{32}$$

Maxwell's relationship from Eqs 32 and 23 gives

$$\left(\frac{\partial\Delta}{R\partial\theta}\right)_{P_t} = 2t\left(\frac{\partial J}{\partial P_t}\right)_\theta \tag{33}$$

For convenience, P_0 in Eq 20 can be written as a function θ only; that is,

$$P_0 = P_0(\theta) \tag{34}$$

A simple differentiation using Eqs 24 to 26 gives

$$\left(\frac{\partial J}{\partial\theta}\right)_{P_t} = J_e\frac{C''(\theta)}{C'(\theta)} + J_p\left(\frac{P_0''(\theta)}{P_0'(\theta)} - (n+1)\frac{P_0'(\theta)}{P_0(\theta)}\right) \tag{35}$$

and

$$\left(\frac{\partial J}{\partial P_t}\right)_\theta = 2\frac{J_e}{P_t} + (n+1)\frac{J_p}{P_t} \tag{36}$$

where $C''(\theta)$ is the second derivative of the function $C(\theta)$ with respect to θ, $P_0'(\theta)$ is the derivative of the limit-load $P_0(\theta)$ with respect to θ, and $P_0''(\theta)$ is the second derivative of the limit-load $P_0(\theta)$ with respect to θ.

The compliance of the cracked pipe when the crack angle θ is fixed is

$$\left(\frac{\partial\Delta}{\partial P_t}\right)_\theta = C_{nc} + C(\theta) + \frac{n}{P_t}\left(\frac{P_t}{c(n)P_0(\theta)}\right)^n \tag{37}$$

Combining Eqs 31, 33, 35, 36, and 37 gives $(\partial J/\partial a)_{\Delta_T}$ as

$$\left(\frac{\partial J}{\partial a}\right)_{\Delta_T} = \frac{J_e}{R}\frac{C''(\theta)}{C'(\theta)} + \frac{J_p}{R}\left(\frac{P_0''(\theta)}{P_0'(\theta)} - (n+1)\frac{P_0'(\theta)}{P_0(\theta)}\right)$$

$$- 2t\left(\frac{2J_e}{P_t} + \frac{(n+1)J_p}{P_t}\right)^2\left(C_s + C_{nc} + C(\theta) + \frac{n}{P_t}\left(\frac{P_t}{c(n)P_0(\theta)}\right)^n\right)^{-1} \tag{38}$$

Equation 38 can be rewritten as

$$\left(\frac{\partial J}{\partial a}\right)_{\Delta_T} = \frac{J_e}{R}\frac{C''(\theta)}{C'(\theta)} + \frac{J_p}{R}\left(\left(\frac{F''(\theta)}{F'(\theta)}\right) - (n+1)\frac{F'(\theta)}{F(\theta)}\right)$$

$$-\frac{2t}{P_t^2}(2J_e + (n+1)J_p)^2\left(C_s + C_{nc} + C(\theta) + \frac{n}{P_t}\left(\frac{P_t}{c(n)P_0(\theta)}\right)^n\right)^{-1} \quad (39)$$

If the elastic contributions except Δ_{nc} are neglected, Eq 39 becomes

$$\left(\frac{\partial J}{\partial a}\right)_{\Delta_T} = \frac{J_p}{R}\frac{F''(\theta)}{F'(\theta)} + \frac{P_t^2}{2R^2t}\left(\frac{F'(\theta)}{F(\theta)}\right)^2$$

$$\times\left(\frac{C_s + C_{nc}}{(C_s + C_{nc})\dfrac{P_t}{n}\left(\dfrac{C(n)P_0(\theta)}{P_t}\right)^n + 1}\right) \quad (40)$$

Equation 40 can also be written as

$$\left(\frac{\partial J}{\partial a}\right)_{\Delta_T} = \frac{J_p}{R}\frac{F''(\theta)}{F'(\theta)} + \frac{P_t^2}{2R^2t}\left(\frac{F'(\theta)}{F(\theta)}\right)^2$$

$$\times\left(\frac{C_s + C_{nc}}{(C_s + C_{nc})\left(\dfrac{\partial P_t}{\partial \Delta_{cp}}\right)_\theta + 1}\right) \quad (41)$$

Aside from the missing factor of 2, Eq 41 is consistent with that of Ref 5 in which the equation for $(\partial J/\partial a)_{\Delta_T}$ is obtained from an assumption of a more general functional form of $P_t = F(\theta)q(\Delta_{cp})$.

When $n \to \infty$ for the nonhardening case, Eq 26 becomes

$$J_p = -\frac{8\sigma_0 R}{Z - L}F'(\theta)\Delta_{cp} \quad (42)$$

Equation 39 becomes

$$\left(\frac{\partial J}{\partial a}\right)_{\Delta_T} = \frac{J_e}{R}\frac{C''(\theta)}{C'(\theta)} + \frac{J_p}{R}\frac{F''(\theta)}{F'(\theta)} + \frac{4J_e}{R}\frac{F'(\theta)}{F(\theta)}$$

$$+ \frac{P_0^2}{2R^2t}\left(\frac{F'(\theta)}{F(\theta)}\right)^2(C_s + C_{nc} + C(\theta)) \quad (43)$$

and Eq 40 becomes

$$\left(\frac{\partial J}{\partial a}\right)_{\Delta_T} = \frac{J_p}{R}\frac{F''(\theta)}{F'(\theta)} + \frac{P_0^2}{2R^2t}\left(\frac{F'(\theta)}{F(\theta)}\right)^2(C_s + C_{nc}) \tag{44}$$

Estimation of the J-Resistance Curve

The *J*-estimation scheme for a cracked body with an extending crack was proposed by Zahoor and Kanninen [5]. Their estimation scheme is summarized here. From Eqs 14, 16, and 23, *J* can be written as

$$J = J_e + J_p \tag{45}$$

where

$$J_e = \frac{1}{2Rt}\frac{\partial}{\partial\theta}\int_0^{P_t}\Delta_{ce}\,dP_t \tag{46}$$

and

$$J_p = \frac{1}{2Rt}\frac{\partial}{\partial\theta}\int_0^{P_t}\Delta_{cp}\,dP_t \tag{47}$$

With the assumption of

$$P_t = F(\theta)q(\Delta_{cp}) \tag{48}$$

or

$$\Delta_{cp} = q^{-1}\left(\frac{P_t}{F(\theta)}\right) \tag{49}$$

J_p can be written as

$$J_p = -\frac{F'(\theta)}{2Rt}\int_0^{\Delta_{cp}}q(\Delta_{cp})\,d\Delta_{cp} \tag{50}$$

The increment of J_p is

$$dJ_p = \left(\frac{\partial J_p}{\partial\theta}\right)_{\Delta_{cp}}d\theta + \left(\frac{\partial J_p}{\partial\Delta_{cp}}\right)_\theta d\Delta_{cp} \tag{51}$$

Therefore J_p for a pipe with an extending crack is

$$J_p = \int_0^{\Delta_{cp}}\left[\frac{-1}{2Rt}\frac{F'(\theta)}{F(\theta)}\right]P_t\,d\Delta_{cp} + \int_{\theta_0}^\theta\left[\frac{F''(\theta)}{F'(\theta)}\right]J_p\,d\theta \tag{52}$$

Equation 52 is consistent with Eq 53 of Ref 5, aside from a factor of 2 inadvertently omitted in that work. The elastic J_e can be evaluated via Eq 25. The power law relation of Eqs 20 and 21 is consistent with the assumption of the functional form of Eq 48 or 49. Therefore Eq 52 is adopted here for the estimation of J_p from the experimental load-displacement curve.

Experiments on Type 304 Stainless Steel Pipe

A number of experiments on 102-mm (4-in.)-diameter through-wall circumferentially cracked pipes subjected to four-point bending were conducted at Battelle [5,12,17]. Two of the experiments (Experiments 1T and 2T) were conducted under displacement-controlled conditions so that a large amount of stable crack growth was achieved. Experiment 3T was conducted with a compliant spring to obtain an instability point that could be compared with the tearing instability prediction. The initial crack lengths for Experiments 1T, 2T, and 3T were 0.229, 0.371, and 0.29 of the pipe circumference, respectively.

The load-displacement curves for Experiments 1T and 2T are shown in Figs. 5 and 6. Note that the crack initiation points marked in Figs. 5 and 6 for Experiments 1T and 2T were detected visually. However, the electric potential method was used to determine crack initiation and crack growth for Ex-

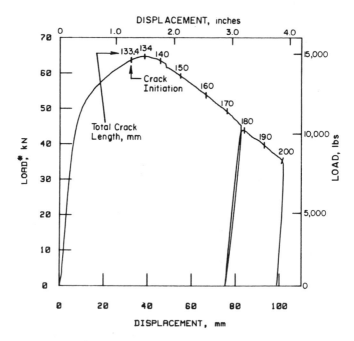

FIG. 5—*Bending load versus load-point displacement from Experiment 1T.*

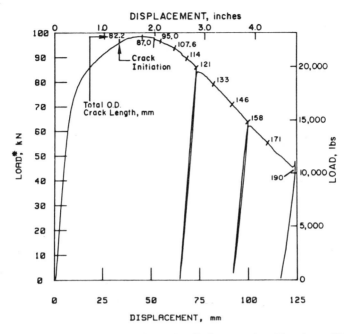

FIG. 6—*Bending load versus load-point displacement from Experiment 2T.*

periment 3T. The load-displacement curve for Experiment 3T is shown in Fig. 7. Note that a part of spring displacement is not shown.

Analysis of the experimental data by simple beam bending theory shows that the outer fiber stress of a pipe with no crack is approximately equal to yield stress when the crack initiates. Therefore the elastic relation between P_t and Δ_{nc} as defined in Eq 15 can be used. According to the simple beam theory,

$$C_{nc} = \frac{(Z - L)^2(Z + 2L)}{48\,EI} \tag{53}$$

where E is Young's modulus and I is the appropriate moment of inertia of the pipe. The quantity Δ_c is then easily obtained by subtracting the contribution of Δ_{nc} from the value of Δ obtained from the experimental data. The elastic Δ_{ce} is calculated from Eq 17 with a $C(\theta)$ derived from experiments [17]. The plastic Δ_{cp} can then be readily obtained. Finally, the J-resistance curves of the cracked pipes in Experiments 1T and 2T are easily obtained via Eqs 25 and 52. The results are shown in Fig. 8.

The J-resistance curves for a three-point bend bar and a center-cracked panel of Type 304 stainless steel are also shown for comparison in Fig. 8 [17]. The geometric dependence of these curves is obvious. It must be kept in mind, however, that the J-resistance curves for pipes shown are inferred from

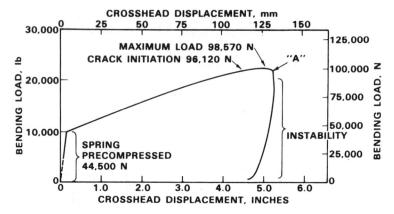

FIG. 7—*Bending load versus displacement of spring and pipe from Experiment 3T.*

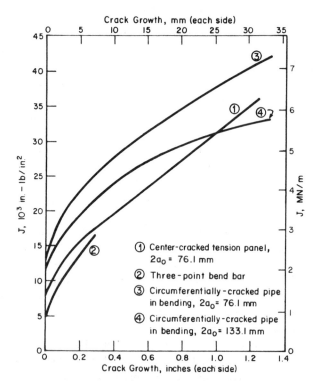

FIG. 8—*Comparison of J-resistance curves.*

a simple estimation scheme which may not be accurate. The J-resistance curve of the three-point bend bar is, as expected, the lowest.

Load-Displacement Curve of the Pipe Test

Rhee used a finite element model to simulate the load-displacement relation of Experiment 3T.[4] In his model, a piece-wise linear representation of the tensile stress-strain curve for Type 304 stainless steel was used as input to the finite element code. The J_2 flow theory of plasticity was employed to model the plastic flow near the crack tip. The results are shown in Fig. 9.

Figure 9 plots the load P_t as a function of Δ_c which was obtained by subtracting Δ_{nc} from Δ. The quantity Δ_{nc} was obtained by using Eqs 15 and 53. The corresponding experimental results are also shown for comparison. It is seen that the finite element results overestimate the stiffness of the cracked pipe at small displacements; however, they agree well at larger displacements. (In the figure, the point at which the crack initiation is marked for Experiment 3T was determined by the electric potential method [17].) Initial investigations [17] showed that the elastic contributions such as Δ_{ce} and J_e are small compared with their plastic counterparts at and after crack initiation. Hence, for a simple estimation, Δ_{ce} can be lumped into Δ_{cp}. A power law $P_t = K\Delta_c(1/6.45)$ where K is a constant was used to simulate the load-displacement relation up to the displacement at crack initiation.

In Experiment 3T, Z was set to 1.346 m (53 in.) as opposed to 1.524 m (60 in.) for Experiments 1T and 2T. In order to test the validity of the proposed relations, Eqs 20 and 21, the load-displacement must be compared on the basis of the same test arrangement (except the crack length). Because a pure bending moment provides the dominant stresses near the crack plane, the load is scaled up by a factor of $(1.524 - L)/(1.346 - L)$ and the displacement is consequently scaled down by the same factor for the experimental data of Experiments 1T and 2T. The results are plotted in Fig. 10.

Power law relations fitted for these curves with a common $n = 6.45$ are also plotted in Fig. 10. It can be seen that that power law relations fit very well in the large displacement range where the crack is about to initiate. However, the power law relation overestimates the stiffness of the cracked pipe considerably in the small displacements range. A more detailed analysis in which Δ_c can be decomposed into a Δ_{ce} and a Δ_{cp} will result in a better fit for the power law relation between Δ_{cp} and P_t. Note that the crack initiation points marked for Experiments 1T and 2T were determined visually.

The θ-dependent coefficients of the power law relation are normalized and plotted against $F(\theta)$ in Fig. 11. The results show that the behavior of the cracked pipe can be well-represented by the power law relation (Eqs 20 and 21) in the deformation range where the cracks initiate.

---

[4]Rhee, C. M., Production Research Division, Conoco Inc., Ponca City, Okla., unpublished results.

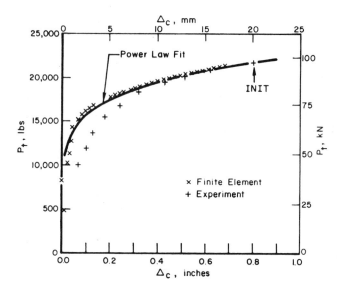

FIG. 9—*Bending load as a function of the displacement due to crack for Experiment 3T.*

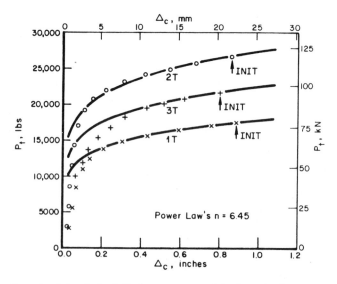

FIG. 10—*Power law fit for the load-displacement curve of the three pipe fracture experiments.*

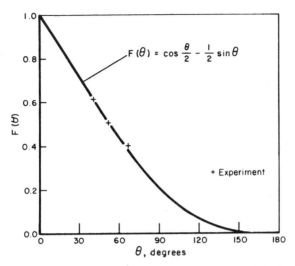

FIG. 11—*Comparison of the power law fit for experimental data with the limit-load function* $F(\theta)$.

Tearing Instability Predictions for a Circumferentially Cracked Pipe Using the Estimation Scheme

Determination of the J-Resistance Curve for Experiment 3T

A significant geometric dependence of the J-resistance curve was shown in Fig. 8. Some of the differences in the J-resistance curves may be due to the simplicity of the J-estimation formula (Eq 52). Since the initial crack length of the cracked pipe in Experiment 3T is between those in Experiments 1T and 2T, a natural choice for the J-resistance curve for Experiment 3T would be one between Experiments 1T and 2T. However, the J_{Ic} value for Experiment 3T is not between the values for Experiments 1T and 2T. (The J_{Ic} values at initiation are easily calculated via Eq 52 as 1.984 MN/m (11 330 in.-lbf/in.^{2}), 2.235 MN/m (12 760 in.-lbf/in.^{2}), and 1.814 MN/m (10 360 in.-lbf/in.^{2}) for Experiments 1T, 2T, and 3T, respectively.) Note that in Experiments 1T and 2T crack initiation was determined by visual observation and by the electric potential energy method in Experiment 3T. Differences in the J_{Ic} values obtained suggest that the pipe material's properties may differ somewhat, or that the J estimation scheme (Eq 52) is not accurate enough, or that the electric potential method may detect the crack initiation earlier. An average value of the J-resistance curve of Experiments 1T and 2T with $J_{Ic} = 1.814$ MN/m (10 360 in.-lbf/in.^{2}) is used as the J-resistance curve of Experiment 3T (Fig. 12a) to serve as the basis for the tearing instability predictions that follow.

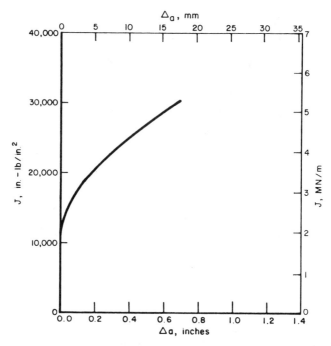

FIG. 12a—*Assumed J-resistance curve for the instability predictions.*

Prediction of the Power Law Approach

The engineering estimation scheme proposed in this paper can be used to predict tearing instability in Experiment 3T. Earlier investigations [*17*] show that the elastic J_e is small. Therefore J_e is neglected and Δ_{ce} is lumped into the plastic Δ_{cp} in this analysis.

With the assumed J_{Ic} value, Eq 26 can be used to calculate the values of P_t at crack initiation with the initial crack length. The calculated P_t is 96 793 N (21 700 lbf); Δ_T is obtained from Eqs 28 and 12 to 15 as 178 mm (7.03 in.). The experimental value of P_t is 96 081 N (21 600 lbf) and Δ_T is 179 mm (7.04 in.). Hence these predictions agree well with the experimental values.

Then, Eq 40 is used to calculate $(\partial J/\partial a)_{\Delta_T}$ to compare with dJ_R/da. A small amount of Δa can be assumed to obtain the $J_R(\Delta a)$ value that is equal to the current applied J value. Next, Eq 26 is used to calculate the value of P_t, and Eqs 28 and 12 to 15 used to calculate the value of Δ_T. Equation 40 is then used to calculate $(\partial J/\partial a)_{\Delta_T}$ to compare with dJ_R/da. The foregoing procedure is repeated to reach some value of J.

The J-(dJ/da) diagram for Experiment 3T is plotted in Fig. 12b. In the figure, the vertical axis denotes J values with several corresponding values of Δa as marked. The solid line represents the material curve, and the dot-dash

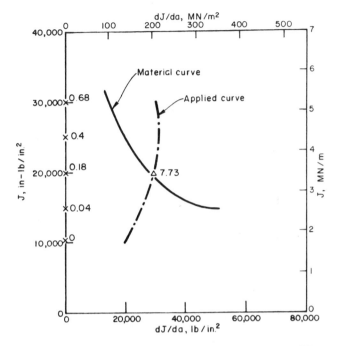

FIG. 12b—*Prediction of tearing instability for Experiment 3T.*

line represents the applied curve calculated from the estimation scheme. The J value at the intersection of the material curve and the applied curve is approximately 3.503 MN/m (20 000 in.-lbf/in.^{2}); it corresponds to a crack growth of 4.572 mm (0.18 in.) and a maximum total displacement of 196 mm (7.73 in.).

Figure 12c plots J as a function of Δ_T. This figure shows vividly that the tearing instability point is at the critical $\Delta_T = 196$ mm (7.73 in.) beyond which no solution exists for this rate-independent deformation plasticity approach. The predictions of $\Delta_T^{cr} = 196$ mm (7.73 in.) compare well with the experimental value of 203 mm (8.0 in.); however, the prediction of crack growth at instability is 4.572 mm (0.18 in.) as compared to a much higher value of 16 to 19 mm (0.63 to 0.75 in.) deduced from the electric potential method. The discrepancy may be due to inaccuracies in modelling of the load-displacement curve and the J-estimation scheme, variability of material properties, and inherent deficiency of the rate-independent deformation plasticity.

Note that the estimation scheme predicts a maximum load $P_t = 99\ 729$ N (22 420 lbf) which compares well with the experimental value of 99 564 N (22 383 lbf). The displacement at the maximum load, $\Delta_T = 192$ mm (7.56 in.), also compares well with the experimental value of 196 mm (7.73 in.). But, the predicted crack growth of 1.02 mm (0.04 in.) at the maximum load is

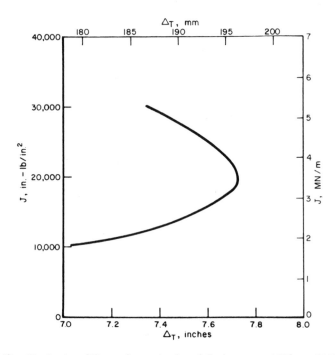

FIG. 12c—*Tearing instability at the maximal total displacement of 196 mm (7.73 in.).*

much smaller than the experimental value of 5.79 mm (0.228 in.). Since the crack growth at the maximum load determined visually in Experiment 1T is 0.51 mm (0.02 in.) and 4.8 mm (0.19 in.) in Experiment 2T, one would expect that the crack growth at the maximum load in Experiment 3T should be within this range. However, the electric potential method gives a higher value of 5.79 mm (0.228 in.) of growth for Experiment 3T. The electric potential method includes any tunneling effects, since it gives the average crack length over the thickness direction. This could explain the fact that the crack growth determined by the electric potential method is larger. Regardless, the predicted value of crack extension at instability is too small compared with the experimental data.

Prediction of the Limit-Load Approach

In the limit-load approach, Eqs 42 and 44 are used instead of Equations 26, 28, and 40 to calculate J and $(\partial J/\partial a)_{\Delta_T}$. The applied J-curve with σ_0 regarded as the flow stress, 503 MPa (73 ksi), is plotted as Curve 2 in Fig. 12*d*. (Note that Curve 1 represents the results for the power hardening approach.) The flow stress is the experimental value of the maximum net section stress of the central notched plate for 304 stainless steel [*17*]. The prediction of insta-

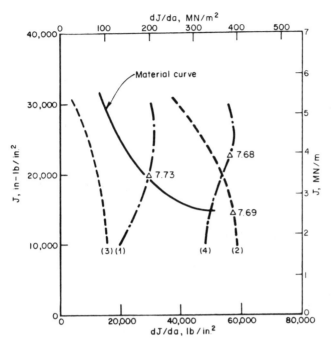

FIG. 12d—*Comparison of prediction schemes.* (1) *Power hardening.* (2) *Limit-load* ($\sigma_0 = 503$ *MPa [73 ksi]).* (3) *Limit-load* ($\sigma_0 = 268$ *MPa [38.92 ksi]).* (4) *Zahoor and Kanninen [17].*

bility is at $\Delta_T^{cr} = 195$ mm (7.69 in.) with $\Delta a^{cr} = 1.02$ mm (0.04 in.). The limit-load approach apparently underestimates crack extension at instability. It predicts the instability shortly after the crack initiation. However, this analysis gives a good estimation of the total displacement at instability. Note that this approach predicts that the maximum load occurs at crack initiation. After crack initiation, the load drops.

The applied J-(dJ/da) curve for σ_0, regarded as the yield stress of a tension test, 268 MPa (38.9 ksi), is plotted as Curve 3 in Fig. 12d. This curve shows that the pipe seems to be "fracture proof" because the material curve and applied curve have no tendency to intersect each other. This, of course, is contradicted by the experimental result.

A parametric study shows that the applied J-(dJ/da) curve will be tangential to the material curve for $\sigma_0 = 358$ MPa (52.0 ksi). This is a case where the cracked body is still stable even though $dJ_R/da = (\partial J/\partial a)_{\Delta_T}$ in the deformation history (Fig. 3). Therefore the limit-load approach will predict the system is "fracture proof" if σ_0 is chosen equal to or less than 358 MPa (52.0 ksi), and will predict instability if σ_0 is chosen between 358 MPa (52.0 ksi) and 503 MPa (73.0 ksi). Higher values of J and Δa and a lower Δ_T value at instability correspond to a lower value of σ_0 chosen between 358 MPa (52.0 ksi) and 503 MPa

(73.0 ksi). The reason for a lower Δ_T value is that a lower elastic displacement results from the lower load-carrying capacity at a lower assumed σ_0.

Clearly, the limit-load approach to predict tearing instability should be made with caution. In particular, experimental data are needed to guide the choice of σ_0. This choice may depend on which of the quantities, J, Δa, or Δ_T, is to be best approximated.

Zahoor and Kanninen [5] developed an approach which includes the material hardening effects.[5] In the prediction of instability for Experiment 3T, however, they had to neglect the term $(\partial P_t / \partial \Delta_{cp})_\theta$ in Eq 41. Instead, they calculated $(\partial J/\partial a)_{\Delta_T}$ via Eq 44 by inserting the experimental values of the total load for P_0 as a function of the amount of the stable crack growth inferred from the electric potential method. The results are plotted as Curve 4 in Fig. 12d. It is seen that J at the intersection of the applied J-$(\partial J/\partial a)_{\Delta_T}$ curve and the material curve is 2.626 MN/m (15 000 in.-lbf/in.²), whereas the maximum displacement of the system is reached at $J = 3.940$ MN/m (22 500 in.-lbf/in.²).

Note that Δ_T is calculated via Eqs 42 and 12 to 15. Tearing instability occurs at the maximum total displacement for a system under a monotonic increasing displacement condition. The intersection of the applied J-(dJ/da) curve and the material curve determines the point of the maximum displacement. Apparently, this is not the case for Curve 4; the maximum total displacement happens at a higher value of J, rather than at the J value of the intersection of the applied J-(dJ/da) curve and the material curve. Comparison of Curve 1 and Curve 4 shows that an overestimation of $(\partial J/\partial a)_{\Delta_T}$ for Curve 4 is due to the fact that $(C_s + C_{nc})(\partial P_t/\partial \Delta_{cp})_\theta$ in Eq 41 is not negligible. Therefore the prediction of tearing instability according to the method used in Ref 5 can be incorrect for a hardening material like Type 304 stainless steel under compliant loading. Table 1 summarizes the comparison of the experimental data and the predictions of the various approaches considered here.

Concluding Remarks

A J and $(\partial J/\partial a)_{\Delta_T}$ estimation scheme for a circumferentially cracked pipe subjected to four-point bending, with material hardening effects included, was developed. The tearing instability predictions compared well with experi-

---

[5]Note that the J-(dJ/da) diagram in Ref 17 has a factor of 2 error in J and an approximate factor of 4 error in $(\partial J/\partial a)_{\Delta_T}$. The reason is that, for Experiment 3T, the contribution of the first term on the right-hand side of Eq 41 is insignificant compared with the second term on the right-hand side of Eq 41. In Ref 17 the term $(C_s + C_{nc})(\partial P_t/\partial \Delta_{cp})_\theta$ is neglected. Therefore $(\partial J/\partial a)_{\Delta_T}$ is underestimated approximately by a factor of 4 due to the factor of 2 error in P_t since Eq 41 contains P_t^2.

The estimated value of $(C_s + C_{nc})(\partial P_t/\partial \Delta_{cp})_\theta$ is approximately equal to 1 for Experiment 3T. This is why the correct value of $(\partial J/\partial a)_{\Delta_T}$ using the method of Ref 17 is approximately twice as much as that of the strain-hardening analysis reported here.

TABLE 1—Comparison of predictions of various approaches and experimental data.[a]

	Maximum Load, lbf	Crack Extension at Maximum Load, in.	Total Displacement at Instability, in.	Crack Extension at Instability, in.
Experiment	22 383	0.228[b]	8.0	0.63 to 0.75[b]
Power hardening approach (this paper) (Curve 1 in Fig. 12d)	22 420	0.04	7.73	0.18
Limit-load approach ($\sigma_0 = 73$ ksi) (Curve 2 in Fig. 12d)	23 926	at initiation	7.69	0.04
Limit-load approach ($\sigma_0 = 38.92$ ksi) (Curve 3 in Fig. 12d)	12 756	at initiation	no instability prediction	no instability prediction
Hardening approach [5,17] (Curve 4 in Fig. 12d)	not predicted[c]	not predicted[c]	7.68[d]	0.28[d]

[a] 1 in. = 0.0254 m, 1 ksi = 6.89 MPa, 1 lbf = 4.45 N.
[b] Deduced from electric potential method.
[c] Experimental data used as input for tearing instability predictions.
[d] Instability does not occur at the intersection of the material curve and the applied curve in the J-(dJ/da) plot.

mental data. However, a more refined analysis is still needed to gain further insight into the final fracture of the cracked pipe. It was also demonstrated that caution is needed in making instability predictions using the limit-load approach, since erroneous conclusions might be made without a judicious choice of "flow stress."

As shown in Fig. 8, the J-resistance curves inferred from the experimental data show a considerable geometric dependence. However, the prediction of tearing instability by using a lower bound J-resistance curve, such as that deduced from a three-point bend bar, will be at a lower total displacement. This suggests that using a lower-bound-resistance curve will result in a conservative estimation of the tearing instability point.

Acknowledgments

This research was supported by the U.S. Nuclear Regulatory Commission, Metallurgy and Materials Research Branch, under Contract NRC-04-81-178. The authors would like to express their appreciation to Jack Strosnider and Milton Vagins of the NRC for their encouragement of this work. The experimental pipe fracture data cited in this paper were developed by G. M. Wilkowski.

References

[1] Kanninen, M. F., Popelar, C. H., and Broek, D., *Nuclear Engineering and Design*, Vol. 67, 1981, pp. 27–55.

[2] Paris, P. C. and Johnson, R. E. in *Elastic-Plastic Fracture: Second Symposium, Volume II—Fracture Resistance Curves and Engineering Applications, ASTM STP 803*, C. F. Shih and J. P. Gudas, Eds., American Society for Testing and Materials, 1983, pp. II-5–II-40.

[3] Paris, P. C., Tada, H., Zahoor, A., and Ernst, H. in *Elastic-Plastic Fracture, ASTM STP 668*, J. D. Landes, J. A. Begley, and G. A. Clarke, Eds., American Society for Testing and Materials, 1979, pp. 5–36.

[4] Tada, H., Paris, P. C., and Gamble, R. M. in *Fracture Mechanics: Twelfth Conference, ASTM STP 700*, American Society for Testing and Materials, 1980, pp. 296–313.

[5] Zahoor, A. and Kanninen, M. F., *Journal of Pressure Vessel Technology*, Vol. 103, 1981, pp. 352–358.

[6] Zahoor, A. and Kanninen, M. F. in *Elastic-Plastic Fracture: Second Symposium, Volume II—Fracture Resistance Curves and Engineering Applications, ASTM STP 803*, C. F. Shih and J. P. Gudas, Eds., American Society for Testing and Materials, 1983. pp. II-291–II-308.

[7] Popelar, C. H., Pan, J., and Kanninen, M. F., this publication, pp. 699–720.

[8] Rice, J. R., "Mathematical Analysis in the Mechanics of Fracture," in *Fracture: An Advanced Treatise*, H. Liebowitz, Ed., Vol. 2, Academic Press, New York, 1968, pp. 191–311.

[9] Hutchinson, J. W. and Paris, P. C. in *Elastic-Plastic Fracture, ASTM STP 668*, J. D. Landes, J. A. Begley, and G. A. Clarke, Eds., American Society for Testing and Materials, 1979, pp. 37–64.

[10] Rice, J. R., Drugan, W. J., and Sham, T. L. in *Fracture Mechanics: Twelfth Conference, ASTM STP 700*, American Society for Testing and Materials, 1980, pp. 189–221.

[11] Paris, P. C., "Tearing Instability Analysis and Applications," presented at Proceedings of U.S.-Japan Cooperative Seminar on Fracture Tolerance Evaluation, Honolulu, 7–11 Dec. 1981.

[12] Wilkowski, G. M., Zahoor, A., and Kanninen, M. F., *Journal of Pressure Vessel Technology*, Vol. 103, 1981, pp. 359–365.

[*13*] Shih, C. F. in *Mechanics of Crack Growth, ASTM STP 590*, American Society for Testing and Materials, 1976, pp. 3–26.
[*14*] Shih, C. F. and Hutchinson, J. W., "Fully Plastic Solutions and Large Scale Yielding Estimates for Plane Stress Crack Problems," *Journal of Engineering Materials and Technology, Transactions of ASME*, 1976, pp. 289–295.
[*15*] Merkle, J. G. and Corten, H. T., "A J-Integral Analysis for the Compact Specimen, Considering Axial Force as Well as Bending Effects," *Transactions of ASME*, 1974, pp. 286–292.
[*16*] Kanninen, M. F. et al, "The Development of a Plan for the Assessment of Degraded Nuclear Piping by Experimentation and Tearing Instability Fracture Mechanics Analysis," NRC-04-81-178, Final Report, Battelle-Columbus Laboratories, Columbus, Ohio, Sept. 1982.
[*17*] Kanninen, M. F. et al, "Instability Predictions for Circumferentially Cracked Type 304 Stainless Steel Pipe under Dynamic Loading—Vols. I and II," EPRI NP-2347, Project T118-2, Final Report, Electric Power Research Institute, Palo Alto, Calif., April 1982.

Summary

Summary

The National Symposium on Fracture Mechanics is intended as a forum for the exchange of ideas related to the fracture of engineering materials. Topics presented at the Fifteenth Symposium included theoretical and numerical analysis of cracks in finite geometries, dynamic fracture and crack arrest, fatigue behavior, fracture at elevated temperature, and elasto-plastic fracture mechanics. As an aid to the reader the papers have been organized into four sections based on the main theme of each paper. No classification can be absolute, however, and topics within the field of interest to the reader may be found throughout the volume. The reader is encouraged to consult the Index for the location of topics of interest.

Linear Elastic Fracture Mechanics

The theme of the first section is linear elastic fracture mechanics (LEFM). Papers on the theoretical, numerical, and experimental aspects of this classical approach to the mechanics of fracture are presented. The first five papers address the topic of determining the stress intensity factor in geometries of practical interest. In a combined numerical and experimental study *Grandt et al* investigate the transition of part-through cracks at holes into through-the-thickness cracks under fatigue loading. They conclude that the crack growth rate (and stress intensity factor) varies along the flaw boundary so as to encourage the trailing edge to "catch up" with the leading edge until a uniform through-thickness condition is reached. *Saff and Sanger* take an alternative approach to the same problem and develop an approximate analytical procedure for determining the variation in stress intensity factor along the flaw border. They represent the geometry by a set of two-dimensional slices and impose displacement continuity conditions between slices to obtain a solution to the three-dimensional problem. The analysis of cracks in thin sheets (skins) with variable thickness (lands) is of particular interest for aircraft structures. *Ratwani and Kan* provide a solution to this class of problem in the form of an integral equation, using Fourier transform techniques, which they solve numerically to obtain the stress intensity factor. Experimental fatigue crack growth results agree well with the crack growth behavior predicted by the analytical model. *Parker and Andrasic* use a modified mapping collocation numerical technique to determine the weight functions for cylindrical specimens with either one or two radial cracks. Extensive graphical results for

749

the stress intensity factor are presented for a range of diameter ratios, crack locations, and crack lengths. The last paper in this set, by *Kathiresan et al*, uses finite element methods to study the stress state and fracture potential in tapered attachment lugs subjected to off-axis loading.

In the next paper in this section, *Newman* describes the results of a series of numerical experiments using the finite element method to predict failure loads under elasto-plastic conditions. Using a crack growth criterion based on the critical crack-tip-opening displacement (CTOD) or, equivalently, the crack-tip-opening angle (CTOA) hypothesis, he presents a comparison of his numerical predictions with experimental results for three specimen geometries, including a specially designed three-hole-crack tension specimen. Results on two aluminum alloys and 304 stainless steel indicate that a single critical CTOD (or CTOA) value, obtained from compact tension specimens, can be used to provide reasonable predictions of crack initiation, stable crack growth, and instability in other geometric shapes.

In contrast to the variety of numerical/analytical techniques presented in earlier papers in the section, the next two papers utilize photoelastic techniques to study stress intensity factor variations under widely differing conditions. *Smith and Kirby*, in a continuation of the work presented at the Fourteenth National Symposium,[1] use frozen stress photoelastic methods to determine the stress intensity factor and width correction factors for flat plates with elliptical cracks in tension. Except for very deep flaws, the experimental results agree favorably with the numerical results of Newman and Raju, provided the influence of the high Poisson's ratio in the experimental models is taken into account. At the other end of the spectrum, *Ramulu et al* re-examine previous dynamic photoelastic results in the light of a new dynamic crack branching criterion. They conclude that the necessary and sufficient conditions for crack branching are the attainment of a critical branching stress intensity factor, K_{Ib}, and a critical radial distance, r_c, which is an additional material property. The proposed model predicted the crack branching angle in the SEN and wedge-loaded RDCB specimens studied.

The next paper in this section summarizes the activities of researchers at Battelle-Columbus Laboratories on crack-arrest toughness testing as part of an ASTM task group on this topic. The results, reported by *Rosenfield et al*, provide information for the establishment of acceptable size requirements. An approach for handling the problem of combined thermomechanical loading of ductile pressure vessels within the framework of fracture mechanics is presented by *Bloom and Malik* in the final paper in this section. The proposed procedure is based on a failure assessment curve which accounts for the ther-

---

[1] Smith, C. W. and Kirby, G. C., "Stress-Intensity Distributions for Natural Cracks Approaching Benchmark Crack Depths in Remote Uniform Tension," *Fracture Mechanics: Fourteenth Symposium—Volume I: Theory and Analysis, ASTM STP 791*, J. C. Lewis and G. Sines, Eds., American Society for Testing and Materials, 1983, pp. I-269–I-280.

mal plus mechanical stresses in terms of the elastic stress intensity factor with a plastically corrected crack length.

Fatigue Crack Growth

The second section of this volume collects together those papers that address the topic of high cycle fatigue. *Sahli and Albrecht* examine the fatigue life of weldments in bridge girders, particularly emphasizing fillet-welded transverse stiffeners. Predicted fatigue lives using average crack growth rates and simplified stress intensity factors for these complex geometries compare favorably with the results of fatigue experiments on plates with welded transverse stiffeners typical of those used in bridge construction. *Schwartz et al* propose a fatigue crack propagation model that uses a hyperbolic sine function with four unknown coefficients. Fitting the model to fatigue data from Incoloy 901, the authors conclude that one of the coefficients is a material constant and the other three are continuous functions of frequency, temperature, and stress ratio. Because of this continuity, the constants can be interpolated to conditions for which no data exist. The fatigue crack growth behavior of four 7XXX aluminum alloys subjected to single cycle overloads every 8000 cycles is described by *Bretz et al*. Utilizing fatigue data and fractographic examination, the authors identify two mechanisms, one mechanical and the other metallurgical, affecting the retardation process. The next paper in this section is by *Lang et al*. They investigate the effect of specimen configuration and frequency on the fatigue behavior of a nonmetallic material, Nylon 66. The differences in propagation rates are attributed to hysteretic heat generation which was found to be frequency and specimen shape dependent.

The preceding papers in the section on fatigue crack growth were directed mainly towards characterizing the behavior in specific materials. The remaining two papers in this section are more abstract and propose models to explain general behavior observed in fatigue. *Socie et al* propose a simple model for predicting fatigue life of notched members. The fatigue life cycle is considered as the sum of the initiation portion which is controlled by notch plasticity and the crack growth phase governed by the applied nominal stress. The final paper in this section, by *Jolles and Tortoriello*, examines the stress field triaxiality along the border of a surface flaw in relation to crack closure at low stress ratios. If the stress ratio is high enough, crack closure does not occur. Experimental results on aluminum plates with surface cracks at varying stress ratios agree well with predictions.

Material Influences on Fracture

The theme of the third section is material influences on fracture. The papers in this section describe the fracture behavior of specific materials in response to external variables or microstructural changes. In the first five papers the external variable is elevated temperature. *Swaminathan and Landes* present the results of elastic and elasto-plastic fracture tests on three advanced melting processed steel forgings for rotor applications. When compared with conventionally processed forgings, the advanced technology forgings demonstrated improved fracture toughness by a factor of two or more. *Garwood's* results on Type 316 stainless steel at elevated temperature show a clear orientation dependence on CTOD for ductile tearing. Using a HRR crack tip stress field representation, *Huang and Pelloux* develop a theoretical model for creep crack growth in a powder-metallurgy nickel-based superalloy. The next two papers examine the behavior in the transition temperature regime. *Landes and Mc-Cabe* show that the temperature associated with J_c instability is dominated by specimen thickness, with additional effects related to ligament length. The University of Maryland researchers, *Ogawa et al*, directed their attention to fracture morphology in order to gain an understanding of cleavage behavior in the transition range. Strong influences of ferrite and prior-austenite grain size were revealed by proper etching. In addition, an explanation for the large scatter in K_{Ic}, typical of tests in this regime, due to inhomogenity in dendritic solidification is proposed.

The influence of nonmetallic inclusions on a variety of fracture properties is described by *Wilson*. He compared test results on conventional and calcium-treated A588 Grade A steels and showed that control of inclusions by calcium treatment has beneficial effects in most orientations. *Landes and Leax* examined the influence of load history on the elasto-plastic fracture toughness of 4340 steel. In the preliminary results reported in this paper, monotonic and cyclic bulk prestraining appear to have opposite effects on measured R-curve toughness. In the next paper, *Leis* describes an anomalous fatigue behavior at a notch root in a 1080 pearlitic steel. *Underwood and Leger* present the results of a correlation study of fracture toughness and Charpy energy or reduction-in-area for A723 steel. The remaining two papers in this section, by *Mills* and *Saxena et al*, examine the effects of fast-neutron irradiation on superalloy A-286 and high temperature creep crack growth in A470 Class 8 steel, respectively.

Elasto-Plastic Fracture Mechanics

In recent years there has been considerable emphasis on the extension of the principles of fracture mechanics to materials undergoing large-scale plastic deformation. This branch of mechanics, referred to collectively as elasto-plastic fracture mechanics (EPFM), is not as well defined as LEFM, and the current research thrust is to further define the underlying principles and procedures governing elasto-plastic fracture. The papers in the last section of this volume continue the progress reported in previous volumes of this series.

The first two papers, by *Vassilaros and Hackett* and *Wilkowski et al*, address the problem of monitoring crack initiation and stable ductile crack growth under large-scale deformation conditions. Both papers demonstrate that, if suitable precautions are taken, the d-c potential drop method can be used as a crack length meter in automated data acquisition systems. In the next paper, *Hellmann* and *Schwalbe* compare *R*-curve results from center-cracked tension specimens, compact specimens, and bend specimens, and conclude that center-cracked tension specimens have longer measurement capacity than either of the bending configurations. Continuing on the topic of test conditions for valid fracture toughness measurements are two papers by researchers at IRSID (St. Germain en Laye, France). In the first paper, *De Roo et al* probe the effect of specimen dimensions and reach conclusions similar to those by Landes and McCabe. *Marandet et al*, in the second paper, describe the influence of loading rate on the transition behavior of structural steels. The results show that an increase in strain rate shifts the K_{Ic}-temperature curve towards higher temperatures.

McCabe and Ernst and *Cheng et al* examine the relation between crack opening displacement and elasto-plastic fracture criteria. Both papers report specimen influences on the measured parameter that limit the usefulness of this parameter as a fracture property. The measurement of compliance is an established method for determining crack length in long-term fracture tests, but *Nicholas et al* report a lack of one-to-one correspondence between crack length and compliance in several nickel-based superalloys. Possible causes are presented; however, a definitive explanation requires a level of analysis capability not yet available.

The final two papers are interrelated. The first of these is by *Popelar, Pan, and Kanninen* and develops a theoretical model for tearing instability that includes the effects of strain hardening. The second paper, by *Ahmad* and the same authors, presents an application of the proposed model to circumferentially cracked pipes and compares the results of experiments with the model predictions.

R. J. Sanford

Department of Mechanical Engineering, University of Maryland, College Park, Maryland; symposium chairman and editor

Index